# Intelligent Design Creationis

# Intelligent Design Creationism and Its Critics

Philosophical, Theological, and Scientific Perspectives

edited by Robert T. Pennock

A Bradford Book
The MIT Press
Cambridge, Massachusetts
London, England

© 2001 Massachusetts Institute of Technology

All rights reserved. No part of this book may be reproduced in any form by any electronic or mechanical means (including photocopying, recording, or information storage and retrieval) without permission in writing from the publisher.

This book was set in Sabon in 3B2 by Asco Typesetters, Hong Kong and was printed and bound in the United States of America.

Library of Congress Cataloging-in-Publication Data

Intelligent design creationism and its critics : philosophical, theological, and scientific perspectives / edited by Robert T. Pennock.
    p.  cm.
"A Bradford book."
Includes bibliographical references and index.
ISBN 0-262-16204-0 (alk. paper) — ISBN 0-262-66124-1 (pbk. : alk. paper)
1. Creationism. 2. Evolution (Biology) 3. Evolution (Biology)—Religious aspects—Christianity. 4. Religion and science. 5. Science—Philosophy.
I. Title.
BS652.P44 2001
231.7'652—dc21                                                    2001031276

10 9 8 7 6 5 4 3

# Contents

# Preface

Most of the classic conflicts between science and religion have been resolved, but the old controversy over evolution and creationism continues to erupt anew, as witnessed most recently in Kansas when creationists on the State Board of Education voted to remove evolution from the state's science curriculum standards. While most mainline Christian theologians and denominations have reconciled their faith and Darwinian evolution, a large number of fundamentalist and evangelical Christians continue to believe that there are irreconcilable differences, primarily because they hold to a literal or inerrant interpretation of the Bible. The American courts have consistently ruled against various forms of so-called scientific creationism, but creationists regularly mutate into new forms and try new tactics to get their views into the schools.

The last decade of the millennium saw the arrival of a new player in the creation/evolution debate—what has become known as the "intelligent design (ID) movement" or "the wedge." This new movement aims to overthrow not only Darwinian evolution, but also the supposedly dogmatic materialist/naturalist philosophy of the modern age that it believes has dethrowned Christian theism from its proper role of moral and intellectual authority.

The most important early publications that heralded the wedge movement included Berkeley law professor Phillip Johnson's book *Darwin on Trial* (Regnery, 1991) and Lehigh biochemist Michael Behe's book *Darwin's Black Box: The Biochemical Challenge to Evolution* (The Free Press, 1995). Another significant publication is the biology textbook by Percival Davis and Dean H. Kenyon titled *Of Pandas and People: The Central Question of Biological Origins* (Foundation for Thought and Ethics, 1993), which creationists have proposed for use in secondary school. The textbook poses a choice for students between those who accept Darwinian evolution and those who believe in intelligent design. Other important leaders of the ID movement include William A. Dembski, Paul Nelson, Stephen Meyer, and Jonathan Wells. Philosopher of religion Alvin Plantinga has also been highly influential.

I discussed and criticized the arguments of the intelligent design creationists (as well as other new varieties of creationism) in detail in my book *Tower of Babel: The Evidence against the New Creationism* (MIT Press, 1999). This current anthology serves as a companion volume that brings together many of the important original sources as well as new articles that highlight the ongoing debate.

This book should be useful to philosophers, theologians, and scientists, in their teaching, and to others, including educators and public policymakers, who are interested in contemporary expressions of anti-evolutionism. It would be appropriate as a textbook in any of those fields for courses that deal with the relation between science and religion. As I discussed in *Tower of Babel*, while the vast majority of creationists still hold the standard "young-earth" view, there are now many significant doctrinal schisms among them. The ID movement aims to unite the various factions by promoting a minimal notion of "mere creation," and in large measure they are succeeding, but it is important not to think of creationists stereotypically as a monolithic movement. This volume is by no means exhaustive; it focuses just on what is most novel about the new movement. For a course on the creationism controversy as a whole, it would need to be used in combination with *Tower of Babel* and with books that examine other forms of creationism.

Obviously, I am not a disinterested party to this debate. I have been one of the most vocal critics of intelligent design creationism (IDC), and members of the movement have criticized me in return. I include three of those exchanges, but have tried as editor to emphasize other voices, allowing the proponents and opponents of IDC to speak for themselves.

This anthology does not provide a "balanced treatment" in the sense of *equal time* that many creationists still desire for their views in public schools; articles from opponents of intelligent design creationism outnumber those from proponents by a margin of two to one. Nevertheless, in a more important sense, the selections are balanced in that they are representative. I picked the most central and significant of IDC's published articles plus a couple that fill in some of the blanks to give a fair sample of the key elements of their position, particularly as articulated by its most influential leaders. If anything, the anthology is somewhat less representative on the side of critics. As many critics point out, they typically find too much that is wrong in IDC writings to cover in any given article, and so focus on just a few salient points. Still, the opponents' articles give a fair sense of the range and intensity of the critical reaction.

About two-thirds of the articles have been previously published, mostly in specialized or small-circulation journals. Anthologizing these serves the purpose of bringing together hard-to-find material to get a more complete view of the debate. I emphasized articles that appeared in academic journals such as *Biology & Philosophy*, *Christian Scholars Review*, and *Philosophy of Science*. However, since most IDC articles appear in their own publications or in general-interest Christian magazines and since some of the important critics have also published in general-interest periodicals or on the internet, in several cases I drew from those sources.

There are no IDC articles from peer-reviewed science journals, since none has been published there so far. Several scientists have published reviews of Johnson's and Behe's books or responses to their arguments, but have mostly dismissed their criticisms as based upon misunderstandings or a caricature of evolutionary theory. I have reprinted a couple of these, but to be able to include some more detailed scientific criticisms of at least some of the their specific arguments against evolution I have included some new articles as well. Several of the new articles, comprising the remainder of the anthology, were solicited to provide direct replies to challenges that intelligent design creationists made in some specific reprinted article, while others deal with more general issues. New and reprinted articles are mixed in the philosophical, theological, and scientific sections. Bibliographic information for the reprinted articles appears at the bottom of the first page of each. Throughout, numbers in square brackets refer to page numbers in this volume.

Each section begins with an introduction that describes the articles and provides some background information about the issues to be discussed, so here I will give only a very brief overview of the sections themselves. The first section provides a broad overview of IDC as a political movement, focusing on its strategic revolutionary plan—"the wedge"—to overturn Darwinism and scientific naturalism. The main thrust of the IDC attack on evolution is philosophical, and section II deals with that issue, particularly with regard to Johnson's central argument that science assumes a dogmatic metaphysics that props up evolution. Section III switches to the theological debate regarding the apparent conflict between evolution and the Bible, and Alvin Plantinga's question of what to do when reason seems to clash with the dictates of Christian faith. Section IV deals with the main scientific challenges that IDCs have made. Michael Behe's contention that the "irreducible complexity" of biomolecules refutes Darwinism takes center stage, but the critical responses also address a wide variety of other scientific challenges that IDCs have made against evolutionary theory. Section V covers Plantinga's arguments against both metaphysical and methodological naturalism as they relate to evolution and the possibility of knowledge. Section VI deals with responses from mainstream Christian theologians to the IDC claim that Darwinian evolution is inconsistent with theism. Section VII includes both philosophical and scientific perspectives on the question of whether genetic information refutes Darwinism and materialism. Section VIII looks at two cases in which IDCs try to turn the tables by charging that it is actually evolutionists who rely upon magical or theological presuppositions. The final section considers the policy question about whether and how creationism or evolution should be taught in the public schools.

The sections can be read independently, but several fit together, with later ones taking up and developing issues begun in earlier ones. For instance, to focus on the philosophical issues involving the relation of evolution, scientific naturalism, and epistemology, sections II and V could be read together. Sections III and VI could be read in tandem to get a range from Catholic to various Protestant theological positions on how evolution may or may not be judged compatible with Christianity.

I now want to mention a few ways in which the anthology is not representative. As noted above, I have focused on articles that deal with the more novel aspects of the ID movement, and have omitted discussions about young-earth versus old-earth creationist views. There are IDC members in both camps but they choose not to discuss such factional differences and to unite on common basic points of agreement. Also missing are other creationist perspectives on such issues, which would have been useful to provide a sense of how the ID position fits and is viewed within the overall movement. For instance, I had planned to include a recent article by Henry Morris titled "Design Is Not Enough." Henry Morris, founder of the young-earth Institute for Creation Research (ICR) writes that the idea behind the intelligent design movement is to begin with the design argument and postpone talking about the bedrock biblical doctrine. "Any discussion of a young earth, 6-day creation, a world-wide flood and other biblical records of early history will turn off scientists and other professionals, they say, so we should simply use the evidence of intelligent design as a 'wedge' to pry them loose from their naturalistic premises. Then, later, we can follow up this opening by presenting the gospel, they hope" (*Back to Genesis*, no. 127a, July 1999). After reviewing arguments for design that have been offered historically, Morris concludes that such an approach is insufficient. "It is obvious," he writes, "that neither 'intelligent design' nor 'irreducible complexity' nor any other such euphemism for creation will suffice to separate a thorough-going Darwinian naturalist from his atheistic religion, in favor of God and special creation" (ibid). Morris declined permission for me to include his article in this collection, explaining that did not intend his article to be a criticism of the intelligent design movement.

Another important way in which the anthology is not representative is that it short-changes the moral and theological concerns that lie behind and power creationists' opposition to evolution. As I discussed in *Tower of Babel*, it is in large part because they believe that morality and meaning are at stake that they see evolution as so problematic. Phillip Johnson and other IDCs have argued that accepting the evolutionary naturalist view leads in the end to ethical relativism. Like other creationists before them, they blame the "evolutionary worldview" for what they take to be the moral degeneracy of contemporary American society. The moral issues are of course closely linked in their view with the theological questions. The theological articles in the anthology mostly focus on rebutting the IDC claim that Darwinian evolution is irreconcilable with true Christian theism, but much more is involved for those on both sides of the issue. For instance, no consideration of the design argument can be complete without a concomitant discussion of the theodicy problem—if it is taken that we may infer God's design of the world from observing its features, then how can we reconcile the clear cases of natural evil in the world with a supposedly benevolent deity? Intelligent design creationists hold that intelligent agency is needed to explain the complexity of the world down to even the level of biomolecules. Robert Frost illustrated the problem with such a view in his poem "Design," and I will close my preface with that for readers to ponder should they choose to pursue such questions on their own.

I found a dimpled spider, fat and white,
On a white heal-all, holding up a moth
Like a white piece of rigid satin cloth—
Assorted characters of death and blight
Mixed ready to begin the morning right,
Like the ingredients of a witch's broth—
A snow-drop spider, a flower like a froth,
And dead wings carried like a paper kite.

What had that flower to do with being white,
The wayside blue and innocent heal-all?
What brought the kindred spider to that height,
Then steered the white moth thither in the night?
What but design of darkness to appall?—
If design govern in a thing so small.

# Acknowledgments

I wish to offer special thanks to Matthew J. Brauer, Daniel R. Brumbaugh, Roy Clouser, Barbara Forrest, Peter Godfrey-Smith, Philip Kitcher, Alvin Plantinga, Michael Ruse, and Kelly C. Smith who contributed previously unpublished articles to this volume. Thanks also to Arthur Peacocke for contributing his Idreos lecture.

The poem "Design" is from *The Poetry of Robert Frost*, edited by Edward Connery Lathem. Copyright 1936 by Robert Frost, © 1964 by Lesley Frost Ballantine, © 1969 by Henry Holt and Co. Reprinted by permission of Henry Holt and Company, LLC.

I would also like to thank the following publishers and copyright holders for permission to reprint previously published articles: Michael Behe; Blackwell Publishers Ltd/University of Southern California; *Christian Scholar's Review*; *First Things*; Stephen Jay Gould; Kluwer Academic Publishers; the META lists; *Natural History*; Arthur Peacocke; *Perspectives on Science & Christian Faith*; *The Skeptic*; *Theology Today*; University of Chicago Press; and University Press of America.

# Contributors

**Michael J. Behe** is Professor of Biological Sciences at Lehigh University in Bethlehem, Pennsylvania, where he lives with his wife and eight children. He is the author of *Darwin's Black Box: The Biochemical Challenge to Evolution*, which has been reviewed by the *New York Times*, *Nature*, *Philosophy of Science*, *Christianity Today*, and over eighty other publications.

**Matthew J. Brauer** received his Ph.D. in Zoology from the University of Texas at Austin, where he studied the dynamics of adaptation in a bacterial virus. He is currently doing postdoctoral research in the Department of Genetics at the Stanford University School of Medicine. His main research interests include the evolution of modular metabolic and regulatory pathways.

**Daniel R. Brumbaugh** received his Ph.D. in Zoology from the University of Washington, where he studied mechanisms of adaptation in natural populations of a marine hydroid species. Currently, as the Marine Program Manager at the American Museum of Natural History's Center for Biodiversity and Conservation, his research includes work on the speciation dynamics of marine bryozoans using molecular phylogenetics and the integration of physical, biological, and social sciences for the design and implementation of marine reserves.

**Roy Clouser** is Professor of Philosophy and Religion at The College of New Jersey. In addition to the usual A.B., M.A., and Ph.D. in philosophy (University of Pennsylvania), he holds a B.D. in theology from Reformed Episcopal seminary and has done graduate work at Harvard Divinity School. Besides numerous articles, he is the author of *The Myth of Religious Neutrality* (Notre Dame Press, 1991) and *Knowing with the Heart: Religious Experience and Belief in God* (InterVarsity Press, 1999).

**Richard Dawkins** is the Charles Simonyi Professor of the Public Understanding of Science at Oxford University. He won the International Cosmos Prize (Japan) for 1997. His six books include *The Selfish Gene* (OUP, 1976, 1989), *The Extended Phenotype* (OUP, 1982), *The Blind Watchmaker* (Norton, 1986), *River Out of Eden* (Basic, 1995) and *Climbing Mount Improbable* (Norton, 1996). His most recent book is *Unweaving the Rainbow: Science, Delusion, and the Appetite for Wonder* (Houghton Mifflin, 1998).

**William A. Dembski** is a Senior Fellow of the Discovery Institute's Center for the Renewal of Science and Culture and is Associate Research Professor at Baylor

University. He is author of *The Design Inference* (Cambridge, 1998) and *Intelligent Design: The Bridge Between Science and Theology* (InterVarsity Press, 1999), and *No Free Lunch: Why Complex Specified Information Requires Intelligence* (Rowman and Littlefield, 2001).

**Evan Fales** is an associate professor at the University of Iowa. He is the author of two books, *Causation and Universals* and *A Defense of the Given*, and is currently at work on a third, *Divine Intervention*. He has contributed articles on the evidential value of mystical experience for theism and on the problem of evil.

**Branden Fitelson** is a Ph.D. candidate in philosophy at the University of Wisconsin. His research in philosophy focuses mainly on the foundations and applications of probability and statistics in both philosophy and science. He has also published papers in logic and computer science in the area of automated reasoning. Branden is also a research associate at the Argonne National Laboratory in Chicago.

**Barbara Forrest** is Associate Professor of Philosophy at Southeastern Louisiana University. She received her Ph.D. in philosophy from Tulane University, writing her dissertation on the naturalism of Sidney Hook. Having previously published papers on critical thinking and the philosopher's role in Holocaust studies, her two most recent articles are "Methodological Naturalism and Philosophical Naturalism: Clarifying the Connection" in the journal *Philo* and "The Possibility of Meaning in Human Evolution" in *Zygon*.

**Peter Godfrey-Smith** received his Ph.D. in philosophy from University of California, San Diego, in 1991. Currently he is Associate Professor of Philosophy at Stanford University. He works on various issues in philosophy of biology and philosophy of mind, including explanation and causation in genetics, the evolution of cognition, and the role of informational and semantic concepts within biology. Godfrey-Smith is author of *Complexity and the Function of Mind in Nature* (Cambridge, 1996).

**Stephen Jay Gould** is Alexander Agassiz Professor of Zoology, Professor of Geology, and Curator in Invertebrate Paleontology in the Museum of Comparative Zoology at Harvard University. He is author of numerous articles and books, including most recently *Full House* (1996), *Questioning the Millennium* (1997), *Leonardo's Mountain of Clams and the Diet of Worms* (1998), *Rock of Ages: Science and Religion in the Fullness of Life* (1999), and *The Lying Stones of Marrakech* (2000).

**Phillip E. Johnson** is the Jefferson E. Peyser Professor of Law at the University of California, Berkeley. He served as Chief Law Clerk to United States Chief Justice Earl Warren, and has taught criminal law and other subjects at Berkeley since 1967. He is the author of five books related to Darwinism, including *Darwin on Trial* (1993), *Reason in the Balance* (1995), and *The Wedge of Truth* (2000).

**Philip Kitcher** is Professor of Philosophy at Columbia University. He is the author of five books on topics in the philosophy of science, including *Abusing Science: The Case Against Creationism*. He is a past President of the Pacific Division of the American Philosophical Association, and a former editor-in-chief of *Philosophy of Science*.

**Ernan McMullin** is Director Emeritus of the Program in History of Science at the University of Notre Dame; Fellow of the American Academy of Arts and Sciences, of the American Association for the Advancement of Science, and of the International Academy of the History of Science; and Past President of the Philosophy of Science Association, and of the American Philosophical Association (Central Division). He has published numerous articles in contemporary philosophy of science, history of the philosophy of science, and the relations of theology and science.

**Nancey Murphy** is Professor of Christian Philosophy at Fuller Theological Seminary. She has a Ph.D. in philosophy of science (University of California at Berkeley) and a Th.D. in theology and philosophy of religion (Graduate Theological Union). She is the author of *Theology in the Age of Scientific Reasoning* (Cornell, 1991), which received the American Academy of Religion Award of Excellence, and was runner-up for *Christianity Today's* Critics' Choice Award. She has written or edited six additional books on the relations between science and theology.

**Paul A. Nelson** is a Senior Fellow of the Discovery Institute's Center for the Renewal of Science and Culture, and editor of the journal *Origins & Design*. His monograph, "On Common Descent," *Evolutionary Monographs* (University of Chicago, 2001), analyzes the theory of common descent.

**Arthur Peacocke** taught and did research on biochemistry at the Universities of Birmingham and Oxford for over twenty-five years, before turning his attention to the relation of science to theology. He is now a priest of the Church of England, an Honorary Canon of Christ Church, Oxford, and founder of several professional societies devoted to the topic. He is the author of *Theology for a Scientific Age* (Fortress Press, 1993), which received a Templeton Prize; *God and Science: The quest for Christian credibility* (SCM Press, 1996); and *Paths from Science towards God: Forging an Open Theology for a Scientific Age* (Oneworld, 2000).

**Robert T. Pennock** is Associate Professor in the Lyman Briggs School and Philosophy Department at Michigan State University. A philosopher and historian of science (Ph.D. University of Pittsburgh, 1991), his research focuses on the relationship of evidential and value issues in science. Among other honors, he has received Templeton awards for his work on science and religion, and was named a national Sigma Xi Distinguished Lecturer (2000). He is the author of *Tower of Babel: The Evidence against the New Creationism* (MIT Press, 1999).

**Alvin Plantinga** is the John A. O'Brien Professor of Philosophy at the University of Notre Dame. He is the author of several books in metaphysics, epistemology, and the philosophy of religion, including *God and Other Minds*, *The Nature of Necessity*, *Warrant: the Current Debate*, *Warrant and Proper Function*, and *Warranted Christian Belief*. He is past president of the Society of Christian Philosophers and the Central Division of the APA.

**Michael Ruse** is Professor of Philosophy and Zoology at the University of Guelph, Ontario. He is the author of many books, including *Taking Darwin Seriously*, *Monad to Man: The Concept of Progress in Evolutionary Biology*, and

*Mystery of Mysteries: Is Evolution a Social Construction?* He is also the founder and editor of the journal *Biology & Philosophy*.

**Kelly C. Smith** is Assistant Professor in the Department of Philosophy and Religion at Clemson University. He received his M.S. in zoology in 1992 and his Ph.D. in philosophy in 1994, both from Duke University. He has published on various issues in philosophy of biology and applied ethics, including biological complexity, genetic causation and genetic disease concepts, and the creation/evolution debate.

**Elliott Sober** is Hans Reichenbach Professor of Philosophy and Henry Vilas Research Professor at the University of Wisconsin-Madison, where he has taught since 1974. His publications include *The Nature of Selection* (MIT Press, 1984); *Reconstructing the Past—Parsimony, Evolution, and Inference* (MIT Press, 1988); and *Philosophy of Biology* (Westview Press, 1993). Most recently, he and the biologist David Sloan Wilson published *Unto Others—The Evolution and Psychology of Unselfish Behavior* (Harvard University Press, 1998).

**Chris Stephens** is Assistant Professor of Philosophy at the University of Oklahoma. He earned his Ph.D. in philosophy at the University of Wisconsin-Madison, where he wrote his dissertation on the evolution of rationality. His research interests are primarily in philosophy of biology, philosophy of science, and philosophy of mind. He has published in these areas in the *British Journal for the Philosophy of Science*, *Philosophy of Science*, and *Philosophical Studies*.

**Howard J. Van Till** is Professor (Emeritus) of Physics and Astronomy at Calvin College. Dr. Van Till's research experience includes both solid-state physics and millimeter-wave astronomy. During the past two decades he has devoted a considerable portion of his writing and speaking efforts to topics regarding the relationship of science and theology. In 1999 he was honored by the Calvin Alumni Association with the Faith and Learning Award.

**George C. Williams** is Professor Emeritus, Department of Ecology and Evolution, State University, Stony Brook, New York. He has been President of the Society for the Study of Evolution and is the recipient of numerous awards, including Eminent Ecologist Award, Ecological Society of America, 1989; Elliot Award, National Academy of Sciences, 1990; Raymond Pearl Award, American Society of Human Biology, 1997; and the Crafoord Prize from the Swedish Royal Academy, 1999. He recently published *Why We Get Sick: The New Science of Darwinian Medicine*, coauthored with R. M. Nesse.

# I

# Intelligent Design Creationism's "Wedge Strategy"

The intelligent design movement views itself as a truly revolutionary movement that aims to overthrow not just Darwinian evolution, but also the pernicious philosophical worldview of materialism and naturalism that its members believe has been built upon it. The essay in this introductory section, "The Wedge at Work: How Intelligent Design Creationism is Wedging Its Way into the Cultural and Acedemic Mainstream" by philosopher Barbara Forrest, gives an overview of the recent intellectual history of the movement, its central goals, and its guiding principles.

Drawing on a wide range of published and unpublished articles, interviews, and documents, Forrest describes how the movement to undermine "evolutionary naturalism" and promote "theistic realism" has coalesced under the leadership of Philip Johnson and others who are now centered at the Center for the Renewal of Science and Culture (CRSC), a branch of the Discovery Institute. She also focuses on an internal CRSC document, titled "The Wedge Strategy," that surfaced from an anonymous source in March, 1999. The document outlines their three-phase plan for achieving their revolution. She compares the ideas and plans therein with the work that IDCs have actually been doing, showing that they are actively engaged in developing intelligent design on every front *except* as a real scientific research program. As Forrest explains, her study "does not analyze the philosophical and scientific arguments (such as they are) of the intelligent design proponents ... [but rather] analyzes the nature of the wedge strategy, providing a framework from which to move at any point into the philosophical and scientific analyses." Her discussion of the IDC Wedge helps provide a context for the specific critical discussions in the sections that follow.

# 1

## The Wedge at Work: How Intelligent Design Creationism Is Wedging Its Way into the Cultural and Academic Mainstream

Barbara Forrest

### Introduction

If we understand our own times, we will know that we should affirm the reality of God by challenging the domination of materialism and naturalism in the world of the mind. With the assistance of many friends I have developed a strategy for doing this.... We call our strategy the "wedge."
—Phillip E. Johnson[1]

With the simplest of metaphors, Phillip Johnson describes the "wedge" strategy adopted in order to advance "intelligent design" theory, the most recent—and most dangerous—manifestation of creationism. Yet the simplicity of the metaphor is deceptive. In reality, the wedge strategy is being aggressively and systematically executed by the Discovery Institute's (DI) Center for the Renewal of Science and Culture (CRSC) through an extensive, constant, and sometimes dizzying range of activities which, as Johnson says, are intended ultimately to "affirm the reality of God." This religious goal, advanced chiefly by means of the Discovery Institute's anti-evolution agenda—and by politics—is the heart of the wedge movement.

This essay does not analyze the philosophical and scientific arguments (such as they are) of DI's intelligent design proponents. Others are doing that quite capably; rather, the present study analyzes the nature of the wedge strategy, providing a framework from which to move at any point into the philosophical and scientific analyses.

This study of the Discovery Institute's wedge strategy consists of three parts:

Part I. A chronological history of the wedge strategy and the authentication of the "Wedge Document." The development of the wedge movement and its strategy can be chronicled by consulting the Discovery Institute's own publications, as well as through articles in friendly sources. The Wedge's chronological development includes the production of a strategy outline known informally outside DI as the "Wedge Document," which is authenticated here. An important point, however, is that DI's activities would speak for themselves *even if the document were not genuine*. These activities betray an aggressive, systematic agenda for promoting not only intelligent design creationism, but the religious worldview that undergirds it.

Part II. A survey of wedge activities.   A survey of the activities which DI's Center for the Renewal of Science and Culture has undertaken to advance the wedge strategy shows that the systematic program has been very successfully carried out *except for its most important component*: the production of scientific data to support intelligent design. Yet despite the *scientific* failure of the wedge, the CRSC is tireless in advancing the rest of its strategy—(1) establishing a beachhead in higher education; (2) influencing public opinion by a steady stream of popular publications; and (3) most insidiously, insinuating "intelligent design theory" into the public education curriculum.

Part III. An analysis of the nature of the wedge strategy and its advance into the mainstream.   The Wedge consists of a tightly knit core of people at DI's Center for the Renewal of Science and Culture who have worked together for almost a decade to advance the wedge strategy; the same people are always active in the CRSC's major events. The movement is fueled by a religious vision which, although it varies among the members in its particulars, is predicated on the shared conviction that America is in need of "renewal" which can be accomplished only by instituting religion as its cultural foundation.

## Part I: Chronological History of the Wedge

Discovery Institute's Center for the Renewal of Science and Culture seeks nothing less than the overthrow of materialism and its cultural legacies....
—Center for the Renewal of Science and Culture[2]

The "Wedge," a movement—aimed at the court of public opinion—which seeks to undermine public support for teaching evolution while cultivating support for intelligent design theory, was not born in the mind of a scientist, or in a science class, or in a laboratory, or from any kind of scientific research, but out of personal difficulties after a divorce which led to Phillip Johnson's conversion to born-again Christianity. The wedge movement thus began, in a very real sense, as a religious epiphany in the life of Phillip Johnson. In accounts given by Johnson himself, he says that "the experience of having marriage and family life crash under me, and of achieving a certain amount of academic success and seeing the meaninglessness of it, made me ... give myself to Christ at the advanced age of 38. And that aroused a particular level of intellectual interest in the question of why the intellectual world is so dominated by naturalistic and agnostic thinking."[3] Nancy Pearcey, a CRSC fellow and Johnson associate, sees enough of a connection between Johnson's leadership of the intelligent design movement and his religious conversion to link both events in two of her most recent publications. In an interview with Johnson for *World* magazine, Pearcey says, "It is not only in politics that leaders forge movements. Phillip Johnson has developed what is called the 'Intelligent Design' movement.... Mr. Johnson is a Berkeley law professor who, spurred by the crisis of a failed marriage, converted to Christianity in midlife."[4] In *Christianity Today*, she made an even sharper connection which reveals the link between Johnson's religious beliefs and his animosity toward evolution: "The unofficial spokesman for ID is Phillip E. Johnson, a Berkeley law professor who converted to Christianity in his late 30s, then turned his sharp lawyer's eyes on the theory of evolution."[5]

Having begun with his religious conversion, Johnson's quest for personal meaning culminated in another epiphany during a sabbatical in England: "In 1987, when UC Berkeley law professor Phillip Johnson asked God what he should do with the rest of his life, he didn't know he'd wind up playing Toto to the ersatz winds of Darwinism. But a fateful trip by a London bookstore hooked Mr. Johnson on a comparative study of evolutionary theory."[6] Johnson purchased Richard Dawkins's *The Blind Watchmaker* and "devoured it and then another book, Michael Denton's *Evolution: A Theory in Crisis*." Such was Johnson's second epiphany: "I read these books, and I guess almost immediately I thought, *This is it. This is where it all comes down to, the understanding of creation*."[7] The Wedge's gestation period had begun.

According to Johnson, the wedge *movement*, if not the term, began in 1992: "The movement we now call the Wedge made its public debut at a conference of scientists and philosophers held at Southern Methodist University in March 1992, following the publication of my book *Darwin on Trial* [1991]. The conference brought together as speakers some key

Wedge figures, particularly Michael Behe, Stephen Meyer, William Dembski, and myself."[8] Johnson had initiated contact with a "cadre of intelligent design (ID) proponents for whom Mr. Johnson acted as an early fulcrum. ... Mr. Johnson made contact, exchanged flurries of e-mail, and arranged personal meetings. He frames these alliances as a 'wedge strategy,' with himself as lead blocker and ID scientists carrying the ball behind him."[9] In 1993, a year after the SMU conference, "the Johnson-Behe cadre of scholars met at Pajaro Dunes. ... Here, Behe presented for the first time the seed thoughts that had been brewing in his mind for a year—the idea of 'irreducibly complex' molecular machinery."[10]

When, in July 1992, *Scientific American* published Stephen Jay Gould's review of *Darwin on Trial*, in which Gould called the book "full of errors, badly argued, based on false criteria, and abysmally written," Johnson's supporters formed the "Ad Hoc Origins Committee" and wrote a letter (probably in 1992 or 1993) on Johnson's behalf: "This letter was mailed to thousands of university professors shortly after Stephen Jay Gould wrote his vitriolic bashing of ... *Darwin on Trial*. Included with it was Johnson's essay 'The Religion of the Blind Watchmaker,' replying to Gould, which *Scientific American* refused to publish."[11] Among the thirty-nine signatories were nine (listed below with their then-current affiliations) who a few years later became fellows of the Discovery Institute's Center for the Renewal of Science and Culture:

**Henry F. Schaefer III, Ph.D.**
Quantum Computational
Chemistry
University of Georgia

**Stephen Meyer, Ph.D.**
Philosophy of Science
Whitworth College

**Michael Behe, Ph.D.**
Biochemistry
Lehigh University

**William Dembski, Ph.D.**
Philosophy
Northwestern University

**Robert Kaita, Ph.D.**
Plasma Physics
Princeton University

**Robert Koons, Ph.D.**
Philosophy
UT, Austin

**Walter Bradley, Ph.D.**
Chairman, Mechanical Engineering
Texas A&M University

**Paul Chien, Ph.D.**
Biology
University of San Francisco

**John Angus Campbell, Ph.D.**
Speech Communication
University of Washington

The signatories describe themselves in the letter as "a group of fellow professors or academic scientists who are generally sympathetic to Johnson and believe that he warrants a hearing.... Most of us are also Christian Theists who like Johnson are unhappy with the polarized debate between biblical literalism and scientific materialism. We think a critical re-evaluation of Darwinism is both necessary and possible without embracing young-earth creationism."[12] A critical mass of supporters had begun to coalesce around Johnson.

By 1995, Johnson's mission had crystallized and he had a loyal contingent of like-minded people to help carry it out. That summer they held another conference, "The Death of Materialism and the Renewal of Culture," which served as the matrix of the Center for the Renewal of Science and Culture, organized the following year.[13] Johnson produced another book, *Reason in the Balance: The Case Against Naturalism in Science, Law, and Education* (InterVarsity Press, 1995), in which he positioned himself as a "theistic realist" against dogmatic "methodological naturalism":

First, here is a definition of MN [methodological naturalism], followed by a contrasting definition of my own position, which I label "theistic realism" (TR)....

1. A methodological naturalist defines science as the search for the best naturalistic theories. A theory would not be naturalistic if it left something (such as the existence of genetic information or consciousness) to be explained by a supernatural cause. Hence all events in evolution (before the evolution of intelligence) are assumed to be attributable to unintelligent causes. The question is not *whether* life (genetic information) arose by some combination of chance and chemical laws ... but merely *how* it did so.

... The Creator belongs to the realm of religion, not scientific investigation.

2. A theistic realist assumes that the universe and all its creatures were brought into existence for a purpose by God. Theistic realists expect this "fact" of creation to have empirical, observable consequences that are different from the consequences one would observe if the universe were the product of nonrational causes.... God always has the option of working through regular secondary mechanisms, and we observe such mechanisms frequently. On the other hand, many important questions—including the origin of genetic information and human consciousness—may not be explicable in terms of unintelligent causes, just as a computer or a book cannot be explained that way.[14]

The opposition between naturalism and theistic realism has become a hallmark of Johnson's thinking.

Now that ID's metaphysical terrain was clearly mapped, Johnson and his allies needed a formal strategy for executing their mission. By 1996, the most crucial development in the wedge strategy had occurred: the Center for the Renewal of Science and Culture was established under the auspices of the conservative Seattle think tank, the Discovery Institute.[15] In its Summer 1996 *Journal*, "a periodic publication that keeps Discovery members and friends up to date on Discovery's programs and events," the Discovery Institute announced the CRSC's formation, which "grew out of last summer's [1995] 'Death of Materialism' conference."[16] According to DI president Bruce Chapman, "The conference pointed the way and helped us mobilize support to attack the scientific argument for the 20th century's ideology of materialism and the host of social 'isms' that attend it." Larry Witham's December 1999 *Washington Times* column reveals the CRSC's topmost position on its parent organization's roster of priorities:

The eight-year-old Discovery Institute is a Seattle think tank where research in transportation, military reform, economics and the environment often takes on the easygoing tenor of its Northwest hometown. But it also sponsors a group of academics in science affectionately called "the wedge." ... The wedge is part of the institute's four-year-old Center for Renewal of Science and Culture (CRSC), a research, publishing and conference program that challenges what it calls an anti-religious bias in science and science education. "I would say it's our No. 1 project," said Bruce Chapman, Discovery's president and founder.[17]

With the formation of the CRSC, the Wedge's core working group was in place: Stephen Meyer and John G. West, Jr., as co-directors; William Dembski, Michael Behe, Jonathan Wells, and Paul Nelson as 1996–97 full-time research fellows; and Phillip Johnson as advisor.[18] Once the movement was securely housed with the Discovery Institute, the execution of the wedge strategy began to pick up speed.

In November 1996, Johnson and his associates convened the "Mere Creation" conference at Biola University in California.[19] The importance of this conference cannot be underestimated; indeed, in the Foreword to the book which issued from it, its importance was explicitly spelled out by Henry Schaefer, the University of Georgia chemist who had supported Phillip Johnson as a signatory to the Ad Hoc Origins Letter: "An unprecedented intellectual event occurred in Los Angeles on November 14–17, 1996. Under the sponsorship of Christian Leadership Ministries, Biola University hosted a major research conference bringing together scientists and scholars who reject naturalism as an adequate framework for doing science and who seek a common vision of creation united under the rubric of intelligent design."[20] (Christian Leadership Ministries has continued to actively assist the Wedge both logistically and in its provision of "virtual" office space to wedge members on its "Leadership University" web site.)[21]

Contrary to Schaefer's labeling the Mere Creation conference a research conference, it did not actually produce any scientific research.[22] It did, however, produce the needed strategy. The movement's goal at this conference was already clear to third-party observers, such as Scott Swanson, who wrote about the conference for *Christianity Today*:

The fledgling "intelligent-design" movement, which says Darwinian explanation of human origins are inadequate, is aiming to shift from the margins to the mainstream.... The first major gathering of intelligent-design proponents took place in November at Biola University in La Mirada, California.... If the turnout at the conference is any indication, intelligent design is gaining a following. More than 160 academics, double what organizers had envisioned, attended from 98 universities, colleges, and organizations. The majority represented secular universities.[23]

Although, according to Swanson, the organizers "chose not to use the conference as a forum to develop a statement of belief for the movement," he learned that "leaders are planning a spring conference at the University of Texas and have begun publishing a journal, *Origins and Design*, edited by Paul Nelson...." This is a reference to the "Naturalism, Theism, and the Scientific Enterprise" conference, held in February 1997 and organized by CRSC fellow Robert Koons, a philosopher and University of Texas faculty member.[24] With a core of supporters who had now been able to convene and strategize, the Wedge's gestation period was over: "Prior to the conference, the intelligent-design movement comprised a loose coalition of scholars from a wide variety of disciplines. The conference brought together like-minded scholars 'to get them thinking in the same range of questions,' says ... Phillip Johnson...."[25]

William Dembski edited a book of conference presentations entitled *Mere Creation: Science, Faith and Intelligent Design* (such books, like the conferences themselves, being an important component in the wedge strategy). Henry Schaefer wrote the Foreword, in which he reveals unmistakably that the wedge strategy had now solidified in important ways:

Wonderful ideas left under a bushel do no good. The conference should produce tangible results that will accelerate the growth of scientific research programs unencumbered by naturalism, encouraging and disseminating scholarship both at the highest level and at the popular level via such activities as preparing a book for publication, with chapters drawn from the conference papers (this goal has been met with the publication of the present volume);

planning a major origins conference at a large university to engage scientific naturalists (this goal remains in the offing);

outlining a research program to encourage the next generation of scholars to work on theories beyond the confines of naturalism;

exploring the need for establishing fellowship programs, and encouraging joint research (Seattle's Discovery Institute is the key player here—see www.discovery. org);

providing resources for the new journal *Origins & Design* as an ongoing forum and a first-rate interdisciplinary journal with contributions by conference participants (see www.arn.org/arn);

preparing information usable in the campus environment of a modern university, such as expanding a World Wide Web origins site (see www.leaderu.com, www.origins.org, www.iclnet.org) and exploring video and other means of communication (see www.daystar.org).

Schaefer also lists the members of the steering committee for the conference:

Michael Behe
Walter Bradley
William Dembski
Phillip Johnson
Sherwood Lingenfelter
Stephen Meyer
J. P. Moreland
Paul Nelson
Pattle Pun
John Mark Reynolds
Henry F. Schaefer III
Jeffrey Schloss[26]

The activities Schaefer lists in his Foreword prefigure most of the activities which are now actually being executed, and the steering committee metamorphosed into some of the Wedge's most active members. All steering committee members except Johnson, who is the CRSC's advisor, and Sherwood Lingenfelter, Biola University provost who hosted the conference, have become CRSC fellows.

By 1997, Johnson was talking openly about the wedge strategy in his book, *Defeating Darwinism by Opening Minds* (dedicated "To Roberta and Howard, who understood 'the wedge' because they love the Truth").[27] Johnson devotes Chapter 6 to "The Wedge: A Strategy for Truth," calling upon a familiar metaphor of using a wedge to make a small opening which then splits a huge log: "We call our strategy 'the wedge.' A log is a seeming solid object, but a wedge can eventually split it by penetrating a crack and gradually widening the split. In this case the ideology of scientific materialism is the apparently solid log."[28] Johnson's 1998 book, *Objections Sustained: Subversive Essays on Evolution, Law and Culture* is dedicated "To the members of the Wedge, present and future."[29] His most recent book is *The Wedge of Truth: Splitting the Foundations of Naturalism* (InterVarsity Press, 2000).

Although Johnson had begun thinking and speaking of the wedge strategy in 1997, there had been no detailed elaboration of the form its execution would take. Such elaborations were stated in a strategy document which has come to be known informally as the "Wedge Document."[30] It surfaced anonymously and was posted on the Internet in March 1999; various aspects of the document indicate that it was written in 1998. This document is the "Five Year Plan" of the Center for the Renewal of Science and Culture, although it includes goals which stretch into the next twenty years, indicating the CRSC's view of the strategy as a long-term commitment. Although Johnson has talked openly about the existence of the strategy, he has not publicly elaborated upon its logistics, and the logistics are ambitious. The document, entitled "The Wedge Strategy," with the name of the organization, "Center for the Renewal of Science and Culture" beneath the title, explains what the CRSC is doing now as well as where they want to go; therefore, it is crucially important.

The authenticity of the Wedge Document has been neither affirmed nor denied by the Discovery Institute; however, a strong case can be made for its authenticity. The Introduction reads as follows:

THE WEDGE STRATEGY
CENTER FOR THE RENEWAL OF SCIENCE & CULTURE
INTRODUCTION

The proposition that human beings are created in the image of God is one of the bedrock principles on which Western civilization was built. Its influence can be detected in most, if not all, of the West's greatest achievements, including representative democracy, human rights, free enterprise, and progress in the arts and sciences.

Yet a little over a century ago, this cardinal idea came under wholesale attack by intellectuals drawing on the discoveries of modern science. Debunking the traditional conceptions of both God and man, thinkers such as Charles Darwin, Karl Marx, and Sigmund Freud portrayed humans not as moral and spiritual beings, but as animals or machines who inhabited a universe ruled by purely impersonal forces and whose behavior and very thoughts were dictated by the unbending forces of biology, chemistry, and environment. This materialistic conception of reality eventually infected virtually every area of our culture, from politics and economics to literature and art.

The cultural consequences of this triumph of materialism were devastating. Materialists denied the existence of objective moral standards, claiming that environment dictates our behavior and beliefs. Such moral relativism was uncritically adopted by much of the social sciences, and it still undergirds much of modern economics, political science, psychology and sociology.

Materialists also undermined personal responsibility by asserting that human thoughts and behaviors are dictated by our biology and environment. The results can be seen in modern approaches to criminal justice, product liability, and welfare. In the materialist scheme of things, everyone is a victim and no one can be held accountable for his or her actions.

Finally, materialism spawned a virulent strain of utopianism. Thinking they could engineer the perfect society through the application of scientific knowledge, materialist reformers advocated coercive government programs that falsely promised to create heaven on earth.

Discovery Institute's Center for the Renewal of Science and Culture seeks nothing less than the overthrow of materialism and its cultural legacies. Bringing together leading scholars from the natural sciences and those from the humanities and social sciences, the Center explores how new developments in biology, physics and cognitive science raise serious doubts about scientific materialism and have re-opened the case for a broadly theistic understanding of nature. The Center awards fellowships for original research, holds conferences, and briefs policymakers about the opportunities for life after materialism.

The Center is directed by Discovery Senior Fellow Dr. Stephen Meyer. An Associate Professor of Philosophy at Whitworth College, Dr. Meyer holds a Ph.D. in the History and Philosophy of Science from Cambridge University. He formerly worked as a geophysicist for the Atlantic Richfield Company.[31]

The most compelling evidence for the Wedge Document's authenticity was located on DI's own web site, on pages which contained *exactly* the same wording as this Introduction. These pages appeared to date from the establishment of the Center for the Renewal of Science and Culture and are no longer accessible.[32] Yet even if the *document* were not authentic, the ambitious spate of activities being carried out by the CRSC proves the existence of a well orchestrated strategy. The document simply provides a specific sketch of the strategy and can be used as a reference point to examine the CRSC's progress in executing its phases.

An ambitious strategy like the Wedge would have been useless, however, without money. The CRSC has been generously funded by a number of benefactors, the most forthcoming of whom is Howard Ahmanson through his organization Fieldstead and Company. A rather ominous aspect of Ahmanson's identity is his long-time membership (until 1995) on the board of the Christian reconstructionist Chalcedon Foundation, one of the most extreme right-wing fundamentalist organizations in the country.[33] Ahmanson's contribution of "crucial start-up funding" is acknowledged in the Discovery Institute's announcement of the CRSC's establishment in its summer 1996 *Journal*.[34] In the 1999 *Journal*, DI announced major funding increases:

... three enlarged grants to the Center for the Renewal of Science and Culture have enabled it to expand the number of fellowships it is supporting for scholarly work on the theory of 'intelligent design.' ... Crucial decisions in the fall of 1998 at the Fieldstead & Co. ... increased its grant to Discovery to $300,000 per year for the next five years. The Maclellan Foundation ... also increased its grant to $400,000 for 1999, while the ... Stewardship Foundation ... voted to increase its CRSC grant to $200,000 per year for the next five years. Special grants are likely to bring the overall CRSC budget to over $1 million for 1999.[35]

According to Larry Witham in the *Washington Times*, all three of the above funding sources have "Christian roots."[36] Their contribution of so much money indicates that they recognize and support the CRSC's mission. DI president Bruce Chapman affirms such support: "We are not going through this exercise just for the fun of it. We think some of these ideas are destined to change the intellectual—and in time the political—world. Fieldstead & Company and the Stewardship Foundation agree, or they would not have given us such substantial funding."[37]

Now, entering the year 2001, with its program of action spelled out in the Wedge Document and ample funding secured, the strategy is at work and gaining steam. Having begun with only four research fellows, the CRSC presently consists of forty-one fellows, thirteen of whom have senior status. Phillip Johnson is still the advisor, along with George Gilder.[38] Their pursuit of the Wedge's goals continues unabated. The split in the log widened, and the Wedge was lodged more firmly in place, with the establishment in October 1999 of the Michael Polanyi Center at Baylor University, run by CRSC fellows William Dembski and Bruce Gordon. As will be seen in the subsequent parts of this study, with the Discovery Institute having provided a secure home, the Center for the Renewal of Science and Culture has grown from infancy to robust adolescence, and it is impatiently racing toward adulthood.

## Part II: A Survey of Wedge Activities

The social consequences of materialism have been devastating. As symptoms, those consequences are certainly worth treating. However, we are convinced that in order to defeat materialism, we must cut it off at its source.... Design theory promises to reverse the stifling dominance of the materialist world view, and to replace it with a science consonant with Christian and theistic convictions.

—"The Wedge Strategy"

In the July/August 1999 *Touchstone* Magazine, Phillip Johnson declares that "it is time to review how the Wedge has grown and progressed, to evaluate how far we have come, and to forecast what we expect to accomplish in the next decade."[39] The Wedge Document lists specific goals by means of which DI's progress in pursuing its wedge strategy can be calculated. Wedge members maintain a dizzying schedule of activities—new ones show up on the Internet constantly, making it difficult to keep track of all of them. Not all of them are of equal significance; some are relatively minor, while others, such as conferences, are more ambitious and wider in their impact. Neither has DI itself organized all of these activities. Some were organized by others—both their allies and their opponents. However, when wedge members participate in events organized by others—whether allies or opponents—the Wedge's goals are advanced even more pronouncedly: Johnson's goal of getting a place at the table is met, and the Wedge registers as a force which must be taken into account.

At least one activity for virtually every major goal in the Wedge Document is identified below. *A notable exception, however, is their very first phase—the goal of scientific research.* Yet when added together, the other activities demonstrate that the strategy—consisting of a great deal of political and public relations work, if not scientific research—is a well funded, aggressive, systematic program which has considerably advanced the goals in the Wedge Document:

THE WEDGE PROJECTS
Phase I. Scientific Research, Writing and Publication
• Individual Research Fellowship Program
• Paleontology Research Program ...
• Molecular Biology Research Program ...

Phase II. Publicity and Opinion-making
- Book Publicity
- Opinion-Maker Conferences
- Apologetics Seminars
- Teacher Training Programs
- Op-ed Fellow
- PBS (or other TV) Co-production
- Publicity Materials/Publications

Phase III. Culture Confrontation and Renewal
- Academic and Scientific Challenge Conferences
- Potential Legal Action for Teacher Training
- Research Fellowship Program: shift to social sciences and humanities[40]

The Wedge Document also states that DI does not consider the chronological order of these phases to be unchangeable and is optimistic about the Wedge's success: "The Wedge strategy can be divided into three distinct but interdependent phases, which are roughly but not strictly chronological. We believe that, with adequate support, we can accomplish many of the objectives of Phases I and II in the next five years (1999–2003), and begin Phase III...."

### Phase I of the Wedge Strategy: "Scientific Research, Writing and Publication"

By the Discovery Institute's own description above, Phase I—the production of scientific research, along with writing and publicity—is the foundation of the wedge strategy. In support of "significant and original research in the natural sciences, the history and philosophy of science, cognitive science and related fields," the CRSC has a generous fellowship program, providing "Full-year research fellowships between $40,000 and $50,000" and "Short-term research fellowships between $2,500 and $15,000 for either summer research, release time from teaching or book promotion activities, or other research-related activities."[41] During CRSC's first year of operation alone, they awarded more than $270,000 in research grants.[42] Such lucrative support should enable industrious young scientists to develop scientific research programs and compile data to support intelligent design.

Yet, in this most important of all the Wedge's goals—the only one that can truly win them the credibility they crave—their record is conspicuously unsuccessful. Ironically, the CRSC boasts of its scientific research program, while Phillip Johnson has admitted the lack of scientific data which would substantiate their boasting. The CRSC's web site declares, "The Center for the Renewal of Science and Culture seeks, therefore, to challenge materialism on specifically scientific grounds. Yet Center Fellows do more than critique theories that have materialistic implications. They have also pioneered alternative scientific theories and research methods that recognize the reality of design.... This new research program—called "design theory"—is based upon recent developments in the information sciences and many new evidences of design."[43] But in 1996, when the wedge strategy was being formalized at the Mere Creation conference at Biola University, Johnson acknowledged that design proponents did not yet have the science to accomplish their goals:

What we need for now is people who want to get thinking going in the right direction, not people who have all the answers in advance. In good time new theories will emerge, and science will change. *We shouldn't try to shortcut the process by establishing some new theory of origins until we know more about exactly what needs to be explained.* Maybe there will be a new theory of evolution, but it is also possible that the basic concept will collapse and science will acknowledge that those elusive common ancestors of the major biological groups never existed. If we get an unbiased scientific process started, we can have confidence that it will bring us closer to the truth.

For the present I recommend that we also put the Biblical issues to one side. The last thing we should want to do, or seem to want to do, is to threaten the freedom of scientific inquiry. Bringing the Bible anywhere near this issue just raises the "Inherit the Wind" stereotype, and closes minds instead of opening them.

*We can wait until we have a better scientific theory*, one genuinely based on unbiased empirical evidence and not on materialist philosophy, before we need to worry about whether and to what extent that theory is consistent with the Bible. *Until we reach that better science, it's just best to live with some uncertainties and incongruities*, which is our lot as human beings—in this life, anyway.[44] [Emphasis added.]

Despite this notable lack of "some new theory of origins," CRSC fellow Nancey Pearcey wrote in 1997 that "The design movement offers more than new and improved critiques of evolutionary theory.... Its goal is to show that intelligent design also functions as a positive research program." The Discovery Institute boasts of CRSC's scientific achievements in its 1999 *Journal*:

Today ... Darwinist dogma is being challenged by new science. It isn't easy getting a hearing, but it is happening more and more. Science's grand tradition of self-examination is leading to new theories based on better evidence, and pointing away from materialism.

Defenders of Darwinian orthodoxy are quarreling among themselves as never before as disturbing evidence against Darwinism appears in such fields as Big Bang cosmology, paleontology (especially in Cambrian era fossils) and molecular biology. Moreover, an alternative to Darwinism—within science—is emerging in the theory of "Intelligent Design." The Center for the Renewal of Science and Culture at Discovery Institute is a major factor in the new scientific debate and the examination of its implications for culture and public policy.[45]

The wedge movement's desired entré into secular academia is impossible without a scientific research program, buttressed by the production of peer-reviewed scientific data. Yet fellows of the CRSC have been successful only in the Phase I goals of writing and publication.[46] They have produced no original scientific data, nor even a genuine scientific research plan, which would mark the successful accomplishment of this most crucial phase.

The CRSC's 1997 "Year End Update" chronicled its activities for that year. Its "Consultation on Intelligent Design" brought together "CRSC fellows and friends from around the world." Featured as "highlights of the weekend" were wedge scientists whose work purportedly holds promise for confirmation of design's scientific viability: Paul Chien, a University of San Francisco biology professor, and Michael Behe, a biochemistry professor at Lehigh University.[47] Research on the scientific output of these wedge members reveals the lack of success in the goal with which they have been entrusted.

**Paul K. Chien**　Paul K. Chien is charged in the Wedge Document with conducting the CRSC's paleontology research. He has cultivated connections with Chinese scientists in Kunming, China, where the famous Chengjiang fossils, dating back to the Cambrian period, have aroused intense international interest. Chien and the Discovery Institute helped organize the June 1999 "International Symposium on the Origins of Animal Body Plans and Their Fossil Records," a conference on the Chengjiang fossils which scientists from around the world attended in Kunming.[48]

The Discovery Institute is exploiting Chien's connection with the Chengjiang discovery in several ways; one example is their argument for teaching intelligent design in the nation's public schools in an article entitled "Intelligent Design in Public School Science Curricula: A Legal Guidebook":

In recent years the fossil record has also provided new support for design. Fossil studies reveal a "biological big bang" near the beginning of the Cambrian period 530 million years ago. At that time roughly fifty separate major groups of organisms or "phyla" (including most all the basic body plans of modern animals) emerged suddenly without evident precursors. Although neo-Darwinian theory requires vast periods of time for the step-by-step development of new biological organs and body plans, fossil finds have repeatedly confirmed a pattern of explosive appearance followed by prolonged stability of living forms. Moreover, the fossil record shows a "top-down" hierarchical pattern of appearance in which major structural themes or body plans emerge before minor variations on those themes.... Not only does this pattern directly contradict the "bottom-up" pattern predicted by neo-Darwinism, but as University of San Francisco marine paleobiologist Paul Chien and several colleagues have argued,... it also strongly resembles the pattern evident in the history of human technological design, again suggesting actual (i.e., intelligent) design as the best explanation for the data.[49]

The Wedge is clearly using Chien's expertise as a "paleobiologist" to shore up its pro-intelligent design stance. However, although Chien does have distinguished credentials, paleontological expertise—a necessary credential for studying the Chinese Cambrian fossils—is not among them. On the University of San Franciso's Department of Biology home page, Chien's degrees are listed: "B.S., Chung Chi College, N.T., Hong Kong, Chemistry, 1962; B.S., Chung Chi College, N.T., Hong Kong, Biology, 1964; Ph.D., University of California, Irvine, 1971." His Ph.D. field is not listed, but according to his bio on the CRSC web site, it is biology. Dr. Chien's research interests are also listed on the USF page: "Prof. Chien is interested in the physiology and ecology of inter-tidal organisms. His research has involved the transport of amino acids and metal ions across cell membranes and the detoxification mechanisms of metal ions."[50] Clearly, Chien has no formal credentials in paleontology; moreover, he is not really interested in acquiring them, as he revealed in a 1997 *Real Issue* interview:

*RI:*  Do you intend to go back to Chengjiang, the Chinese Cambrian site?
*Chien:*  I would very much like to do that. Somehow I would like to get more involved in fossil work. Although I have lectured so many years in my own area of marine biology and pollution, I think I would like to concentrate on this aspect. This was an opportunity presented to me which nobody else has.

*RI:* Perhaps you could add "paleontologist" to your credentials.

*Chien:* Not really; that's not my purpose. I am more interested in working on the popular level....[51]

A survey of the scientific literature reveals that Chien's study of the Chengjiang fossils has progressed no farther than a hobby.[52] A survey in May 2000 on *SciSearch* (an electronic database containing both the *Science Citation Index* and *Current Contents*), which contains citations dating to 1988, using the name "P. K. Chien," with no date restrictions, yielded only six articles, none of which were about either the Chengjiang fossils or intelligent design. A Medline search on June 26, 2000, for "P. K. Chien" yielded the same results. A combined search of *Biological and Agricultural Index*, *Medline*, and *Zoological Record* on June 26, 2000, for "P. Chien" (which also picked up anything by "P. K. Chien") yielded a total of forty-five articles, but none about either the Chengjiang fossils or intelligent design.

Chien has expended considerable effort to advance the cause of intelligent design in China:

More important for me is to tell the Chinese people about [the Cambrian Explosion]. The Chengjiang Biota is a "Treasure" discovered by our own scientists on our own land! In Taiwan, China and Hong Kong, very few people know about it. Many Chinese have been taught the wrong theory, namely Darwinism. When I told them about this new scientific finding, some were very angry because they had been told the wrong story all their lives. Of course some thought I was telling a lie. But after I showed them the evidence, the real fossils from Chengjiang, they turned around and blamed the education they had received.[53]

This kind of work serves the Discovery Institute's agenda of undermining evolution and advancing the cause of intelligent design beyond the U.S. The Wedge Document predicts, as one of the CRSC's Five Year Objectives, "An active design movement in Israel, the UK and other influential countries outside the US." However, Chien's work does nothing anywhere in the world to provide genuine scientific support for intelligent design.

**Michael Behe** Michael Behe is the wedge member whom the Discovery Institute presents as its most formidable scientist. His book *Darwin's Black Box*, which the Wedge Document lauds for being published in paperback after "nine print runs in hard cover," is credited in the document with helping to increase the momentum of the wedge strategy that began with Johnson's book *Darwinism on Trial*. Yet Behe's failure to produce original intelligent design research and to publish on intelligent design in scientific journals proves that publicity, not real scientific accomplishment, is DI's primary goal—Behe serves a vital function for the organization, but not a scientific one.

An inspection of professional information about Behe on his departmental web site at Lehigh University yielded nothing which could be taken as scientific research supporting intelligent design.[54] Behe's own faculty page is about his professional research; interestingly, he has nothing at all to say there about his avocation of promoting intelligent design. He has posted a picture of himself at what appears to be a bookstore, holding a copy of *Darwin's Black Box*. He lists four representative publications, none of which are about intelligent design. He lists his favorite links, one of which is the creationist web site, Access Research Network, where he maintains a schedule of his engagements.[55] However, other than these two rather subtle references, there is nothing related to intelligent design in his professional postings.

A survey of the scientific literature shows that Behe, like Chien, has to date published no peer-reviewed research on intelligent design in any scientific journal. *Darwin's Black Box*, published by a respectable publishing company, The Free Press, and aimed at a popular audience, has been thoroughly critiqued by scientists, but the ideas in the book have not been published in a scientific journal. A May 2000 *SciSearch* survey using "Behe, M." yielded ten articles in scientific journals, none of which are about intelligent design. The only listed titles attributed to Behe referring to either evolution or intelligent design were five letters: "Embryology and Evolution," *Science*, 1998, V. 281, N. 5375; "Defining Evolution," *Scientist*, 1997, V. 11, N. 22; "Defining Evolution," *Scientist*, V. 11, N. 12; "Darwinism and Design," *Trends in Ecology and Evolution*, 1997, V. 12, N. 6; and "Understanding Evolution," *Science*, 1991, V. 253, N. 5023. None is more than one page long.

Behe responded to an article in the July 1999 issue of *Philosophy of Science*, "Redundant Complexity: A Critical Analysis of Intelligent Design in Biochemistry," by Niall Shanks and Karl H. Joplin.[56] His response, "Self-Organization and Irreducibly Complex Systems: A Reply to Shanks and Joplin," appears in the March 2000 issue of *Philosophy of Science*. However, the only work of his own that he cites in this response is *Darwin's Black Box*, published in 1996. He cites CRSC fellow William Dembski's *The Design Inference* (Cambridge University Press, 1998). He also cites scientific articles and books written by others, but he cites no articles bearing his own name in scientific journals.[57]

Behe has produced several articles in response to his critics. One of them, "Correspondence with Science Journals: Response to Critics Concerning Peer-Review," purports to provide evidence of his attempt to publish "a full-length reply-to-critics paper" in "a journal in the field of evolution."[58] However, responses to one's critics published on the web site of a political think tank do not qualify as scientific publications. In light of Behe's failure to publish anything about intelligent design in scientific journals—which would gain him some measure of respect, if not agreement—it is ironic that he issues a warning to proponents of "the theory of Darwinian molecular evolution": "'Publish or perish' is a proverb that academicians take seriously. If you do not publish your work for the rest of the community to evaluate, then you have no business in academia and, if you don't already have tenure, you will be banished."[59]

**General Survey of Intelligent Design Publications**    The utter absence of any scientific publications supporting intelligent design by Chien and Behe is characteristic of intelligent design as a whole. The dearth of scientific data supporting intelligent design was confirmed in 1997 by George W. Gilchrist in a survey of the scientific literature: "This search of several hundred thousand scientific reports published over several years failed to discover a single instance of biological research using intelligent design theory to explain life's diversity."[60] In the May/June 1997 *Reports of the National Center for Science Education*, Gilchrist reports his survey up to 1997 in five computerized databases—*BIOSIS*, the *Expanded Academic Index*, the *Life Sciences Collection*, *Medline*, and the *Science Citation Index*—for any scientific publications on intelligent design as a biological theory. His search yielded a total of only thirty-seven references, of which "none report scientific research using intelligent design as a biological theory."[61] The situation has not improved since 1997.

A similar search conducted for the present study, supplementing Gilchrist's survey by looking for intelligent design articles published since 1997, had the same results: no scientific research supporting intelligent design as a biological theory has been published. In order to survey other databases for anything the earlier described *SciSearch* survey might have missed, surveys were conducted, with no date restrictions, of the *BIOSIS* and *Medline* databases, using both "intelligent design" and "design theory" as keywords. The "intelligent design" search in *BIOSIS* yielded four articles, only one of which was about intelligent design (the July 1999 Shanks-Joplin article). Using "design theory" as the keyword, a *BIOSIS* survey yielded sixteen articles, but none about design theory as it relates to intelligent design creationism. The *Medline* search using *both* "intelligent design" and "design theory" yielded fourteen articles, none of which were about intelligent design creationism. A *SciSearch* survey using "intelligent design" yielded sixty-one titles. All except four were related to industrial technology, engineering, computers, shipbuilding, etc. Of the remaining four, only two were on intelligent design as a biological theory: Shanks-Joplin's and Behe's *Philosophy of Science* articles cited above. The other two were letters entitled "Intelligent Design" in *Geotimes* and "Intelligent Design Reconsidered" in *Technology Review* —ambiguous titles given the fact that articles with "intelligent design" in their titles were also listed, but were clearly not about intelligent design as a biological theory (e.g., "HyperQ Plastics: An Intelligent Design Aid for Plastic Material Selection").

The surveys reveal that the wedge strategy is failing miserably in its most important goal: the production of scientific research data to support intelligent design creationism and the publication of such data in scientific journals. Not only have Chien and Behe failed to produce such work, but so has every other CRSC fellow—if other wedge members' work were being presented at scientific meetings and published in scientific journals, it should have been found in the survey of the databases. In the only part of its strategy that could genuinely win the wedge acceptance into the academic and cultural mainstream, the Discovery Institute's Center for the Renewal of Science and Culture is so far an utter failure.

### Phases II and III of the Wedge Strategy

Despite the total absence of scientific productivity on which Phase I depends, the Wedge is tirelessly engaged in the Phase II and III activities. Even a brief list of these activities for each major category paints a convincing picture of the organized, systematic nature of the Wedge's advance.

## Phase II.   Publicity and Opinion-making

• Book Publicity
Wedge members are advancing quickly toward the wedge document's stated goal of "Thirty published books on design and its cultural implications...." Phillip Johnson alone has now written five books. Their books are available at major online retail outlets such as Barnes and Noble and Amazon.com.[62] Books, videotapes, audiotapes, and their journal, *Origins and Design*, are aggressively marketed on the creationist web site, Access Research Network, which is largely operated by wedge members.[63] They use their science education web site's "Science Education Bookstore" to sell their books via a direct link to the "Bookstore" on the Discovery Institute web site.[64] Book publicity is therefore a constant activity, exemplified by Phillip Johnson's promoting his books at the February 2000 National Religious Broadcasters convention in Anaheim, California.[65]

• Opinion-Maker Conferences
The CRSC has attended at least one conference which it explicitly labeled an "Opinion-Maker Conference," recounted in its November/December 1997 "Year End Update." This event was clearly a networking opportunity for wedge members:

Opinion-Maker Conference:   At the invitation of Ed Atsinger, President of Salem Communications, Inc., Steve Meyer and Phillip Johnson recently addressed a national conference of radio talk-show hosts. The talk-show hosts were extremely enthusiastic in response to Steve and Phil, and their presentation of the case for Intelligent Design. Afterward, Howard Freedman, National Program Director of Salem Communications Inc., and many of the talk-show hosts invited Steve, Phil, and other scientists to appear on their programs to discuss the evidence for design,...[66]

• Apologetics Seminars
The Wedge Document states that the CRSC seeks "to build up a popular base of support among our natural constituency, namely, Christians. We will do this primarily through apologetics seminars." William Dembski's Fieldstead & Company–supported seminar, "Design, Self-Organization, and the Integrity of Creation," would fit into this category. The course description shows that the June 19–July 28, 2000, "Summer Seminar" at Calvin College was designed to attract Christian, pro-intelligent design participants: "The aim of this seminar is to see whether a rapprochement between design and self-organization is possible that pays proper due both to the divine wisdom in creation and to the integrity of the world as

an act of creation. . . . [S]cholars with expertise in the following disciplines are especially encouraged to apply: complex systems theory, information/design theory, history and philosophy of science, philosophy of religion, philosophical theology, and any special sciences dealing with complex systems." Each applicant was required to submit "a one-page description of his/her vocation as Christian scholar and teacher."[67]

• Teacher Training Programs

Of all the Discovery Institute's goals designed to advance the wedge, this is the most insidious, since it is aimed at getting intelligent design into the public school classroom. The Wedge's most prominent tactic toward this end is a CRSC web site offering "Science Education Resources."[68] When the site first went up early in 2000, DI allowed public access to a set of learning (i.e., *teaching*) objectives and a lesson plan on the "Cambrian Explosion."[69] After only a brief period of accessibility, DI restricted access by requiring a user name and a password to obtain this "pilot curriculum"—no doubt because of its dubious constitutionality (the rest of the site is accessible). The now-restricted "Cambrian Explosion Objectives" include learning "Current evidence for the Cambrian explosion" based on "Recent Chinese fossils," with a reference to "P. Chen" (apparently a misspelling of "Chien") so as to present him as an authority; as has been shown, Paul Chien's scientific work on the Cambrian fossils is nonexistent.[70] Listed in "Current competing explanations of the Cambrian explosion" is "Design theory (P. Chen [Chien] and S. [Stephen] Meyer)"—along with "Punctuated equilibrium (N. Eldredge and S. J. Gould)," implying the scientific legitimacy of design theory. The also-restricted "Cambrian Explosion Lesson Plans" actually call for students to "Role Play the Lives of Scientists, Science Educators, and Philosophers of Science," with students pairing off as opposing characters: "[Phillip] Johnson/[William] Provine," "[Stephen] Meyer/[Eugenie] Scott," and "[Michael] Behe/[Michael] Ruse." On a page entitled "Related Websites," listed under "Other Progressive Supplementary Science Curricula," is a link to the creationist Access Research Network. Recognizing the power of the World Wide Web in promoting their teaching resources, the CRSC's web site development schedule includes the plan to "Present [the web curriculum] to NABT, NSTA, and other science education professional groups (1999–2001)."[71] On a page entitled "Web Curriculum Lowers Political Hurdles," CRSC asserts that its "Web curriculum can be appropriated without textbook adoption wars"—meaning, of course, that teachers who want to use it are encouraged to do an end run around textbook adoption procedures.[72]

• Op-ed Fellow

Many wedge members have published editorials in major newspapers. Phillip Johnson has published op-ed pieces in *The Wall Street Journal*

(August 16, 1999) and *The Chronicle of Higher Education* (November 12, 1999). Michael Behe has had pieces in *The New York Times* (October 29, 1996, and August 13, 1999), and Stephen Meyer has written for *The Wall Street Journal* (December 6, 1993) and *The Washington Times* (July 4, 1996). Jay Richards wrote a column for *The Washington Post* (August 21, 1999). CRSC fellow Nancy Pearcey produces a steady stream of op-eds for journals and magazines, most prominently the religious *World* Magazine and *Christianity Today*.[73] The list here is by no means exhaustive.

• PBS (or other TV) Co-production
The Discovery Institute has not yet achieved its goal of a PBS production on intelligent design, but the Wedge has logged quite a bit of television time. Michael Behe and Phillip Johnson were featured on two segments of PBS's *Technopolitics* series, for which DI provided funding.[74] Phillip Johnson was invited by PBS's NOVA Online, as one of the "leading spokesmen in the evolution/creation debate," to share an online discussion with biologist Kenneth Miller in 1996.[75] Stephen Meyer appeared on the PBS program *Freedom Speaks* in March 1997.[76] Johnson, Behe, and CRSC fellow David Berlinski formed the pro-intelligent design side of a debate on PBS's *Firing Line* in December 1997.[77] Johnson was featured as an authority on the Scopes trial on the History Channel's *In Search of History*.[78] Whether they arrange these TV appearances themselves or receive invitations, the air time raises the profile of wedge members.

• Publicity Materials/Publications
Some of the above information is relevant to this category. Discovery Institute also cultivates publicity by announcing CRSC fellows' availability and including contact information for interviews in its U.S. Newswire press releases.[79] In addition, the Wedge has taken masterful advantage of the Internet for publicity. One example is a banner which ran in June 2000 on the conservativeWorldNetDaily.com, advertising the videotape "TheTriumph of Design and the Demise of Darwin," featuring Phillip Johnson. According to PCDataOnline, WorldNetDaily received 4.235 million page views in one week.[80] The videotape was produced by conservative writer and producer Jack Cashill (see www.cashill.com) and advertised by Video Post Productions on a web site, www.triumphofdesign.com, devoted exclusively to its promotion.

## Phase III.   Cultural Confrontation and Renewal

• Academic and Scientific Challenge Conferences
Conferences are supremely important to the wedge strategy: "Once our research and writing have had time to mature, and the public prepared for the reception of design theory, we will move toward direct

confrontation with the advocates of materialist science through challenge conferences in significant academic settings.... The attention, publicity, and influence of design theory should draw scientific materialists into open debate with design theorists, and we will be ready" (Wedge Document, Phase III). CRSC has obviously not waited for intelligent design research to mature before holding conferences. As stated by Phillip Johnson, the wedge movement actually started in 1992 with a conference at Southern Methodist University. Instead of waiting for their research to mature, the Wedge has used conferences since its inception in its attempt to become a "player" in American academia. The wedge strategy crystallized at the "Mere Creation" conference at Biola University in 1996, a conference which, according to CRSC fellow Ray Bohlin, constituted "the backbone of the future direction of the fledgling intelligent design movement."[81] CRSC fellows also attend many conferences held by others—in short, conferencing is a full-time concern for the Wedge. Taken together, just the Wedge's *own* major conferences—six in only eight years, four of them in *very* "significant academic settings"—can be clearly identified as a primary component of the wedge strategy:

(1) Darwinism: Scientific Inference or Philosophical Preference?
Southern Methodist University, Dallas Texas, March 26–28, 1992
(www.leaderu.com/orgs/fte/darwinism/)

(2) Mere Creation: Reclaiming the Book of Nature
Conference on Design and Origins
Biola University, La Mirada, California, November 14–17, 1996
(www.origins.org/mc/menus/sched.html)

(3) Naturalism, Theism and the Scientific Enterprise
University of Texas-Austin, February 20–23, 1997
(www.leaderu.com/offices/koons/menus/conference.html)

(4) The Nature of Nature: An Interdisciplinary Conference on the Role of Naturalism in Science
Michael Polanyi Center, Baylor University, Waco, Texas, April 12–15, 2000
(www.baylor.edu/~polanyi/natconf.htm)

(5) Design and Its Critics: Conference on Intelligent Design—A Critical Appraisal
Concordia University, Mequon, Wisconsin, June 22–24, 2000
(www.cuw.edu/cranach/schedule.htm)

(6) Science and Evidence for Design in the Universe
Yale University, New Haven, Connecticut, November 2–4, 2000
(www.idurc.org/yale.html    and    www.rivendellinstitute.org/business/featurea.htm)

Although these conferences may be officially sponsored or co-sponsored by non-wedge entities, their identity as major wedge events can be discerned by the constant presence of a core of wedge members who, given their relationship as a tightly knit group with carefully orchestrated activities, stand out as the dominant presence at these events.

• Potential Legal Action for Teacher Training
Given the certainty of constitutional challenges should a teacher introduce intelligent design into a public school classroom, the CRSC has taken measures to meet this challenge. Senior fellows David K. DeWolf, a law professor at Gonzaga University, and Stephen Meyer, a philosophy professor at Whitworth College, along with Mark E. DeForrest (not a CRSC fellow), have written *Intelligent Design in Public School Science Curricula: A Legal Guidebook* (Foundation for Thought and Ethics, 1999).[82] On its science education web site, CRSC assures those who register for access to their restricted curriculum that "Email listserv for registered web curriculum users can include legal advice." With the assurance that "Our Curriculum is Legally Permissible in Public Schools," the CRSC urges, "Don't let legal intimidation squash classroom innovation."[83]

• Research Fellowship Program: Shift to Social Sciences and Humanities
The re-emphasis on fellowships in Phase III appears to have figured in the Wedge's plan for the controversial and now defunct Michael Polanyi Center at Baylor University, given the MPC's plan, stated on its "Events and Programs" web page, for "A research fellowship program so that the MPC can sponsor a steady stream of top scientists and scholars ... at Baylor."[84] Run by CRSC fellows William Dembski (Director) and Bruce Gordon (Associate Director), the MPC was clearly established to advance the study of intelligent design theory: "Design ... may serve to elucidate various phenomena that prove intractable from the standpoint of neo-Darwinian and self-organizational approaches. Present design-theoretic research holds much promise, but the ultimate significance of design theory remains to be seen. Nonetheless, the MPC sees design-theoretic ideas as a promising resource for understanding the complexity we observe in nature, and is committed to pursuing this avenue of research to see what fruit it will bear."[85] More importantly with respect to Phase III was the MPC's explicit reflection of the Wedge's goal of extending design theory to the social sciences and humanities: "The first [goal of the Polanyi Center] is to promote and pursue research in the historical development and conceptual foundations of the natural and social sciences.... The impact of science on the humanities and the arts is the second focus of research at Michael Polanyi Center." One of the Wedge Document's Twenty Year Goals is "To see design theory application in specific fields, including ... psychology, ethics, politics, theology and philosophy in the humanities; to see its [influence] in the fine arts." Yet the most telling

connection between the Polanyi Center and the Wedge was this statement of the Polanyi Center's purpose: "The successful achievement of these goals, therefore, is a task that the Michael Polanyi Center shares with a network of individual scholars and other established Centers around the world that have similar research projects." The most prominent center with a similar research project is the Discovery Institute's Center for the Renewal of Science and Culture.

## Part III: Analysis of the Nature of the Wedge Strategy and Its Advance into the Mainstream

Christians in the 20th century have been playing defense.... They've been fighting a defensive war to defend what they have, to defend as much of it as they can.... It never turns the tide. What we're trying to do is something entirely different. We're trying to go into enemy territory, their very center, and blow up the ammunition dump. What is their ammunition dump in this metaphor? It is their version of creation.[86]

—Phillip E. Johnson, February 6, 2000, at a meeting of the National Religious Broadcasters in Anaheim, California

Since Darwin, we can no longer believe that a benevolent God created us in His image.... Intelligent Design opens the whole possibility of us being created in the image of a benevolent God.... The job of apologetics is to clear the ground, to clear obstacles that prevent people from coming to the knowledge of Christ.... And if there's anything that I think has blocked the growth of Christ as the free reign of the Spirit and people accepting the Scripture and Jesus Christ, it is the Darwinian naturalistic view.... It's important that we understand the world. God has created it; Jesus is incarnate in the world.[87]

—William Dembski, February 6, 2000, at a meeting of the National Religious Broadcasters in Anaheim, California

The key to understanding the CRSC's activities is to understand fully the true nature of the wedge movement. The important point is that the wedge strategy—the intelligent design movement as a whole—really has nothing to do with science, despite its proponents' affirmations to the contrary. Johnson actually admitted this in 1996: "This isn't really, and never has been, a debate about science.... It's about religion and philosophy."[88] Not a single area of science has been affected in any way by intelligent design theory. In actuality, this "scientific" movement which seeks to permeate the American academic and cultural mainstream is religious to its core.

In March 1994, Johnson attended a conference on "Regaining a Christian Voice in the University." He delivered a lecture entitled "The Real Issue: Is God Unconstitutional?" in which he lamented the increasing prevalence in American universities of "scientific naturalism," which, according to Johnson, is "the established religious philosophy of America."[89] This lament is a constant theme in Johnson's campaign to promote "intelligent design"; it fuels the mission by Johnson and his CRSC associates to get "intelligent design theory" into the academic world and into public life as the chief competitor of the theory of evolution. Johnson's words at this conference reveal just how important this mission is to them: "The bitter debate over whether 'creation' or 'intelligent design' may be considered as a possibility in scientific discourse is no minor matter. Behind it lies one of the most important questions of human existence: Did God create Man, or did Man create God?"[90]

In 2000, Johnson was still lamenting what he sees as the departure of God from both secular and religious universities. In *The Wedge of Truth*, he recounts the story of Philip Wentworth, who, according to Wentworth's essay "What College Did to My Religion," entered Harvard in 1924 as a faithful Presbyterian youth and emerged several years later as a disillusioned convert to "scientific naturalism," having been taught, as Johnson puts it, by "infidels."[91] Johnson sees parallels between Wentworth's and his own experience at Harvard more than thirty years later. They were both victims of an "elite" who "are particularly skilled at inventing ways to tame God because they desire either to ignore God or to use him for their own purposes." They were defenseless young people in a university which "offered no instructions in how to recognize idolatry."[92]

According to Johnson, Wentworth's—and Johnson's own—experience of "apostasy" at Harvard are "representative of the experience of an entire culture of educated people over more than a century" because of scientific naturalism (*Wedge of Truth*, 20). Hence, in Johnson's mind, the only remedy for such apostasy is to institute a completely new scientific paradigm and methodology: "Phillip Johnson's idea of revolution is not ... a struggle to control one corner of the ivory tower. He is playing for all the marbles for the governing paradigm of the entire thinking world. He believes evolution's barren rule can be overturned, that it is rip[e] for revolution...."[93] Yet Johnson and his wedge associates are only using science as the façade behind which to stage their revolution, which, according to their plan, will establish their religious worldview as the foundation of American cultural and academic life. Paradoxically, they are pursuing their remedy for "scientific naturalism" from *outside science*. They are not attempting to change the way science is currently done by introducing a *better* methodology or more viable hypotheses; if they were, they would actually be doing scientific research and presenting it at scientific conferences to be vetted by scientific peers. Rather, they are trying to change the way the public and influential policy-makers *perceive* science through their aggressive program of public relations activities. This is crucial to their strategy.[94]

In May 2000, the wedge strategy took another crucial turn—toward implementing the Discovery Institute's overtly political goal to "cultivate and convince . . . congressional staff . . ." (Wedge Document). In a May 8 press release, DI announced that "Discovery Institute will bring top scientists and scholars to Washington D.C. to brief Congressional Representatives and Senators and their staffs on the scientific evidence of intelligent design and its implications for public policy and education, Wednesday, May 10, in the U.S. Capitol Building and Rayburn Office Building."[95] Seven members of the U.S. House of Representatives, both Republicans and Democrats, co-hosted the briefing, which was attended by about fifty people. One congressman, Rep. Thomas Petri of Wisconsin, was at one time a Discovery Institute Adjunct Fellow, according to the Summer 1996 Discovery Institute *Journal*.[96] Another, Rep. Mark Souder of Indiana, published a defense of intelligent design after the briefing in the June 14, 2000, *Congressional Record* (H4480). David Applegate, director of the American Geological Institute's Government Affairs Program, in an AGI "Special Update" sent to AGI member societies to alert them about the briefing, noted Rep. Souder's membership on the House Education Committee and Rep. Petri's expected upcoming chairmanship of the House Education and Workforce Committee.[97] This is the most convincing evidence to date of the political ambitions of the Wedge, and this ambition is aimed primarily at an important target: American public schools. The possibility also exists that Discovery Institute will attempt to secure government funding for intelligent design research. The briefing was a small but significant advance for the Wedge into the political arena.

In the Discovery Institute's August 1996 *Journal*, DI president Bruce Chapman explicitly connects the Center for the Renewal of Science and Culture with not only a religious mission but also a political one:

The new Center for the Renewal of Science and Culture is an exciting and ambitious exemplar of Discovery Institute's role as a futurist think tank with prudential principles.

... [It] challenges policy makers—and even our members and sponsors—to stretch their own thinking, ... It calls upon their imagination to see the world not just as we received it, but as it is becoming and can become....

... We think some of these ideas are destined to change the intellectual—and in time the political—world....

... The more you read about this program ... the more you will realize the radical assault it makes on the tired and depressing materialist culture and politics of our times, as well as the science behind them. Then, when you start to ponder what society and politics might become under a sounder scientific dispensation, you will become truly inspired.

... There is great comfort, courage and resolve in the moral and political legacy of our civilization as formulated in the Bible, history and the writing of the American Founders. So it is fitting that our news of the Center for Renewal of Science and Culture this month is accompanied by publication of Discovery Senior Fellow John West's superb book, *The Politics of Revelation and Reason.* Before you opine about the place of religion in politics, (or why there shouldn't be any), use this scholarly, but very readable, account of religion in early American politics. It will surprise you—and perhaps, it will inspire, too.[98]

The Center for Renewal of Science and Culture fits well with Discovery's existing programs in high technology and religion.[99]

With the political wheels of the Wedge having been set in motion, Johnson, as is obvious in his most recent book, *The Wedge of Truth*, is no longer trying to disguise the religious nature of the wedge strategy. He reveals this with a biblical reference in a recent interview about the book:

*[Interviewer]:*   How would you describe the main purpose of *The Wedge of Truth* in comparison to your other books?

*[Johnson]:*   Each of my books builds upon the logic that was erected in my previous ones. My prior books argued that the real discoveries of science—as opposed to the materialist philosophy that has been imposed upon science—point straight towards the reality of intelligent causes in biology.... [T]here are two definitions of "science" in our culture. One definition says that scientists follow the evidence regardless of the philosophy; the other says that scientists must follow the (materialist) philosophy regardless of the evidence. The "Wedge of Truth" is driven between those two definitions, and enables people to recognize that "In the beginning was the Word" is as true scientifically as it is in every other respect.[100]

Moreover, not only is the wedge strategy founded on and fueled by religious zeal, but it is merely the newest "evolution" of good old-fashioned American creationism. According to Johnson, the scientific "creation myth" must be replaced by the true account of human existence:

That God created us is part of God's general revelation to humanity, built into the fabric of creation. This foundational truth is something which, in the words of the political philosopher [and Wedge member] J. Budziszewski, *we can't not know....*

... The proper metaphysical basis for science is not naturalism or materialism but the fact that the creator of the cosmos not only created an intelligible universe but also created the powers of reasoning which enable us to conduct scientific investigations.... True science will also remember that only some aspects of reality can be understood through observation and experiments....

... [T]he materialist story thrives only as long as it does not confront the biblical story directly. In a direct conflict, where the public perceives the issues clearly, the biblical story will eventually prevail over the materialist story....

... What we need is for God himself to speak, to give us a secure foundation on which we can build.... So it is of the greatest importance that we ask the question: "Has God *done* something to give us a start in the right direction, or has he left us alone and on our own?"

When we have reached that point in our questioning, we will inevitably encounter the person of Jesus Christ, the one who has been declared the incarnate Word of God, and through whom all things came into existence. This time *he* will be asking the question that is recorded in the Gospels: "Who do men say that I am?" ...

... When the naturalistic understanding of reality finally crashes and burns ... the great question Jesus posed will come again to the forefront of consciousness. Who should we say that he is? Is he the one who was to come, or should we look for another?

As a Christian I have answers to those questions, and of course other people will have different answers. The Wedge philosophy is that the important thing is to get the right questions on the table, and that task requires that we invite any and all answers for a fair hearing. For now my point is merely that a question which was long assumed to be off the table will become important again if the cultural debate over Darwinism and naturalism goes in the direction I am predicting. We are not talking about some mere revision of a particular scientific theory. *We are talking about a fatal flaw in our culture's creation myth, and therefore in the standard of reasoning that culture has applied to all questions of importance....*

... The basic story of the Incarnation—that God has taken human form ... is more equivalent to the scientific truth that apples fall down rather than up....[101]

There is no doubt that the message Johnson wants his readers to hear is that "theistic" science, properly built only upon a metaphysics of supernatural creation, is marching toward Jesus—and straight into the academic and cultural mainstream.

Establishing a presence in American higher education is one of the Wedge's most ambitious goals. CRSC fellow Nancy Pearcey is optimistic about its success: "The new strategy centers on a concept labeled intelligent design. The design movement shows promise of winning a place at the table in secular academia, while uniting Christians concerned about the role science plays in the current culture wars."[102] Wedge strategists do not expect to establish a large presence—indeed, as stated in the Wedge Document, they do not even believe it is necessary: "A lesson we have learned from the history of science is that it is unnecessary to outnumber the opposing establishment. Scientific revolutions are usually staged by an initially small and relatively young group of scientists who are not blinded by the prevailing prejudices and who are able to do creative work at the pressure points, that is, on those critical issues upon which whole systems of thought hinge."[103] What is most important is that they establish a presence *inside* the academic establishment and the cultural mainstream. This is slowly but surely taking place.

Nothing is more important to the Wedge than the academic respectability that comes from earning degrees and securing teaching positions at well known, respectable universities. One of the goals of the wedge strategy is to have ten CRSC fellows teaching at major universities by 2003. They have already more than realized that goal in the following CRSC fellows:

Michael Behe—Lehigh University
Walter Bradley—Texas A&M (until his retirement)
J. Budziszewski—University of Texas-Austin
John Angus Campbell—University of Memphis
Robert Koons—University of Texas-Austin
Paul Chien—University of San Francisco
David K. DeWolf—Gonzaga University
Guillermo Gonzales—University of Washington-Seattle
Bruce Gordon—Baylor University (former part-time instructor and current assistant research professor)[104]
Phillip E. Johnson—University of California-Berkeley (until retirement)
Robert Kaita—Princeton University
Dean H. Kenyon—San Francisco State University (California State University system)
Scott Minnich—University of Idaho

Henry F. Schaefer—University of Georgia
Richard Weikart—California State University-Stanislaus

Mary Beth Marklein's comment in *USA Today*, "From the intelligent-design movement, advanced by scholars at respected universities, is emerging what could become a battle in science research," creates in the minds of its readers exactly the impression that the Discovery Institute wants the American public to have.[105]

By carving a niche for themselves in university life, both secular and religious, DI's intelligent design creationists are in a position to accomplish several things:

(1) To cultivate a facade of academic legitimacy.
Johnson realizes that academic legitimacy is the first hurdle the Wedge must overcome:

> The conference [at Southern Methodist University in March 1992] brought together as speakers some key Wedge figures, particularly Michael Behe, Stephen Meyer, William Dembski, and myself. It also brought a team of influential Darwinists, headed by Michael Ruse, to the table to discuss this proposition: "Darwinism and neo-Darwinism as generally held in our society carry with them an *a priori* commitment to metaphysical naturalism, which is essential to making a case on their behalf." ... the amazing thing was that a respectable academic gathering was convened to discuss so inherently subversive a proposition.[106]

In an interview with *Communiqué: A Quarterly Journal*, Johnson also acknowledges the difficulty of acquiring this legitimacy, but is resolutely committed to achieving it:

> *CJ:*   That seems to be behind the idea of driving "the wedge" into the scientific community—that you'd just encourage them [students and faculty] to get behind guys like Behe and join that momentum.
>
> *Phil:*   Yes, the idea is that you get a few people out promoting a new way of thinking and new ideas, it's very shocking, and they take a lot of abuse.... [Y]ou have to have people that talk a lot about the issue and get it up front and take the punishment and take all the abuse, and then you get people used to talking about it. It becomes an issue they are used to hearing about, and you get a few more people and a few more, and then eventually you've legitimated it as a regular part of the academic discussion. And that's my goal: to legitimate the argument over evolution ... as a mainstream scientific and academic issue.... We're bound to win.... We just have to normalize it, and that takes patience and persistence, and that's what we are applying.[107]

(2) To influence college students, too many of whom are ignorant of genuine science, thus recruiting them into the wedge movement.
In *Touchstone* Magazine (July/August 1999), Johnson updates the progress of the wedge. His remarks indicate that universities are fertile recruit-

ing ground for the wedge: "[M]any ... college students are reading our literature, and are responding very favorably.... The most talented of these will be the Wedge members of the future."[108] Colleges and universities are the logical source of the "future talent" which, according to the Wedge Document, the CRSC seeks "to cultivate and convince."

(3) To cultivate the support of university administrators and financial donors.

We believe that, with adequate support, we can accomplish many of the objectives ... in the next five years (1999–2003).... For this reason we seek to cultivate and convince influential individuals,... college and seminary presidents and faculty,... and potential academic allies. (Wedge Document)

As shown earlier, the Wedge has secured the financial support of benefactors such as Howard Ahmanson. The Wedge's stability may depend upon the continuation of this lucrative support, as indicated by Stephen Meyer after they received increased funding in 1999: "We not only have a larger program than before, the existence of 'outyear' funding means greater long term stability."[109] William Dembski and Bruce Gordon were successful in securing Baylor University president Robert Sloan's support for the establishment, if not the indefinite continuation, of the Michael Polanyi Center. This support manifested itself the following year in Sloan's strenuous defense of their presence at Baylor during a controversy over its establishment, recounted in the *Houston Chronicle*: "[Sloan] said alumni, students and parents have 'overwhelmingly' supported the goals of the Polanyi Center, but he would still back the center even without such support."[110] Sloan at the time buttressed his moral support of the Polanyi Center with financial backing for Dembski and Gordon: "Baylor spokesman Larry Brumley said the university will pay for Dembski's salary after the [John Templeton Fund] grant expires next year, and that it is paying Gordon's salary." As the *Chronicle* also revealed, the John Templeton Fund was also a source of support for these activities: "[Dembski's] salary at the Polanyi Center is paid by a $75,000 grant from the John Templeton Fund, distributed through the Discovery Institute."[111]

(4) To acquire physical bases of operation, with access to all the advantages this brings.

The Polanyi Center was established at Baylor in October 1999, giving the CRSC its first physical base outside the Discovery Institute in Seattle. As of this writing, the Wedge has not established anything like the Polanyi Center in any secular university. However, both at the Polanyi Center and other universities where they have held conferences, wedge members have accomplished the next best thing: they are, in effect, bringing the secular universities to *them* by inviting mainstream scholars and scientists to participate in their conferences. The "Nature of Nature" conference featured major academic figures in April 2000, and Discovery Institute publicized this fact in an April 7, 2000, news release, pointing out that "Among the participants are two Nobel Prize winners, Steven Weinberg and Christian de Duve, as well as noted scientists Alan Guth, Simon Conway Morris and others.... 'This is going to be the greatest collection of minds on the subject of directionality versus contingency in the natural sciences,' said [William] Dembski."[112] After the conference, DI celebrated the achievement of the goal of making "the role of naturalism in science an acceptable topic of academic discussion, and to create a non-confrontational forum for rival scholars to interact on the issue."[113] Phillip Johnson explicitly linked himself to the Baylor conference when he boasted that "we had a conference at Baylor University in April 2000 to discuss whether the evidence of nature points towards or away from the need for a supernatural creator. It was probably the most distinguished conference in Baylor history, with two Nobel Prize winners and many of the country's most distinguished professors in science, philosophy, and history."[114]

(5) They can exploit their presence in higher education, using their credentials to "snow" the public.

Academic credentials are the ticket to success for the Wedge, and members take every opportunity to publicize their own. An example is a short article by CRSC fellow Ray Bohlin, executive director of Probe Ministries, entitled "Mere Creation: Science, Faith & Intelligent Design." In no more than roughly five pages, he never mentions a fellow CRSC member (and he refers to six of them) without also stating the fellow's academic credentials and accomplishments, as in the case of Henry Schaefer: "So said Dr. Henry F. Schaefer III, professor of chemistry at the University of Georgia, author of over 750 scientific publications, director of over fifty successful doctoral students, and five-time Nobel nominee...."[115]

The accomplishment of these goals is especially important to the CRSC's strategy to advance their brand of creationism; indeed, it is critical because they are the only creationists who stand a chance of pulling it off. The old-style creationism represented by Henry Morris, Duane Gish, and others is unlikely to be tolerated on mainstream campuses, even religious ones like Baylor. The CRSC creationists have taken the time and trouble to acquire legitimate degrees, providing them a degree of cover both while they are students and after they join university faculties. Johnson alludes to this in the interview with *Communiqué*:

*CJ:*   Along those lines, what encouragement would you offer to a young student of science—let's say a young lady beginning a Ph.D. program in microbiology at a major university?

*Phil:*   We have a wonderful example here in Michael Behe ... in what he is able to do while retaining a well funded lab and standing in the scientific world.... The fact is that there are a lot of people in science who just don't want to be bothered with the whole Darwinian ideological agenda. It doesn't have anything to do with the scientific work that they do, so they are patient with it. I think if we're clever enough in quoting the arguments and keeping people in the conversation and so on, and reassuring them that they can doubt Darwinism and still practice science just as well as ever—that it doesn't mean they are going to give up science and, you know, start thumping bibles instead or whatever—I think there'll just be a growing number of people who will get used to that conversation in that element. Behe has so far been able to maintain his standing, and he's getting invitations everywhere. Once you get someone like that [who] breaks the ice, then there are opportunities for more people. So, I don't think you need to be in despair, *but you need to use a lot of tact and judgment and keep your head down while you're getting your Ph.D. in a lot of places—because there is dogmatism, but there are ways to overcome that.*[116] [Emphasis added.]

By keeping their "head down" at the universities where they teach and study, intelligent design creationists blend more smoothly into the academic population. They can do this either by compartmentalizing their creationism—separating their involvement from what they do professionally on their respective campuses—or by cloaking it in technical, esoteric, and therefore more palatable, language. They thereby present less risk of embarrassment to their universities and increase their chances of being tolerated, at least by administrators who are either sympathetic to them, unaware of their agendas, or scientifically unsophisticated. An example of this is Robert Koons' hosting of the "Naturalism, Theism and the Scientific Enterprise" conference in 1997 at the University of Texas. Koons acknowledges the advantage of a sympathetic department head: "I ... spoke to my department head [Daniel Bonevac] about making the department the official host. My chairman is a good friend of mine (who also happens to be a Christian and is very sympathetic to this sort of thing) and he agreed to attach the department's name to the conference. We didn't get any money from the university, but we did get clerical and administrative support."[117]

William Dembski plays an essential role in the advancement of the wedge strategy in academia; the proof of this was his directorship of the Michael Polanyi Center. An essential point to understand, moreover, is that Dembski and Phillip Johnson are inseparable. Each cites the other as a key figure in the intelligent design movement. Johnson refers to Dembski as one of the "key Wedge figures."[118] Dembski cites Johnson as one of the people with whom the movement begins and whose book *Darwin on Trial* was a "key text" in the movement.[119] Moreover, Johnson has acknowledged as recently as August 1999 his own role as a representative of the movement and its role in carrying the intelligent design debate into higher education, as well as public discussion: "[Evidence for intelligent design] is given in books published by the academic publishers, like Cambridge University Press, and by other scholars, scientists, philosophers in the intelligent design movement, which I represent, and which is carrying this issue into the universities and into the mainstream public discussion."[120]

Targeting academia and public opinion is intended to advance the Wedge's goal of undermining evolutionary theory, thus creating an opening for CRSC's "new" paradigm of "theistic science." The fact that "theistic science" will never overthrow mainstream science is irrelevant to the strategy. At present, just getting the subject into the academic and cultural mainstream—even when it is attacked—is an advancement. As early as 1996, in a review of Del Ratzsch's book, *The Battle of the Beginnings: Why Neither Side Is Winning the Creation/Evolution Debate*, Johnson acknowledged that even carrying the discussion into the Christian academic world is a "scandal" but "exciting":

Our movement is something of a scandal in some sections of the Christian academic world for the same reason that it is exciting: we propose actually to engage in a serious conversation with the mainstream scientific culture on fundamental principles, rather than to submit to the demand that naturalism be conceded as the basis for all scientific discussions. That raises the alarming possibility, as one of Ratzsch's colleagues put it in criticizing me, that "the gulf between the academy and the sanctuary will only grow wider." The bitter feeling that has been spawned in some quarters by that possibility may explain why Ratzsch discusses our group so tentatively, but no matter. What matters for the present is to open up discussion....[121]

He acknowledged it again in 1997 in a quote by the *New York Times*: "Mr. [Kenneth] Miller also skewered Mr. Behe's book in a recent review. But that the book was even reviewed is progress in Mr. Johnson's view: 'This issue is getting into the mainstream. People realize they can deal with it the way they deal with other intellectual issues.... My goal is not

so much to win the argument as to legitimate it as part of the dialogue.'"[122] Two years later, in Spring 1999, Johnson was still describing the intelligent design movement as primarily "destructive" in its
function—admitting that the intelligent design movement so far has
produced no "answers" of its own, despite its hope to have some in the
future:

*CJ:* So, would it be fair to say that the goal is to undermine or call into question
what has generally been accepted in the scientific community rather than purporting your own answers to all of the questions?

*Phil:* Yes, the starting point is to understand what in the official answers is just
dead wrong, because you can't get anywhere until you've made that step. Now,
obviously at some time in the future you hope to get to better answers which are
actually true, and that's a positive program, but you can't begin to work in that
direction until you have an acknowledgement that the existing answers are false.
You have to get the questions right before you can even determine the falsity of
the answers. So, for the time being, it's primarily a destructive work that's aimed
at opening up a closed dogmatic field to new insights.[123]

Despite the difficulties, however, the movement continues unabated, and
getting a foothold in the academic world is crucial to the strategy, as
Johnson stressed in February 1999 at D. James Kennedy's "Reclaiming
America for Christ" conference: "Johnson added that he is happy to be
working with university professors, such as Michael Behe of Lehigh
University in Pennsylvania. . . . This strategy, he said, 'enables us to get a
foothold in the academic world and the academic journals. . . .'"[124]

Clearly, Johnson does not see this effort as having short-term results;
rather, he sees the promotion of intelligent design as a long-term project
which will bear fruit after present wedge members are gone: ". . . I hope
we'll be remembered as the pioneers who opened up the criticism and
made it possible for the change to occur. It'll take decades . . . and we
won't be around to see the final days, but maybe we'll be remembered as
among those who started the ball rolling, and that'll be a great satisfaction."[125] In the meantime, the goal is to stay on the offensive and wear
down the opposition: "Johnson speaks of a wedge strategy, with himself
the leading edge. 'I'm like an offensive lineman in pro football,' he says.
. . . 'My idea is to clear a space by legitimating the issue, by exhausting
the other side, by using up all their ridicule.'"[126]

In Johnson's mid-1999 assessment of the success of the wedge strategy
(*Touchstone* Magazine, July/August 1999), he remains convinced that all

the wedge strategists have to do is to be patient and eventually the academic world will come around: "As the discussion proceeds, the intellectual world will become gradually accustomed to treating materialism and naturalism as subjects to be analyzed and debated, rather than as tacit foundational assumptions that can never be criticized. Eventually the answer to our prime question will become too obvious to be in doubt."[127] Asserting his confidence that this strategy will work, he uses Dembski as an example: "I attended a seminar on Dembski's ideas recently at a major university philosophy department where I saw from the reactions how common it is for clever people to deploy their mental agility in the service of obscurity. But Dembski put the concept of intelligent design on their mental maps, and eventually they will get used to it."[128] Clearly, the resistance to Dembski's ideas does not deter Johnson, and just as clearly, he thinks the key is to just dig in and ride out the controversy. Notably, he does *not* say wedge strategists must improve their arguments or present hard scientific data to bolster their arguments.

Also clear is the fact that Johnson views this movement as religious at its core. Speaking in February 1999 at the "Reclaiming America for Christ" conference convened by D. James Kennedy, who has become one of the Religious Right's leading figures through his Coral Ridge Ministries, Johnson again revealed the true religious nature of the movement, which is aimed at creating divisions in "the other side":

The objective, he said, is to convince people that Darwinism is inherently atheistic, thus shifting the debate from creationism vs. evolution to the existence of God vs. the non-existence of God. From there people are introduced to "the truth" of the Bible and then "the question of sin" and finally "introduced to Jesus."

"You must unify your own side and divide the other side," Johnson said. He added that he wants to temporarily suspend the debate between the young-Earth creationists, who insist that the planet is only 6,000 years old, and old-Earth creationists, who accept that the Earth is ancient. This debate, he said, can be resumed once Darwinism is overthrown.[129]

Apparently this view is shared by Johnson's colleagues in the movement and is considered its "defining concept":

My colleagues and I speak of "theistic realism"—or sometimes, "mere creation" —as the defining concept of our movement. That means that we affirm that God is objectively real as Creator, and that the reality of God is tangibly recorded in evidence accessible to science, particularly in biology. We avoid the tangled arguments about how or whether to reconcile the Biblical account with the present state of scientific knowledge, because we think these issues can be much more constructively engaged when we have a scientific picture that is not distorted by naturalistic prejudice.[130]

## Conclusion

Until the present study, there was no comprehensive survey of the activities in which the Discovery Institute's Center for the Renewal of Science and Culture has engaged in order to execute the wedge strategy. Such a survey is highly instructive. The head-spinning pace of wedge activity is an indication of the urgency with which wedge strategists at the CRSC view their mission. It is unlikely that even the recent election losses of the creationist candidates in the August 2000 Kansas school board primaries will slow the Wedge's momentum. For such a movement—fueled by religious zeal, funded by sympathetic benefactors, and aided by political alliances—this defeat was only a momentary setback. The stream of public relations events, conferences, books, and "educational materials" for public schools continues energetically. The Wedge continues its advance, guided by its vision of a "promised land" in which Darwin's powerful legacy has lost its hard-won place in the scientific enterprise.

## Notes

1. Phillip E. Johnson, *Defeating Darwinism by Opening Minds* (Downers Grove, Illinois: InterVarsity Press, 1997), 91–92.

2. Center for the Renewal of Science and Culture, "What Is the Center for the Renewal of Science and Culture All About? The Mission of the Center" [online]. Accessed 18 March 2000 at http://www.discovery.org/w3/discovery.org/crsc/aboutcrsc.html. This document was found in a directory which is no longer accessible.

3. Stephen Goode, "Johnson Challenges Advocates of Evolution," *Insight on the News*, October 25, 1999 [online]. Accessed at Access Research Network on 9 July 2000 at http://www.arn.org/docs/johnson/insightprofile1099.htm.

4. Nancy Pearcey, "Wedge Issues: An Intelligent Discussion with Intelligent Design's Designer," *World*, July 29, 2000 [online]. Accessed on 13 April 2001 at http://www.worldmag.com/world/issue/07-29-00/closing_2.asp.

5. Nancy Pearcey, "We're Not in Kansas Anymore: Why Secular Scientists and Media Can't Admit that Darwinism Might Be Wrong," *Christianity Today*, May 22, 2000 [online]. Accessed on 3 August 2000 at http://christianityonline.com/ct/2000/006/1.42.html.

6. Lynn Vincent, "Science vs. Science," *World*, February 26, 2000 [online]. Accessed on 13 April 2001 at http://www.worldmag.com/world/issue/02-26-00/national_1.asp.

7. Tim Stafford, "The Making of a Revolution," *Christianity Today*, December 8, 1997 [online]. Accessed on 13 April 2001 at http://www.christianitytoday.com/ct/7te/7te016.html.

8. Phillip E. Johnson, "The Wedge: Breaking the Modernist Monopoly on Science," *Touchstone* Magazine [online], July/August 1999. Accessed 9 March 2000 at http://www.touchstonemag.com/docs/issues/12.4docs/12-4pg18.html.

9. Vincent, "Science vs. Science," at http://www.worldmag.com/world/issue/02-26-00/national_1.asp.

10. Tom Woodward, "Meeting Darwin's Wager," Part 2, *Christianity Today*, April 28, 1997 [online]. Accessed on 27 March 2000 at http://www.christianitytoday.com/ct/7t5/7t514b.html.

11. "Ad Hoc Origins Committee: Scientists Who Question Darwinism" [online]. Accessed on 24 May 2000 at http://www.apologetics.org/news/adhoc.html. Johnson's reply to Gould was printed in *Origins Research* 15:1 (the forerunner of the creationist journal *Origins and Design*) and is online at http://www.arn.org/docs/orpages/or151/151johngould.htm.

12. "Ad Hoc Origins Committee." The complete list of signatories is available at http://www.apologetics.org/news/adhoc.html. Notre Dame philosopher Alvin Plantinga was also a signatory to this letter, which is early evidence of his continuing support of the intelligent design movement. Nancy Pearcey refers to Plantinga as a "design proponent." See Nancy Pearcey, "We're Not in Kansas Anymore: Why Secular Scientists and Media Can't Admit that Darwinism Might Be Wrong," *Christianity Today*, May 22, 2000 [online]. Accessed on 13 April 2001 at http://christianityonline.com/ct/2000/006/1.42.html.

13. "Major Grants Help Establish Center for Renewal of Science and Culture," Discovery Institute *Journal*, summer (August) 1996 [online]. Accessed on 24 July 2000 at http://wwwdiscovery.org/w3/discovery.org/journal/center.html. See published presentations from this conference in *The Intercollegiate Review*, spring 1996.

14. Phillip E. Johnson, *Reason in the Balance: The Case against Naturalism in Science, Law and Education* (Downers Grove, Ilinois: InterVarsity Press, 1995), 208–209.

15. Even though the CRSC—the creationist arm of the larger Discovery Institute—is the subject of this study, "CRSC" and "Discovery Institute" are often used interchangeably, as will occasionally be done here.

16. "Major Grants Help Establish Center for Renewal of Science and Culture," Discovery Institute *Journal*, summer (August) 1996 [online]. Accessed on 24 July 2000 at http://wwwdiscovery.org/w3/discovery.org/journal/center.html.

17. Larry Witham, "Contesting Science's Anti-Religious Bias," *The Washington Times*, December 29, 1999. Accessed online on 24 May 2000 on the Discovery Institute web site at http://www.discovery.org/viewDB/index.php3?program=CRSCstories&command=view&id=65.

18. This information was on a web page at http://www.discovery.org/w3/ discovery.org/crsc/crsc96fellows.html but is no longer accessible. Another page from this directory, which appears to somewhat later than 1996–97, although the year is not specified, also lists additional people as fellows: Walter Bradley, Chair, Dept. of Mechanical Engineering, Texas A&M; John Angus Campbell, Professor of Speech Communications, University of Memphis; William Lane Craig, Research Professor, Talbot School of Theology; Jack Harris, Ph.D. candidate, University of Washington; Dean H. Kenyon [co-author, *Of Pandas and People*], San Francisco State University; Nancy Pearcey, Wilberforce Forum; and Charles Thaxton, Charles University, Prague. George Gilder is listed as an advisor along with Phillip Johnson.

19. See the "Mere Creation" web site for this conference at http://www.origins. org/mc/menus/index.html. See also the article by Jay Grelen, "Witnesses for the Prosecution," *World*, November 30, 1996 [online]. Accessed 1 March 2000 at http://www.worldmag.com/world/issue/11-30-96/national_2.asp.

20. Henry F. Schaefer III, "Foreword," in *Mere Creation: Science, Faith and Intelligent Design*, ed. William Dembski (Downers Grove, Illinois: InterVarsity Press, 1998), 9.

21. See http://www.leaderu.com/menus/aboutus.html. Accessed on 13 April 2001.

22. Neither this conference nor any other CRSC activities have produced any scientific research, as will be shown in Part II of this study.

23. Scott Swanson, "Debunking Darwin?" *Christianity Today*, January 6, 1997 [online]. Accessed on 25 July 2000 at http://www.christianitytoday.com/ct/7t1/ 7t1064.html.

24. The NTSE conference is discussed in the analysis of wedge activities in Part III.

25. Swanson, "Debunking Darwin?" *Christianity Today*.

26. Schaefer, "Foreword," in *Mere Creation: Science, Faith and Intelligent Design*, 10–11.

27. This is a reference to Roberta and Howard Ahmanson, whose financial support of the CRSC will be discussed later.

28. Johnson, *Defeating Darwinism*, 92.

29. Phillip E. Johnson, *Objections Sustained: Subversive Essays on Evolution, Law and Culture* (Downers Grove, Illinois: InterVarsity Press), 1998.

30. Center for the Renewal of Science and Culture, "The Wedge Strategy," [online]. Available at http://www.humanist.net/skeptical/wedge.html [accessed 17 April 2000].

31. Center for the Renewal of Science and Culture, "The Wedge Strategy," at http://www.humanist.net/skeptical/wedge.html.

32. Accessed 18 March 2000 at http://www.discovery.org/w3/discovery.org/crsc/ aboutcrsc.html. These documents were no longer available as of 17 May 2000, when access was attempted and an error message saying "File Not Found!" was received. The directory had been available at http://www.discovery.org/w3/ discovery.org/crsc as of 18 March 2000, but access was denied as of 17 April

2000 ("Forbidden. You don't have permission to access w3/discovery.org/crsc/ on this server").

33. See the Capitol Research Center's "After Henry Salvatori: California's 'Most Generous' Conservative Philanthropists," October 1998 [online]. Accessed on 4 April 2000 at http://www.capitalresearch.org/fw/fw-1098.html. See also Steve Benen, "From Genesis to Dominion," *Church & State*, July/August 2000 [online]. Accessed on 21 July 2000 at http://www.au.org/churchstate/cs7003.htm.

34. "Major Grants Help Establish Center for the Renewal of Science and Culture," *Discovery Institute Journal*, summer 1996 [online]. Accessed on 3 July 2000 at http://www.discovery.org/w3/discovery.org/journal/center.html.

35. "Major Grants Increase Programs, Nearly Double Discovery Budget," Discovery Institute *Journal*, 1999 [online]. Accessed on 13 June 2000 at http://www.discovery.org/w3/discovery.org/journal/1999/grants.html.

36. Larry Witham, "Contesting Science's Anti-Religious Bias," *Washington Times*, December 29, 1999. Accessed online on 22 May 2000 at http://www.discovery.org/viewDB/index.php3?program=CRSCstories&command=view&id=65.

37. Bruce Chapman, "Ideas Whose Time Is Coming," Discovery Institute *Journal*, summer 1996 [online]. Accessed on 28 August 2000 at http://www.discovery.org/w3/discovery.org/journal/president.html.

38. See the current list of CRSC fellows at http://www.discovery.org/crsc/fellows/index.html. Accessed on 13 April 2001.

39. Johnson, "The Wedge: Breaking the Modernist Monopoly on Science," at http://www.touchstonemag.com/docs/issues/12.4docs/12-4pg18.html.

40. Center for the Renewal of Science and Culture, "The Wedge Strategy," at http://www.humanist.net/skeptical/wedge.html.

41. Center for the Renewal of Science and Culture, "The Research Fellowship Program" [online]. Accessed on 28 August 2000 at http://www.crsc.org/fellows/fellowshipInfo.html.

42. "CRSC Innovates in Media and Academia," Discovery Institute *Journal*, spring 1998. Accessed on 13 April 2001 at http://www.discovery.org/w3/discovery.org/journal/spring98.html.

43. Center for the Renewal of Science and Culture, "Design Theory: A New Science for a New Century." Accessed on 27 April 2000 at http://www.discovery.org/crsc/index.php3.

44. Phillip E. Johnson, "How to Sink a Battleship: A Call to Separate Materialist Philosophy from Empirical Science," *The Real Issue*, November/December 1996 [online]. Accessed on 31 August 2000 at http://www.leaderu.com/real/ri9602/johnson.html. This article is located on the Christian Leadership Ministries' "Leadership University" site, which CLM describes on the site as part of its "Telling the Truth project."

45. Center for the Renewal of Science and Culture, "As the Millennium Ends: The Promise of Better Science and a Better Culture" [online]. Accessed on 13 June 2000 at http://www.discovery.org/w3/discovery.org/journal/1999/crsc.html.

46. Most of their books have been published by religious presses—Zondervan and InterVarsity Press. However, that is changing. William Dembski's book *The Design Inference* was published by Cambridge University Press [1998], and Michael Behe's book *Darwin's Black Box* was published by The Free Press [1996]. As time goes on, the CRSC's goal of wedging into the cultural and academic mainstream is being facilitated by their entry into the publishing mainstream.

47. Center for the Renewal of Science and Culture, "Year End Update," November/December 1997. Accessed on 18 March 2000 at http://www.discovery.org/w3/discovery.org/crsc/crscnotes2.html. This document was in a directory which is no longer accessible. The "Consultation on Intelligent Design" is mentioned in the Discovery Institute spring 1998 *Journal*. Available at http://www.discovery.org/w3/discovery.org/journal/spring98.html.

48. Chien's affiliation with the Discovery Institute (CRSC) was not disclosed in his organizing of the China symposium. Scientists from around the world who participated learned of Chien's identity as a creationist only after they arrived in China and were alerted by David Bottjer, Professor of Earth Sciences (Paleobiology and Evolutionary Paleoecology) in the Department of Earth Sciences at the University of Southern California, who had arrived early and was assisting Chien with preparation of an abstract book for participants. (David Bottjer to Barbara Forrest, telephone interview, May 31, 2000)

49. David K. DeWolf, Stephen C. Meyer, and Mark E. DeForrest, "Intelligent Design in Public School Science Curricula: A Legal Guidebook" [online]. Accessed on 4 June 2000 at http://www.baylor.edu/~William_Dembski/docs_resources/guidebook.htm. Also available at http://law.gonzaga.edu/people/dewolf/fte.htm. Accessed on 13 April 2001.

50. The Department of Biology, University of San Francisco, "Faculty." Accessed on 25 May 2000 at http://www.usfca.edu/biology/faculty.htm#chien.

51. "The Explosion of Life," *The Real Issue*, March/April 1997 [online]. Accessed on 7 July 2000 at http://www.clm.org/real/ri9701/chien.html.

52. Chien actually characterizes his interest in the Chengjiang fossils as just a hobby. See Cecilia Yau, "The Twilight of Darwinism at the Dawn of a New Millennium: An Interview with Dr. Paul Chien," *Challenger*, February/March 2000. Accessed on 14 June 2000 at http://www.ccmusa.org/challenger/000203/doc1.html.

53. Yau, "The Twilight of Darwinism at the Dawn of a New Millennium: An Interview with Dr. Paul Chien."

54. See Behe's web site at the Department of Biological Sciences, Lehigh University [online]. Accessed on 13 April 2001 at http://www.lehigh.edu/~inbios/behe.html.

55. Michael J. Behe: "Spring 2001 Schedule," Access Research Network [online]. Accessed on 13 April 2001 at http://www.arn.org/behe/mb_schedule.htm.

56. Available online at http://www.etsu.edu/philos/faculty/niall/complexi.htm. Accessed on 28 May 2000.

57. The most thorough scientific criticism of Behe's work is the book *Finding Darwin's God* (Cliff Street Books, 1999), by Brown University biologist Kenneth Miller. For articles and reviews of *Darwin's Black Box*, see the web site of David Ussery at http://www.cbs.dtu.dk/dave/Behe_links.html. See also John Catalano's web site, "Behe's Empty Box," at http://www.world-of-dawkins.com/box/behe. htm#intro and theTalk.Origins Archive at http://www.talkorigins.org/faqs/behe. html. All accessed on 13 April 2001.

58. Available online at http://www.discovery.org/crsc/index.php3. Accessed on 25 August 2000. See also Behe's other responses to his critics published on the CRSC web site.

59. Michael J. Behe, "Evidence for Intelligent Design from Biochemistry (From a Speech Delivered at Discovery Institute's God & Culture Conference)," Discovery Institute [online]. Accessed on 13 April 2001 at http://www.discovery.org/ viewDB/index.php3?program=CRSC&command=view&id=51.

60. George W. Gilchrist, "The Elusive Scientific Basis of Intelligent Design Theory," *Reports of the National Center for Science Education* 17:3 (May/June 1997), 14–15. Also available online at http:/natcenscied.org/newsletter.asp?curiss =3. Accessed on 13 April 2001.

61. Gilchrist, "The Elusive Scientific Basis of Intelligent Design Theory," 15.

62. They also enjoy serendipitous publicity, an example of which occurred on August 29, 2000, when Behe's *Darwin's Black Box* and Johnson's *Defeating Darwinism by Opening Minds* appeared as "Quick Picks" on the Amazon.com web site.

63. See http://www.arn.org/arnproducts/catalog.htm. See also "About Access Research Network" at http://www.arn.org/infopage/info.htm. Accessed on 13 April 2001.

64. See http://www.discovery.org/crsc/scied/bookstore/index.html. Accessed on 13 April 2001.

65. Joseph L. Conn, "God's Air Force: How the National Religious Broadcasters Provide Troops and Ammo for the Religious Right's Christian Nation Crusade," *Church & State*, April 2000 [online]. Accessed on 13 April 2001 at http://www. au.org/churchstate/cs4002.htm.

66. Center for the Renewal of Science and Culture, "Year End Update," November–December 1997. Accessed on 18 March 2000 at http://www.discovery.org/ w3/discovery.org/crsc/crscnotes2.html. This document is no longer accessible.

67. Seminars in Christian Scholarship, "Design, Self-Organization, and the Integrity of Creation," William Dembski. Accessed on 10 February 2000 at http://www.calvin.edu/fss/fieldstd.htm. For application requirements for this "Fieldstead Seminar" see http://www.calvin.edu/fss/fldinfo.htm. Accessed on 13 April 2001.

68. See http://www.discovery.org/crsc/scied/. Accessed on 13 April 2001.

69. The objectives and the lesson plan were available at that time at http://www. discovery.org/crsc/scied/evol/cambrian/object/index.html and http://www.discovery. org/crsc/scied/evol/cambrian/resource/lesson/index.html, respectively.

70. The CRSC's science education web site contains a lengthy bibliography of sources on the Cambrian fossils. Consistent with this study's findings regarding Chien's lack of publication on this subject, the CRSC's own bibliography lists not a single work by Chien. Yet the learning objectives present him as an authority.

71. See http://www.discovery.org/crsc/scied/present/other/index.html. Accessed on 13 April 2001.

72. See http://www.discovery.org/crsc/scied/present/topics/political.htm. Accessed on 13 April 2001. The Science Education site also has a page listing "Other Web Curriculum Examples" on which there is a link entitled "Infectious AIDS: Have We Been Misled?" This is a link to the site of Berkeley scientist Peter Duesberg, who has been criticized for his claim that "there is no virological, nor epidemiological, evidence to back-up the HIV-AIDS hypothesis." See http://www.duesberg.com/duesberg.html. Accessed on 13 April 2001.

73. See http://www.arn.org/pearcey/nphome.htm. Accessed on 13 April 2001.

74. Transcripts are available at http://www.arn.org/technohome.htm. Accessed on 13 April 2001. For the announcement about funding, see the Discovery Institute *Journal*, spring 1998, p. 14, at http://www.discovery.org/w3/discovery.org/journal.

75. "How Did We Get Here? (A Cyber Debate)," NOVA [online]. Accessed on 29 August 2000 at http://www.pbs.org/wgbh/nova/odyssey/debate/index.html.

76. This announcement was on a page in a directory which is no longer available at http://www.discovery.org/w3/discovery.org/crsc/freedom.html.

77. See Access Research Network's page referring to this debate at http://www.arn.org/fline1297.htm#anchor22822. Accessed on 13 April 2001.

78. This program aired in the author's locality on November 10, 1998.

79. Previous press releases are not archived online and are no longer accessible at www.usnewswire.com.

80. Joseph Farah, "WorldNetDaily's Explosive Growth," WorldNetDaily, June 1, 2000 [online]. Accessed on 13 April 2001 at http://www.worldnetdaily.com/news/article.asp?ARTICLE_ID=14981.

81. Dr. Ray Bohlin, "Mere Creation: Science, Faith and Intelligent Design," Probe Ministries [online]. Accessed on 27 August 2000 at http://www.probe.org/docs/mere.html.

82. See the description of the book at Access Research Network's purchasing page at http://www.arn.org/arnproducts/catalog.htm. A long article of the same title is available on David Dewolf's Gonzaga University web site at http://law.gonzaga.edu/people/dewolf/fte.htm. Accessed on 13 April 2001.

83. See www.discovery.org/crsc/scied/present/topics/political.htm and www.discovery.org/crsc/scied/what/topics/missiontopics/legal.htm, respectively. Accessed on 13 April 2001.

84. See Ron Nissimov, "Baylor Professors Concerned Center Is Front for Promoting Creationism," *Houston Chronicle*, July 1, 2000. See also "Dembski Relieved of Duties as Polanyi Center Director," Baylor Public Relations, October 19,

2000. Accessed on 12 April 2001 at http://pr.baylor.edu/rel.fcgi?2000.10.19.05. The research function of the Polanyi Center was absorbed into Baylor's Institute for Faith and Learning. William Dembski's current title is "Associate Research Professor in the Conceptual Foundations of Science," while Bruce Gordon is "Interim Director, The Baylor Center for Science, Philosophy and Religion" and "Assistant Research Professor, Institute for Faith and Learning." See http://www.baylor.edu/~William_Dembski/biosketch.htm and http://www.baylor.edu/~Bruce_Gordon/vita.htm. Accessed on 12 April 2001.

85. "Purpose of Center," Michael Polanyi Center [online]. Accessed on 27 August 2000 at http://www.baylor.edu/~polanyi/purpose.htm. The MPC web site has been removed.

86. Steve Benen, "Science Test," *Church & State*, July/August, 2000 [online]. Accessed on 13 April 2001 at http://www.au.org/churchstate/cs7002.htm.

87. Benen, "Science Test."

88. Jay Grelen, "Witnesses for the Prosecution," *World*, November 30, 1996. Accessed on 12 April 2001 at http://www.worldmag.com/world/issue/11-30-96/national_2.asp. Johnson repeated this assertion in 1999 in an interview with *Insight on the News*: "Naturally, I get asked all the time, 'How can you do this when you're not a scientist?' The answer is that it is not mainly about science. It is about a certain way of thinking." Accessed on 9 July 2000 at http://www.arn.org/docs/johnson/insightprofile1099.htm.

89. Phillip Johnson, "Is God Unconstitutional?" Part I, in *The Real Issue* [online]. Accessed 20 February 2000 at http://www.leaderu.com/real/ri9403/johnson.html.

90. Johnson, "Is God Constitutional?"

91. Phillip E. Johnson, *The Wedge of Truth: Splitting the Foundations of Naturalism* (Downers Grove, Illinois: InterVarsity Press, 2000). See Chapter 1, "Philip Wentworth Goes to Harvard." For Phillip Wentworth's essay, see http://www.theatlantic.com/issues/95nov/warring/whatcoll.htm. Accessed on 13 April 2001.

92. Johnson, *The Wedge of Truth*, 38.

93. Stafford, "The Making of a Revolution," at http://www.christianitytoday.com/ct/7te/7te016.html.

94. The importance to the Wedge of reaching the public through publicity and publications is emphasized in the Wedge Document: "The primary purpose of Phase II [Publicity and Opinion-making] is to prepare the popular reception of our ideas. The best and truest research can languish unread and unused unless it is properly publicized."

95. This announcement was made in a May 8, 2000, U.S. Newswire press release which is not archived.

96. See "Petri Ideas Attracting Respect and Attention," in Discovery Institute *Journal*, summer 1996, at http://www.discovery.org/w3/discovery.org/journal/petri.html. Accessed on 13 April 2001.

97. See David Applegate, "Special Update: Evolution Opponents Hold Congressional Briefing" at http://www.agiweb.org/gap/legis106/id_update.html. Accessed 13 April 2001. See also Applegate's *Geotimes* article, "Creationists Open a New Front," at http://www.geotimes.org/july00/scene.html. Accessed 13 April 2001. Petri did not assume the position of chair, although he is vice chair of the committee. See the committee member list at http://edworkforce.house.gov/members/mem-fc.htm. Accessed on 17 March 2001.

98. Bruce Chapman, "From the President, Bruce Chapman: Ideas Whose Time Is Coming," Discovery Institute *Journal*, summer (August) 1996 [online]. Accessed on 24 July 2000 at http://www.discovery.org/w3/discovery.org/journal/president.html.

99. "Major Grants Help Establish Center for the Renewal of Science and Culture," Discovery Institute *Journal*, at http://www.discovery.org/w3/discovery.org/journal/center.html.

100. "Interview with Phillip Johnson about *The Wedge of Truth*," Christianbook.com, August 14, 2000 [online]. Accessed on 13 April 2001 at http://www.christianbook.com/Christian/Books/dpep/interview.pl/16559901?sku=22674. "In the beginning was the Word" is the first line in "The Gospel According to St. John" in the New Testament. Johnson cites this passage in support of supernatural "mere" creation in order to avoid the disputes which arise when young earth, biblical literalist creationists cite the book of Genesis.

101. Johnson, *The Wedge of Truth*, 152–162.

102. Nancey Pearcey, "Opening the 'Big Tent' in Science: The New Design Movement," Access Research Network [online]. Accessed on 6 June 2000 at http://www.arn.org/docs/pearcey/np_bigtent30197.htm. Originally published as "The Evolution Backlash," *World*, March 1, 1997. Accessed on 13 April 2001 at http://www.worldmag.com/world/issue/03-01-97/cover_1.asp.

103. "The Wedge Strategy: Five Year Strategy Plan Summary—Phase I."

104. Gordon states in his curriculum vitae (as of April 2001) that he had "adjunct faculty" status at Baylor (1999–2000). His web site lists two courses he taught in fall 1999 and spring 2000. Gordon had no current teaching assignments as of April 2001, although he still had assistant research professor status. This status allows him to carry on his research activities in Baylor's Institute for Faith and Learning, but it includes no teaching duties. See http://www.baylor.edu/~Bruce_Gordon/ for both his c.v. and his previous course listings. Accessed on 13 April 2001.

105. Mary Beth Marklein, "Evolution's Next Step in Kansas," *USA Today*, July 19, 2000. Accessed on 13 April 2001 at the Discovery Institute at http://www.discovery.org/viewDB/index.php3?program=CRSCstories&command=view&id=393.

106. Johnson, "The Wedge: Breaking the Modernist Monopoly on Science," http://www.touchstonemag.com/docs/issues/12.4docs/12-4pg18.html.

107. Jeff Lawrence, "Communiqué Interview: Phillip E. Johnson," *Communiqué*, spring 1999 [online]. Accessed on 13 April 2001 at http://www.communiquejournal.org/q6/q6_johnson.html.

108. Johnson, "The Wedge: Breaking the Modernist Monopoly on Science."

109. "Major Grants Increase Programs, Nearly Double Discovery Budget," Discovery Institute *Journal*, 1999, at http://www.discovery.org/w3/discovery.org/journal/1999/grants.html.

110. Ron Nissimov, "Baylor Professors Concerned Center Is Front for Promoting Creationism," *Houston Chronicle*, July 1, 2000. Regarding the controversy over the Polanyi Center's presence at Baylor, the *Chronicle* points out that although Sloan refused a request from the Faculty Senate to dissolve the Center, he had established a "nine-member committee of scholars primarily from outside Baylor to examine whether the Polanyi Center can contribute to constructive dialogue." See "Baylor Releases Committee Report" at http://pr.baylor.edu/feat.fcgi?2000.10.17.polanyi. The report is also available here. Accessed on 17 April 2001.

111. In the same article is a denial by Jay Richards, CRSC program director, that the Discovery Institute was directing Dembski's work at the Polanyi Center. Richards said, however, that the Discovery Institute hoped intelligent design would be taught along with evolution.

112. "Nobel prize winners, international scientists and scholars meet to discuss the nature of the universe," Discovery Institute News, April 7, 2000. Accessed 16 April 2000 at http://www.discovery.org/news/baylor.html. News stories are not archived by DI, so this document is no longer accessible.

113. "Baylor Naturalism Conference Focused on Scientific Differences," Discovery Institute News, May 1, 2000. Accessed on 7 May 2000 at http://www.discovery.org/news/baylorConfUpdate.html. This file is no longer available.

114. "Interview with Phillip Johnson about *The Wedge of Truth*," Christianbook.com [online], at http://www.christianbook.com/Christian/Books/dpep/interview.pl/16559901?sku=22674.

115. Bohlin, "Mere Creation: Science, Faith, and Intelligent Design," at http://www.probe.org/docs/mere.html.

116. Lawrence, "Communiqué Interview: Phillip E. Johnson," at http://www.communiquejournal.org/q6/q6_johnson.html.

117. "Great Beginnings: UT Origins Conference Opens Doors to Dialogue," *The Real Issue*, March/April 1997 [online]. Accessed on 25 March 2000 at http://www.leaderu.com/real/ri9701/koons.html.

118. Johnson, "The Wedge: Breaking the Modernist Monopoly on Science," at http://www.touchstonemag.com/docs/issues/12.4docs/12-4pg18.html.

119. William A. Dembski, "The Intelligent Design Movement" [online]. Reprinted from *Cosmic Pursuit*, spring 1998. Accessed 19 May 2000 at http://www.origins.org/offices/dembski/docs/bd-idesign.html.

120. "Kansas Deletes Evolution from State Science Test," *Talkback Live* [online]. Aired August 16, 1999, 3:00 p.m. ET. Accessed on 19 May 2000 at http://www.arn.org/docs/kansas/talkback81699.htm.

121. Phillip E. Johnson, "Starting a Conversation About Evolution," review of *The Battle of the Beginnings: Why Neither Side Is Winning the Creation-*

*Evolution Debate*, by Del Ratzsch. Accessed on 31 August 2000 at Access Research Network: Phillip Johnson Archives, http://www.arn.org/docs/johnson/ratzsch.htm.

122. Laurie Goodstein, "Christians and Scientists: New Light for Creationism," *New York Times*, December 21, 1997. Accessed on 25 May 2000 at Access Research Network at http://www.arn.org/docs/fline1297/fl_goodstein.htm.

123. Lawrence, "Communiqué Interview: Phillip E. Johnson," at http://www.communiquejournal.org/q6/q6_johnson.html.

124. Rob Boston, "Missionary Man," *Church & State*, April 1999 [online]. Accessed on 8 February 2000 at http://www.au.org/churchstate/cs4995.htm.

125. Lawrence, "Communiqué Interview: Phillip E. Johnson." Evidence of the long-term nature of the wedge strategy consists in the fact that, except for Johnson, the most important wedge members, such as William Dembski and Stephen Meyer, are relatively young.

126. Stafford, "The Making of a Revolution," http://www.christianitytoday.com/ct/7te/7te016.html.

127. Johnson, "The Wedge: Breaking the Modernist Monopoly on Science." Johnson's "prime question" is this: "What should we do if empirical evidence and materialist philosophy are going in different directions? Suppose, for example, that the evidence suggests that intelligent causes were involved in biological creation. Should we follow the evidence or the philosophy?" For Johnson, this question is tantamount to asking the academic establishment, "If the evidence suggests intelligent design, should we be genuinely scientific and admit this, or should we be unscientific and refuse to admit it?"

128. Johnson, "The Wedge: Breaking the Modernist Monopoly on Science," at http://www.touchstonemag.com/docs/issues/12.4docs/12-4pg18.html.

129. Boston, "Missionary Man," at http://www.au.org/churchstate/cs4995.htm.

130. Johnson, "Starting a Conversation about Evolution," at http://www.arn.org/docs/johnson/ratzsch.htm.

# II

# Johnson's Critique of Evolutionary Naturalism

The first article in this section, Phillip E. Johnson's "Evolution as Dogma: The Establishment of Naturalism" originally appeared in the periodical *First Things*, at the end of 1990. In this influential manifesto, Johnson laid out the ideas and themes that have since become his hallmark and that are now echoed by his followers in the intelligent design movement. He argues that the "official story" of what is wrong with creationism "contains just enough truth to mislead persuasively." He proposes that all the discussion about Noah's Flood and other "side issues" have obscured the central issue, namely, that the so-called fact of evolution "is based not upon any incontrovertible empirical evidence, but upon a highly controversial philosophical presupposition." Only the "doctrinaire naturalism" of evolutionary scientists, he claims, prevents them from recognizing the gaping holes in evolutionary theory and admitting that how complex animal groups came into existence may be one of those "mysteries beyond our comprehension." He claims that by "skillful manipulation of categories and definitions, the Darwinists have established philosophical naturalism as educational orthodoxy" while simultaneously working officially to deny that there is a real conflict between Darwinism and religion. The way to win the battle against evolution is to challenge the dogmatic "rules of discourse" that Darwinist naturalists have set up, such as the "arbitrary" rule against negative argument. Darwinists will continue to try to blur the issues as they try to "convert the nation's school children to a naturalist outlook," but their "strategy of indoctrination" will never be enough to make their view true. Since this article was published, Johnson has published five books further developing these themes.

My article "Naturalism, Evidence, and Creationism: The Case of Phillip Johnson" challenges the central elements of Johnson's philosophical attack as he presented them in the article above, and in his book *Darwin on Trial* (Regnery Gateway, 1991), that appeared the next year. I wrote this while a Fellow at an NEH-funded five-week summer Institute devoted to the topic of naturalism. In the article, I show that Johnson fails to distinguish *metaphysical* or *ontological naturalism* from *methodological naturalism*. It is not the former but the latter that science makes use of, and I explain how it is not dogmatic but follows from sound requirements for empirical evidential testing. I also show that he has no serious alternative type of positive evidence to offer for creationism, and why his attempt to legitimate purely negative argument fails.

I mentioned that this article was forthcoming in *Biology & Philosophy* when I gave an invited talk on the topic at Ohio State University. Apparently someone in the audience alerted Johnson to this, for within a few days he had called the editor of *Biology & Philosophy*, Michael Ruse, to say he knew an article was coming out and asking for the right to reply. Ruse granted Johnson's request and then allowed me to write a response. Those two articles—"Reply to Pennock" and "Johnson's Reason in the Balance"—are reprinted as the third and fourth items of this section. In his reply, Johnson writes that if only "innocent methodological naturalism" were at issue there would be little to argue about, but he claims that I engage in the classic game of bait-and-switch and turn it into "the real thing." He claims that he is advocating "intelligent causes," not "irrational causes," and that my "caricature" of design would imply that no science was done in the age of Newton. In mine, I respond to his allegations while extending my earlier critique to his second book *Reason in the Balance* (InterVarsity, 1995).

The discussion has since continued. I expanded my two articles and incorporated them as part of chapter 4 of *Tower of Babel: The Evidence against the New Creationism* (MIT Press, 1999). Johnson and I had a subsequent exchange of papers about my book in the September/October 1999 issue of *Books and Culture*, and Johnson has recently criticized me further in his recently published book *The Wedge of Truth* (InterVarsity, 2000). Others in the intelligent design movement continue to attack evolution by means of attacks upon naturalism. Section IV of this volume deals with Alvin Plantinga's argument that even the methodological form of naturalism is problematic.

# 2

# Evolution as Dogma: The Establishment of Naturalism

Phillip E. Johnson

The orthodox explanation of what is wrong with creationism goes something like this:

Science has accumulated overwhelming evidence for evolution. Although there are controversies among scientists regarding the precise mechanism of evolution, and Darwin's particular theory of natural selection may have to be modified or at least supplemented, there is no doubt whatsoever about the *fact* of evolution. All of today's living organisms including humans are the product of descent with modification from common ancestors, and ultimately in all likelihood from a single microorganism that itself evolved from nonliving chemicals. The only persons who reject the fact of evolution are biblical fundamentalists, who say that each species was separately created by God about 6,000 years ago, and that all the fossils are the products of Noah's Flood. The fundamentalists claim to be able to make a scientific case for their position, but "scientific creationism" is a contradiction in terms. Creation is inherently a religious doctrine, and there is no scientific evidence for it. This does not mean that science and religion are necessarily incompatible, because science limits itself to facts, hypotheses, and theories and does not intrude into questions of value, such as whether the universe or mankind has a purpose. Reasonable persons need have no fear that scientific *knowledge* conflicts with religious *belief*.

Like many other official stories, the preceding description contains just enough truth to mislead persuasively. In fact, there is a great deal more to the creation-evolution controversy than meets the eye, or rather than meets the carefully cultivated media stereotype of "creationists" as Bible-quoting know-nothings who refuse to face up to the scientific evidence. The creationists may be wrong about many things, but they have at least one very important point to argue, a point that has been thoroughly obscured by all the attention paid to Noah's Flood and other side issues.

Originally published in *First Things* (1990, no. 6, pp. 15–22).

What the science educators propose to teach as "evolution," and label as fact, is based not upon any incontrovertible empirical evidence, but upon a highly controversial philosophical presupposition. The controversy over evolution is therefore not going to go away as people become better educated on the subject. On the contrary, the more people learn about the philosophical content of what scientists are calling the "fact of evolution," the less they are going to like it.

To understand why this is so, we have to define the issue properly, which means that we will have to redefine some terms. Nobody doubts that evolution occurs, in the narrow sense that certain changes happen naturally. The most famous piece of evidence for Darwinism is a study of an English peppered-moth population consisting of both dark- and light-colored moths. When industrial smoke darkened the trees, the percentage of dark moths increased, due to their relative advantage in hiding from predators. When the air pollution was reduced, the trees became lighter and more light moths survived. Both colors were present throughout, and so no new characteristics emerged, but the percentage of dark moths in the population went up and down as changing conditions affected their relative ability to survive and produce offspring.

Examples of this kind allow Darwinists to assert as beyond question that "evolution is a fact," and that natural selection is an important directing force in evolution. If they mean only that evolution of a sort has been known to occur, and that natural selection has observable effects upon the distribution of characteristics in a population, then there really is nothing to dispute. The important claim of "evolution," however, is not that limited changes occur in populations due to differences in survival rates. It is that we can extrapolate from the very modest amount of evolution that can actually be observed to a grand theory that explains how moths, trees, and scientific observers came to exist in the first place.

Orthodox science insists that we can make the extrapolation. The "neo-Darwinian synthesis" (hereafter Darwinism) begins with the assumption that small random genetic changes (mutations) occasionally have positive survival value. Organisms possessing these favorable variations should have a relative advantage in survival and reproduction, and they will tend to pass their characteristics on to their descendants. By differential survival a favorable characteristic spreads through a population, and the population becomes different from what it was. If sufficient favorable mutations show up when and where they are needed, and if natural selection allows them to accumulate in a population, then it is conceivable that by tiny steps over vast amounts of time a bacterial ancestor might produce descendants as complex and varied as trees, moths, and human beings.

That is only a rough description of the theory, of course, and there are all sorts of arguments about the details. Some Darwinists, such as Harvard Professor Steven Jay Gould, say that new mechanisms are about to be discovered that will produce a more complicated theory, in which strictly Darwinian selection of individual organisms will play a reduced role. There is also a continuing debate about whether it is necessary to "decouple macroevolution from microevolution." Some experts do not believe that major changes and the appearance of new forms (i.e., macroevolution) can be explained as the products of an accumulation of tiny mutations through natural selection of individual organisms (microevolution). If classical Darwinism isn't the explanation for macroevolution, however, there is only speculation as to what sort of alternative mechanisms might have been responsible. In science, as in other fields, you can't beat something with nothing, and so the Darwinist paradigm remains in place.

For all the controversies over these issues, however, there is a basic philosophical point on which the evolutionary biologists all agree. Some say new mechanisms have to be introduced and others say the old mechanisms are adequate, but nobody with a reputation to lose proposes to invoke a supernatural creator or a mystical "life force" to help out with the difficulties. The theory in question is a theory of *naturalistic* evolution, which means that it absolutely rules out any miraculous or supernatural intervention at any point. Everything is conclusively presumed to have happened through purely material mechanisms that are in principle accessible to scientific investigation, whether they have yet been discovered or not.

That there is a controversy over how macroevolution could have occurred is largely due to the increasing awareness in scientific circles that the fossil evidence is very difficult to reconcile with the Darwinist scenario. If all living species descended from common ancestors by an accumulation of tiny steps, then there once must have existed a veritable universe of transitional intermediate forms linking the vastly different organisms of today (e.g., moths, trees, and humans) with their hypothetical common ancestors. From Darwin's time to the present, paleontologists have hoped to find the ancestors and transitional intermediates and trace the course of macroevolution. Despite claims of success in some areas, however, the results have been on the whole disappointing. That the fossil record is in important respects hostile to a Darwinist interpretation has long been known to insiders as the "trade secret of paleontology," and the secret is now coming out in the open. New forms of life tend to be fully formed at their first appearance as fossils in the rocks. If these new forms actually evolved in gradual steps from pre-existing forms, as Darwinist science insists, the numerous intermediate forms that once must have existed have not been preserved.

To illustrate the fossil problem, here is what a particularly vigorous advocate of Darwinism, Oxford Zoology Professor (and popular author) Richard Dawkins, says in *The Blind Watchmaker* about the "Cambrian explosion," i.e., the apparently sudden appearance of the major animal forms at the beginning of the Cambrian era:

The Cambrian strata of rocks, vintage about 600 million years, are the oldest ones in which we find most of the major invertebrate groups. And we find many of them in an advanced state of evolution, the very first time they appear. It is as though they were just planted there, without any evolutionary history. Needless to say, this appearance of sudden planting has delighted creationists. Evolutionists of all stripes believe, however, that this really does represent a very large gap in the fossil record, a gap that is simply due to the fact that, for some reason, very few fossils have lasted from periods before about 600 million years ago.

The "appearance of sudden planting" in this important instance is not exceptional. There is a general pattern in the fossil record of sudden appearance of new forms followed by "stasis" (i.e., absence of basic evolutionary change). The fossil evidence in Darwin's time was so discouraging to his theory that he ruefully conceded: "Nature may almost be said to have guarded against the frequent discovery of her transitional or linking forms." Leading contemporary paleontologists such as David Raup and Niles Eldredge say that the fossil problem is as serious now as it was then, despite the most determined efforts of scientists to find the missing links. This situation (along with other problems I am passing over) explains why many scientist would dearly love to confirm the existence of natural mechanisms that can produce basically new forms of life from earlier and simpler organisms without going through all the hypothetical intermediate steps that classical Darwinism requires.

Some readers may wonder why the scientists won't admit that there are mysteries beyond our comprehension, and that one of them may be how those complex animal groups could have evolved directly from pre-exisiting bacteria and algae without leaving any evidence of the transition. The reason that such an admission is out of the question is that it would open the door to creationism, which in this context means not simply biblical fundamentalism, but *any* invocation of a creative intelligence or purpose outside the natural order. Scientists committed to philosophical naturalism do not claim to have found the precise answer to every problem, but they characteristically insist that they have the important problems sufficiently well in hand that they can narrow the field of possibilities to a set of naturalistic alternatives. Absent that insistence, they would have to concede that their commitment to naturalism is based upon faith rather than proof. Such a concession could be exploited by promoters of rival sources of knowledge, such as philosophy and religion, who would be quick to point out that faith in naturalism is no more "scientific" (i.e. empirically based) than any other kind of faith.

Immediately after the passage above about the Cambrian explosion, Dawkins adds the remark that, whatever their disagreements about the tempo and mechanism of evolution, scientific evolutionists all "despise" the creationists who take delight in pointing out the absence of fossil transitional intermediates. That word "despise" is well chosen. Darwinists do not regard creationist as sincere doubters but as dishonest propagandists, persons who probably only pretend to disbelieve what they must know in their hearts to be the truth of naturalistic evolution. The greater their apparent intelligence and education, the greater their fault in refusing to acknowledge the truth that is staring them in the face. These are "dark times," Dawkins noted last year in the *New York Times* because nearly half of the American people, including many "who should know better," refuse to believe in evolution. That such people have any rational basis for their skepticism is out of the question, of course, and Dawkins tells us exactly what to think of them: "It is absolutely safe to say that if you meet somebody who claims not to believe in evolution, that person is ignorant, stupid, or insane (or wicked, but I'd rather not consider that)."

Darwinists disagree with creationists as a matter of definition, of course, but the degree of contempt that they express for creationism in principle requires some explanation beyond the fact that certain creationists have used unfair tactics such as quoting scientists out of context. It is not just the particular things that creationists do that infuriate the Darwinists; the creationists' very existence is infuriating. To understand why this is so, we must understand the powerful assumptions that mainstream scientists find it necessary to make, and the enormous frustration they feel when they are asked to take seriously persons who refuse to accept those assumptions.

What Darwinists like Dawkins despise as "creationism" is something much broader than biblical fundamentalism or even Christianity and what they proclaim as "evolution" is something much narrower than what the word means in common usage. All persons who affirm that "God creates" are in an important sense creationists, even if they believe that the Genesis story is a myth and that God created gradually through evolution over billions of years. This follows from the fact that the theory of evolution in question is *naturalistic* evolution, meaning evolution that involves no intervention or guidance by a creator outside the world of nature.

Naturalistic evolution is consistent with the existence of "God" only if by that term we mean no more than a first cause which retires from further activity after establishing the laws of nature and setting the natural mechanism in motion. Persons who say they believe in evolution, but who have in mind a process guided by an *active* God who purposely intervenes or controls the process to accomplish some end, are using the same term that the Darwinists use, but they mean something very different by it. For example, here is what Douglas Futuyma, the author of a leading college evolutionary biology textbook, finds to be the most important conflict between the theory of evolution and what he thinks of as the "fundamentalist" perspective:

Perhaps most importantly, if the world and its creatures developed purely by material, physical forces, it could not have been designed and has no purpose or goal. The fundamentalist, in contrast, believes that everything in the world, every species and every characteristic of every species, was designed by an intelligent, purposeful artificer, and that is was made for a purpose. Nowhere does this contrast apply with more force than to the human species. Some shrink from the conclusion that the human species was not designed, has no purpose, and is the product of mere material mechanisms—but this seems to be the message of evolution. (*Science on Trial: The Case for Evolution*)

It is not only "fundamentalists," of course, but theists of any description who believe that an intelligent artificer made humanity for a purpose, whether through evolution or otherwise. Futuyma's doctrinaire naturalism is not just some superfluous philosophical addition to Darwinism that can be discarded without affecting the real "science" of the matter. If some powerful conscious being exists outside the natural order, it might use its power to intervene in nature to accomplish some purpose, such as the production of beings having consciousness and free will. If the possibility of an "outside" intervention is allowed in nature at any point, however, the whole naturalistic worldview quickly unravels.

Occasionally, a scientist discouraged by the consistent failure of theories purporting to explain some problem like the first appearance of life will suggest that perhaps supernatural creation is a tenable hypothesis in this one instance. Sophisticated naturalists instantly recoil with horror, because they know that there is no way to tell God when he has to stop. If God created the first organism, then how do we know he didn't do the same thing to produce all those animal groups that appear so suddenly in the Cambrian rocks? Given the existence of a designer ready and willing to do the work, why should we suppose that random mutations and natural selection are responsible for such marvels of engineering as the eye and the wing?

Because the claims of Darwinism are presented to the public as "science," most people are under the impression that they are supported by direct evidence such as experiments and fossil record studies. This impression is seriously misleading. Scientists cannot observe complex biological structures being created by random mutations and selection in a laboratory or elsewhere. The fossil record, as we have seen, is so unhelpful that the important steps in evolution must be assumed to have occurred within its "gaps." Darwinists believe that the mutation-selection mechanism accomplishes wonders of creativity not because the wonders can be demonstrated, but because they cannot think of a more plausible explanation for the existence of wonders that does not involve an unacceptable *creator*, i.e., a being or force outside the world of nature. According to Gareth Nelson, "evidence, or proof, of origins—of the universe, of life, of all the major groups of life, of all the minor groups of life, indeed of all the species—is weak or nonexistent when measured on an absolute scale." Nelson, a senior zoologist at the American Museum of Natural History, wrote that statement in the preface to a recent book by Wendell Bird, the leading attorney for the creationist organizations. Nelson himself is no creationist, but he is sufficiently disgusted with Darwinist dogmatism that he looks benignly upon unorthodox challengers.

Philosophical naturalism is so deeply ingrained in the thinking of many educated people today, including theologians, that they find it difficult even to imagine any other way of looking at things. To such people, Darwinism seems so logically appealing that only a modest amount of confirming evidence is needed to prove the whole system, and so they point to the peppered-moth example as virtually conclusive. Even if they do develop doubts whether such modest forces can account for large-scale change, their naturalism is undisturbed. Since there is nothing outside of nature, and since *something* must have produced all the kinds of organisms that exist, a satisfactory naturalistic mechanism must be waiting to be discovered.

The same situation looks quite different to people who accept the possibility of a creator outside the natural order. To such people, the peppered-moth observations and similar evidence seem absurdly inadequate to prove that natural selection can make a wing, an eye, or a brain. From their more skeptical perspective, the consistent pattern in the fossil record of sudden appearance followed by stasis tends to prove that there is something wrong with Darwinism, not that there is something wrong with the fossil record. The absence of proof "when measured on an absolute scale" is unimportant to a thoroughgoing naturalist, who feels that science is doing well enough if it has a plausible explanation that maintains the naturalistic worldview. The same absence of proof is highly significant to any person who thinks it possible that there are more things in heaven and earth than are dreamt of in naturalistic philosophy.

Victory in the creation-evolution dispute therefore belongs to the party with the cultural authority to establish the ground rules that govern the discourse. If creation is admitted as a serious possibility, Darwinism cannot win, and if it is excluded *a priori* Darwinism cannot lose. The point is illustrated by the logic which the Natural Academy of Sciences employed to persuade the Supreme Court that "creation-scientists" should not be given an opportunity to present their case against the theory of evolution in science classes. Creation-Science is not science, said the Academy, because

it fails to display the most basic characteristic of science: reliance upon naturalistic explanations. Instead, proponents of "creation-science" hold that the creation of the universe, the earth, living things, and man was accomplished through supernatural means inaccessible to human understanding.

Besides, the Academy's brief continued, creationists do not perform scientific research to establish the mechanism of supernatural creation, that being by definition impossible. Instead, they seek to discredit the scientific theory of evolution by amassing evidence that is allegedly consistent with the relatively recent, abrupt appearance of the universe, the earth, living things, and man in substantially the same form as they now have.

"Creation-science" is thus manifestly a device designed to dilute the persuasiveness of the theory of evolution. The dualistic mode of analysis and the negative argumentation employed to accomplish this dilution is, moreover, antithetical to the scientific method.

The Academy's brief went on to cite evidence for evolution, but evidence was unnecessary. Creationists are disqualified from making a positive case, because science by definition is based upon naturalism. The rules of science also disqualify any purely negative argumentation designed to dilute the persuasiveness of the theory of evolution. Creationism is thus out of court—and out of the classroom—before any consideration of evidence. Put yourself in the place of a creationist who has been silenced by that logic, and you may feel like a criminal defendant who has just been told that the law does not recognize so absurd a concept as "innocence."

With creationist explanations disqualified at the outset, it follows that the evidence will always support the naturalistic alternative. We can be absolutely certain that the Academy will not say, "The evidence on the whole supports the theory of evolution, although we concede that the apparent abrupt appearance of many fully formed animal groups in the Cambrian rocks is in itself a point in favor of the creationists." There are *no* scientific points in favor of creation and there never will be any as long as naturalists control the definition of science, because creationist explanations by definition violate the fundamental commitment of science to naturalism. When the fossil record does not provide the evidence that naturalism would like to see, it is the fossil record, and not the natural-istic explanation, that is judged to be inadequate.

When pressed about the unfairness of disqualifying their opponents *a priori*, naturalists sometimes portray themselves as merely insisting upon a proper definition of "science," and not as making any absolute claims about "truth." By this interpretation, the National Academy of Sciences did not say that it is *untrue* that "the creation of the universe, the earth, living things and man was accomplished through supernatural means inaccessible to human understanding," but only that this statement is *unscientific*. Scientific naturalists who take this line sometimes add that they do not necessarily object to the study of creationism in the public schools, provided it occurs in literature and social science classes rather than in science class.

This naturalist version of balanced treatment is not a genuine attempt at a fair accommodation of competing worldviews, but a rhetorical maneuver. It enables naturalists effectively to label their own product as fact and its rival as fantasy, without having to back up the decision with evidence. The dominant culture assumes that science provides *knowl-edge*, and so in natural science classes fundamental propositions can be proclaimed as objectively true, regardless of how many dissenters believe them to be false. That is the powerful philosophical meaning of the claim that "evolution is a fact." By contrast, in literature class we read poetry and fiction, and in social science we study the subjective *beliefs* of various cultures from a naturalistic perspective. If you have difficulty seeing just how loaded this knowledge-belief distinction is, try to imagine the reac-tion of Darwinists to the suggestion that their theory should be removed from the college biology curriculum and studied instead in a course de-voted to nineteenth-century intellectual history.

By skillful manipulation of categories and definitions, the Darwinists have established philosophical naturalism as educational orthodoxy in a nation in which the overwhelming majority of people express some form of theistic belief inconsistent with naturalism. According to a 1982 Gallup poll aimed at measuring nationwide opinion, 44 percent of res-pondents agreed with the statment that "God created man pretty much in his present form at one time within the last 10,000 years." That would seem to mark those respondents as creationists in a relatively narrow sense. Another 38 percent accepted evolution as a process guided by

God. Only 9 percent identified themselves as believers in a naturalistic evolutionary process not guided by God. The philosophy of the 9 percent is now to be taught in the school as unchallengeable truth.

Cornell University Professor William Provine, a leading historian of Darwinism, concluded from Gallup's figures that the American public simply does not understand what the scientists means by evolution. As Provine summarized the matter, "The destructive implications of evolutionary biology extend far beyond the assumptions of organized religion to a much deeper and more pervasive belief, held by the vast majority of people, that non-mechanistic organizing designs or forces are somehow responsible for the visible order of the physical universe, biological organisms, and human moral order." Provine blamed the scientific establishment itself for misleading the public about the absolute incompatibility of contemporary Darwinism with any belief in God, designing forces, or absolute standards of good and evil. Scientific leaders have obscured the conflict for fear of jeopardizing public support for their funding, and also because some of them believe that religion may still play a useful role in maintaining public morality. According to Provine, "These rationalizations are politic but intellectually dishonest."

The organizations that speak officially for science continue to deny that there is a conflict between Darwinism and "religion." This denial is another example of the skillful manipulation of definitions, because there are evolution-based religions that embrace naturalism with enthusiasm. Stephen Jay Gould holds up the geneticist Theodosius Dobzhansky, "the greatest evolutionist of our century and a lifelong Russian Orthodox," as proof that evolution and religion are compatible. The example is instructive, because Dobzhansky made a religion out of evolution. According to a eulogy by Francisco Ayala, "Dobzhansky was a religious man, although he apparently rejected fundamental beliefs of traditional religion, such as the existence of God and of life beyond physical death. His religiosity was grounded on the conviction that there is meaning in the universe. He saw that meaning in the fact that evolution has produced the stupendous diversity of the living world and has progressed from primitive forms of life to mankind. . . . He believed that somehow mankind would eventually evolve into higher levels of harmony and creativity." In short, Dobzhansky was what we would today call a New Age pantheist. Of course evolution is not incompatible with religion when the religion is evolution.

Dobzhansky was one of the principal founders of the neo-Darwinian synthesis. Another was Julian Huxley, who promoted a religion of "evolutionary humanism. A third was the paleontologist George Gaylord Simpson. Simpson explained in his book *The Meaning of Evolution* that "there are some beliefs still current, labeled as religious and involved with religious emotions, that conflict with evolution and are therefore intellectually untenable in spite of their emotional appeal." Simpson added that it is nonetheless "self-evident ... that evolution and *true* religion are compatible." By true religion he meant naturalistic religion, which accepts that "man is the result of a purposeless and natural process that did not have him in mind." Because efforts have been made to obscure the point, it should be emphasized that Simpson's view is not some personal opinion extraneous to the real "science" of Darwinism. It is an expression of the same naturalism that gives Darwinists confidence that mutation and natural selection, Darwinism's "blind watchmaker," can do all the work of a creator.

Against this background readers may perceive the cruel irony in Justice Brennan's opinion for the Supreme Court majority, holding the Louisiana "balanced treatment" statute unconstitutional because the creationists who promoted it had a "religious purpose." Of course they had a religious purpose, if by that we mean a purpose to try to do something to counter the highly successful efforts of proponents of naturalism to have their philosophy established in the public schools as "fact." If creationists object to naturalistic evolution on religious grounds, they are admonished that it is inappropriate for religion to meddle with science. If they try to state scientific objections, they are disqualified instantly by definitions devised for that purpose by their adversaries. Sisyphys himself, eternally rolling his stone up that hill in Hades, must pity their frustration.

The Darwinists are also frustrated, however, because they find the resurgence of creationism baffling. Why can't these people learn that the evidence for evolution is overwhelming? Why do they persist in denying the obvious? Above all, how can they be so dishonest as to claim that scientific evidence supports their absurd position? Writing the introduction to a collection of polemics titled *Scientists Confront Creationism*, Richard Lewontin attempted to explain why creationism is doomed by its very nature. Because he is a dedicated Marxist as well as a famous geneticist, Lewontin saw the conflict between creation and evolution as a class struggle, with history inevitably awarding the victory to the naturalistic class. The triumph of evolution in the schools in the post-Sputnik era signaled that "the culture of the dominant class had triumphed, and traditional religious values, the only vestige of control that rural people had over their own lives and the lives of their families, had been taken away from them." In fact, many creationists are urban professionals who make their living from technology, but Lewontin's basic point is valid. The "fact of evolution" is an instrument of cultural domination, and it is only to be expected that people who are being consigned to the dustbin of history should make some protest.

Lewontin was satisfied that creationism cannot survive because its acceptance of miracles puts it at odds with the more rational perception of the world as a place where all events have natural causes. Even a creationist "crosses seas not on foot but in machines, finds the pitcher empty when he has poured out its contents, and the cupboard bare when he has eaten the last of the loaf." Lewontin thus saw creationism as falsified not so much by any discoveries of modern science as by universal human experience, a thesis that does little to explain either why so absurd a notion has attracted so many adherents or why we should expect it to lose ground in the near future.

Once again we see how the power to define can be used to distort, especially when the critical definition is implicit rather than exposed to view. (I remind the reader that to Lewontin and myself, a "creationist" is not necessarily a biblical literalist, but rather any person who believes that God creates.) If creationists really were people who live in an imaginary world of continual miracles, there would be very few of them. On the contrary, from a creationist point of view, the very fact that the universe is on the whole orderly, in a manner comprehensible to our intellect, is evidence that we and it were fashioned by a common intelligence. What is truly a miracle, in the pejorative sense of an event having no rational connection with what has gone before, is the emergence of a being with consciousness, free will, and a capacity to understand the laws of nature in a universe which in the beginning contained only matter in mindless motion.

Once we understand that biologists like Lewontin are employing their scientific prestige in support of a philosophical platform, there is no longer any reason to be intimidated by their claims to scientific expertise. On the contrary, the inability of most biologists to make any sense out of creationist criticisms of their presuppositions is evidence of their own philosophical naivete. The "overwhelming evidence for naturalistic evolution" no longer overwhelms when the naturalistic worldview is itself called into question, and that worldview is as problematical as any other set of metaphysical assumptions when it is placed on the table for examination rather than being taken for granted as "the way we think today."

The problem with scientific naturalism as a worldview is that it takes a sound methodological premise of natural science and transforms it into a dogmatic statement about the nature of the universe. Science is committed by definition to empiricism, by which I mean that scientists seek to find truth by observation, experiment, and calculation rather than by studying sacred books or achieving mystical states of mind. It may well be, however, that there are certain questions—important questions, ones to which we desperately want to know the answers—that cannot be answered by the methods available to our science. These may include not only broad philosophical issues such as whether the universe has a purpose, but also questions we have become accustomed to think of as empirical, such as how life first began or how complex biological systems were put together.

Suppose, however, that some people find it intolerable either to be without answers to these questions or to allow the answers to come from anyone but scientists. In that case science must provide answers, but to do this, it must invoke *scientism*, a philosophical doctrine which asserts arbitrarily that knowledge comes only through the methods of investigation available to the natural sciences. The Soviet Cosmonaut who announced upon landing that he had been to the heavens and had not seen God was expressing crudely the basic philosophical premise that underlies Darwinism. Because we cannot examine God in our telescopes or under our microscopes, God is unreal. It is meaningless to say that some entity exists if in principle we can never have knowledge of that entity.

With the methodology of scientism in mind, we can understand what it means to contrast scientific "knowledge" with religious "belief," and what follows from the premise that natural science is not suitable for investigating whether the universe has a purpose. Belief is inherently subjective, and includes elements such as fantasy and preference. Knowledge is in principle objective, and includes elements such as facts and laws. If science does not investigate the purpose of the universe, then the universe effectively has no purpose, because a purpose of which we can have no knowledge is meaningless to us. On the other hand, the universe does exist, and all its features must be explicable in terms of forces and causes accessible to scientific investigation. It follows that the best naturalistic explanation available is effectively true, with the proviso that it may eventually be supplanted by a better or more inclusive theory. Thus naturalistic evolution is a fact, and the fact implies a critical guiding role for natural selection.

Scientism itself is not a fact, however, nor is it attractive as a philosophy once its elements and consequences are made explicit. Persons who want naturalistic evolution to be accepted as unquestioned fact must therefore use their cultural authority to enact rules of discourse that protect the purported fact from the attacks of unbelievers. First, they can identify science with naturalism, which means that they insist as a matter of first principle that no consideration whatever be given to the possibility that mind or spirit preceded matter. Second, they can impose a rule of procedure that disqualifies purely negative argument, so that a theory which obtains some very modest degree of empirical support can become immune to disproof until and unless it is supplanted by a better naturalistic theory. With these rules in place, Darwinists can claim to have proved that natural selection crafted moths, trees, and people, and point to the peppered-moth observation as proof.

The assumption of naturalism is in the realm of speculative philosophy, and the rule against negative argument is arbitrary. It is as if a judge were to tell a defendant that he may not establish his innocence unless he can produce a suitable substitute to be charged with the crime. Such vulnerable rules of discourse need protection from criticism, and two distinct rhetorical strategies have been pursued to provide it. First, we have already seen that the direct conflict between Darwinism and theism has been blurred, so that theists who are not committed to biblical inerrancy are led to believe that they have no reason to be suspicious of Dar-

winism. The remaining objectors can be marginalized as fundamentalists, whose purportedly scientific objections need not be taken seriously
because "everybody knows" that people like that will believe, and say,
anything.

The second strategy is to take advantage of the prestige that science
enjoys in an age of technology, by asserting that anyone who disputes
Darwinism must be an enemy of science, and hence of rationality itself.
This argument gains a certain plausibility from the fact that Darwinism is
not the only area within the vast realm of science where such practices
as extravagant extrapolation, arbitrary assumptions, and metaphysical
speculation have been tolerated. The history of scientific efforts to explain
human behavior provides many examples, and some aspects of cosmology, such as its Anthropic Principle, invite the label "cosmo-theology."
What makes the strategy effective, however, is not the association of
Darwinism with the more speculative aspects of cosmology, but its purported link with technology. Donald Johanson put the point effectively, if
crudely: "You can't accept one part of science because it brings you good
things like electricity and penicillin and throw away another part because
it brings you some things you don't like about the origin of life."

But why can't you do exactly that? That scientists can learn a good
deal about the behavior of electrons and bacteria does not prove that
they know how electrons or bacteria came into existence in the first
place. It is also possible that contemporary scientists are insightful upon
some matters and, like their predecessors, thoroughly confused about
others. Twentieth-century experience demonstrates that scientific technology can work wonders, of course. It also demonstrates that dubious
doctrines based upon philosophy can achieve an undeserved respectability
by cloaking themselves in the mystique of science. Whether Darwinism
is another example of pseudoscience is the question, and this question
cannot be answered by a vague appeal to the authority of science.

For now, things are going well for Darwinism in America. The
Supreme Court has dealt the creationists a crushing blow, and state
boards of education are beginning to adopt "science frameworks." These
policy statements are designed to encourage textbook publishers to proclaim boldly the fact of evolution—and therefore the naturalistic philosophy that underlies the fact—instead of minimizing the subject to avoid
controversy. Efforts are also under way to bring under control any individual teachers who express creationist sentiments in the classroom,

especially if they make use of unapproved materials. As ideological authority collapses in other parts of the world, the Darwinists are successfully swimming against the current.

There will be harder times ahead, however. The Darwinist strategy depends upon a certain blurring of the issues, and in particular upon maintaining the fiction that what is being promoted is an inoffensive "fact of evolution," which is opposed only by a discredited minority of religious fanatics. As the Darwinists move out to convert the nation's school children to a naturalistic outlook, it may become more and more difficult to conceal the religious implications of their system. Plenty of people within the Darwinist camp know what is being concealed, and cannot be relied upon to maintain a discreet silence. William Provine, for example, has been on a crusade to persuade the public that it has to discard either Darwinism or God, and not only God but also such non-materialistic concepts as free will and objective standards of morality. Provine offers this choice in the serene confidence that the biologists have enough evidence to persuade the public to choose Darwinism, and to accept its philosophical consequences.

The establishment of naturalism in the schools is supposedly essential to the improvement of science education, which is in such a dismal state in America that national leaders are truly worried. It is not likely, however, that science education can be improved in the long run by identifying science with a worldview abhorred by a large section of the population, and then hoping that the public never finds out what is being implied. The project requires that the scientific establishment commit itself to a strategy of indoctrination, in which the teachers first tell students what they are supposed to believe and then inform them about any difficulties only later, when it is deemed safe to do so. The weakness that requires such dogmatism is evident in Philip Kitcher's explanation of why it is "insidious" to propose that the creationists be allowed to present their negative case in the classroom:

There will be ... much dredging up of misguided objections to evolutionary theory. The objections are spurious—but how is the teacher to reveal their errors to students who are at the beginning of their science studies? ... What Creationists really propose is a situation in which people without scientific training—fourteen-year-old students, for example—are asked to decide a complex issue on partial evidence.

A few centuries ago, the defenders of orthodoxy used the same logic to explain why the common people needed to be protected from exposure to the spurious heresies of Galileo. In fairness, the creationists Kitcher had in mind are biblical fundamentalists who want to attack orthodox scientific doctrine on a broad front. I do not myself think that such advocacy groups should be given a platform in the classroom. In my experience, however, Darwinists apply the same contemptuous dismissal to any suggestion, however well-informed and modestly stated, that in constructing their huge theoretical edifice upon a blind commitment to naturalism, they may have been building upon the sand. As long as the media and the courts are quiescent, they may retain the power to marginalize dissent and establish their philosophy as orthodoxy. What they do not have the power to do is to make it true.

# 3

# Naturalism, Evidence, and Creationism: The Case of Phillip Johnson

Robert T. Pennock

## I. Introduction

Creationism has a new champion. He is Phillip E. Johnson, professor of law at the University of California, Berkeley. In his book *Darwin on Trial* (1991, hereafter abbreviated JDT), Johnson renewed the Creationist attack against evolution, and has since become a popular speaker on the lecture circuit, regularly drawing capacity crowds. He has debated scientists, such as physicist and Nobel Prize winner Steven Weinberg, and philosophers of science, such as Michael Ruse, who stated at a recent conference of the American Association for the Advancement of Science that Johnson's challenge deserves to be taken seriously (Ruse 1993). Christian Creationist groups have been quick to recognize Johnson as an important asset and they sponsor forums for him to present his arguments against evolution. The Ad Hoc Origins Committee, a group of professors and academic scientists from universities including Princeton and the University of Texas who describe themselves as Christian Theists, claims that Johnson has given a "penetrating and fundamental critique of modern Darwinism" (Bocarsly *et al.* 1993) and distributes free videotapes of one of his speeches. Creationists have become increasingly well funded and well organized in the last two decades, but they have lacked an articulate spokesman with a high-profile institutional affiliation. Johnson fills this role and provides the movement with a measure of credibility it has longed for. Of course, Johnson is a lawyer and not a biologist, so his credibility is not on the scientific side—indeed,

Originally published in *Biology & Philosophy* (1996, vol. 11, no. 4, pp. 543–549).

William Provine and other biologists have called his descriptions of evolutionary theory a "crude caricature" (Provine 1990, p. 20)—but he knows how to draw upon his strengths and makes a classic courtroom move of shifting the locus of argument in a way that seeks to undermine the expert testimony of his scientist adversaries. His key argument is broadly philosophical, but Johnson also uses his considerable rhetorical skills to try to turn the tables on scientists by portraying them as naively doctrinaire and intolerant, and Creationists as rational and fair-minded skeptics. To meet Johnson's challenge we must not only show how his argument fails on logical grounds, but also deconstruct his rhetoric.

One of Johnson's titles—"Evolution as Dogma: The Establishment of Naturalism" (Johnson 1990, hereafter abbreviated JED)—neatly captures both his argumentative and rhetorical strategies. Unlike the Creation-Scientists, who try to put Creationism on a par with evolution by claiming that Creationism is scientific, Johnson tries to put them on a par by alleging that evolution is ideological. Darwinian evolution, he claims, "is based not upon any incontrovertible empirical evidence, but upon a highly controversial philosophical presupposition" (JED p. 1 [60]). That presupposition is Naturalism. Johnson argues that naturalistic evolution is not scientific but is rather a dogmatic belief system held in place by the authority of a scientific priesthood, and that without the naturalist assumption evolutionary theory would be rejected in favor of Creationism. The charge that science is a "secular religion" is not new, but Johnson is the first to locate a basis for the charge in specific philosophical assumptions made by science, and to try to exploit this as a point of weakness in evolutionary theory to the advantage of Creationism. Johnson's attack contains a kernel of truth—it is true that science makes use of a naturalistic philosophy—but Johnson has misunderstood Naturalism's role in science in general and its implications in this instance. To see this, we begin with a review of Johnson's main argument and discussion of its key concepts. The next section contains a discussion of Naturalism. I will argue that Johnson has conflated two varieties of Naturalism—Ontological Naturalism and Methodological Naturalism. If science assumed the former then Johnson's charge of scientific dogmatism might have some merit, but we shall see that science relies on the latter, which is innocent of the charge. We examine the import of Methodological Naturalism and show why science assumes it as part of its commitment to empirical evidence. Next, we will ask what positive evidence Creationism has to offer that Johnson claims has been suppressed, and shall see that Creationism does not have a case even if the requirements of Methodological Naturalism are relaxed. Finally, we show why purely negative argument against evolution, a Creationist tactic Johnson wants to justify, will not establish Creationists' preferred conclusion. In the course of discussion I will also highlight Johnson's prejudicial and misleading rhetoric, which serves to polarize the debate and undermine the possibility of peaceful co-existence between science and religion.

## II. Johnson's Argument

Johnson offers variations of the usual Creationist arguments that try to poke holes in the broad fabric of scientific evidence for evolution, but we will focus upon his novel and strongest challenge, which is the whole-cloth charge that evolution is metaphysical dogma. We may summarize Johnson's main argument in the following three-step form: Evolution is a naturalistic theory that denies by fiat any supernatural intervention. The scientific evidence for evolution is weak, but the philosophical assumption of Naturalism dogmatically disallows consideration of the Creationist's alternative explanation of the biological world. Therefore, if divine interventions were not ruled out of court, Creationism would win over evolution.

This is not laid out formally as a deductive argument, but one recognizes at once a version of the familiar Creationist "Dual Model" tactic; the argument is presented as though evolution and Creationism are the only alternatives, so if evolution gets knocked out, Creationism wins by default. Creation-Scientists, requesting "balanced treatment" of the issue in the public schools, used a very crude form of this type of argument structure, with Darwinian evolution on the one side and a thinly disguised Biblical literalism on the other; let the children judge the evidence and decide for themselves which one is right, they asked in the name of fairness. Of course, they did not plan to mention Mayan or Hindu or Asanti creation stories as alternatives. Johnson is more sophisticated. He, too, wants to get his conclusion by means of a negative argument against evolution[1] but he tries harder to set up the dichotomy to logically exclude other alternatives by attempting to define the key terms of the debate—

"Darwinism" and "Creationism"—so they are mutually exclusive and jointly exhaustive.

Johnson takes pains to distinguish his brand of Creationism from the specific scripture-based commitments of Creation-Science[2] and to define Creationism broadly. Here is the way he puts the definition in *Darwin on Trial*:

"Creationism" means belief in creation in a . . . general sense. Persons who believe that the earth is billions of years old, and that simple forms of life evolved gradually to become more complex forms including humans, are "creationists" if they believe that a supernatural Creator not only initiated this process but in some meaningful sense *controls* it in furtherance of a purpose. (JDT p. 4)

Elsewhere he reiterates this with a slightly different emphasis.

The essential point of creation has nothing to do with the timing or the mechanisms the Creator chose to employ, but with the element of design or purpose. In the broadest sense, a "creationist" is simply a person who believes that the world (and especially mankind) was *designed*, and exists for a *purpose*. (JDT p. 113)

A significant feature of Johnson's definitions is that they put no explicit restrictions on the manner of creation as long as God is involved in a significant way; guided evolution, special creation or any other mode of divine creation seems allowed. The definitions make no reference to the Bible, making it appear that Johnson countenances as Creationist the cosmogonies of any other religious or cultural tradition. In *Evolution as Dogma* Johnson is even more general:

[A] "creationist" is . . . any person who believes that God creates. (JED p. 13 [71])

Such apparent open-mindedness makes the defender of evolution look narrow-minded in contrast to the tolerant Creationist. It also serves to enlarge Johnson's constituency, for most people will identify themselves as Creationist in the minimal sense of commitment to the idea that God creates.[3] Additionally, the broad definition helps bolster Johnson's claim that evolution is necessarily at odds with religion, for he contrasts this mild-mannered Creationism with a view of evolutionary theory that makes the latter essentially atheistic.

Johnson defines "evolution" very narrowly. He does not deny that evolution by natural selection occurs if all one means by that is that "limited changes occur in populations due to differences in survival rates" (JED p. 2 [60]). Even Creation-Science allows micro-evolution, he claims—God created "kinds" but thereafter individuals may diversify within the limits of the kind.[4] On the other hand, the important thesis of evolutionary theory, he says, is the further one about macro-evolution: that evolutionary processes also explain "how moths, trees, and scientific observers came to exist in the first place" (JED p. 2 [60]). Most of *Darwin on Trial* attacks this claim, but Johnson narrows his sights still further to set up his general argument. Since it is possible that God did not create creatures suddenly, but instead used a gradual evolutionary process, even macro-evolution does not contradict Creationism unless it is "explicitly or tacitly defined as *fully naturalistic evolution*—meaning evolution that is not directed by any purposeful intelligence" (JDT p. 4). This is the form of evolution that Johnson sets up as his target. Here is his positive definition:

By "Darwinism" I mean fully naturalistic evolution, involving chance mechanisms guided by natural selection. (JDT p. 4)

Johnson's main argument will hang on his conception of the role of Naturalism in this scheme, which we will examine shortly, but his central point is that in naturalistic evolution God's intervention is excluded.

Taking these two definitions together, we see how the argument is supposed to work. Creationism holds that God plays a role in Creation (however it occurs) and Darwinism denies the same. Though closer inspection makes it clear that Johnson's definitions do not establish the logical dichotomy he needs (as we will discuss in the last section), on the surface it looks as though he has set up the major premise of a valid disjunctive syllogism that will then allow Creationists to rely solely upon negative argumentation. This is Johnson's first innovation. His second is that his characterization of the terms of the debate allows evolution to be attacked not only on scientific grounds, but also on philosophical grounds. He spends seven chapters in *Darwin on Trial* on the first task, trying to cast doubt upon the wide range of empirical evidence for evolutionary theory so that he can claim that Creationism is a better theory that would be accepted if not for the "powerful" and "doctrinaire" naturalistic assumption (JED p. 6 [64–65]) that rules it out by definition. As we shall see, however, the philosophical charge does the real work.

Johnson is well aware that scientists have not and will not now find Creationist criticisms of the evidence for evolution to be persuasive,[5] but this matters little for he is playing to the jury. When the scientific expert witness rebuts his negative appraisals of the evidence for evolution, Johnson will argue that biologists are "[unable] to make any sense out of creationist criticisms of their presuppositions" because of their "philosophical naiveté" (JED p. 13 [72]) and their "blind commitment to naturalism" (JED p. 17 [76]). The evidence cited for evolution, Johnson claims,

looks quite different to people who accept the possibility of a creator outside the natural order. To such people, the peppered-moth observations and similar evidence seem absurdly inadequate to prove that natural selection can make a wing, an eye, or a brain. From their more skeptical perspective, the consistent pattern in the fossil record of sudden appearance followed by stasis tends to prove that there is something wrong with Darwinism, not that there is something wrong with the fossil record. The absence of proof "when measured on an absolute scale" is unimportant to a thoroughgoing naturalist, who feels that science is doing well enough if it has a plausible explanation that maintains the naturalistic worldview. The same absence of proof is highly significant to any person who thinks it possible that there are more things in heaven and earth than are dreamt of in naturalistic philosophy. (JED p. 8 [66])

Again we see how Johnson's rhetoric tries to make the Creationist appear to be the rational "skeptic" who merely accepts the "possibility" of a Creator, and the biologist the "blind" and "naive" ideologue who dogmatically rejects that possibility and thereby misjudges the evidence. Thus, he concludes, it is not the evidence, but the ideology that supports evolution. Here is the conclusion the reader is supposed to draw:

Victory in the creation-evolution dispute therefore belongs to the party with the cultural authority to establish the ground rules that govern the discourse. If creation is admitted as a serious possibility, Darwinism cannot win, and if it is excluded *a priori* Darwinism cannot lose. (JED p. 8 [66])

The claim that evolution is held up solely by "metaphysical assumptions" (JED p. 13 [72]) and "speculative philosophy" (JED p. 15 [73]) allows Johnson to ignore the weakness of his negative scientific arguments. In his public lectures, Johnson follows the same pattern, usually taking a few token swipes at the empirical evidence for evolution and then quickly moving to his philosophical indictment of its naturalistic metaphysics.

Although this philosophical criticism of Naturalism has to carry the weight of his conclusion, Johnson fails to provide any philosophical analysis of the concept that he charges scientists have uncritically accepted. Neither does he support his thesis that the concept is inherently dogmatic, or provide evidence that scientists do subscribe to it in the way he claims. Let us now briefly review the history of Naturalism, and then evaluate Johnson's characterization and his application of the concept to the biological case.

## III. Varieties of Naturalism

The generic meaning of "Naturalism" is a philosophical view based upon study of the natural world, with an implicit contrast to the supernatural world, but this leaves room for a wide range of specific variations. Since the time of the ancient Greeks, Naturalism has often been associated with various forms of secularism, especially epicureanism and materialism, but it has also been used as a label for religious views such as pantheism, as well as the theological doctrine that we learn religious truth not by revelation but by study of natural processes. In the centuries leading up to the 20th century, concomitant with the rise of the natural sciences, the term became associated more directly with the methods and fruits of the scientific study of nature. One spin-off at the turn of the century was the Naturalist movement in literature, epitomized by Zola but continuing in a form through Steinbeck, which featured "scientific" portrayals of human characters playing out predetermined roles as amoral creatures governed by natural law. Another extreme expression was Auguste Compte's philosophy of Positivism, the scientific stage of philosophical development which society purportedly reached after progressing beyond theological and metaphysical conceptions of the world. Positivism concerned itself only with regularities of observable phenomena, so Naturalism at that time became associated with Phenomenalism. This version of Naturalism was carried forward into the philosophy of science in the early 20th-century by the influential Logical Positivists who restricted knowledge to propositions with a determinable truth-value—if a proposition was not verifiable then it was taken to be meaningless. The so-called verifiability criterion of meaning turned out to be unworkable and its collapse was one of several reasons for the demise of the Logical Positivist view in mid-century. Since then, in philosophy at least, the Naturalist view of the world has become coincident with the scientific view of the world, whatever that may turn out to be. Many people continue to think of the scientific world view as being exclusively materialist and deterministic, but if science discovers forces and fields and indeterministic causal processes, then these too are to be accepted as part of the Naturalistic worldview.[6] The key point is that Naturalism is not necessarily tied to specific ontological claims (about what sorts of being do or don't exist); its base commitment is to a method of inquiry.

Of course one may choose to take some set of basic ontological categories from science at a particular time and then claim that only these things exist. The seventeenth century Mechanistic Materialists, who held that the world consists of nothing but material particles in motion, did just this, and there are any number of other ways that one could decide to fix base ontological commitments. Let us call this type of view *Ontological Naturalism*. The Ontological Naturalist makes a commitment to substantive claims about what exists in nature, and then adds a closure clause stating "and that is all there is." A thorough historical review of positive formulations of Ontological Naturalism could fill an article in itself, but amidst this variety many do agree on a common negative claim: because God standardly is assumed to be supernatural, the Ontological Naturalist usually denies God's existence. It is possible, however, for an Ontological Naturalist to allow God in the picture, provided God's attributes are appropriately constrained to conform to the regimen of the given natural ontology. Hobbes and Spinoza were Ontological Naturalists who thought they found room for God (indeed, for a Judeo-Christian God) in this way. Some traditional theists, however, were not willing to countenance their naturalized conceptions of the deity; Hobbes was branded an atheist and Spinoza a pantheist. The problem of trying to naturalize theology is that traditionalists want God to be able to control nature from outside nature; they take God to be supernatural by definition. Probably the main reason for the strong secularist strand among the varieties of Naturalism is that many Naturalists also have tended to take for granted this traditional conception of God and have found it difficult to square it with their other ontological commitments.

Ontological Naturalism should be distinguished from the more common contemporary view, which we can call *Methodological Naturalism*. The Methodological Naturalist does not make a commitment directly to a picture of what exists in the world, but rather to a set of methods as a reliable way to find out about the world—typically the methods of the natural sciences, and perhaps extensions that are continuous with them—and indirectly to what those methods discover. An important feature of science is that its conclusions are defeasible on the basis of new evidence, so whatever tentative substantive claims a Methodological Naturalist makes are always open to revision or abandonment on the basis of new, countervailing evidence. Because the base commitment of a Methodological Naturalist is to a mode of investigation that is good for finding out about the empirical world, even the specific methods themselves are open to change and improvement; science may adopt promising new methods and refine existing ones if doing so would provide better evidential warrant. Understanding the nature of scientific evidence will be critical for answering Johnson's charge, but let us postpone examination of that concept and how it relates to the question of God's existence and creativity until we have seen the details of Johnson's philosophical claims that Naturalism is assumed dogmatically and that its ideology alone supports evolutionary theory.

Although it is the linchpin of his argument, Johnson provides only a cursory discussion of the concept of Naturalism. Taken individually, his few statements do pick out versions of Naturalism, but taken together they suggest a biased and misleading picture. In *Darwin on Trial*, Johnson defines Naturalism as follows:

Naturalism assumes the entire realm of nature to be a closed system of material causes and effects, which cannot be influenced by anything from "outside." Naturalism does not explicitly deny the mere existence of God, but it does deny that a supernatural being could in any way influence natural events, such as evolution, or communicate with natural creatures like ourselves. (JDT pp. 114–115)

This is a good definition of a common form of Ontological Naturalism; the "causal closure of the physical" is another way this idea is expressed. The acknowledgment that Naturalism does not "explicitly" deny the "mere existence" of God, however, is significant, for it is another indication that Johnson is not as tolerant and ecumenical as his definition of Creationism might initially lead one to believe. The clear implication here is that, because Naturalism rejects continuing divine intervention, it does *implicitly* deny God's existence, but this conclusion follows only if one has a particular conception of divine power. We see this view expressed again as Johnson immediately follows the above definition by introducing a specific form of Naturalism that he calls "Scientific Naturalism."

Scientific naturalism makes the same point by starting with the assumption that science, which studies only the natural, is our only reliable path to knowledge. A God who can never do anything that makes a difference, and of whom we can have no reliable knowledge, is of no importance to us. (JDT p. 115)

In such statements Johnson is dismissing views such as Deism that do allow God to influence natural events, to make a difference and conceivably even to communicate with us, by setting up the world in the appropriate way at Creation but thereafter not intervening in the natural order. He is also rejecting views that hold that God is concerned with our spiritual rather than our material being and intervenes only at a spiritual level. He is also ignoring religious views that do not posit a personal God, but conceive of God as a universal life force or a mystical unity. Also unimportant, apparently, are views that say we can have "no reliable knowledge" of God; this restriction leaves out even many Judeo-Christian thinkers who hold that the nature of God is unknowable to the human mind. These spiritual views Johnson excludes are prevalent world-wide, so we should not be misled by his attempt to portray his form of Creationism as generically tolerant. Such views, however, *are* compatible with varieties of both Ontological and Methodological Naturalism and belie Johnson's attempts to conflate Naturalism and atheism.

Returning to the definitions, one may think at first that "Scientific Naturalism" is Johnson's term for Methodological Naturalism, but in light of his other comments we see that he mixes in elements of Ontological Naturalism. He says, for example, that in the present context he considers scientific naturalism to be equivalent to evolutionary naturalism, scientific materialism, and scientism.

All these terms imply that scientific investigation is either the exclusive path to knowledge or at least by far the most reliable path, and that only natural or material phenomena are real. In other words, what science can't study is effectively unreal. (JDT p. 114)

By ignoring distinctions among such positions,[7] Johnson again is able to associate evolution with (godless) materialism and to portray Naturalism as monolithically dogmatic. "Scientism," for example, is a term of derision coined by Hermeneutic critics of science to label those who wanted to apply the methods of the natural sciences "inappropriately" to the human sciences, for which they though the literary model of Hermeneutic *interpretation* should reign as the proper method. Their target was specifically the followers of the Logical Positivists, but, as was noted, the exclusionary Positivist view that only the scientifically verifiable was meaningful has not held currency for several decades. Contemporary Methodological Naturalists would not recognize themselves in this description, yet it is just this sort of view that Johnson insistently portrays as the essence of Scientific Naturalism.[8]

When he applies Naturalism to evolution Johnson says that one gets:

... a theory of naturalistic evolution, which ... absolutely rules out any miraculous or supernatural intervention at any point. Everything is conclusively presumed to have happened through purely material mechanisms that are in principle accessible to scientific investigation, whether they have yet been discovered or not. (JED p. 3 [61])

Here it is clear that Johnson is describing a form of Ontological Naturalism—besides the reference to Mechanistic Materialism, the terms "absolutely" and "conclusively" emphasize the supposed dogmatic commitment to the substantive ontological claims. Johnson claims that evolutionary biologists assume this sort of positivistic philosophy, but certainly evolutionary biology as a science does not have to do so, and it is hard to believe even that any scientist who has kept abreast of developments in philosophy of science would affirm this form of Ontological Naturalism.

Indeed, it seems clear that the two biologists that Johnson most often decries—George Gaylord Simpson and Stephen Jay Gould—do not endorse such a view, but are Methodological Naturalists. Simpson discussed Naturalism as part of his review of the principle of Uniformitarianism in geology and biology and he is explicit that the scientific postulate of Naturalism is "a necessity of method" and that the rejection of appeal to preternatural factors must be made on "heuristic grounds" (Simpson 1970, p. 61). When one looks for Gould's view on the matter one finds in his discussion of Uniformitarianism that he used precisely the distinction reviewed above to disambiguate "Substantive Uniformitarianism" (a descriptive hypothesis holding that the history of life was uniform) from "Methodological Uniformitarianism" (Gould 1965, p. 226). Gould uses the latter term to label the assumption in geology that natural laws are invariable—a position that implies absence of supernatural invention. The name Gould gives to this presupposition tells us just how he views it. He recommends that the special term be dropped because it follows from the fact that geology is a science. Clearly, both Simpson and Gould understand that science does not affirm Naturalism as a substantive ontological claim but rather as a methodological assumption.

## IV.    Methodological Naturalism and Evidence

We have seen how Johnson misleadingly inserts terminology with con-
notations of dogmatism into the very definition of Naturalism. He regularly refers to Naturalism using such terms as "extravagant extrapolation,
arbitrary assumptions, and metaphysical speculation" (JED p. 15 [74]),
but such name-calling is no argument. Johnson provides no analysis to
show that science assumes the Naturalistic principle dogmatically; he
simply asserts this. We have now seen that Naturalism is not properly
put forward as an ontological claim about what conclusively does or
does not exist, but rather as a methodological rule that states a valid way
for investigation to proceed, so clearly it is not dogmatic in the sense
Johnson claimed. But is the methodological rule itself dogmatic? To say
that a belief or principle is dogmatic is to say that it is opinion put forward as true or valid on the grounds of authority rather than reason.
Does science put forward the methodological principle not to appeal to
supernatural powers or divine agency simply on authority? Is it just an
extravagant, arbitrary, speculative assumption? Certainly not. There is a
simple and sound rationale for the principle based upon the requirements
of scientific evidence.

Empirical testing relies fundamentally upon use of the lawful regularities of nature that science has been able to discover and sometimes
codify in natural laws. For example, telescopic observations implicitly
depend upon the laws governing optical phenomena. If we could not rely
upon these laws—if, for example, even when under the same conditions,
telescopes occasionally magnified properly and at other occasions produced various distortions dependent, say, upon the whims of some supernatural entity—we could not trust telescopic observations as evidence.
The same problem would apply to any type of observational data. Lawful regularity is at the very heart of the naturalistic world view and to
say that some power is supernatural is, by definition, to say that it can
violate natural laws.[9] So, when Johnson argues that science should allow
in supernatural powers and intelligences he is in effect saying that it
should allow beings that are above the law (a rather strange position for
a lawyer to take). But without the constraint of lawful regularity, inductive evidential inference cannot get off the ground. Controlled, repeatable
experimentation, for example, which Johnson explicitly endorses in his

video "Darwinism on Trial" (Johnson 1992), would not be possible without the methodological assumption that supernatural entities do not intervene to negate lawful natural regularities.

Of course science is based upon a philosophical system, but not one that is extravagant speculation. Science operates by empirical principles of observational testing; hypotheses must be confirmed or disconfirmed by reference to empirical data. One supports a hypothesis by showing consequences obtained that would follow if what is hypothesized were to be so in fact. Darwin spent most of the *Origin of Species* applying this procedure, demonstrating how a wide variety of biological phenomena could have been produced by (and thus explained by) the simple causal processes of the theory. Supernatural theories, on the other hand, can give no guidance about what follows or does not follow from their supernatural components. For instance, nothing definite can be said about the processes that would connect a given effect with the will of the supernatural agent—God may simply say the word and zap anything into or out of existence. Furthermore, in any situation, any pattern (or lack of pattern) of data is compatible with the general hypothesis of a supernatural agent unconstrained by natural law. Because of this feature, supernatural hypotheses remain immune from disconfirmation.[10] Johnson's form of Creationism is particularly guilty on this count. Creation-Science does include supernatural views at its core that are not testable and it was rightly dismissed as not being scientific because of these in the Arkansas court case, but it at least was candid about a few specific non-supernatural claims that are open to disconfirmation (and indeed that have been disconfirmed), such as that the earth is less than 10,000 years old and that many geological and paleontological features were caused by a universal flood (the Noahian Deluge). Johnson, however, does not provide any Creationist claim beyond his generic one that "God creates for some purpose," and as a purely supernatural hypothesis this is not open to empirical test. Science assumes Methodological Naturalism because to do otherwise would be to abandon its empirical evidential touchstone.

Finally, allowing appeal to supernatural powers in science would make the scientist's task just too easy, because one would always be able to call upon the gods for quick theoretical assistance. Johnson wants us to accept "God creates for some purpose" as an explanation of the biological world, but there would be no reason to stop there. Once such super-natural explanations are permitted they could be used in chemistry and physics as easily as Creationists have used them in biology and geology. Indeed, all empirical investigation beyond the purely descriptive could cease, for scientists would have a ready-made answer for everything. Clearly science must reject this kind of one-size-fits-all explanation. By disqualifying such short-cuts, the Naturalist principle also has the virtue of spurring deeper investigation. If one were to find some phenomenon that appeared inexplicable according to some current theory one might be tempted to attribute it to the direct intervention of God, but Method-ological Naturalism prods one to look further for a natural explanation. Clearly, it is not just because such persistence has proved successful in the past that science should want to encourage this attitude.

Johnson claims that, "If the possibility of an 'outside' intervention is allowed in nature at any point ... the whole naturalistic worldview quickly unravels" (JED p. 7 [65]). He intends by this only that atheistic Darwinism will lose in a head-on comparison with theistic Creationism once the "ideological" restrictions are removed, but as we have seen the consequences would be far more serious. Johnson wants to make an ex-ception to the law in this one area but it would infect the entire enter-prise. Methodological Naturalism is not a dogmatic ideology that simply is tacked on to the principles of scientific method; it is essential for the basic standards of empirical evidence.

## V.   Creationism's Evidence

With his attack upon Naturalism, Johnson is arguing that science aban-don a sound methodological principle and reintroduce miraculous "ex-planations." We have seen that science has good reasons for retaining this principle—without it standard inductive evidential inferences would be undermined—but we have also admitted that rules of scientific in-quiry are themselves open to change or modification if a better method of evidential warrant is found. Does Johnson have something better in mind? It seems that he does, for he regularly claims that the supernatural theory of Creationism is a better theory than Darwinism, and constantly complains that "Creationists are disqualified from making a positive case, because science by definition is based upon Naturalism" (JED p. 9 [67]).

Such statements lead one to expect that Johnson will supply what he says the scientific priesthood has suppressed, but one will look in vain for this positive evidence. Amidst all the negative arguments one finds only two small hints of what type of positive evidence the Creationists would have to offer—revelation and the Design Argument.

The first occurs only as a passing remark following an (inadvertently self-undermining) acknowledgment that empiricism is a "sound methodological premise" (JED p. 14 [72]). Johnson writes:

Science is committed by definition to ... find[ing] truth by observation, experiment, and calculation rather than by studying sacred books or achieving mystical states of mind. It may well be, however, that there are certain questions ... that cannot be answered by the methods available to our science. These may include not only broad philosophical issues such as whether the universe has a purpose, but also questions what we have become accustomed to think of as empirical, such as how life first began or how complex biological systems were put together. (JED p. 14 [72])

The sly implication here is that the "sacred books" and "mystical states of mind" may indeed be appropriate ways to answer empirical as well as teleological questions. Is this Johnson's new source of positive evidence for Creationism? I asked Johnson just this question following one of his public lectures and he replied that he was not defending this position. However, neither did he deny that such appeal to scriptural authority or mystical experience would count as positive empirical evidence. Johnson seems to be pleading the Fifth on this important issue. He cannot reject these methods without alienating his constituency, for the Biblical account, perhaps supplemented by religious experiences, is the prime motivation for Christian Creationists. On the other hand, he cannot endorse the "method" of supernatural revelation without abdicating his claim of expertise as a lawyer, for anyone would be laughed out of court who argued that one could help establish an empirical fact (say, that the defendant set off the explosion) by reference to the authority of psychic or spiritual testimony.

The second hint of a source of positive evidence for Creationism is in the following statement:

[F]rom a creationist point of view, the very fact that the universe is on the whole orderly, in a manner comprehensible to our intellect, is evidence that we and it were fashioned by a common intelligence. (JED p. 13 [71])

This is the only instance where Johnson makes an explicit commitment to any type of positive evidence the Creationist can provide, but what we have here is nothing but a version of the old Argument from Design—the world appears to exhibit a designed arrangement so we may infer the existence of a designer—which provides, at best, only a weak analogy from the human case to the divine. This article is not the place to review the vast literature on the design argument, so I will confine myself to a few remarks on Johnson's specific formulation of it.

What sort of order do we find in the universe, and what can it tell us? Examples of design to which we infer to a human designer are such as a house, or a formal well-manicured garden, or Paley's famous pocket-watch, but I would argue that we draw the inference in these cases precisely because the kind of order we see in them is so *unlike* what we typically found in nature; their simple geometrical forms and periodicities are strikingly different from the complexities and irregularities that the surface structure of the world presents. And when we do discover an underlying order in nature that is "comprehensible to our intellect" it is order to which we have been able to give natural, scientific explanations. Such order thus does not provide good reason to infer a supernatural creator. It is rather those features of the world that, for the time being at least, are *incomprehensible* to our intellect that are more likely to lead us to think of a higher power. At one time this may have been the awful power of a thunderstorm, leading us to suppose an angry god. Or we may have been struck by the wondrous beauty of a rainbow after the storm and interpreted it as a sign that god was appeased. We appeal to supernatural agency to explain that which we cannot explain otherwise. But when these phenomena are eventually accounted for in terms of natural electrical and optical properties they lose their persuasiveness as indications of the literal presence of god and at most retain only an emotional or symbolic force.

Similarly, the version of the Design Argument that appealed to the adaptedness of organisms was persuasive only when this adaptation was mysterious and the idea of purposeful creation seemed the only possible explanation. But Darwin showed how simple natural processes could explain such adaptations. Johnson's "God creates for a purpose" view can say nothing about the supernatural processes by which the creation was accomplished or what divine ends it serves; such an "explanation" starts and stops with the will of God. To pick just one of any number of common examples, Darwin's theory also accounts for those organisms that are not properly adapted to their environment—random variation produces both fit and unfit individuals, and natural selection is more likely to eliminate those that cannot compete as well in the given environment—but why would God create the world in such a way that the vast majority of individual organisms die because they are maladapted? The Design Argument has always been criticized on this sort of point even before Darwin; if God designed the world, why did He do such a poor job of it? Isn't the evident waste and sloppiness actually an argument *against* the existence of God? Such questions show the flip side of the Design Argument and highlight its weakness, for if it is applicable at all it is applicable for both the theist and the atheist. The Creationist answer to such impious questions is that God must have had His reasons. Period.

And of course Creationists are right that we are not privy to God's reasons and purposes. The appeal to supernatural forces, whether divine or occult, is always available because we can cite no necessary constraints upon the powers of supernatural agents. This is just the picture of God that Johnson presents. He says that God could create out of nothing or use evolution if He wanted (JDT p. 14, 113); God is "omnipotent" (JDT p. 113). He says God creates in the "furtherance of a purpose" (JDT p. 4), but that God's purposes are "inscrutable" (JDT p. 71) and "mysterious" (JDT p. 67). A god that is all-powerful and whose will is inscrutable may be called upon to explain *any* event in any situation, and, as we saw, this is one reason for the methodological prohibition against such appeals in science.

There are just two ways a Creationist might become more specific about the will and methods of God. The first would be to appeal to revelation through mystical experience or scripture, but we have already seen that this does not in itself count at all as empirical evidence. The other is to revert to an earlier form of Naturalism we mentioned— natural theology—and try to judge the nature of God by looking at the book of nature. Johnson indulges this approach just once, concluding that the elaborate-tailed peacock and peahen are "just the kind of creatures a whimsical Creator might favor, but that an "uncaring mechanical process' like natural selection would never permit to develop" (JDT

p. 31). If we are to take Johnson seriously, such "Creationist explanations" in terms of divine "whimsy" postulated on the basis of peacock tails, are better accounts than those given by evolutionary theory and are thereby supposed to favor Creationism.

If this is the "positive case" that purportedly has been suppressed, it is no wonder that Creationists rely upon exclusively negative argumentation, and why Johnson labors to legitimize it. The Creationists' insistence upon viewing the issue as a simple either/or choice mistakenly leads them to think that their negative arguments against Darwinian evolutionary mechanisms directly prove Creationism. However, negative argument will not suffice to establish the Creationists' desired conclusion even using Johnson's non-standard definitions of "Darwinism" and "Creationism" that were quoted earlier. Johnson's definition of Darwinism mentioned only the classical evolutionary mechanism of natural selection operating upon tiny chance variations, and omitted other possible processes. So even if negative argument were to rule out this sort of mechanism (something that Johnson certainly has not done), one could not thereby accept Creationism since there are other alternatives that do not rely upon divine intervention. His definitions of Creationism are similarly problematic. For example, one definition seems to rule out the sort of Deist who believes that God created the world and set it going in the way He wanted, but then no longer intervenes. Another definition rules out supernatural but non-theistic views that would stand in contrast to naturalistic evolution. Johnson would have us believe that the logical situation is the following,

Creationism ($C_1$) or Darwinism ($D_1$)

as though these were the only candidates. If this were the case then if negative argument disproved the latter then the former would follow as a deductive logical conclusion, but the logical situation is rather:

$C_1$ or $C_2$ or ... or $C_n$ or $D_1$ or $D_2$ or ... or $D_n$ or $X_n$

and in this case even if $D_1$ were rejected a variety of options remain beside Johnson's preferred $C_1$, so purely negative argument is not sufficient to establish it. Furthermore, these positions are not necessarily mutually exclusive. Direct conflict occurs when evolutionary theory is confronted with *specific* Creationist stories (like the Creation-Scientists' literal 6-day instantaneous creation) but Johnson *claims* not to defend any such view.

"Darwinism" as Johnson defines it is not a single thesis but rather the conjunction of at least two different theses: ($D_m$) the specific evolutionary mechanism of the New Synthesis, and ($D_a$) the atheistic denial of divine intervention. But all of Johnson's and other Creationists' negative arguments are directed at undermining ($D_m$), so even if they were successful they would leave the possibility of ($D_a$) untouched. The existence or nonexistence of some particular evolutionary process is independent of the question of whether or not there is a creative deity. Johnson himself must grant this for he claims to allow the possibility that "He might have created things instantaneously in a single week or through gradual evolution over billions of years" (JDT p. 113). Thus, negative arguments against evolutionary processes are irrelevant to the key question of divine creative power when stated in this general way. Indeed, if Johnson cared just about his broad, ecumenical sort of Creationism as he claims, then he should have no reason at all to argue against ($D_m$) and could confine himself to arguing against ($D_a$). Only someone who has a specific conflicting Creationist scenario in mind such as the one-week instantaneous creation story need worry about the evolutionary mechanism. In any case, negative argumentation is not going to establish the general or the specific Creationist thesis. As prosecuting attorney for the new Creationists, Johnson needs to provide positive evidence for his and his clients' preferred conclusion, but, as we have seen, he has none to offer.

## Notes

This paper was written while I was a Fellow at the 1993 National Endowment for the Humanities (N.E.H.) Summer Institute on Naturalism, and I would like to thank the N.E.H. for its financial support. I would also like to thank my fellow N.E.H. Fellows, J. F. Austin, and an anonymous reviewer for *Biology & Philosophy* for their comments on an earlier draft.

1. One of Johnson's main complaints against science, which he illustrates by reference to the National Academy of Science's friend of the court brief in the trial one the 1981 Arkansas "Balanced Treatment" Act, is that it does not allow Creationism to use merely negative arguments (JDT p. 8).

2. Politically organized creationists are mostly Christian literalists who believe, for example, in Biblical inerrancy, in Special Creation by God of all biological kinds, and in a historical global flood. The Bible-Science Association, the Creation Science Fellowship, Inc., and the Creation Research Society, a few of the main Creationist organizations, all require that one sign a statement of belief in such fundamentalist Christian principles to become a member.

3. (Culliton 1989) reports that only *half* of Americans accepted the statement that "Human beings as we know them today developed from earlier species of animals." Of these many will agree that this evolutionary devolopment was guided by God.

4. Johnson notes the idea that the different races of human beings are all descendants of the original ancestral pair—Adam and Eve. (JDT p. 68) He calls this an example of allowable micro-evolution, but clearly the Creationist notion of diversification "within the limits" of created "kinds" is not the same as micro-evolution on the biological model. Furthermore, it is doubtful that Creation-Scientists, given their theological beliefs, could consistently accept that races differentiated on the basis of natural selection of random genetic variation.

5. The focus of this paper is upon Johnson's philosophical criticism, so I will not review these scientific criticisms. Johnson rehearses the standard Creationist arguments against the power of mutation and natural selection, carpings on gaps in the fossil record, and doubts about the possibility of a molecular origin of life. These and other such arguments have previously been addressed by scientists and philosophers of science such as Philip Kitcher (1982), Michael Ruse (1982), Arthur N. Strahler (1987), and Tim Berra (1990).

6. To recognize the enrichment of the scientific ontology beyond the classical form of Materialism, it is now more common to speak of "Physicalism," where the reference is to the ontology of current physics, or sometimes "Physicalistic Materialism." Admittedly, some scientists and philosophers continue to use simply "Materialism," but it is understood in the broader sense of the *fabric* of the universe, which includes space-time and electro-magnetic fields and so on, rather than in the old sense of mere *matter*. Johnson, however, explicitly links Natural-ism and the old Mechanistic Materialism throughout his works, with the rhetor-ical effect of conflating naturalistic evolution with atheistic materialism.

7. Johnson provides no bibliographic references for any of these terms so we cannot evaluate the specifics of the definitions he may have in mind.

8. Another example is this passage: "The Soviet Cosmonaut who announced upon landing that he had been to the heavens and had not seen God was expressing crudely the basic philosophical premise that underlies Darwinism. Be-cause we cannot examine God in our telescopes or under our microscopes, God is unreal. It is meaningless to say that some entity exists if in principle we can never have knowledge of that entity" (JED p. 14 [72]). Not only does Johnson once again link Naturalism to atheism, by way of Communism, but he explicitly characterizes it in Positivist terms, with the reference to the verifiability criterion of meaning.

9. This is not to say, however, that things we now think of as supernatural nec-essarily are so. It could turn out, for example, that ghosts exist but that, unlike our fictional view of them, they are subject to natural law. In such a case we would have learned something new about the natural world (which may require revising current theories), and would not have truly found anything supernatural.

10. We should note a possible red herring on this point. Creationists often erro-neously claim that evolutionary theory itself is not disconfirmable, and so charge

that it is in the same boat as their view. To his credit, Johnson does not make this error. He understands that the theory is disconfirmable, and all of his negative argumentation purports to show that in fact it has already been disconfirmed. Johnson's positive claim is that biologists' philosophical prejudices prevent them from recognizing the disconfirming evidence.

## References

Berra, T. M.: 1990, *Evolution and the Myth of Creationism*, Stanford, Stanford Univ. Press.

Bocarsly, A. *et al.*: 1993, "Ad Hoc Origins Committee's Letter to Colleagues."

Culliton, B. J.: 1989, "The Dismal State of Scientific Literacy", *Science* 243, 600.

Gould, S. J.: 1965, "Is Uniformitarianism Necessary?" *American Journal of Science* 265, 223–228.

Johnson, P. E.: 1990, *Evolution as Dogma: The Establishment of Naturalism*, U.S.A.: Haughton Publishing Company.

Johnson, P. E.: 1991, *Darwin on Trial*, Washington, D.C., Regnery Gateway.

Johnson, P. E.: 1992, "Darwinism on Trial" (Video), Pasadena, CA, Reasons to Believe.

Kitcher, P.: 1982, *Abusing Science: The Case Against Creationism*, U.S.A., The MIT Press.

Provine, W. B.: 1990, "Response to Phillip Johnson," in P. E. Johnson, *Evolution as Dogma: The Establishment of Naturalism*, U.S.A., Houghton Publishing Company, pp. 19–22.

Ruse, M.: 1982, *Darwinism Defended*, Reading, Mass., Addison-Wesley.

Ruse, M.: "Nonliteralist Antievolution," *American Association for the Advancement of Science Annual Meeting. Symposium on "The New Antievolutionism,"* February 13, 1993.

Simpson, G. G.: 1970, "Uniformitarianism. An Inquiry into Principle, Theory, and Method in Geohistory and Biohistory," in Kecht, M. K. and Steere, W. C. (eds.), *Essays in Evolution and Genetics in Honor of Theodosius Dobzhansky*, New York, Appleton-Century-Crofts, pp. 43–96.

Strahler, A. N.: 1987, *Science and Earth History: The Evolution/Creation Controversy*, Buffalo, N.Y., Prometheus Books.

# 4

# Response to Pennock

Phillip E. Johnson

Robert Pennock argues that evolutionary theory embraces only "methodological," and not "ontological," naturalism. That distinction implies that Darwinists do not claim to make ontologically true statements about the history of life, but only statements about what inferences can be drawn from a naturalistic starting point. If the Darwinists were really as modest as that, there would be little to argue about. For example, I agree with Richard Dawkins that the "blind watchmaker" mechanism is the most plausible naturalistic hypothesis for how complex organisms might have come into existence. We disagree only over whether the theory is true.

Of course, Darwinists do not claim that their theory is merely the best among the naturalistic alternatives. They insist that it is *true* ("fact," in the vernacular). Similarly, their counterparts in prebiotic chemistry do not merely say that they would like to find a naturalistic process that can transform non-living chemicals into living organisms: they insist that they are (at the very least) well on the way to solving the problem. Such claims are false. While Darwinism is better than any available naturalistic alternative, it is clearly against the evidence outside the narrow confines of "finch beaks" and "industrial melanism." Yet, because of the premise of naturalism, the theory survives despite the devastating criticism it has received from scientists who know that the theory is, in the language of our "P.C." culture, "empirically challenged."

We who know how this game of bait-and-switch is played just look for the "switch" that turns innocent "methodological" naturalism into the

Originally published in *Biology & Philosophy* (1996, vol. 11, no. 4, pp. 561–563).

real thing. In Pennock's version, the switch is the argument that natural-ism and rationality are virtually identical—because he thinks that attrib-uting the design of organisms to an intelligence would imply that all events occur at the whim of capricious gods, so that there would be no regularities for scientists to observe. No doubt this caricature explains why no science was done in the century of Newton. What theists like myself are urging biologists to consider, however, is not irrational causes, but intelligent causes. Computers are intelligently designed, but this does not mean that they cannot be the objects of scientific study.

The main point at issue is very simple. Richard Dawkins states, on page 1 of *The Blind Watchmaker*, that "Biology is the study of com-plicated things that give the appearance of having been designed." That is positive evidence that they *were* designed—if you leave it there. Of course, Dawkins does not leave it there. He says that natural selection really did the "designing," and natural selection is an unintelligent, pur-poseless material process. It is on that basis that George Gaylord Simp-son (Pennock's example of a mere "methodological" naturalist) could proclaim that "Man is the result of a purposeless and natural process that did not have him in mind."

Is the "blind watchmaker thesis" true, or false? Prior to that, is the thesis empirical, or is it philosophical? If science disqualifies any consid-eration of intelligent design, and if the most plausible unintelligent mechanism is protected by a rule against negative argument, then some-thing at least roughly like Darwinism follows as a matter of logic, re-gardless of the evidence. If that is Pennock's point, then I thoroughly agree with it. Even the most crushing disconfirmation of the Darwinian story (like the Cambrian fossil record) is met with a shrug. The Cambrian animals had to have evolved from single-celled ancestors by the Darwin-ian mechanism, no matter what the fossils say. What else could have happened? Paleontologists like Gould may flirt with saltationism, but saltation without a mechanism is a lot like special creation. As Daw-kins puts it, you can call the creation of man from the dust of the earth a saltation.

One reason that it is appropriate for a Professor of Law to com-ment upon the philosophy of biology is that so many of the philosophers and biologists want to be litigators. They have taken on evolutionary naturalism as a client, and they employ their rhetorical skills to protect their client's hidden assumptions from unfriendly investigators. The philosopher-litigators were successful for a while, but now the public increasingly understands that a naturalistic worldview is being promulgated in the name of "evolution," and that the metaphysicians of science (e.g., Dawkins, Sagan, Gould, Weinberg, Maynard Smith) employ a terminology that makes their philosophy appear to be a conclusion of unbiased scientific investigation. Perhaps it is time for the philosophers to get ahead of the curve, and start helping the non-professionals to think critically about this subject, instead of trying to mystify them with the bogus notion that Dawkins et al. respect a firm distinction between methodology and ontology.

For those who want to carry the discussion further, I recommend that you:

1. Read (as Pennock apparently did not) my book about naturalism—*Reason in the Balance*, Intervarsity Press, 1995.

2. Read the magnificent book by Michael Behe, titled *Darwin's Black Box* (Free Press, 1996). Ask all the molecular biologists you know if they can answer Behe's argument that molecular mechanisms exhibit "irreducible complexity." What you will learn is that most have never thought about the problem, because they were taught that Darwinism had solved it long ago.

3. Subscribe to *Origins and Design*, a new journal which is based on the heresy that the scientific study of biological origins should be divorced from its current precommitment to naturalistic philosophy. The first issue is now available.

# 5

# Reply: Johnson's *Reason in the Balance*

Robert T. Pennock

Phillip Johnson correctly infers that I did not read his *Reason in the Balance* (RIB), but since my article was written in 1993 and accepted for publication in 1994, and his book did not appear until 1995, I hope I may be forgiven that omission. I will take this opportunity to extend my criticisms to his new book while replying to the challenges in his Response. As it turns out, the main thesis and rhetorical strategy of Johnson's RIB are formed of the same stuff as his earlier body of work—he reiterates that "... Darwinism is not really based on empirical evidence. Its true basis is in philosophy ..." (RIB 16). He continues to describe that philosophy as "dogmatic metaphysical naturalism" (RIB 90) or "dogmatic materialism" (RIB 203) and to call scientists "the ruling priesthood" (RIB 198) of this "established religion" (RIB 35). We still get no consistent analysis of the concept of naturalism, which in RIB is made virtually synonymous with or is blamed for atheism, modernism, reductionism, pragmatism, relativism and liberal rationalism. However, unlike earlier works, RIB does mention the methodological form of naturalism (MN), though it relegates the discussion of MN mostly to the Appendix, and the body of the work either ignores it or conflates it with ontological ("metaphysical") naturalism. I do not have the space here to document the ambiguities, and will address the clearest statement of Johnson's position on MN: "The key question raised by the qualifier *methodological* is this: What is being limited—science or reality?" (RIB 212)

This links to the first point in Johnson's Response and shows why it is misleading, as are related points he makes in RIB about truth and ratio-

Originally published in *Biology & Philosophy* (1996, vol. 11, no. 4, pp. 565–568).

nality. Johnson wants truth about reality—indeed he wants absolute truth (RIB 196)—but he neglects the more basic issue that truth claims must be justified by some method. He says it is irrational to act in conflict with true reality, and asserts that God is truly real, but he fails to provide a method by which we can justify claims about God. Instead, he challenges scientists to say that they know, rather than just dogmatically assume, that God did not create us. But we have seen that biologists do not assert by fiat that God played no role in the development of life forms; they simply proceed, as all scientists must, to search for purely natural mechanisms. When they find evidence for a natural explanation (and Johnson admits that the Darwinian mechanism is not only possible but somewhat plausible) they may legitimately say that they have discovered something true about the natural world. To be sure, this is an approximate and tentative scientific truth, not an ontological (metaphysical) truth in the sense that it cannot rule out the possibility that a supernatural Creator is involved in the process. (On the other hand, it does rule out one version of the teleological argument: that God is necessary to explain this development.) Consider for comparison the geneticist who, applying MN, searches for a natural explanation for hypertrichosis. People with hypertrichosis grow hair all over their faces and upper bodies, and were once thought of as werewolves. Finding evidence for the X-linked gene and evolutionary explanation of the trait, the geneticist might reassure a patient that his disorder is "the result of a purposeless and natural process that did not have him in mind." Surely we may accept that statement as true, even though, as a merely naturalistic scientific truth, it does not rule out the possibility of an intelligent supernatural cause—a "curse of the werewolf," say—so it cannot be said to be absolutely true in the ontological (metaphysical) sense. Similarly, the Creationists' supernatural story may be a metaphysical truth—God may have created the world 6,000 years ago but made it look older as "Appearance of Age" Creationists hold—but it is not a scientific truth.

As for Johnson's related challenge, I do not see how I am guilty of a bait-and-switch since I admit, nay insist, that scientific method is a way to find empirical truths about the world (so long as we remain clear that we may have to revise our assessments of what is true if countervailing evidence is found, and we recognize that there are other sorts of truths that science simply cannot address), and I also do not hold that rationality is limited to science. It is Johnson who is trying to pull the switch, or perhaps the wool over our eyes. Recall that his Creationist hypothesis (which remains essentially unchanged in RIB—e.g., p. 74) is that "God creates for a purpose" not, as he now implies in his rebuttal, that "Organisms were designed by an [unspecified] intelligence." (Is Johnson now officially aligning with Creationists who promote "Intelligent Design Theory"?) In my paper I never said that "attributing the design of organisms to an intelligence would imply that all events occur at the whim of capricious gods" but simply noted that scientists would have no way to test which, if any, events were caused by supernatural entities because, by definition, they are above natural law. The picture would look quite different if one "naturalized" the Deity by restricting God's powers and attributes (biologists combating Creationism sometimes inadvertently do this, arguing, in natural terms, that a "perfect" or "wise" or "benevolent" God would not have created "imperfect" organs or "wasteful" fecundity or "cruel" adaptations), but this is not a "god" in the standard, Creationist sense, which is what we must confront. If there is a "caricature" it is Johnson's own, for I simply quoted his characterization of God as an omnipotent, omniscient supernatural entity whose willed purposes are inscrutable, mysterious and, yes, whimsical. Neither do I deny that there can be, in Herbert Simon's parlance, a "science of the artificial"—we may study computers that were made by natural intelligences because neither computers nor their human creators are above natural law. However, organisms are not artificial and Johnson is not really interested in natural intelligences; his real Creationist hypothesis is about a divine, supernatural intelligence, but that is beyond the purview of science.

In the most important section of RIB Johnson proposes "Theistic Realism" and tries to support the Creationist hypothesis from the viewpoint of its "theory of knowledge" (RIB 107). He begins by citing John 1: 1–13 as "the essential, bedrock position of Christian theism about creation" (RIB 107), and goes on to make his central argument that it is "obvious" to all who have not had their reason clouded by the "drug" of naturalism that living beings are the products of intelligent creation: "Because in our universal experience unintelligent material processes do not create life ..." (RIB 108). Weigh that reason in the balance! Genetic engineering may one day allow humans to create life, but so far we do not have a single case of intelligent creation of life; rather, our universal experience to date is that *only* unintelligent material processes do so.

Johnson's "theological science" (RIB 105), of which the above argument is the prime example, may emphasize different Scripture, but it is not too different from the standard "Creation Science" of Henry Morris' Bible-based Young Earth Creationism. We also hear echoes of Morris in Johnson's comment that mentions Newtonian science, in which he follows a trend among Creationists to pick up on studies in the history of science that showed how some natural philosophers of the scientific revolution supported their notion of natural law by reference to a conception of God as an orderly law-giver. Creationists are now eager to credit Christianity with the origin of modern science (or at least of the "true science" that supports their views) and to anachronistically call theistic scientists like Newton fellow "Creationists." However, they want Newton as one of their own because of his scientific successes, not because of his less enduring theological work on "ancient wisdom," interpretation of the Book of Revelations, and numerology regarding "the number of the beast." Having been fortunate to study with J. E. McGuire, the Newton scholar who spearheaded the historical research into Newton's non-scientific interests, I am well aware of Newton's religious motivations, but also of the complexities and tensions in his views. God may have underwritten the active principles that govern the world described in the *Principia* and the *Opticks*, but He did not interrupt any of the equations or regularities therein. Johnson would do well to consult Newton' own rules of reasoning, especially his first that says not to admit unnecessary causes when explaining phenomena, and his fourth that says to regard the conclusions of inductive methods as "accurately or very nearly true" (Newton 1962, p. 399) and to eschew contrary hypotheses until new evidence requires them. Are such rules metaphysical dogma? Commentators note that Newton's *theology* sometimes led him to regard these dogmatically, but that in his scientific passages he took them as ". . . a matter of method merely, to be used tentatively as a principle of further inquiry" (Burtt 1925, p. 215). I hold that there are good reasons for science's methods and for having confidence in its conclusions, but methods may be improved, and I still await Johnson's positive alternative.

In Newton's day many Christians thought atomism was tantamount to atheism in much the same way that Johnson and other Creationists now say that evolution is, and Newton engaged in spirited exegesis to combat this view. He traced the atomic idea back through Pagans like Lucretius, Epicurus, Democritus, Thales, Pythagoras and finally to Mochus, whom he identified with Moses to make the link to Christianity. Today Christian theists rightly find such contrivances unnecessary, for theism is not threatened by atomic theory. Nor do Christians still feel troubled by the heliocentric and geokinetic theories of the solar system, though an earlier generation went through a great turmoil over that conflict with the Bible. Such theories have been confirmed under the methodological assumptions of naturalistic science and we may properly call them true and factual in as strong a sense as empirical justification allows. Evolutionary theory is of a kind and, again after some turmoil, most religious groups have come to accept it as a scientific truth about reality that is fully compatible with their faith and with a mature understanding of Scripture. Johnson's attempt to turn back the clock does a disservice to both religion and science. I certainly agree that believers can be good scientists; what I cannot agree with is that they must be believers of the Creationist variety who want not just absolute truth, but their unjustified, antiscientific version of it.

### References

Burtt, E. A.: 1925, *The Metaphysical Foundations of Modern Science*, Harcourt, Brace & Co., New York.

Johnson, Phillip E.: 1995, *Reason in the Balance: The Case Against Naturalism in Science, Law & Education*, InterVarsity Press, Downers Grove, IL.

Newton, Isaac: 1962, *Mathematical Principles of Natural Philosophy and His System of the World*, University of California Press, Berkeley.

# III

## A Theological Conflict? Evolution vs. the Bible

Non-Christian creationists oppose evolution because they see it as conflicting with their own various religious creation stories, but here we are focusing upon Christian creationists, for whom it is an apparent conflict with the Bible that spurs their opposition. In the first essay in this section, "When Faith and Reason Clash: Evolution and the Bible," originally published in 1991 in *Christian Scholars Review*, philosopher of religion Alvin Plantinga articulates the nature of the problem. He asks "how shall we Christians deal with apparent conflicts between faith and reason, between what we know as Christians and what we know in other ways, between teaching of the Bible and the teachings of science?"—particularly when Reformed Christianity takes the Bible to be "a special revelation from God himself, demanding our absolute trust and allegiance." After arguing that Scripture can also correct reason and that reason can sometimes correct *misinterpretations* of Scripture, he turns to the question of evolution. He argues first that evolution is truly in conflict with regard to the teaching of the Bible, and next that the core evolutionary theses, such as common ancestry and the Darwinian mechanism, are "dubious" from a Christian perspective and that the empirical evidence for them is far from certain. (Plantinga cites Philip Johnson's then unpublished 1989 manuscript *Science and Scientific Naturalism in the Evolution Controversy* and gives other standard creationist objections to evolution.) He concludes that "There is indeed a battle between the Christian community and the forces of unbelief" and that, because science is not neutral with respect to the conflict, "in all the areas of academic endeavor, we Christians must think about the matter at hand from a Christian perspective: we need Theistic Science." He commends to Christians the task of creating such a "Christian science" because these issues are "a matter of absolutely crucial importance to the health of the Christian community."

Of course, not all Christians agree with Plantinga's position. Plantinga writes from the perspective of Reformed Christianity, but so does physicist Howard J. Van Till who, in the second essay in this section, takes issue with many of Plantinga's arguments. Van Till rejects the conflict metaphor and argues that Plantinga relies on "folk exegesis" in his assessment of the Bible's teachings on creation. He argues that the problem for the Christian should not be with evolution, but with those few scientists who misuse it and engage in "naturalistic apologetics."

The third article is by Plantinga's Notre Dame colleague, historian and philosopher of science Ernan McMullin. While he agrees with Plantinga about the importance of finding an integrated worldview that includes both science and Christian faith, McMullin seriously disagrees with Plantinga's call for a Theistic Science, which he argues should not be described *as* science. McMullin is also a Catholic priest, and he goes on to liken the clash over evolution to the seventeenth-century conflict between the Church and Galileo over the geocentric versus heliocentric cosmology, which was in the period taken equally to be a matter of central importance to the health of Christianity. He argues that a literal or quasi-literal reading of Genesis misses its point and that "the cosmological references in the Old Testament ought be understood as conveying fundamental theological truths about the dependence of the natural and human worlds on their Creator, rather than explaining how exactly these worlds first took shape." He notes that there is "little support for the thesis of special creation on the part of contemporary biblical scholars." McMullin also takes Plantinga to task for seriously understating the evidential support for evolution. He concludes by showing how common ancestry "gives a meaning to the history of life that it previously lacked," and disputing Plantinga's claims about what is or is not "antecedently probable" from a Christian perspective.

In the final piece in this section, Plantinga replies to Van Till and McMullin and sorts out his points of agreement and disagreement with them. (In his introduction, Plantinga also thanks "Professor Pun" for a third commentary that I have not reprinted here. Pattle Pun is an early member of the intelligent design Wedge and is now a Fellow at the Center for the Renewal of Science and Culture [see Barbara Forrest's article in section I]; in his article he enthusiastically endorsed Plantinga's arguments. Plantinga's one disagreement with Pun involves the molecular clock hypothesis, against which he cites the work of Sigfried Scherer, who is also an intelligent design creationist, though in the "young-earth" camp.)

# 6

# When Faith and Reason Clash: Evolution and the Bible

Alvin Plantinga

My question is simple: how shall we Christians deal with apparent conflicts between faith and reason, between what we know as Christians and what we know in other ways, between teaching of the Bible and the teachings of science? As a special case, how shall we deal with apparent conflicts between what the Bible initially seems to tell us about the origin and development of life, and what contemporary science seems to tell us about it? Taken at face value, the Bible seems to teach that God created the world relatively recently, that he created life by way of several separate acts of creation, that in another separate act of creation, he created an original human pair, Adam and Eve, and that these our original parents disobeyed God, thereby bringing ruinous calamity on themselves, their posterity and the rest of creation.

According to contemporary science, on the other hand, the universe is exceedingly old—some 15 or 16 billion years or so, give or take a billion or two. The earth is much younger, maybe $4\frac{1}{2}$ billion years old, but still hardly a spring chicken. Primitive life arose on earth perhaps $3\frac{1}{2}$ billion years ago, by virtue of processes that are completely natural if so far not well understood; and subsequent forms of life developed from these aboriginal forms by way of natural processes, the most popular candidates being perhaps random genetic mutation and natural selection.

Now we Reformed Christians are wholly in earnest about the Bible. We are people of the Word; *Sola Scriptura* is our cry; we take Scripture to be a special revelation from God himself, demanding our absolute trust and allegiance. But we are equally enthusiastic about *reason*, a

Originally published in *Christian Scholar's Review* (1991, vol. 21, no. 1, pp. 8–32). © 1991 by *Christian Scholar's Review*. Reprinted by permission.

God-given power by virtue of which we have knowledge of ourselves, our world, our past, logic and mathematics, right and wrong, and God himself; reason is one of the chief features of the image of God in us. And if we are enthusiastic about reason, we must also be enthusiastic about contemporary natural science, which is a powerful and vastly impressive manifestation of reason. So this is my question: given our Reformed proclivities and this apparent conflict, what are we to do? How shall we think about this matter?

## I.    When Faith and Reason Clash

If the question is simple, the answer is enormously difficult. To think about it properly, one must obviously know a great deal of science. On the other hand, the question crucially involves both philosophy and theology: one must have a serious and penetrating grasp of the relevant theological and philosophical issues. And who among us can fill a bill like that? Certainly I can't. (And that, as my colleague Ralph McInerny once said in another connection, is no idle boast.) The scientists among us don't ordinarily have a sufficient grasp of the relevant philosophy and theology; the philosophers and theologians don't know enough science; consequently, hardly anyone is qualified to speak here with real authority. This must be one of those areas where fools rush in and angels fear to tread. Whether or not it is an area where angels fear to tread, it is obviously an area where fools rush in. I hope this essay isn't just one more confirmation of that dismal fact.

But first, a quick gesture towards the history of our problem. Our specific problem—faith and evolution—has of course been with the church since Darwinian evolution started to achieve wide acceptance, a little more than a hundred years ago. And this question is only a special case of two more general questions, questions that the Christian Church has faced since its beginnings nearly two millennia ago: first, what shall we do when there appears to be a conflict between the deliverances of faith and the deliverances of reason? And another question, related but distinct: how shall we evaluate and react to the dominant teachings, the dominant intellectual motifs, the dominant commitments of the society in which we find ourselves? These two questions, not always clearly distinguished, dominate the writings of the early church fathers from the second century on.

Naturally enough, there have been a variety of responses. There is a temptation, first of all, to declare that there really can't be any conflict between faith and reason. The no-conflict view comes in two quite different versions. According to the first, there is no such thing as truth *simpliciter*, truth just as such: there is only truth from one or another perspective. An extreme version of this view is the medieval two-truth theory associated with Averroes and some of his followers: some of these thinkers apparently held that the same proposition can be true according to philosophy or reason, but false according to theology or faith; true as science but false as theology. Thinking hard about this view can easily induce vertigo: the idea, apparently, is that one ought to affirm and believe the proposition as science, but deny it as theology. How you are supposed to do that isn't clear. But the main problem is simply that truth isn't merely truth with respect to some standpoint. Indeed, any attempt to explain what *truth from a standpoint* might mean inevitably involves the notion of truth *simpliciter*.

A more contemporary version of this way of thinking—the truth-from-a-standpoint way of thinking—takes its inspiration from contemporary physics. To oversimplify shamelessly, there is a problem: light seems to display both the properties of a wave in a medium and also the properties of something that comes in particles. And of course the problem is that these properties are not like, say, *being green* and *being square*, which can easily be exemplified by the same object; the problem is that it looks for all the world as if light *can't* be both a particle and a wave. According to Nils Bohr, the father of the Copenhagen interpretation of quantum mechanics, the solution is to be found in the idea of *complementarity*. We must recognize that there can be descriptions of the same object or phenomenon which are both true, and relevantly complete, but nonetheless such that we can't see how they could both hold. From one point of view light displays the particle set of properties; from another point of view, it displays the wave properties. We can't see how both these descriptions can be true, but in fact they are. Of course the theological application is obvious: there is the broadly scientific view of things, and the broadly religious view of things; both are perfectly acceptable, perfectly correct, even though they appear to contradict one another.[1] And the point of the doctrine is that we must learn to live with and love this situation.

But this view itself is not easy to learn to love. Is the idea that the properties in question *really are* inconsistent with each other, so that it isn't possible that the same thing have both sets of properties? Then clearly enough they *can't* both be correct descriptions of the matter, and the view is simply false. Is the idea instead that while the properties are *apparently* inconsistent, they aren't really inconsistent? Then the view might be correct, but wouldn't be much by way of a *view*, being instead nothing but a redescription of the problem.

Perhaps a more promising approach is by way of territorial division, like that until recently between East and West Germany, for instance. We assign some of the conceptual territory to faith and Scripture, and some of it to reason and science. Some questions fall within the jurisdiction of faith and Scripture; others within that of reason and science, but none within both. These questions, furthermore, are such that their answers can't conflict; they simply concern different aspects of the cosmos. Hence, so long as there is no illegal territorial encroachment, there will be no possibility of contradiction or incompatibility between the teachings of faith and those of science. Conflict arises only when there is trespass, violation of territorial integrity, by one side or the other. A limited version of this approach is espoused by our colleague Howard van Till in *The Fourth Day*. Science, he says, properly deals only with matters *internal* to the universe. It deals with the properties, behavior and history of the cosmos and the objects to be found therein; but it can tell us nothing about the *purpose* of the universe, or about its *significance*, or its *governance*, or its *status*; that territory has been reserved for Scripture. The Bible addresses itself only to questions of external relationships, relationships of the cosmos or the things it contains to things beyond it, such as God. Scripture deals with the status, origin, value, governance and purpose of the cosmos and the things it contains, but says nothing of their properties, behavior or history.

Now van Till means to limit these claims to the *prehistory* (i.e., history prior to the appearance of human beings) of the cosmos; he does not hold that science and Scripture cannot both speak on matter of *human* history, for example.[2] This means that his view doesn't give us a general approach to *prima facie* conflicts between science and Scripture; for it says nothing about such apparent conflicts that pertain to matters of human history, or to matters concerning how things have gone in the cosmos since the appearance of human beings. Van Till limits his approval of this approach for very good reason; taken as a *general* claim, the contention that Scripture and science never speak on the same topic is obviously much too simple. First, there are many questions such that both science (taken broadly) and the Bible purport to answer them: for example, *Was there such a person as Abraham? Was Jesus Christ crucified? Has anyone ever caught fish in the Sea of Galilee? Do ax heads ever float?* Indeed, even if we restrict or limit the claim, in van Till's way, to the prehistory of the cosmos, we still find questions that both Scripture and science seem to answer: for example, *Has the cosmos existed for an infinite stretch of time?*

Further, it is of the first importance to see that when we remove that limitation (and here, of course, van Till would agree), then it isn't true at all that the Bible tells only about status, value, purpose, origin, and the like. It tells us about Abraham, for example, and not only about his status and purpose; it tell us he lived in a certain place, made the long journey from Ur to Canaan, had a wife Sarah who had a son when she was really much too old, proposed at one time to sacrifice Isaac in obedience to the Lord, and so on. Even more important, the Bible tells us about Jesus Christ, and not simply about his origin and significance. It *does* tell us about those things, and of course they are of absolutely crucial importance to its central message; but it also tells us much else about Christ. We learn what he did: he preached and taught, drew large crowds, performed miracles. It tells us that he was crucified, that he died, and that he rose from the dead. Some of the teachings most central to Scripture and to the Christian faith tell us of concrete historical events; they therefore tell us of the history and properties of things within the cosmos. Christ died and then rose again; this tells us much about some of the entities within the cosmos. It tells us something about the history, properties, and behavior of his body, for example: namely, that it was dead and then later on alive. It thus tells us that some of the things in the cosmos behaved very differently on *this* occasion from the way in which they *ordinarily* behave. The same goes, of course, for the Ascension of Christ, and for the many other miracles reported in Scripture.

So we can't start, I think, by declaring that the teachings of contemporary science cannot conflict with the deliverances of the faith; obviously they can. We can't sensibly decide in advance what topics Scripture can or does speak on: instead we must look and see. And in fact it speaks on an enormous variety of topics and questions—some having to do with origin, governance, status and the like, but many more having to do with what happened within the cosmos at a particular place and time, and hence with what also falls within the province of science. It speaks of history, of miracles, of communications from the Lord, of what people did and didn't do, of battles, healings, deaths, resurrections, and a thousand other things.

Let's look a little deeper. As everyone knows, there are various intellectual or cognitive powers, belief-producing mechanisms or powers, various sources of belief and knowledge. For example, there are perception, memory, induction, and testimony, or what we learn from others. There is also reason, taken narrowly as the source of logic and mathematics, and reason taken more broadly as including perception, testimony and both inductive and deductive processes; it is reason taken this broader way that is the source of science. But the serious Christian will also take our grasp of Scripture to be a proper source of knowledge and justified belief Just how does Scripture work as a source of proper belief? An answer as good as any I know was given by John Calvin and endorsed by the Belgic Confession: this is Calvin's doctrine of the Internal Testimony of the Holy Spirit. This is a fascinating and important contribution that doesn't get nearly the attention it deserves; but here I don't have time to go into the matter. Whatever the mechanism, the Lord speaks to us in Scripture.

And of course what the Lord proposes for our belief is indeed what we should believe. Here there will be enthusiastic agreement on all sides. Some conclude, however, that when there is a conflict between Scripture (or our grasp of it) and science, we must reject science; such conflict automatically shows science to be wrong, at least on the point in question. In the immortal words of the inspired Scottish bard William E. McGonagall, poet and tragedian,

When faith and reason clash,
Let reason go to smash.

But clearly this conclusion doesn't follow. *The Lord* can't make a mistake: fair enough; but *we* can. Our grasp of what the Lord proposes to teach us can be faulty and flawed in a thousand ways. This is obvious, if only because of the widespread disagreement among serious Christians as to just what it is the Lord *does* propose for our belief in one or another portion of Scripture. Scripture is indeed perspicuous: what it teaches with respect to the way of salvation is indeed such that she who runs may read. It is also clear, however, that serious, well-intentioned Christians can disagree as to what the teaching of Scripture, at one point or another, really is. Scripture is inerrant: the Lord makes no mistakes; what he proposes for our belief is what we ought to believe. Sadly enough, however, our grasp of what he proposes to teach is fallible. Hence we cannot simply identify the teaching of Scripture with our grasp of that teaching we must ruefully bear in mind the possibility that we are mistaken. "He sets the earth on its foundations; it can never be moved," says the Psalmist.[3] Some sixteenth-century Christians took the Lord to be teaching here that the earth neither rotates on its axis nor goes around the sun; and they were mistaken.

So we can't identify our understanding or grasp of the teaching of Scripture with the teaching of Scripture; hence we can't automatically assume that conflict between what we see as the teaching of Scripture, and what we seem to have learned in some other way must always be resolved in favor of the former. Sadly enough, we have no guarantee that on every point our grasp of what Scripture teaches is correct; hence it is possible that our grasp of the teaching of Scripture be corrected or improved by what we learn in some other way—by way of science, for example.

But neither, of course, can we identify either the current deliverances of reason or our best contemporary science (or philosophy, or history, or literary criticism, or intellectual efforts of any kind) with the truth. No doubt what reason taken broadly, teaches is by and large reliable; this is, I should think, a consequence of the fact that we have been created in the image of God. Of course we must reckon with the fall and its noetic effects; but the sensible view here, overall, is that the deliverances of reason are for the most part reliable. Perhaps they are most reliable with respect to such common everyday judgments as that there are people here, that it is cold outside, that the pointer points to 4, that I had breakfast this morning, that $2 + 1 = 3$, and so on; perhaps they are less reliable when it comes to matters near the limits of our abilities, as with certain questions in set theory, or in areas for which our faculties don't seem to be primarily designed, as perhaps in the world of quantum mechanics. By and large, however, and over enormous swatches of cognitive territory, reason is reliable.

Still, we can't simply embrace current science (or current anything else either) as the truth. We can't identify the teaching of Scripture with our grasp of it because serious and sensible Christians disagree as to what Scripture teaches; we can't identify the current teachings of science with truth, because the current teachings of science change. And they don't change just by the accumulation of new facts. A few years back, the dominant view among astronomers and cosmologists was that the universe is infinitely old; at present the prevailing opinion is that the universe began some 16 billion years ago; but now there are straws in the wind suggesting a step back towards the idea that there was no beginning.[4] Or think of the enormous changes from nineteenth- to twentieth-century physics. A prevailing attitude at the end of the nineteenth century was that physics was pretty well accomplished; there were a few loose ends here and there to tie up and a few mopping up operations left to do, but the fundamental lineaments and characteristics of physical reality had been described. And we all know what happened next.

As I said above, we can't automatically assume that when there is a conflict between science and our grasp of the teaching of Scripture, it is science that is wrong and must give way. But the same holds *vice versa*; when there is a conflict between our grasp of the teaching of Scripture and current science, we can't assume that it is our interpretation of Scripture that is at fault. It *could* be that, but it doesn't *have* to be; it could be because of some mistake or flaw in current science. The attitude I mean to reject was expressed by a group of serious Christians as far back as 1832, when deep time was first being discovered; "If sound science appears to contradict the Bible," they said, "we may be sure that it is our interpretation of the Bible that is at fault."[5] To return to the great poet McGonagall,

When faith and reason clash,
'Tis faith must go to smash.

This attitude—the belief that when there is a conflict, the problem must inevitably lie with our interpretation of Scripture, so that the correct course is always to modify that understanding in such a way as to accommodate current science—is every bit as deplorable as the opposite error. No doubt science can correct our grasp of Scripture; but Scripture can also correct current science. If, for example, current science were to return to the view that the world has no beginning, and is infinitely old, then current science would be wrong.

So what, precisely, must we do in such a situation? Which do we go with: faith or reason? More exactly, which do we go with, our grasp of Scripture or current science? I don't know of any infallible rule, or even any pretty reliable general recipe. All we can do is weigh and evaluate the relative warrant, the relative backing or strength, of the conflicting teachings. We must do our best to apprehend both the teachings of Scripture and the deliverances of reason; in either case we will have much more warrant for some apparent teachings than for others. It may be hard to see just what the Lord proposes to teach us in the Song of Solomon or Old Testament genealogies; it is vastly easier to see what he proposes to teach us in the Gospel accounts of Christ's resurrection from the dead. On the other side, it is clear that among the deliverances of reason is the proposition that the earth is round rather than flat; it is enormously harder to be sure, however, that contemporary quantum mechanics, taken realistically, has things right.[6] We must make as careful an estimate as we can of the degrees of warrant of the conflicting doctrines; we may then make a judgment as to where the balance of probability lies, or alternatively, we may suspend judgment. After all, we don't *have* to have a view on all these matters.

Let me illustrate from the topic under discussion. Consider that list of apparent teachings of Genesis: that God has created the world, that the earth is young, that human beings and many different kinds of plants and animals were separately created, and that there was an original human pair whose sin has afflicted both human nature and some of the rest of the world. At least one of these claims—the claim that the universe is young—is very hard to square with a variety of types of scientific evidence: geological, paleontological, cosmological and so on. Nonetheless a sensible person might be convinced, after careful and prayerful study of the Scriptures, that what the Lord teaches there implies that this evidence is misleading and that as a matter of fact the earth really *is* very young. So far as I can see, there is nothing to rule this out as automatically pathological or irrational or irresponsible or stupid.

And of course this sort of view can be developed in more subtle and nuanced detail. For example, the above teachings may be graded with respect to the probability that they really are what the Lord intends us to learn from early Genesis. Most clear, perhaps, is that God created the world, so that it and everything in it depends upon him and neither it nor anything in it has existed for an infinite stretch of time. Next clearest, perhaps, is that there was an original human pair who sinned and through whose sinning disaster befell both man and nature; for this is attested to not only here but in many other places in Scripture. That humankind was separately created is perhaps less clearly taught; that many other kinds of living beings were separately created might be still less clearly taught; that the earth is young, still less clearly taught. One who accepted all of these theses ought to be much more confident of some than of others—both because of the scientific evidence against some of them, and because some are much more clearly the teachings of Scripture than others. I do not mean to endorse the view that all of these propositions are true: but it isn't just silly or irrational to do so. One need not be a fanatic, or a Flat Earther, or an ignorant Fundamentalist in order to hold it. In my judgment the view is mistaken, because I take the evidence for an old earth to be strong and the warrant for the view that the Lord teaches that the earth is young to be relatively weak. But these judgments are not simply *obvious*, or inevitable, or such that anyone with any sense will automatically be obliged to agree.

## II.  Faith and Evolution

So I can properly correct my view as to what reason teaches by appealing to my understanding of Scripture; and I can properly correct my understanding of Scripture by appealing to the teachings of reason. It is of the first importance, however, that we correctly *identify* the relevant teachings of reason. Here I want to turn directly to the present problem, the apparent disparity between what Scripture and science teach us about the origin and development of life. Like any good Christian Reformed preacher, I have three points here. First, I shall argue that the theory of evolution is by no means religiously or theologically neutral. Second, I want to ask how we Christians should in fact think about evolution; how probable is it, all things considered, that the Grand Evolutionary Hypothesis is true? And third, I want to make a remark about how, as I see it, our intellectuals and academics should serve us, the Christian community, in this area.

## A. Evolution Religiously Neutral?

According to a popular contemporary myth, science is a cool, reasoned, wholly dispassionate attempt to figure out the truth about ourselves and our world, entirely independent of religion, or ideology, or moral convictions, or theological commitments. I believe this is deeply mistaken. Following Augustine (and Abraham Kuyper, Herman Dooyeweerd, Harry Jellema, Henry Stob and other Reformed thinkers), I believe that there is conflict, a battle between the *Civitas Dei*, the City of God, and the City of the World. As a matter of fact, what we have, I think, is a three-way battle. On the one hand there is Perennial Naturalism a view going back to the ancient world, a view according to which there is no God, nature is all there is, and mankind is to be understood as a part of nature. Second, there is what I shall call "Enlightenment Humanism": we could also call it "Enlightenment Subjectivism" or "Enlightenment Antirealism": this way of thinking goes back substantially to the great eighteenth-century enlightenment philosopher Immanuel Kant. According to its central tenet, it is really we human beings, we men and women, who structure the world, who are responsible for its fundamental outline and lineaments. Naturally enough, a view as startling as this comes in several forms. According to Jean Paul Sartre and his existentialist friends, we do this world-structuring freely and individually; according to Ludwig Wittgenstein and his followers we do it communally and by way of language; according to Kant himself it is done by the transcendental ego which, oddly enough, is neither one nor many, being itself the source of the one-many structure of the world. So two of the parties to this three-way contest are Perennial Naturalism and Enlightenment Humanism; the third party, of course, is Christian theism. Of course there are many unthinking and ill-conceived combinations, much blurring of lines, many cross currents and eddies, many halfway houses, much halting between two opinions. Nevertheless I think these are the three basic contemporary

Western ways of looking at reality, three basically *religious* ways of viewing ourselves and the world. The conflict is real, and of profound importance. The stakes, furthermore, are high; this is a battle for men's souls.

Now it would be excessively naive to think that contemporary science is religiously and theologically neutral, standing serenely above this battle and wholly irrelevant to it. Perhaps *parts* of science are like that: mathematics, for example, and perhaps physics, or parts of physics— although even in these areas there are connections.[7] Other parts are obviously and deeply involved in this battle: and the closer the science in question is to what is distinctively human, the deeper the involvement.

To turn to the bit of science in question, the theory of evolution plays a fascinating and crucial role in contemporary Western culture. The enormous controversy about it is what is most striking, a controversy that goes back to Darwin and continues full force today. Evolution is the regular subject of courtroom drama; one such trial—the spectacular Scopes trial of 1925—has been made the subject of an extremely popular film. Fundamentalists regard evolution as the work of the Devil. In academia, on the other hand, it is an idol of the contemporary tribe; it serves as a shibboleth, a litmus test distinguishing the ignorant and bigoted fundamentalist goats from the properly acculturated and scientifically receptive sheep. Apparently this litmus test extends far beyond the confines of this terrestrial globe: according to the Oxford biologist Richard Dawkins, "If superior creatures from space ever visit earth, the first question they will ask, in order to assess the level of our civilization, is: 'Have they discovered evolution yet?'" Indeed many of the experts—for example, Dawkins, William Provine, Stephen Gould—display a sort of revulsion at the very idea of special creation by God, as if this idea is not merely not good science, but somehow a bit obscene, or at least unseemly; it borders on the immoral; it is worthy of disdain and contempt. In some circles, confessing to finding evolution attractive will get you disapproval and ostracism and may lose you your job; in others, confessing doubts about evolution will have the same doleful effect. In Darwin's day, some suggested that it was all well and good to discuss evolution in the universities and among the *cognoscenti*: they thought *public* discussion unwise, however; for it would be a shame if the lower classes found out about it. Now, ironically enough, the shoe is sometimes on the other foot; it is the devotees of evolution who sometimes express the fear that public discussion of doubts and difficulties with evolution could have harmful political effects.[8]

So why all the furor? The answer is obvious: evolution has deep religious connections; deep connections with how we understand ourselves at the most fundamental level. Many evangelicals and fundamentalists see in it a threat to the faith; they don't want it taught to their children, at any rate as scientifically established fact, and they see acceptance of it as corroding proper acceptance of the Bible. On the other side, among the secularists, evolution functions as a *myth*, in a technical sense of that term: a shared way of understanding ourselves at the deep level of religion, a deep interpretation of ourselves to ourselves, a way of telling us why we are here, where we come from, and where we are going.

It was serving in this capacity when Richard Dawkins (according to Peter Medawar, "one of the most brilliant of the rising generation of biologists") leaned over and remarked to A. J. Ayer at one of those elegant, candle-lit, bibulous Oxford dinners that he couldn't imagine being an atheist before 1859 (the year Darwin's *Origin of Species* was published); "although atheism might have been logically tenable before Darwin," said he, "Darwin made it possible to be an intellectually fulfilled atheist."[9] (Let me recommend Dawkins' book to you: it is brilliantly written, unfailingly fascinating, and utterly wrongheaded. It was second on the British best-seller list for some considerable time, second only to Mamie Jenkins' *Hip and Thigh Diet*.) Dawkins goes on:

All appearances to the contrary, the only watchmaker in nature is the blind forces of physics, albeit deployed in a very special way. A true watchmaker has foresight: he designs his cogs and springs, and plans their interconnections, with a future purpose in his mind's eye. Natural selection, the blind, unconscious automatic process which Darwin discovered, and which we now know is the explanation for the existence and apparently purposeful form of all life, has no purpose in mind. It has no mind and no mind's eye. It does not plan for the future. It has no vision, no foresight, no sight at all. If it can be said to play the role of watchmaker in nature, it is the *blind* watchmaker. (p. 5)

Evolution was functioning in that same mythic capacity in the remark of the famous zoologist G. G. Simpson: after posing the question "What is man?" he answers, "The point I want to make now is that all attempts to answer that question before 1859 are worthless and that we will be better off if we ignore them completely."[10] Of course it also functions in that capacity in serving as a litmus test to distinguish the ignorant

fundamentalists from the properly enlightened *cognoscenti*; it functions
in the same way in many of the debates, in and out of the courts, as to
whether it should be taught in the schools, whether other views should be
given equal time, and the like. Thus Michael Ruse: "the fight against
creationism is a fight for all knowledge, and that battle can be won if we
all work to see that Darwinism, which has had a great past, has an even
greater future."[11]

The essential point here is really Dawkins' point: Darwinism, the
Grand Evolutionary Story, makes it possible to be an intellectually fulfilled atheist. What he means is simple enough. If you are Christian, or a
theist of some other kind, you have a ready answer to the question, how
did it all happen? How is it that there are all the kinds of floras and
faunas we behold; how did they all get here? The answer, of course, is
that they have been created by the Lord. But if you are not a believer in
God, things are enormously more difficult. How did all these things get
here? How did life get started and how did it come to assume its present
multifarious forms? It seems monumentally implausible to think these
forms just popped into existence; that goes contrary to all our experience.
So how did it happen? Atheism and Secularism need an answer to this
question. And the Grand Evolutionary Story gives the answer: somehow
life arose from nonliving matter by way of purely natural means and in
accord with the fundamental laws of physics; and once life started, all the
vast profusion of contemporary plant and animal life arose from those
early ancestors by way of common descent, driven by random variation
and natural selection. I said earlier that we can't automatically identify
the deliverances of reason with the teaching of current science because
the teaching of current science keeps changing. Here we have another
reason for resisting that identification: a good deal more than reason goes
into the acceptance of such a theory as the Grand Evolutionary Story.
For the nontheist, evolution is the only game in town; it is an essential
part of any reasonably complete nontheistic way of thinking; hence the
devotion to it, the suggestions that it shouldn't be discussed in public,
and the venom, the theological odium with which dissent is greeted.

## B.   The Likelihood of Evolution

Of course the fact the evolution makes it possible to be a fulfilled atheist
doesn't show either that the theory isn't true or that there isn't powerful

evidence for it. Well then, how likely is it that this theory is true? Suppose we think about the question from an explicitly theistic and Christian perspective; but suppose we temporarily set to one side the evidence, whatever exactly it is, from early Genesis. From this perspective, how good is the evidence for the theory of evolution?

The first thing to see is that a number of *different* large-scale claims fall under this general rubric of evolution. First, there is the claim that the earth is very old, perhaps some 4.5 billion years old: The *Ancient Earth Thesis*, as we may call it. Second, there is the claim that life has progressed from relatively simple to relatively complex forms of life. In the beginning there was relatively simple unicellular life, perhaps of the sort represented by bacteria and blue green algae, or perhaps still simpler unknown forms of life. (Although bacteria are simple compared to some other living beings, they are in fact enormously complex creatures.) Then more complex unicellular life, then relatively simple multicellular life such as seagoing worms, coral, and jelly fish, then fish, then amphibia, then reptiles, birds, mammals, and finally, as the culmination of the whole process, human beings: the *Progress Thesis*, as we humans may like to call it (jelly fish might have a different view as to where to whole process culminates). Third, there is the *Common Ancestry Thesis*: that life originated at only one place on earth, all subsequent life being related by descent to those original living creatures—the claim that, as Stephen Gould puts it, there is a "tree of evolutionary descent linking all organisms by ties of genealogy."[12] According to the Common Ancestry Thesis, we are literally cousins of all living things—horses, oak trees and even poison ivy—distant cousins, no doubt, but still cousins. (This is much easier to imagine for some of us than for others.) Fourth, there is the claim that there is a (naturalistic) *explanation* of this development of life from simple to complex forms; call this thesis *Darwinism*, because according to the most popular and well-known suggestions, the evolutionary mechanism would be natural selection operating on random genetic mutation (due to copy error or ultra violet radiation or other causes); and this is similar to Darwin's proposals. Finally, there is the claim that life itself developed from non-living matter without any special creative activity of God but just by virtue of the ordinary laws of physics and chemistry: call this the *Naturalistic Origins Thesis*. These five theses are of course importantly different from each other. They are also logically independent in pairs, except for the third and fourth theses: the fourth entails the third, in that you can't sensibly propose a mechanism or an explanation for evolution without agreeing that evolution has indeed occurred. The combination of all five of these theses is what I have been calling "The Grand Evolutionary Story"; the Common Ancestry Thesis together with Darwinism (remember, Darwinism isn't the view that the mechanism driving evolution is just what Darwin says it is) is what one most naturally thinks of as the Theory of Evolution.

So how shall we think of these five theses? First, let me remind you once more that I am no expert in this area. And second, let me say that, as I see it, the empirical or scientific evidence for these five different claims differs enormously in quality and quantity. There is excellent evidence for an ancient earth: a whole series of interlocking different kinds of evidence, some of which is marshalled by Howard van Till in *The Fourth Day*. Given the strength of this evidence, one would need powerful evidence on the other side—from Scriptural considerations, say—in order to hold sensibly that the earth is young. There is less evidence, but still good evidence in the fossil record for the Progress Thesis, the claim that there were bacteria before fish, fish before reptiles, reptiles before mammals, and mice before men (or wombats before women, for the feminists in the crowd). The third and fourth theses, the Common Ancestry and Darwinian These, are what is commonly and popularly identified with evolution; I shall return to them in a moment. The fourth thesis, of course, is no more likely than the third, since it includes the third and proposes a mechanism to account for it. Finally, there is the fifth thesis, the Naturalistic Origins Thesis, the claim that life arose by naturalistic means. This seems to me to be for the most part mere arrogant bluster; given our present state of knowledge, I believe it is vastly less probable, on our present evidence, than is its denial. Darwin thought this claim very chancy; discoveries since Darwin and in particular recent discoveries in molecular biology make it much less likely than it was in Darwin's day. I can't summarize the evidence and the difficulties here.[13]

Now return to evolution more narrowly so-called: the Common Ancestry Thesis and the Darwinian Thesis. Contemporary intellectual orthodoxy is summarized by the 1979 edition of the *New Encyclopedia Britannica*, according to which "evolution is accepted by all biologists and natural selection is recognized as its cause.... Objections ... have come from theological and, for a time, from political standpoints" (Vol. 7). It goes on to add that "Darwin did two things; he showed that evolution was in fact contradicting Scriptural legends of creation and that its cause, natural selection, was automatic, with no room for divine guidance or design." According to most of the experts, furthermore, evolution, taken as the Thesis of Common Ancestry, is not something about which there can be sensible difference of opinion. Here is a random selection of claims of certainty on the part of the experts. Evolution is certain, says Francisco J. Ayala, as certain as "the roundness of the earth, the motions of the planets, and the molecular constitution of matter."[14] According to Stephen J. Gould, evolution is an established fact, not a mere theory; and no sensible person who was acquainted with the evidence could demur.[15] According to Richard Dawkins, the theory of evolution is as certainly true as that the earth goes around the sun. This comparison with Copernicus apparently suggests itself to many; according to Philip Spieth, "A century and a quarter after the publication of the Origin of Species, biologists can say with confidence that universal genealogical relatedness is a conclusion of science that is as firmly established as the revolution of the earth about the sun."[16] Michael Ruse, trumpets, or perhaps screams, that "evolution is Fact, FACT, **FACT!**" If you venture to suggest doubts about evolution, you are likely to be called ignorant or stupid or worse. In fact this isn't merely *likely*; you have already *been* so-called: in a recent review in the *New York Times*, Richard Dawkins claims that "It is absolutely safe to say that if you meet someone who claims not to believe in evolution, that person is ignorant, stupid or insane (or wicked, but I'd rather not consider that)." (Dawkins indulgently adds that "You are probably not stupid, insane or wicked, and ignorance is not a crime....")

Well then, how should a serious Christian think about the Common Ancestry and Darwinian Theses? The first and most obvious thing, of course is that a Christian holds that all plants and animals, past as well as present, have been created by the Lord. Now suppose we set to one side what we take to be the best understanding of early Genesis. Then the next thing to see is that God could have accomplished this creating in a thousand different ways. It was entirely within his power to create life in a way corresponding to the Grand Evolutionary scenario: it was within his power to create matter and energy, as in the Big Bang, together with laws for its behavior, in such a way that the outcome would be first, life's coming into existence three or four billion years ago, and then the various higher forms of life, culminating, as we like to think, in humankind. This is a semideistic view of God and his workings: he starts everything off and sits back to watch it develop. (One who held this view could also hold that God constantly *sustains* the world in existence—hence the view is only *semi*deistic—and even that any given causal transaction in the universe requires specific divine concurrent activity.)[17] On the other hand, of course, God could have done things very differently. He has created matter and energy with their tendencies to behave in certain ways—ways summed up in the laws of physics—but perhaps these laws are *not* such that given enough time, life would automatically arise. Perhaps he did something different and special in the creation of life. Perhaps he did something different and special in creating the various kinds of animals and plants. Perhaps he did something different and special in the creation of human beings. Perhaps in these cases his action with respect to what he has created was different from the ways in which he ordinarily treats them.

How shall we decide which of these is initially the more likely? That is not an easy question. It is important to remember, however, that the Lord has not merely left the Cosmos to develop according to an initial creation and an initial set of physical laws. According to Scripture, he has often intervened in the working of his cosmos. This isn't a good way of putting the matter (because of its deistic suggestions); it is better to say that he has often treated what he has created in a way different from the way in which he ordinarily treats it. There are miracles reported in Scripture, for example; and, towering above all, there is the unthinkable gift of salvation for humankind by way of the life, death, and resurrection of Jesus Christ, his son. According to Scripture, God has often treated what he has made in a way different from the way in which he ordinarily treats it; there is therefore no initial edge to the idea that he would be more likely to have created life in all its variety in the broadly deistic way. In fact it looks to me as if there is an initial probability on the other side; it is a bit more probable, before we look at the scientific evidence, that the Lord created life and some of its forms—in particular, human life—specially.

From this perspective, then, how shall we evaluate the evidence for evolution? Despite the claims of Ayala, Dawkins, Gould, Simpson and the other experts, I think the evidence here has to be rated as ambiguous and inconclusive. The two hypotheses to be compared are (1) the claim that God has created us in such a way that (a) all of contemporary plants and animals are related by common ancestry, and (b) the mechanism driving evolution is natural selection working on random genetic variation and (2) the claim that God created mankind as well as many kinds of plants and animals separately and specially, in such a way that the thesis of common ancestry is false. Which of these is the more probable, given the empirical evidence and the theistic context? I think the second, the special creation thesis, is somewhat more probable with respect to the evidence (given theism) than the first.

There isn't the space, here, for more than the merest hand waving with respect to marshalling and evaluating the evidence. But according to Stephen Jay Gould, certainly a leading contemporary spokesman,

> our confidence that evolution occurred centers upon three general arguments. First, we have abundant, direct observational evidence of evolution in action, from both field and laboratory. This evidence ranges from countless experiments on change in nearly everything about fruit flies subjected to artificial selection in the laboratory to the famous populations of British moths that became black when industrial soot darkened the trees upon which the moths rest....[18]

Second, Gould mentions homologies: "Why should a rat run, a bat fly, a porpoise swim, and I type this essay with structures built of the same bones," he asks, "unless we all inherited them from a common ancestor?" Third, he says, there is the fossil record:

> transitions are often found in the fossil record. Preserved transitions are not common, ... but they are not entirely wanting.... For that matter, what better transitional form could we expect to find than the oldest human, *Australopithecus afrarensis*, with its apelike palate, its human upright stance, and a cranial capacity larger than any ape's of the same body size but a full 1000 cubic centimeters below ours? If God made each of the half-dozen human species discovered in ancient rocks, why did he create in an unbroken temporal sequence of progressively more modern features, increasing cranial capacity, reduced face and teeth, larger body size? Did he create to mimic evolution and test our faith thereby?[19]

Here we could add a couple of other commonly cited kinds of evidence: (a) we along with other animals display vestigial organs (appendix, coccyx, muscles that move ears and nose); it is suggested that the best explanation is evolution. (b) There is alleged evidence from biochemistry: according to the authors of a popular college textbook, "All organisms ... employ DNA, and most use the citric acid cycle, cytochromes, and so forth. It seems inconceivable that the biochemistry of living things would be so similar if all life did not develop from a single common ancestral group."[20] There is also (c) the fact that human embryos during their development display some of the characteristics of simpler forms of life (for example, at a certain stage they display gill-like structures). Finally, (d) there is the fact that certain patterns of geographical distribution—that there are orchids and alligators only in the American south and in China, for example—are susceptible to a nice evolutionary explanation.

Suppose we briefly consider the last four first. The arguments from vestigial organs, geographical distribution and embryology are suggestive, but of course nowhere near conclusive. As for the similarity in biochemistry of all life, this is reasonably probably on the hypothesis of special creation, hence not much by way of evidence against it, hence not much by way of evidence for evolution.

Turning to the evidence Gould develops, it too is suggestive, but far from conclusive; some of it, furthermore, is seriously flawed. First, those famous British moths didn't produce a new species; there were both dark and light moths around before, the dark ones coming to predominate when the industrial revolution deposited a layer of soot on trees, making the light moths more visible to predators. More broadly, while there is wide agreement that there is such a thing as microevolution, the question is whether we can extrapolate to macroevolution, with the claim that enough microevolution can account for the enormous differences between, say, bacteria and human beings. These is some experiential reason to think not; there seems to be a sort of envelope of limited variability surrounding a species and its near relatives. Artificial selection can produce several different kinds of fruit flies and several different kinds of dogs, but, starting with fruit flies, what it produces is only more fruit flies. As plants or animals are bred in certain direction, a sort of barrier is encountered; further selective breeding brings about sterility or a reversion to earlier forms. Partisans of evolution suggest that, in nature, genetic mutation of one sort or another can appropriately augment the reservoir of genetic variation. That it can do so sufficiently, however, is not known; and the assertion that it does is a sort of Ptolemaic epicycle attaching to the theory.

Next, there is the argument from the fossil record; but as Gould himself points out, the fossil record shows very few transitional forms. "The extreme rarity of transitional forms in the fossil record," he says, "persists as the trade secret of paleontology. The evolutionary trees that adorn our textbooks have data only at the tips and nodes of their branches; the rest is inference, however reasonable, not the evidence of fossils."[21] Nearly all species appear for the first time in the fossil record fully formed, without the vast chains of intermediary forms evolution would suggest. Gradualistic evolutionists claim that the fossil record is woefully incomplete. Gould, Eldredge and others have a different response to this difficulty: punctuated equilibriumism, according to which long periods of evolutionary stasis are interrupted by relatively brief periods of very rapid evolution. This response helps the theory accommodate some of the fossil data, but at the cost of another Ptolemaic epicycle.[22] And still more epicycles are required to account for puzzling discoveries in molecular biology during the last twenty years.[23] And as for the argument from homologies, this too is suggestive, but far from decisive. First, there are of course many examples of architectural similarity that are not attributed to common ancestry, as in the case of the Tasmanian wolf and the European wolf; the anatomical givens are by no means conclusive proof of common ancestry. And secondly, God created several different kinds of animals; what would prevent him from using similar structures?

But perhaps the most important difficulty lies in a slightly different direction. Consider the mammalian eye: a marvelous and highly complex instrument, resembling a telescope of the highest quality, with a lens, an adjustable focus, a variable diaphragm for controlling the amount of light, and optical corrections for spherical and chromatic aberration. And here is the problem: how does the lens, for example, get developed by the proposed means—random genetic variation and natural selection—when at the same time there has to be development of the optic nerve, the relevant muscles, the retina, the rods and cones, and many other delicate and complicated structures, all of which have to be adjusted to each other in such a way that they can work together? Indeed, what is involved isn't, of course, just the eye; it is the whole visual system, including the relevant parts of the brain. Many different organs and suborgans have to be developed together, and it is hard to envisage a series of mutations which is such that each member of the series has adaptive value, is also a step on the way to the eye, and is such that the last member is an animal with such an eye.

We can consider the problem a bit more abstractly. Think of a sort of space, in which the points are organic forms (possible organisms) and in which neighboring forms are so related that one could have originated from the other with some minimum probability by way of random genetic mutation. Imagine starting with a population of animals without eyes, and trace through the space in question all the paths that lead from this form to forms with eyes. The chief problem is that the vast majority of these paths contain long sections with adjacent points such that there would be no adaptive advantage in going from one point to the next, so that, on Darwinian assumptions, none of them could be the path in fact taken. How could the eye have evolved in this way, so that each point on its path through that space would be adaptive and a step on the way to the eye? (Perhaps it is possible that some of these sections could be traversed by way of steps that were not adaptive and were fixed by genetic drift; but the probability of the population's crossing such stretches will be much less than that of its crossing a similar stretch where natural selection is operative.) Darwin himself wrote, "To suppose that the eye, with all its inimitable contrivances ... could have been formed by natural selection seems absurd in the highest degree." "When I think of the eye, I shudder" he said (3–4). And the complexity of the eye is enormously greater than was known in Darwin's time.

We are never, of course, given the *actual* explanation of the evolution of the eye, the actual evolutionary history of the eye (or brain or hand or whatever). That would take the form: in that original population of eyeless life forms, genes $A_1-A_n$ mutated (due to some perhaps unspecified cause), leading to some structural and functional change which was adaptively beneficial; the bearers of $A_1-A_n$ thus had an advantage and came to dominate the population. Then genes $B_1-B_n$ mutated in an individual or two, and the same thing happened again; then gene $C_1-C_{n'}$ etc. Nor are we even given any possibilities of these sorts. (We couldn't be, since, for most genes, we don't know enough about their functions.) We are instead treated to broad brush scenarios at the macroscopic level:

perhaps reptiles gradually developed feathers, and wings, and warm-bloodedness, and the other features of birds. We are given possible evolutionary histories not of the detailed genetic sort mentioned above, but broad macroscopic scenarios: what Gould calls "just-so stories."

And the real problem is that we don't know how to evaluate these suggestions. To know how to do *that* (in the case of the eye, say), we should have to start with some population of animals without eyes; and then we should have to know the rate at which mutations occur for that population; the proportion of those mutations that are on one of those paths through that space to the condition of having eyes; the proportion of *those* that are adaptive, and, at each stage, given the sort of environment enjoyed by the organisms at that stage, the rate at which such adaptive modifications would have spread through the population in question. Then we'd have to compare our results with the time available to evaluate the probability of the suggestion in question. But we don't know what these rates and proportions are. No doubt we *can't* know what they are, given the scarcity of operable time-machines: still, the fact is we don't know them. And hence we don't really know whether evolution is so much as biologically possible: maybe there is no path through that space. It is *epistemically* possible that evolution has occurred: that is, we don't know that it hasn't; for all we know, it has. But it doesn't follow that it is *biologically* possible. (Whether every even number is the sum of two primes is an open question; hence it is epistemically possible that every even number is the sum of two primes, and also epistemically possible that some even numbers are not the sum of two primes; but one or the other of those epistemic possibilities is in fact mathematically impossible.) Assuming that it *is* biologically possible, furthermore, we don't know that it is not prohibitively improbable (in the statistical sense), given the time available. But then (given the Christian faith and leaving to one side our evaluation of the evidence from early Genesis) the right attitude towards the claim of universal common descent is, I think, one of a certain interested but wary skepticism. It is *possible* (epistemically possible) that this is how things happened; God could have done it that way; but the evidence is ambiguous. That it is *possible* is clear; that it *happened* is doubtful; that it is *certain*, however, is ridiculous.

But then what about all those exuberant cries of certainty from Gould, Ayala, Dawkins, Simpson and the other experts? What about those claims that evolution, universal common ancestry, is a rock-ribbed certainty, to be compared with the fact that the earth is round and goes around the sun? What we have here is at best enormous exaggeration. But then what accounts for the fact that these claims are made by such intelligent luminaries as the above? There are at least two reasons. First, there is the cultural and religious, the mythic function of the doctrine; evolution helps make it possible to be an intellectually fulfilled atheist. From a naturalistic point of view, this is the only answer in sight to the question "How did it all happen? How did all this amazing profusion of life get here?" From a nontheistic point of view, the evolutionary hypothesis is the only game in town. According to the thesis of universal common descent, life arose in just one place; then there was constant development by way of evolutionary mechanisms from that time to the present, this resulting in the profusion of life we presently see. On the alternative hypothesis, different forms of life arose independently of each other; on that suggestion there would be many different genetic trees, the creatures adorning one of these trees genetically unrelated to those on another. From a nontheistic perspective, the first hypothesis will be by far the more probable, if only because of the extraordinary difficulty in seeing how life could arise even once by any ordinary mechanisms which operate today. That it should arise many different times and at different levels of complexity in this way, is quite incredible.

From a naturalist perspective, furthermore, many of the arguments for evolution are much more powerful than from a theistic perspective. (For example, *given* that life arose naturalistically, it is indeed significant that all life employs the same genetic code.) So from a naturalistic, nontheistic perspective the evolutionary hypothesis will be vastly more probable than alternatives. Many leaders in the field of evolutionary biologists, or course, *are* naturalists—Gould, Dawkins, and Stebbins, for example; and according to William Provine, "very few truly religious evolutionary biologists remain. Most are atheists, and many have been driven there by their understanding of the evolutionary process and other science."[24] If Provine is right or nearly right, it becomes easier to see why we hear this insistence that the evolutionary hypothesis is certain. It is also easy to see how this attitude is passed on to graduate students, and, indeed, how accepting the view that evolution is certain is itself adaptive for life in graduate school and academia generally.

There is a second and related circumstance at work here. We are sometimes told that natural science is *natural* science. So far it is hard to object: but how shall we take the term 'natural' here? It could mean that natural science is science devoted to the study of nature. Fair enough. But it is also taken to mean that natural science involves a *methodological naturalism* or provisional atheism:[25] no hypothesis according to which God has done this or that can qualify as a *scientific* hypothesis. It would be interesting to look into this matter: is there really any compelling or even decent reason for thus restricting our study of nature? But suppose we irenically concede, for the moment, that natural science doesn't or shouldn't invoke hypotheses essentially involving God. Suppose we restrict our explanatory materials to the ordinary laws of physics and chemistry; suppose we reject divine special creation or other hypotheses about God as *scientific* hypotheses. Perhaps indeed the Lord has engaged in special creation, so we say, but that he has (if he has) is not something with which natural science can deal. So far as natural science goes, therefore, an acceptable hypothesis must appeal only to the laws that govern the ordinary, day-to-day working of the cosmos. As natural scientists we must eschew the supernatural—although, of course, we don't mean for a moment to embrace naturalism.

Well, suppose we adopt this attitude. Then perhaps it looks as if by far the most probable of all the properly scientific hypotheses is that of evolution by common ancestry: it is hard to think of any other real possibility. The only alternatives, apparently, would be creatures popping into existence fully formed; and that is wholly contrary to our experience. Of all the scientifically acceptable explanatory hypotheses, therefore, evolution seems by far the most probable. But if this hypothesis is vastly more probable than any of its rivals, then it must be certain, or nearly so.

But to reason this way is to fall into confusion compounded. In the first place, we aren't just given that one or another of these hypotheses is in fact correct. Granted: if we *knew* that one or another of those scientifically acceptable hypotheses were in fact correct, then perhaps this one would be certain; but of course we don't know that. One real possibility is that we don't have a very good idea how it all happened, just as we may not have a very good idea as to what terrorist organization has perpetrated a particular bombing. And secondly, this reasoning involves a confusion between the claim that of all of those *scientifically* acceptable hypotheses, that of common ancestry is by far the most plausible, with the vastly more contentious claim that of all the acceptable hypotheses *whatever* (now placing no restrictions on their kind) this hypothesis is by far the most probable. Christians in particular ought to be alive to the vast difference between these claims; confounding them leads to nothing but confusion.

From a Christian perspective, it is dubious, with respect to our present evidence, that the Common Ancestry Thesis is true. No doubt there has been much by way of microevolution: Ridley's gulls are an interesting and dramatic case in point. But it isn't particularly likely, given the Christian faith and the biological evidence, that God created all the flora and fauna by way of some mechanism involving common ancestry. My main point, however, is that Ayala, Gould, Simpson, Stebbins and their coterie are wildly mistaken in claiming that the Grand Evolutionary Hypothesis is *certain*. And hence the source of this claim has to be looked for elsewhere than in sober scientific evidence.

So it could be that the best scientific hypothesis was evolution by common descent—i.e., of all the hypotheses that conform to methodological naturalism, it is the best. But of course what we really want to know is not which hypothesis is the best from some artificially adopted standpoint of naturalism, but what the best hypothesis is *overall*. We want to know what the *best* hypothesis is, not which of some limited class is best—particularly if the class in question specifically excludes what we hold to be the basic truth of the matter. It could be that the best scientific hypothesis (again supposing that a scientific hypothesis must be naturalistic in the above sense) isn't even a strong competitor in *that* derby.

Judgments here, of course, may differ widely between believers in God and non-believers in God. What for the former is at best a methodological restriction is for the latter the sober metaphysical truth; her naturalism is not merely provisional and methodological, but, as she sees it, settled and fundamental. But believers in God can see the matter differently. The believer in God, unlike her naturalistic counterpart, is free to look at the evidence for the Grand Evolutionary Scheme, and follow it where it leads, rejecting that scheme if the evidence is insufficient. She has a freedom not available to the naturalist. The latter accepts the Grand Evolutionary Scheme because from a naturalistic point of view this scheme is the only visible answer to the question *what is the explanation of the presence of all these marvelously multifarious forms of life?* The Christian, on the other hand, knows that creation is the Lord's; and she isn't blinkered by *a priori* dogmas as to how the Lord must have accomplished it. Perhaps it was by broadly evolutionary means, but then again perhaps not. At the moment, 'perhaps not' seems the better answer.

Returning to methodological naturalism, if indeed natural science is essentially restricted in this way, if such a restriction is a part of the very essence of science, then what we need here, of course, is not natural science, but a broader inquiry that can include *all* that we know, including the truths that God has created life on earth and could have done it in many different ways. "Unnatural Science," "Creation Science," "Theistic Science"—call it what you will: what we need when we want to know how to think about the origin and development of contemporary life is what is most plausible from a Christian point of view. What we need is a scientific account of life that isn't restricted by that methodological naturalism.

### C.   What Should Christian Intellectuals Tell the Rest of Us?

Alternatively, how can Christian intellectuals—scientists, philosophers, historians, literary and art critics, Christian thinkers of every sort—how can they best serve the Christian community in an area like this? How can they—and since we are they, how can we—best serve the Christian community, the Reformed community of which we are a part, and, more importantly, the broader general Christian community? One thing our experts can do for us is help us avoid rejecting evolution for stupid reasons. The early literature of Creation-Science, so called, is littered with arguments of that eminently rejectable sort. Here is such an argument. Considering the rate of human population growth over the last few centuries, the author points out that even on a most conservative estimate the human population of the earth doubles at least every 1000 years. Then if, as evolutionists claim, the first humans existed at least a million years ago, by now the human population would have doubled 1000 times. It seems hard to see how there could have been fewer than two original human beings, so at that rate, by the inexorable laws of mathematics, after only 60,000 years or so, there would have been something like 36 quintillion people, and by now there would have to be $2^{1000}$ human beings. $2^{1000}$ is a large number; it is more than $10^{300}$, 1 with 300 zeros after it; if there were that many of us the whole universe would have to be packed solid with people. Since clearly it isn't, human beings couldn't have existed for as long as a million years; so the evolutionists are wrong. This is clearly a lousy argument; I leave as homework the problem of saying just where it goes wrong. There are many other bad arguments against evolution floating around, and it is worth our while to learn that these arguments are indeed bad. We shouldn't reject contemporary science unless we have to, and we shouldn't reject it for the wrong reasons. It is a good thing for our scientists to point out some of those wrong reasons.

But I'd like to suggest, with all the diffidence I can muster, that there is something better to do here—or at any rate something that should be done in addition to this. And the essence of the matter is fairly simple, despite the daunting complexity that arises when we descend to the nitty-gritty level where the real work has to be done. The first thing to see, as I said before, is that Christianity is indeed engaged in a conflict, a battle. There is indeed a battle between the Christian community and the forces of unbelief. This contest or battle rages in many areas of contemporary culture—the courts, in the so-called media and the like—but perhaps most particularly in academia. And the second thing to see is that important cultural forces such as science are not neutral with respect to this conflict—though of course certain parts of contemporary science and many contemporary scientists might very well be. It is of the first importance that we discern in detail just *how* contemporary science—and contemporary philosophy, history, literary criticism and so on—is involved in the struggle. This is a complicated, many-sided matter; it varies from discipline to discipline, and from area to area within a given discipline. One of our chief tasks, therefore, must be that of cultural criticism. We must *test* the spirits, not automatically welcome them in because of their great academic prestige. Academic prestige, wide, even nearly unanimous acceptance in academia, declarations of certainty by important scientists—none of these is a guarantee that what is proposed is true, or a genuine deliverance of reason, or plausible from a theistic point of view. Indeed, none is a guarantee that what is proposed is not animated by a spirit wholly antithetical to Christianity. We must discern the religious and ideological connections; we can't automatically take the word of the experts, because their word might be dead wrong from a Christian standpoint.

Finally, in all the areas of academic endeavor, we Christians must think about the matter at hand from a Christian perspective; we need Theistic Science. Perhaps the discipline in question, as ordinarily practiced, involves a methodological naturalism; if so, then what we need, finally, is not answers to our questions from *that* perspective, valuable in some ways as it may be. What we really need are answers to our questions from the perspective of *all* that we know—what we know about God, and what we know by faith, by way of revelation, as well as what we know in other ways. In many areas, this means that Christians must rework, rethink the area in question from this perspective. This idea may be shocking, but it is not new. Reformed Christians have long recognized that science and scholarship are by no means religiously neutral. In a way this is our distinctive thread in the tapestry of Christianity, our instrument in the great symphony of Christianity. This recognition underlay the establishment of the Free University of Amsterdam in 1880; it also underlay the establishment of Calvin College. Our forebears recognized the need for the sort of work and inquiry I've been mentioning, and tried to do something about it. What we need from our scientists and other academics, then, is both cultural criticism and Christian science.

We must admit, however, that it is our *lack* of real progress that is striking. Of course there are good reasons for this. To carry out this task with the depth, the authority, the competence it requires is, first of all, enormously difficult. However, it is not just the *difficulty* of this enterprise that accounts for our lackluster performance. Just as important is a whole set of historical or sociological conditions. You may have noticed that at present the Western Christian community is located in the twentieth-century Western world. We Christians who go on to become professional scientists and scholars attend twentieth-century graduate schools and universities. And questions about the bearing of Christianity on these disciplines and the questions within them do not enjoy much by way of prestige and esteem in these universities. There are no courses at Harvard entitled "Molecular Biology and the Christian View of Man." At Oxford they don't teach a course called "Origins of Life from a Christian Perspective." One can't write his Ph.D. thesis on these subjects.

The National Science Foundation won't look favorably on them. Working on these questions is not a good way to get tenure at a typical university; and if you are job hunting you would be ill-advised to advertise yourself as proposing to specialize in them. The entire structure of contemporary university life is such as to discourage serious work on these questions.

This is therefore a matter of uncommon difficulty. So far as I know, however, no one in authority has promised us a rose garden; and it is also a matter of absolutely crucial importance to the health of the Christian community. It is worthy of the very best we can muster; it demands powerful, patient, unstinting and tireless effort. But its rewards match its demands; it is exciting, absorbing and crucially important. Most of all, however, it needs to be done. I therefore commend it to you.

## Notes

1. Perhaps the shrewdest contemporary spokesman for this view is the late Donald MacKay in *The Clockwork Image: A Christian Perspective on Science* (London: Intervarsity Press, 1974) and "'Complementarity' in Scientific and Theological Thinking" in *Zygon*, Sept. 1974, pp. 225 ff.

2. *The Fourth Day* (Grand Rapids: Wm. B. Eerdmans Publishing Co., 1986), p. 195.

3. Ps. 104 vs. 5.

4. See Stephen Hawking, *A Brief History of Time* (New York: Bantam Books, 1988), pp. 115 ff.

5. *Christian Observer* 1832, p. 437.

6. Here the work of Bas van Fraassen is particularly instructive.

7. As with intuitionist and constructivist mathematics, idealistic interpretations of quantum mechanics, and Bell theoretical questions about information transfer violating relativistic constraints on velocity.

8. Thus according to Anthony Flew, to suggest that there is real doubt about evolution is to corrupt the youth.

9. Richard Dawkins, *The Blind Watchmaker* (London and New York: W. W. Norton and Co., 1986), pp. 6 and 7.

10. Quoted in Richard Dawkins, *The Selfish Gene* (Oxford: Oxford University Press, 1976), p. 1.

11. *Darwinism Defended*, pp. 326–327.

12. Evolution as Fact and Theory" in *Hen's Teeth and Horse's Toes* (New York: Norton, 1983).

13. Let me refer you to the following books: *The Mystery of Life's Origins*, by Charles Thaxton, Walter Bradley and Roger Olsen; *Origins*, by Robert Shapiro, *Evolution, Thermodynamics, and Information: Extending the Darwinian Program*, by Jeffrey S. Wicken, *Seven Clues to the Origin of Life* and Genetic Takeover and the Mineral Origins of Life, by A. G. Cairns-Smith, and *Origins of Life*, by Freeman Dyson; see also the relevant chapters of Michael Denton, *Evolution: A Theory in Crisis* (further publication data on these books, if desired, is to be found in the bibliography). The authors of the first book believe that God created life specially; the authors of the others do not.

14. The Theory of Evolution: Recent Successes and Challenges," in *Evolution and Creation*, ed. Ernan McMullin (Notre Dame: University of Notre Dame Press, 1985), p. 60.

15. "Evolution as Fact and Theory" in *Hen's Teeth and Horse's Toes* (New York: W. W. Norton and Company, 1980), pp. 254–55.

16. "Evolutionary Biology and the Study of Human Nature," presented at a consultation on Cosmology and Theology sponsored by the Presbyterian (USA) Church in Dec. 1987.

17. The issues here are complicated and subtle and I can't go into them; instead I should like to recommend my colleague Alfred Freddoso's powerful piece, "Medieval Aristotelianism and the Case Against Secondary Causation in Nature," in *Divine and Human Action*, edited by Thomas Morris (Ithaca: Cornell University Press, 1988).

18. *Op. cit.* p. 257.

19. *Op. cit.*, pp. 258–259.

20. Claude A. Villee, Eldra Pearl Solomon, P. William Davis, *Biology*, Saunders College Publishing 1985, p. 1012. Similarly, Mark Ridley (*The Problems of Evolution* (Oxford: Oxford University Press, 1985) takes the fact that the genetic code is universal across all forms of life as proof that life originated only once; it would be extremely improbable that life should have stumbled upon the same code more than once.

21. *The Panda's Thumb* (New York: 1980), p. 181. According to George Gaylord Simpson (1953): "Nearly all categories above the level of families appear in the record suddenly and are not led up to by known, gradual, completely continuous transitional sequences."

22. And even so it helps much less than you might think. It does offer an explanation of the absence of fossil forms intermediate with respect to closely related or adjoining species; the real problem, though, is what Simpson refers to in the quote in the previous footnote: the fact that nearly all categories above the level of families appear in the record suddenly, without the gradual and continuous sequences we should expect. Punctuated equilibriumism does nothing to explain the nearly complete absence, in the fossil record, of intermediates between such major divisions as, say, reptiles and birds, or fish and reptiles, or reptiles and mammals.

23. Here see Michael Denton, *Evolution: A Theory in Crisis* (London: Burnet Books, 1985), chapter 12.

24. *Op. Cit.*, p. 28.

25. "Science must be provisionally atheistic or cease to be itself." Basil Whilley "Darwin's Place in the History of Thought" in M. Banton, ed., *Darwinism and the Study of Society* (Chicago: Quadrangle Books, 1961).

## Brief Bibliography

Ayala, Francis, "The Theory of Evolution: Recent Successes and Challenges," in *Evolution and Creation*, ed. Ernan McMullin (Notre Dame: University of Notre Dame Press, 1985).

Cairns-Smith, A. G., *Genetic Takeover and the Mineral Origins of Life* (Cambridge: Cambridge University Press, 1982).

————, *Seven Clues to the Origin of Life* (Cambridge: Cambridge University Press, 1985).

Darwin, Charles, *The Origin of Species*.

Dawkins, Richard, *The Blind Watchmaker* (London and New York: W. W. Norton and Co., 1986).

————, *The Selfish Gene* (Oxford: Oxford University Press, 1976).

Denton, Michael, *Evolution: A Theory in Crisis* (London: Burnet Books, 1985).

Dyson, Freeman, *Origins of Life* (Cambridge: Cambridge University Press, 1985).

Eldredge, Niles, *Time Frames* (New York: Simon And Schuster, 1985).

Freddoso, Alfred, "Medieval Aristotelianism and the Case Against Secondary Causation in Nature," in *Divine and Human Action*, edited by Thomas Morris (Ithaca: Cornell University Press, 1988).

Gould, Stephen J., "Evolution as Fact and Theory" in *Hen's Teeth and Horses' Toes* (New York: Norton, 1983).

Hawking, Stephen, *A Brief History of Time* (New York: Bantam Books, 1988).

Kitcher, Philip, *Vaulting Ambition* (Cambridge: MIT Press, 1985).

Johnson, Philip, *Science and Scientific Naturalism in the Evolution Controversy*. Unpublished manuscript.

MacKay, Donald, *The Clockwork Image: A Christian Perspective on Science* (London: Intervarsity Press, 1974).

————, "'Complementarity' in Scientific and Theological Thinking" in *Zygon*, Sept. 1974, pp. 225 ff.

Neill, Stephen, *Anglicanism* (Penguin, 1958).

Ridley, Mark, *The Problems of Evolution* (Oxford: OUP, 1985).

Ruse, Michael, *Darwinism Defended* (Reading, Mass.: Addison-Wesley Publishing Co., 1982).

Shapiro, Robert, *Origins* (New York: Summit Books, 1986).

Simpson, George Gaylord, *Fossil and the History of Life* (New York: Scientific American Books and W. H. Freeman and Co., 1983).

———, *The Major Features of Evolution* (New York: Columbia University Press, 1953).

———, *The Meaning of Evolution* (New Haven: Yale University Press, 1949).

———, *This View of Life* (New York: Harcourt Brace and World, 1964).

Spieth, Philip, "Evolutionary Biology and the Study of Human Nature," presented at a consultation on Cosmology and Theology sponsored by the Presbyterian (USA) Church in Dec., 1987.

Stanley, Steven, *The New Evolutionary Timetable* (New York: Basic Books, 1981).

Stebbins, G. Ledyard, *Darwin To DNA, Molecules to Humanity* (San Francisco: W. H. Freeman and Co., 1982).

Thaxton, Charles, Walter Bradley, and Roger Olsen, *The Mystery of Life's Origins* (New York: Philosophical Library, 1984).

van Fraassen, Bas, *The Scientific Image* (Oxford: Clarendon Press; New York: Oxford University Press, 1980).

Van Till, Howard, *The Fourth Day: What the Bible and the Heavens are Telling Us About the Creation* (Grand Rapids: W. B. Eerdmans, 1986).

Villee, Claude A., Eldra Pearl Solomon, and P. William Davis, *Biology* (Philadelphia: Saunders College Publishing, 1985).

Wicken, Jeffrey S., *Evolution, Thermodynamics, and Information: Extending the Darwinian Program* (New York: Oxford University Press, 1987).

Willey, Basil, "Darwin's Place in the History of Thought" in M. Banton, (ed.) *Darwinism and the Study of Society* (London: Tavistock Publication, and Chicago: Quadrangle Books, 1961).

# 7

# When Faith and Reason Cooperate

Howard J. Van Till

As beneficiaries of the same Reformed Christian heritage, Alvin Plantinga and I are likely to articulate our theological positions in similar conceptual vocabularies, view many issues from nearby standpoints, and hold numerous important beliefs in common. With little difficulty I could use my allotted space to reflect appreciatively on the sizeable intersection of our commitments and perspectives. However, at the risk of appearing excessively contentious, most of this response will concern those areas where there is some disagreement and those matters that need further development. Even in this endeavor I shall have to be both selective and brief.

## A. Faith, Reason, and Conflict

I have long been sorely vexed at the frequency with which the warfare metaphor has been employed in the discussion of the relationship of natural science and Christian belief. And my irritation seems to be irreversibly amplified each time I observe the proponents of "creation-science" or the preachers of modern Western naturalism resonantly encourage one another in the perpetuation of this conflict thesis in the service of their own polemical agendas.[1]

Hence, with all due respect for the long history of the faith-versus-reason discussion, I believe it was most inappropriate for Plantinga to employ the conflict metaphor as frequently as he did in his paper. In fact, I would strongly contest the idea that a Christian critique of the scientific

Originally published in *Christian Scholar's Review* (1991, vol. 21, no. 1, pp. 33–45). © 1991 by *Christian Scholar's Review*. Reprinted by permission.

concept of evolution can fruitfully be conducted in the arena of the historical faith-versus-reason tension. As persons already committed to faith in God, our concern is not with faith *versus* reason, nor is it with the unambiguous teaching of the Bible *versus* the "teachings" (Plantinga's term) of contemporary natural science. Rather, it is our earnest desire to bring into consonance our reasoned understanding of the Scriptures and our reasoned understanding of the Creation. At a number of points Plantinga acknowledged this, but all too often the emotion-laden rhetoric of faith *versus* reason appeared, thereby clouding the atmosphere of fruitful response to the question, "How shall we think about this matter?"

Perhaps my complaint is primarily with Plantinga's use of the term *faith*. In most instances in this paper the term does not refer to one's personal commitment to act in the warranted confidence that the object of one's faith is trustworthy (e.g., to have faith that God will provide lovingly for our needs); rather, Plantinga employs the term principally as an abbreviated version of "a deliverance of the faith." And, in the context of the issue at hand, I take this to be a reference to some specific belief concerning what the Scriptures require a Christian to affirm, a belief held mostly for the reason that it constitutes an element in the received Christian tradition (e.g., the belief that the Bible teaches the concept of special creation). In fairness, however, we should note the possibility that belief in such a "deliverance of the faith" might well be strengthened as a result of contemporary reexamination.

As I see it, the real challenge that stands before us is not to resolve some *conflict* between Christian faith and scientific reasoning, but instead to promote a *cooperative effort* of Christian scientists and biblical scholars as together we seek to grow in our understanding of what Scripture requires of us and of what the Creation's formative history was like. My version of McGonagall's rhyme would be:

When faith and reason appear to clash,
'tis the appearance must go to smash.

Plantinga does, however, go well beyond the confines of conflict rhetoric to note the genuine difficulties encountered when Christians seek to draw from *all* relevant resources in formulating or evaluating theories concerning the historical manifestation of God's creative activity. And I must admit that I had Plantinga's concerns in mind when writing the introduction to *Portraits of Creation*, the product of an interdisciplinary study on the topic "Creation and Cosmogony." Because of their direct relevance, I shall quote three full paragraphs from that introduction.

As Christians we rightly seek to grow in our understanding of God and of his works in this world. Focusing our attention here on our efforts to understand the physical universe as the Creation with which God continues to interact, we desire to construct and evaluate our theoretical models by drawing upon all that we know about the world. Specifically we wish to incorporate both what we know (or think that we know) by empirical study of the created world and what we know (or think that we know) by exegetical study of the Scriptures.

In drawing from each resource, however, we face difficult questions regarding both epistemology and hermeneutics. What do we really *know*, for example, from the results of empirical science? In the context of conflicting claims by "experts," some genuine, some bogus, how do we come to *know* that a particular scientific theory should be held with a high degree of confidence? Have the relevant empirical data been competently *interpreted*? Have we made the move from "raw data" to systematizing and interpretive theory with the requisite level of scientific proficiency and integrity? And in making this move, have we made full and proper use of what we know from a faithful and well-informed study of the Scriptures?

Likewise, what do we really *know* from the results of biblical exegesis? In the context of conflicting claims by "experts," some genuine, some bogus, how do we come to *know* that a particular theological theory should be held with a high degree of confidence? Have the relevant biblical data been competently *interpreted*? Have we made the move from "raw text" to systematizing and interpretive doctrine with the requisite level of exegetical proficiency and integrity? And in making this move have we made full and proper use of what we know from an honest and skillful investigation of the Creation?[2]

Plantinga seems well aware of the need for paying careful attention to matters of epistemology and hermeneutics in the arena of the natural sciences. (I may submit this observation as my entry in the "colossal understatements" category of the *Guinness Book of World Records*.) More than half of his paper is devoted to questions in this arena of concern. But where is the evidence for a comparable level of concern for dealing with the equally difficult and relevant issues of epistemology and hermeneutics in the arena of biblical exegesis? Where is the concern to deal with the crucial questions of literary genre and historical context? Where is the concern to distinguish the enduring message for today from its original textual expression in the limited conceptual vocabulary of a much earlier culture? In places where we might have expected Plantinga to pose penetrating questions that would challenge (in the constructive sense) the accuracy of the relevant "deliverances of the faith" (i.e., received traditions concerning what the Bible does or does not teach us regarding the historical particulars of God's creative work in forming the universe and its rich variety of structures and creatures), we find Plantinga instead choosing not to engage the contributions of biblical scholarship. In place of scholarly considerations we find the popular language of unexamined traditional exegesis. Plantinga writes, for example, of what "the Bible initially seems to tell us about the origin and development of life," and about what the Bible "taken at face value ... seems to teach."

But how well does this hermeneutic of "initially seems" and "taken at face value" actually hold up? Granted that the Bible's redemptive message is sufficiently clear (the doctrine concerning the perspicuity of Scripture is not under attack here), the issue at hand concerns complex intellectual questions, and for *these* questions—"head knowledge" as distinct from "heart knowledge"—we need far more than a naive biblical hermeneutic or a simple "folk exegesis." Plantinga did acknowledge this in a later section, but it seems to me that he allowed the results of "folk exegesis" to set the initial tone and to provide the appearance of warrant for the conflict thesis approach.

This brings me back to my criticism regarding Plantinga's use of the faith-versus-reason rhetoric. The "faith" of which he speaks is not our core commitment of *trust* in God but "deliverances of the faith," that is, particular statements that might be believed as *authentic manifestations* of our trust in God and the truth of what he has revealed.[3] However, even if our trust in God's truthfulness were as genuine and complete as humanly possible, our knowledge of what particular beliefs ought thereby follow as manifestations of that trust might be both incomplete and inaccurate. Determining what specific propositions (e.g., proper interpretations of the biblical text) constitute authentic and properly warranted manifestations of Christian faith is an ongoing and never completed human enterprise.

For example, is the concept of special creation required of all persons who profess trust in the Creator-God revealed in Scripture? While most Christians in my acquaintance who are engaged in either scientific or biblical scholarship have concluded that the special creationist picture of the world's formation is *not* a necessary component of Christian belief, the larger North American Christian community is deeply divided over this issue.[4] But questions regarding the relationship between the scientific concept of evolutionary development and biblical proclamations concerning creation can not legitimately be framed in the faith-versus-reason rhetoric unless we are agreed (on the basis of something far more substantial than unexamined tradition or "folk exegesis") either that the concept of special creation is required or that the concept of biological evolution is excluded by faithful biblical exegesis. Furthermore, it could well be argued that to hold the special creationist interpretation of early Genesis is not so much a manifestation of faith in God as it is a manifestation of uncritical acceptance of a particular exegetical tradition. Hence the faith-versus-reason discussion soon degenerates into a contest between rigid traditionalism and open inquiry.

Personally, I heartily agree with those biblical scholars who conclude that the concept of special creation (immediate formation of creatures not genealogically related) is *not* biblically warranted. Furthermore, it is my firm conviction that the Bible does not at all exclude a full evolutionary development of lifeforms in a manner similar to that envisioned by modern biological theory. In fact, I would argue—along the lines of the "categorical complementarity" approach offered in *The Fourth Day*—that the Bible has little of relevance to offer toward either the formulation or evaluation of scientific theories concerning biological history.[5] I say this not by "deciding in advance what Scripture can speak on," but in large part by respecting God's choice for the historical and cultural contexts in which the biblical text was to be written. Was it not God's choice to accommodate this mode of revelation to the historically and culturally limited conceptual vocabularies of the day? The human writers inspired by God had no vocabulary for concepts like galactic redshift, thermonuclear fusion, plate tectonics, spacetime metrics, radiometric dating, stellar evolution, ionizing radiation, chemical reactions, atomic spectra, deoxyribonucleic acid, proteinoid microspheres, genetic drift, molecular clocks, configurational entropy, microevolution, macroevolution, etc., etc., etc.. Hence, to expect the Scriptures to provide us with the kind of statements that would be directly relevant to the evaluation of contemporary scientific theories on the world's formative history strikes me as profoundly misguided.

Plantinga's question, "Just how does Scripture work as a source of proper belief?" is, therefore, one that we must continue to ask. This question regarding the proper epistemological role of the biblical text in the formulation and evaluation of theories—especially of scientific theories—deserves far more attention than Plantinga gives it in this particular paper. One thing, however, seems clear to me: framing the Christian critique of evolutionary theories in the rhetoric of faith-versus-reason offers little hope for growth in our reasoned understanding of either the Scriptures or the Creation. When the concept of special creation is presented in association with *faith* and the concept of evolutionary development is identified with *reason*, many persons within the Christian constituency that both Plantinga and I seek to serve leap with little hesitation to the conclusion that to speak approvingly of evolution is to be unfaithful to God our Creator. Because my personal experience has provided me with more than enough of that rhetoric to make me nauseously weary, I strongly discourage further employment of the warfare metaphor in this context.

## B.  Is the "Grand Evolutionary Story" Religiously Neutral?

The answer to this question depends, of course, on the precise definitions of both "religious neutrality" and the "Grand Evolutionary Story." Let's deal with the neutrality question first. What qualities must a scientific theory possess in order to be called religiously neutral?

Does religious neutrality here require that a scientific theory be immune from employment in the mythology of all religious cultures, including atheistic ones? If so, then *no* scientific theory could possibly qualify. Given the creativity of the human imagination, anything from atomism to zoology could be incorporated into one's mythology. Hence while it may be fitting to recognize that some scientific theories have in fact been woven into the fabric of more comprehensive myths, to equate religious neutrality with mythological immunity would be to trivialize the concept and to render it meaningless.

Suppose, then, that we try to define religious neutrality by saying that a scientific theory is religiously neutral if it entails neither the affirmation nor the denial of religious dogma. Taking "religious dogma" in the broad sense to mean both the formalized creeds of a religion and the set of communally held auxiliary beliefs traditionally associated with that religion, we are once again faced with a problem. Religious dogma taken in this broad sense could easily include beliefs about the physical universe (its age, for instance) that fall within the domain of empirical science. Hence it soon becomes clear that almost any theory of modern science could entail either the denial or affirmation of a "deliverance of the faith" (an element in the received tradition) and thereby fail this test of religious neutrality.

But what if we take "religious dogma" in the narrow sense of formalized creedal statements alone? Suppose, to be specific, we selected as a set of formalized creeds the Heidelberg Catechism, the Belgic Confession and the Canons of Dort? The neutrality test is now clarified substantially, but numerous scientific theories would still likely fail. It is unavoidable that statements referring to the created world would be expressed in the conceptual vocabulary drawn from the world picture that prevailed in the sixteenth and seventeenth centuries when these creeds were written. Hence, wherever contemporary science has replaced that vocabulary with new words and concepts a tension between modern science and the traditional creeds will necessarily arise.

With Plantinga, I would not defend natural science as being religiously neutral in the sense either of mythological immunity or of dogmatic isolation. I would, however, argue for the religious *inconclusiveness* of contemporary theories of natural science in the restricted sense of logical independence from both modern Western naturalism and basic Christian theism. Suppose we define "basic Christian theism" by the Apostles' Creed and modern Western naturalism by Sagan's oft-quoted line, "The Cosmos is all there is or ever was or ever will be." Then I would argue that modern scientific theories concerning the properties, behavior and formative history of the physical universe are logically independent of both theism and naturalism, favoring neither one nor the other. I see no logical entanglement, for instance, between these religious commitments and the theories of big-bang cosmology, stellar evolution, plate tectonics and the like.

Does this apply to the "Grand Evolutionary Story" (GES) as well? That depends on how the GES is told. If the scope of the story is restricted to matters pertaining only to the physical properties, physical behavior and formative history of lifeforms, then the GES may well be logically independent of both theism and naturalism. If, on the other hand, the GES is told in such a way as to include substantive statements regarding the source of the world's existence, the relationship of material behavior to divine governance, or the place of questions about purpose, value and ultimate meaning, then the GES will be inherently religious. My personal conviction is that the scientific core of the story (most likely to appear in the professional journal literature) is logically independent of both theism and naturalism, but that the story told in most modern popularizations of evolutionary theory is embedded in a matrix of naturalistic apologetics, thereby giving many unwary readers the impression that the popularity of the scientific theory grows primarily because of its apologetic usefulness. Monod's *Chance and Necessity*, Dawkins' *The Blind Watchmaker*, Gould's *Wonderful Life* and similar works have served well to promote that impression. Now, although this apologetic employment of evolutionary theory may help us understand some of the things we see these days—the popularity of certain books, the "creation-science" movement, large audiences at creation-evolution debates—we should be extremely cautious in assuming that apologetic attractiveness also accounts for the success of evolutionary theory in the arena of professional science. On the advice of numerous Christian biologists I am led to the conclusion that the scientific success of the concept of biological evolution is the product of proper theory evaluation and that the apologetic employment of evolutionary theory in the "folk-science" of evolutionary naturalism is a regrettable and irritating cultural phenomenon that we must deal with on its own terms—not as science, but as the misemployment of science in a religious agenda.[6]

One aspect of the scientific enterprise brought to our attention by Thomas Kuhn is that our *doing* of science and our *talking* about what we do in science may be quite different from one another, even to the point of incongruity. Hence we have no right uncritically to assume that the popular rhetoric of biologists like Provine or Dawkins constitutes an accurate indicator of how theories of biological evolution have in fact achieved their present state of credibility in the professional arena.

## C.  Is Dawkins' Intellectual Fulfillment Warranted?

In his argument against the idea that the GES could be religiously neutral (neutrality being defined, apparently, as mythological immunity) Plantinga called attention to several examples of the way in which proponents of naturalism have incorporated the concept of evolution into their mythology. The example that played the leading role in the development of this point was the quotation from Richard Dawkins: "although atheism might have been *logically* tenable before Darwin, Darwin made it possible to be an intellectually fulfilled atheist."[7]

Plantinga goes on to explain why he grants Dawkins' point. The question at issue, he says, is, "How is it that there are all the kinds of floras and faunas we behold; how did they all get here?" For the Christian, says Plantinga, the answer is obvious: "They have been created by the Lord." For the atheist, on the other hand, "the Grand Evolutionary Story gives the answer" by describing how life arose and diversified by "purely natural means." But I would argue that both of these answers encourage a mischievous misconstrual of the issue. The central problem with each is a failure to distinguish authentically religious questions from questions accessible to modern empirical science—the common error of treating *creation* and *evolution* as if they were in essence alternative answers to the same question.

If the "how" question posed by Plantinga is meant to focus our attention on the physical, chemical, biological and chronological questions of the formative history of the particular array of flora and fauna we now see, then a Christian's reference to their having been created by the Lord, while religiously very important, is scientifically irrelevant. But if, on the other hand, Plantinga's "how" question is meant to focus our attention on the authentically religious question of whether or not the formative history of the universe and its lifeforms is an expression of the sovereign Creator's intentions and is radically dependent on God's enabling sustenance and directing governance, then an atheist's reference to the concept of biological evolution, while scientifically important, is religiously irrelevant (or at least inconclusive). Consequently, I find Dawkins' claims to intellectual fulfillment extremely shallow and unsatisfying, and I find Plantinga's granting of Dawkins' point very puzzling. Dawkins deserves a far larger dose of Plantinga's analytical, rhetorical and polemical skills than he received in this essay.

Perhaps the point I wish to make could best be illustrated by recounting a little-known episode from the Copernican controversy. Our story centers on the Soltheists, a medieval society of sun-worshipers, whose flourishing was attributed by many observers to the heliocentric theory of the solar system made famous by Copernicus.

The Solthcists took immediate advantage of the Copernican theory. For them it was not merely a theory of terrestrial revolution, but the "Principle of Solar Centrality." As such it appeared to be employable as the ideal apologetic tool in promoting their religion of Soltheism. To the scientific concept of the orbital centrality of the sun they could easily append the religious concept of the divine identity of the Sun. (In ancient Egypt this would have been seen as a Reinterpretation of the heliocentric theory.)

Hardric Snikwad, author of an utterly wrong-headed book that shamelessly exploited the Copernican theory in his Soltheist mythology, was recognized as one of the more colorful and outspoken advocates of the Principle of Solar Centrality and an ardent admirer of Copernicus. In fact, at one of those elegant, candle-lit, bibulous Bullford Academy dinners he leaned over and remarked to a like-minded colleague, "Although Soltheism might have been *logically* tenable before Copernicus, Copernicus has made it possible to be an intellectually fulfilled Soltheist."

Overhearing this specious remark was another dinner guest, World B. Fixed, who used to play roundball with the Alexandria Ptolemaics. Fixed quickly seized the occasion to score a point in favor of his own agenda and responded in a confident voice that all nearby diners could hear, "Snikward's remark confirms a suspicion I've held for a long time. The theory of terrestrial revolution is no more than a thin scientific disguise for the religious Principle of Solar Centrality. The concept of earth's dual motion is not only contradicted by what the Scriptures, taken at face value, seem to teach (cf. Ps. 93:1 and 104:5) but its growing acceptance in our day is merely a consequence of its apologetic usefulness for the sunworshipers who have come to dominate the Medieval Astronomical Society." After a brief but effective pause he continued, "But, of course, for the Soltheist, heliocentrism is the only game in town; it is an essential part of any reasonably complete Soltheistic way of thinking; hence the devotion to it, and the venom, the theological odium with which dissent is greeted."

Although it was not immediately apparent to the diners listening to this exchange, Snikwad was actually quite pleased with the kind of adversarial approach taken by Fixed, because it gave the appearance of making the truth of biblical theism dependent on the falsehood of the Copernican theory regarding planetary motions—a matter not essentially religious but relevant because the Ptolemaic picture had become historically associated with other deliverances of the Christian faith and because a popular exegetical tradition in support of that picture had developed.

One of the listeners, however, was displeased with both of the contenders. A middle-aged astronomer named Sigh Yensma (part Frisian, I believe), a chap with strong interests in both philosophy and theology but with credentials in neither, found himself unable to maintain silence. "Gentlemen," began Sigh, "you have embarked on a debate which is assured to promote nothing but confusion and hostility. Both of you must come to the realization that the Copernican theory of terrestrial revolution is religiously inconclusive. Although Snikwad may find the theory apologetically useful, he hardly has cause to celebrate intellectual fulfillment; the question of the sun's deity is a question categorically distinct from the matter of its orbital centrality in the planetary system. And while World B. Fixed might find that the Copernican theory calls for a re-examination of his traditional reading of the Bible, his faith in God should never be allowed to become dependent on the accuracy of his Ptolemaic picture of the geometrical arrangement or motion of the sun and the planets. Soltheists and biblical geocentrists alike will have to learn to evaluate scientific theories on the basis of a competently and honestly applied system of appropriate criteria."

Sigh interrupted his ponderous monologue just long enough to take a fortifying sip of the local brew and then, turning to Snikwad, said, "You know, Snikwad, I deeply resent what you and your Soltheist friends are doing to the reputation of the scientific enterprise. You see, I seek to serve God just as faithfully as does Mr. Fixed here. At the same time I find the Copernican theory of terrestrial revolution to have considerable scientific merit. Oh, there are numerous questions on which it is silent, and it hasn't been unequivocally established, but it's highly credible nonetheless. However, by your specious and shrewd exploitation of the Copernican theory in your Soltheist mythology many of my fellow believers are moved to denounce the entire concept of a heliocentric solar system as a threat to their faith. I'm trying to get them to distinguish the issues of solar centrality and solar deity as being logically independent, but in the atmosphere created by your pompous preaching of Soltheist folkscience it's not an easy task."

The conversation among Sigh, Fixed and Snikwad is reported to have continued for another eighteen minutes, but unfortunately there is an unexplained gap in the tape recording at this point. We'll have to use our imagination to compensate the loss. Furthermore, although I could make the application of this story to the present issue of biological evolution explicit, that would be belaboring the obvious. As the ocean floor said to the continental plate, "I'm sure you get my drift."

### D.  Should Christian Scholarship Reject Methodological Naturalism?

In the section titled "The Likelihood of Evolution," comprising about half of the paper, Plantinga presents his evaluation of selected empirical evidence relevant to the macroevolutionary scenario and he concludes that as a Christian he finds the concept of special creation more credible. My own inclinations, on the other hand, are toward an evolutionary picture—not only because it provides a coherent means of integrating a broad array of empirical evidence (I will leave Plantinga's handling of specific empirical matters for Christian biologists to evaluate), but also because it comports with the theological position that the world created by the God who reveals himself in Scripture is a world characterized by what I shall call *functional integrity*.

By this term I mean to denote a created world that has no functional deficiencies, no gaps in its economy of the sort that would require God to act immediately, temporarily assuming the role of creature to perform functions within the economy of the creation that other creatures have not been equipped to perform. When the Creator says, "Let the land produce vegetation," or "Let the water teem with living creatures," or "Let the land produce living creatures," a world created with functional integrity will, by the enabling power and directing governance of God, be able to respond obediently and employ its capacities to carry out the intentions of the Creator.

This theological position is stated clearly by John Stek in his contribution to *Portraits of Creation*. Concerning the creation as God's Kingdom Stek says,

*It possesses its own integral and integrating economy.* Each of its components has its own internal economy (e.g., the biological economy of plants and animals), and all of its components were created to fill out and integrate the economy of the whole.... Furthermore, the internal economy of the created realm is neither incomplete nor defective. That is to say, it contains no gaps that have to be filled with continuous or sporadic *immediate* operations of divine power; God is not himself a component within the internal economy of his creaturely realm.[8]

From this theological starting point, drawn from a thorough exegetical study of Scripture, Stek is led to state a number of implications regarding human knowledge of the created world. One especially relevant observation is that:

Since the created realm is replete with its own economy that is neither incomplete (God is not a component within it) nor defective, *in our understanding of the economy of that realm so as to exercise stewardship over it ... we must methodologically exclude all notions of immediate divine causality.* As stewards of the creation we must methodologically honor the principle that creation interprets creation.... In pursuit of a stewardly understanding of the creation, we may not introduce a "God of the gaps...." We may not do so (1) because God is not an internal component within the economy of the created realm, and (2) because to do so would be to presume to exercise power over God—the presumptuous folly of those in many cultures who have claimed to be specialists in the manipulation of divine powers.[9]

In *Christian Belief in a Postmodern World*, a lucid and helpful work on the constructive relationship of Christian faith and careful reasoning, philosopher Diogenes Allen articulates similar concepts of the created world and scientific descriptions of it. Concerning the limits of the competence of scientific investigation he reminds us that

our natural sciences seek to describe and explain the relations *between* the members of the universe, not their origin. The existence of the universe and its basic constituents are taken for granted by our sciences.... When we consider the whole of nature, the relations we find within nature cannot tell us why the universe exists nor why it is the kind of universe that it is. The continuing increase of scientific knowledge, which discovers the relations that exist within our universe, does not get us closer to an answer to either question.[10]

Hence, as Christians we do *not* look for gaps either in the present-day functioning of the universe or in its formative history—gaps into which the immediate action of God might be inserted. Says Allen,

This is theologically improper because God, as creator of the universe, is not a member of the universe. God can never properly be used in scientific accounts, which are formulated in terms of the relations between the members of the universe, because that would reduce God to the status of a creature. According to a Christian conception of God as creator of a universe that is rational through and through, there are no missing relations between the members of nature. If, in our study of nature, we run into what seems to be an instance of a connection missing between members of nature, the Christian doctrine of creation implies that we should keep looking for one.[11]

Applying this approach to questions regarding biological evolution, Allen rejects the view, rooted largely in Aristotle, that species must be fixed—a view that would force Christians to reject any concept of genealogical continuity in favor of special creationism. "But," says Allen, "The Christian conviction that God is the Creator is the claim that nature's order is intended by God, not the claim that present-day life forms arose directly from God's action."[12] To argue, as do some contemporary proponents of special creationism, that belief in God as Creator logically entails the impossibility of genealogical continuity in the formative history of living creatures is, according to Allen, "contrary to the Christian conviction that a rational God creates a universe with members that are coherently connected. Rather than defending Christianity against science, it contradicts a fundamental Christian conviction."[13]

Given my theologically rooted conviction that the created world is characterized by functional integrity, both in its present operation and in its formative history, and given my judgment that speciation within the bounds of genealogical continuity would constitute a remarkably elegant expression of that functional integrity, I expect the scientific search for evolutionary accounts of the history of life forms to be an increasingly fruitful enterprise. As Allen notes, "Christianity rightly endorses the search for such accounts on the ground that the source of the universe is rational and so there are connections to be found between its members."[14]

With Stek, Allen and others, I understand the natural sciences to be properly engaged in, and limited to, a study of the properties, behavior and interrelationships among the diverse members of the created world. As such, science does not make explicit mention of God's creative activity as the source and sustainer of the world's existence, or as the architect of the particular dynamic order exhibited by the world, or as the One whose enabling power and blessing are necessary for all components of the world to act as they do in carrying out his intentions for their being. "There is a discontinuity," says Allen, "between God and our scientific explanations because God's relation to the universe is that of Creator [to the created] and the sciences study [only] the relations between the members of the universe."[15] Science does not have either the competence or the calling to study the creative relationship between the universe and God. Neither do the sciences have the right to bring God's creative activity down to the level of creaturely action and treat God as if he were a component in the economy of the created world.

But the scientific approach that I have here described (as have others elsewhere) is characterized by Plantinga as the strategy of "methodological naturalism" or "provisional atheism"—a strategy that he judges to be inferior because it seems arbitrarily to exclude explicit reference to God's immediate action from its theoretical explanations. "What we need," says Plantinga, "is a scientific account of life that isn't restricted by methodological naturalism.... We need Theistic Science."

My criticism will concern both the terminology and substance of this claim. With Stek and Allen I believe there are good theological reasons for keeping *scientific* accounts within the bounds of creaturely phenomena alone; God's creative activity is not an empirically accessible component within the economy of the created world. Hence I find Plantinga's use of terms like "methodological naturalism" and "provisional atheism" highly pejorative. As I see it, granting the limited competence of natural science is not a concession to naturalism; rather, it is simply a recognition that we have empirical access only to creaturely phenomena. And constructing scientific accounts of the formative history of life forms without explicit appeal to immediate acts of the Creator is not a capitulation to atheism; rather, it is a theologically-based recognition that God is not one component among many others in the economy of the created world—God's creative action, operating at a level different from creaturely action, undergirds *all* that occurs, not only that which eludes our first efforts toward scientific description.

Posing the possibility that all of the present floras and faunas developed over an extended period of time in a genealogically continuous manner from earlier forms, or even proposing that the first living structure formed from inanimate components, in no way denies or even calls into question the creative work of God; rather, it should call us to consider the incomprehensible creativity required of God to give being to a world with such a degree of functional integrity that it could, with God's blessing, so respond in obedience to his "Let the land bring forth...."

Methodological naturalism? Provisional atheism? Highly perjorative terms that function only to cloud the issue. It would be far more fitting, I believe, to think of the approach that I have described simply as natural science performed within the arena permitted by biblical theism—science recognized as providing an incomplete picture of reality because of its inability to probe beyond the creaturely realm (I deplore scientism as much as Plantinga does), science that needs to be placed within the framework of an all-emcompassing, biblically informed, theistic world-view that does indeed draw from *all* that we know about God, his creation and his revelation. Call this broader activity of placing the results of natural science in the framework of theism by the name *Theistic Science* if you like, but contrary to Plantinga's vision for it I propose that it should differ from ordinary science not by the occasional insertion of immediate divine acts into a world whose internal economy is either deficient or defective, but rather by its recognition that every aspect of the world's functionally complete economy is radically dependent on the Creator's ceaseless activity as the world's Originator, Sustainer, Governor and Provider.

Plantinga placed his reflections on evolution and the Bible in the framework of the faith versus reason debate—"when faith and reason clash." I see no hope for progress in that approach. As I see it, progress in this discussion will come only when we work communally toward growth in our reasoned understanding of the Scriptures and in our reasoned understanding of the Creation—when faith and reason cooperate.

## Notes

1. For an extended expression of my irritation relative to this issue see chapter 8 of Van Till, Snow, Stek and Young, *Portraits of Creation: Biblical and Scientific Perspectives on the World's Formation* (Grand Rapids: Eerdmans, 1990), pp. 266–77.

2. Van Till et al., *Portraits of Creation*, p. ix.

3. This useful distinction between *trust* in God and specific beliefs that might constitute *authentic manifestations* of that trust can be found in the introduction to Alvin Plantinga and Nicholas Wolterstorff, editors, *Faith and Rationality: Reason and Belief in God* (Notre Dame: University of Notre Dame Press, 1983).

4. For critiques of how the "creation-science" movement encourages this division see Van Till et al., *Portraits of Creation*, pp. 166–202, and Van Till, Young and Menninga, *Science Held Hostage: What's Wrong with Creation-Science AND Evolutionism* (Downers Grove: InterVarsity Press, 1988), pp. 45–124.

5. Howard J. Van Till, *The Fourth Day: What the Bible and the Heavens Are Telling Us about the Creation* (Grand Rapids: Eerdmans, 1986).

6. See *Science Held Hostage*, pp. 125–78.

7. Richard Dawkins, *The Blind Watchmaker* (New York: W. W. Norton, 1986), p. 6.

8. John Stek, in *Portraits of Creation*, p. 254.

9. Stek, in *Portraits of Creation*, p. 261.

10. Diogenes Allen, *Christian Belief in a Postmodern World* (Louisville: Westminster/John Knox Press, 1989), p. 53.

11. Allen, *Christian Belief*, p. 45.

12. Allen, *Christian Belief*, p. 59.

13. Allen, *Christian Belief*, p. 59.

14. Allen, *Christian Belief*, p. 59.

15. Allen, *Christian Belief*, p. 75.

# 8

# Plantinga's Defense of Special Creation

Ernan McMullin

My colleague, Alvin Plantinga, bids the reader of his essay, "When faith and reason clash: Evolution and the Bible," to take his spirited defence of special creation "with a grain of salt." Perhaps he will forgive me if I take it seriously, because I think that this is how many readers *would* take it, in the context of the continuing controversies about "creation science."

## 1. Theistic Science

His thesis in regard to evolution is that, for the Christian, the claim that God created mankind, as well as many kinds of plants and animals, separately and specially, is more probable than the claim of common ancestry that is central to the theory of evolution. And his larger context is that of an exhortation to Christian intellectuals to join battle against "the forces of unbelief," particularly in academia, instead of always yielding to "the word of the experts." These intellectuals must be brought to "discern the religious and ideological connections ... [they must not] automatically take the word of the experts, because their word might be dead wrong from a Christian standpoint." The implication that worries me is that Christian intellectuals should ally themselves with the critics of evolution, despite the almost universal support it has among experts in the relevant fields of natural science.

The "science" these Christian intellectuals profess will not be of the usual naturalist sort. Their account of the origin of species, for instance, will be at odds with that given by Darwin, on grounds that are dis-

Originally published in *Christian Scholar's Review* (1991, vol. 21, no. 1, pp. 55–70); © 1991 by *Christian Scholar's Review*. Reprinted by permission.

tinctively Christian in content. Despite the fact that claims such as these on the part of the Christian depend on what he or she knows "by faith, by way of revelation," Plantinga believes that they can appropriately be called science, and he suggests as labels for them "theistic science" or "Christian science." An important function of this broader knowledge would be revisionary. He reminds us that "Scripture can correct current science," in regard to whether or not, for example, the universe originated at a particular moment in the past.

Plantinga's "theistic science" bears some similarity to the "creation science" that has commanded the headlines in the U.S. so often in recent decades. Like creation scientists, he maintains that the best explanation of the origin of "many kinds of plants and animals" is an interruption in the ordinary course of natural process, a moment when God treats "what he has created in a way different from the way in which he ordinarily treats it." Like them, he relies on a critique of the theory of evolution, pointing to what he regards as fundamental shortcomings in the Darwinian project of explaining new species by means of natural selection, and emphasizing recent criticisms of one or other facet of the synthetic theory from within the scientific community itself. Like them, he calls for a struggle against prevailing scientific orthodoxy, one which may pit the teachers of Christian youth against the "experts."

But the differences between them are obvious.[1] Most creation-scientists believe in a "young earth" dating back only a few thousand years, and attempt to undermine the many arguments that can be brought against this view. Plantinga allows "the evidence for an old earth to be strong and the warrant for the view that the Lord teaches that the earth is young to be relatively weak." The creation-scientists argue for a whole series of related cosmological theses (that stars and galaxies do not change, that the history of the earth is dominated by the occurrence of catastrophe, and so forth); Plantinga focuses on the single issue of the origins of the kinds of living things, and especially of humankind. And he is in the end more concerned to combat the claims of certainty made by the evolutionists than he is to argue that the Christian is irrevocably committed against a full evolutionary account of origins. He allows (which the creation-scientists, I suspect, would not) that as evolutionary science advances, his own present estimate that special creation is more likely might have to give way.

The creation-scientists attempt to detach their arguments from any sort of reliance on Scripture, or more generally, from theological considerations, whereas Plantinga appeals explicitly to the Scriptural understanding of the manner of God's action in the world. The former make a heroic attempt to qualify their creationism as "scientific," in what they take to be the conventional sense of that term. Their effort, I think it is fair to say, was hopeless right from the start. But it may have been prompted as much by political necessity as by strength of conviction regarding the purely scientific merits of their arguments. The creationists would undoubtedly have preferred to defend a view more explicitly based on the Bible, but the exigencies of the constitutional restrictions on what may be taught in the public schools of the U.S. prevented this. The scientists among them would have wanted to shore up their case with various consonances between the catastrophism of their young-earth account and the geological record. But the inspiration for their account lay, and clearly *had* to lie, in the Bible. Trying to fudge this, though understandable in the circumstances, proved a disastrous strategy.

Plantinga offers a far more consistent theme. True, his "theistic science" will not pass constitutional muster, so it will not serve the purposes for which creation-science was originally advanced. But that is not an argument against it; it is merely a consequence of the unique situation of public education in the U.S., a situation that imposes losses as well as gains. I do not think, however, that "theistic science" should be described *as* science. It lacks the universality of science, as that term has been understood in the Western tradition. It also lacks the sort of warrant that has gradually come to characterize natural science, one that points to systematic observation, generalization, and the testing of explanatory hypothesis. It appeals to a specifically Christian belief, one that lays no claim to assent from a Hindu or an agnostic. It requires faith, and faith (we are told) is a gift, a grace, from God. To use the term "science" in this context seems dangerously misleading; it encourages expectations that cannot be fulfilled, in the interests of adopting a label generally regarded as honorific.

Plantinga objects to the sort of "methodological naturalism" that would deny the label "science" to any explanation of natural process that invokes the special action of God; indeed, he characterizes it, in Basil Willey's phrase, as "provisional atheism." "Is there really any compelling or even decent reason for thus restricting our study of nature?" he asks. But, of course, methodological naturalism does *not* restrict our study of nature; it just lays down which sort of study qualifies as *scientific*. If someone wants to pursue another approach to nature—and there are *many* others—the methodological naturalist has no reason to object. Scientists *have* to proceed in this way; the methodology of natural science gives no purchase on the claim that a particular event or type of event is to be explained by invoking God's creative action directly. Calling this *methodological* naturalism is simply a way of drawing attention to the fact that it is a way of characterizing a particular *methodology*, no more. In particular, it is not an ontological claim about what sort of agency is or is not possible. Dubbing it "provisional atheism" seems to me objectionable; the scientist who does not include God's direct action among the alternatives he or she should test scientifically when attempting to explain some phenomenon is surely not to be accused of atheism!

Let me make myself clear. I do not object (as the concluding section of this essay will make clear) to the use of theological considerations in the service of a larger and more comprehensive world-view in which natural science is only one factor. I would be willing to use the term "knowledge" in an extended sense here (though I am well aware of some old and intricate issues about how faith and knowledge are to be related).[2] But I would *not* be willing to use the term, "science," in this context. Nor do I think it necessary to do so in order to convey the respectability of the claim being made: that theology may appropriately modulate other parts of a person's belief-system, including those deriving from science. I would be much more restrictive than Plantinga is, however in allowing for the situation he describes as "Scripture correcting current science." But before we come to our differences, it may be worth laying out first the large areas where he and I agree.

## 2.  Points of Agreement

What really galls Plantinga are the views of people like Dawkins and Provine who not only insist that evolution is a proven "fact," but suppose that this somehow undercuts the reasonableness of any sort of belief in a Creator. Their argument hinges on the notion of design. The role of the Creator in traditional religious belief was that of Designer; the success of the theory of evolution has shown that design is unnecessary. Hence, there is no longer any valid reason to be a theist. In a recent review of a history of the creationist debate in the U.S., Provine lays out this case, and concludes that Christian belief can be made compatible with evolutionary biology only by supposing that God "works through the laws of nature" instead of actively steering biological process by way of miraculous intervention. But this view of God, he says, is "worthless," and "equivalent to atheism."[3] (On this last point, Plantinga and he might not be so far apart!) He chides scientists for publicly denying, presumably on pragmatic grounds, that evolution and Christian belief are incompatible; they *must*, he says, know this to be nonsense.

Plantinga puts his finger on an important point when he notes that for someone who does not believe in God, evolution is some form or other is the only *possible* answer to the question of origins. Prior to the publication of *The Origin of Species* in 1859, the argument from design was part of biological science itself. The founders of physico-theology two centuries earlier, naturalists like John Ray and William Derham, had shown the pervasive presence in Nature of means-end relationships, the apparently purposive adjusting of structure and instinctive behavior to the welfare of each kind of organism. Someone who rejected the idea of God had, therefore, to face some awkward problems in explaining some of the most obvious features of the living world; it seemed as though science itself testified to the existence of God.[4]

Darwin changed all this. He made it possible to reconcile atheism with biological science; from then onwards, the fortunes of atheism as a form of intellectual belief would depend upon the fortunes of the theory of evolution. No wonder, then, that evolution became a crucial "myth" (as Plantinga puts it) of our secular culture, replacing for many the Christian myth as "a shared way of understanding ourselves at the deep level of religion."[5] No wonder also that an attack on the credentials of evolutionary theory would so often evoke from its defenders a reaction reminiscent in its ferocity of the response to heresy in other days.

Is evolution fact or theory? No other question has divided the two sides in the creation-science controversy more sharply. Plantinga notes that someone who denies the existence of a Creator is left with no other option for explaining the origin of living things than an evolutionary-type account. It thus, equivalently, becomes "fact" not just because of the strength of the scientific evidence in its favor but because for the atheist no other explanation is open. On the other side, the believer in God is going to resent this use of the word, "fact," because it seems to exclude in principle the possibility of a Divine intervention, and hence by implication, the possibility of the existence of a Creator. "Fact" seems to convey not just the assurance of a well-supported theory, but the certainty that no other explanation is open.

The debate may often, therefore, be something other than it seems. Instead of being just a disagreement about the weight to be accorded to a particularly complex scientific theory in the light of the evidence available, the debate may conceal a far more fundamental religious difference, each side appearing to the other to call into question an article of faith. To religious believers, calling the assertion of common ancestry a "fact" appears to violate good scientific usage; no matter how well-supported a theory may be (they argue), it remains a theory. To non-believers, the phrase "merely a theory" comes as a provocation, because it suggests a substantial doubt about a claim that appears to them as being beyond question, a doubt prompted furthermore in their view by an illegitimate intrusion of religious belief.

At one level, then, Plantinga's essay can be read as a plea for a more informed understanding of the real nature of the creation-science debate, and a more sympathetic appreciation of what led the proponents of creation-science to take the stand they did. Even their defense of a "young" earth (the major point of disagreement between his view and theirs) ought not (he says) be regarded as "silly or irrational." One need not be "a fanatic, or a Flat Earther, or an ignorant Fundamentalist" to hold such a view. The claim that the earth is ancient is neither obvious nor inevitable; it has to be argued for, and disagreement may, therefore, easily occur.

Plantinga is right, to my mind, to see more in the creation-science debate than evolutionary scientists (or the media) have been wont to allow. And the sort of challenge he offers to the defenders of evolution, though it is not new, could serve the purposes of science in the long run if it forced a clarification and strengthening of argument on the other side, or if it punctured the sometimes troubling smugness that experts tend to display when dealing with outsiders. Plantinga leans too far in the other direction, however. The charitable reading of creation-science that he urges could easily mislead. A claim does not have to be obvious or inevitable for its rejection to connote fanaticism or ignorance. If the indirect evidence for the great age of the earth is overwhelming (Plantinga himself allows that it is "strong"), if its denial would call into question some of the best-supported theoretical findings of an array of natural sciences (cosmology, astrophysics, geology, biology), then one is entitled to issue a severe judgment on the challenger. Perusal of works like that of Morris would lead one to suspect that no matter *how* strong the scientific case were in favor of an ancient earth, it would make no difference to their authors. Their implicit commitment to a literalist interpretation of *Genesis* is such that it blocks a genuinely rational assessment of the alternative. The term, "fanatic," is a notoriously difficult one to apply fairly, because it conveys moral, as well as epistemic, disapprobation. But I would be willing to defend its appropriateness to such expositions of creation-science as that of Morris.

What bothers Plantinga, I suspect, about the use of the term here, is that from *his* point of view the creation-scientist's heart is in the right place, even if perhaps his head isn't. Anyone who stands up for "*sola Scriptura*" in the modern world, even in contexts as unpromising as the debate about the age of the earth, ought not (he seems to suggest) simply be dismissed. Creation-scientists may be wrong in holding that the earth is only a few thousand years old, but their *motivation* for making this claim ought be regarded with sympathy by their fellow-Christians. I would disagree, but it is because of a deeper disagreement about the merits of the "*sola Scriptura*" premiss and of the remaining major theses of creation-science. Though I would not be as harsh on the creation-scientist as leading evolutionists have been, I would, as a Christian, want to register disapproval of creation-science at least as strong as theirs, though for reasons that go beyond theirs. These reasons will become clear, I hope, in what I have to say about Plantinga's analysis of what happens when "faith and reason clash."

## 3. Galileo and Genesis

In his *Letter to the Grand Duchess Christina* (1615), Galileo gave the most extended account that anyone perhaps had written up to that time of how the Christian should proceed when an apparent conflict between science and Scripture arises. Aided, doubtless, by some of his theologian-friends, he drew upon Augustine, Jerome, Aquinas, and an impressive array of other authorities, in order to show that the use made of Scripture by those who opposed the Copernican theory was illegitimate.[6] There may be some lessons to be drawn from this historic document in the context of the Darwinian debate, apart from the obvious one of the embarrassment that the Church would later suffer because of its ill-advised attempt to make the geocentric cosmology of the Old Testament authors a matter, equivalently, of Christian faith.

What, then, did Galileo hold about the bearing of the Scriptures on our knowledge of the natural world? It does not take long for the reader to discover that two radically different principles are proposed in different parts of the *Letter*, and to realize that Galileo almost certainly was not aware of the resulting incoherence.[7] On the one hand, he cites Augustine in support of the traditional view that in cases of apparent conflict, the literal interpretation of Scripture is to be maintained, unless the opposing scientific claim can be *demonstrated*. In that case, theologians must look for an alternative reading of the Scriptural passage(s), since it is a first principle that faith and natural reason cannot really be in conflict. However, the straightforward interpretation of Scripture is to be preferred in cases where the scientific claim has something less than "necessary demonstration" in its support, because of the inherently greater authority to be attached to the word of God.[8]

On the other hand, Galileo also argues that one should not look to Scripture for knowledge of the natural world: the function of the Bible is to teach us how to go to heaven, not how the heavens go, in the aphorism attributed to Baronius. God has given us reason and the senses to enable us to come to understand the world around us. Attempting to teach the underlying structures of natural process would have baffled the readers of Scripture and defeated its obvious purpose. Galileo produced a number of convergent lines of argument to the effect that Scripture is simply not relevant to the concerns of the sciences to begin with.

The implications of these two quite different hermeneutic principles were, of course, altogether different for the resolution of the Copernican debate.[9] But that is not my concern here. More to my purpose is to note that the first principle has one quite disastrous consequence: it sets theologians evaluating the validity of the arguments of the natural philosophers, and natural philosophers defending themselves by composing theological tracts. Either way, there will be immediate charges of trespass. The theologian challenges the force of technical scientific argument; scientists urge their own readings of Scripture or their own theories as to how Scripture, in general, *should* be read. In both cases, the professionals are going to respond, quite predictably: what right have you to intrude in a domain where you lack the credentials to speak with authority? The assessment of theory-strength is not a simple matter of logic and rule but requires a long familiarity with the procedures, presuppositions, and prior successes of a network of connected domains, and a trained skill in the assessment of particular types of argument.

Transposing from the Galilean to the Darwinian debate demands some care, yet there are obvious morals to be drawn. Neither of Galileo's principles is entirely adequate for regulating disputes of this general kind. One ought not require that a scientific case be demonstrative before it have any standing in the face of an apparently contrary Scriptural assertion. On the other hand, one cannot (as Plantinga correctly says) simply rule out conflicts of this kind by laying down that Christian doctrine can have *no* implications for matters that fall under scientific jurisdiction. Galileo was right to maintain that the Bible was not intended as (in part) a manual of natural knowledge. The biblical writers simply made use of the language and the cosmological beliefs of their own day while telling the story of human salvation. But that story itself does require certain presuppositions about human nature, about freedom and moral responsibility, for example, that would clash with psychological theories affirming the unreality of our claim to freedom of moral action. So it is not as though the two domains are, in principle, so safely walled off from one another that no conflict can possibly arise.

Nevertheless, Scripture scholars would be in a large measure of agreement today that the domain of such potential conflict is quite limited. In particular, the creation narratives of the first two chapters of *Genesis* are not to be read as literal history. The points they are making lie deeper and can only be reliably discerned by investigating the wider literary context of that day, on the one hand, and the later theological appropriations of those narratives, on the other.[10] It is not as though the texts are to be taken literally unless and until a conclusive scientific account of origins can be constructed. Rather, no likelihood is to be attached in the first place to the literal construal of the story, say, of the separate creation of the animals. To interpret it literally or quasi-literally is to misunderstand the point that the writers of those narratives were trying to make, the great majority of contemporary Scripture scholars would agree.[11]

An interesting feature of Plantinga's argument is that he explicitly brackets the reference to *Genesis* that one would expect to find, given the primacy he accords to Scripture. Did he originally plan to return to the text of *Genesis* at the end of his paper to clinch his case? ("Suppose we temporarily set to one side the evidence, whatever exactly it is, from early *Genesis*.") I kept waiting for him to turn finally to the biblical narratives of creation as the strongest reason for distrusting the evolutionary story. Historically, these narratives provided the main warrant for the traditional Christian belief that God intervened in a special way to bring to be the first members in the lineage of each natural kind. Those who, like Plantinga, have urged the superiority of special creation over evolution have almost always relied upon a direct appeal to *Genesis*, unless (as we have seen) they were prevented, as the recent creation-science advocates were, by political constraints.

Plantinga is under no such constraints, however. Since he evidently believes that a case can be made for special creation without any overt appeal to *Genesis*, he may have felt it better to avoid the controversies that surround the literalist approach to the *Genesis* text. He already has the scientists on his hands. Why open a second front and take on the theologians too? There is no rhetorical advantage to be gained by making explicit that he can expect little support for the thesis of special creation on the part of contemporary biblical scholars.

Despite his silence in regard to *Genesis*, I do not think that a linkage between his argument and the more traditional *Genesis*-based argument can be denied. Without the latter, would anyone think it a priori more likely that the God of the salvation story would intervene to originate natural kinds instead of allowing them to appear gradually and in a "natural" way? And if biblical theologians are right in holding that the cosmological references in the Old Testament ought be understood as conveying fundamental theological truths about the dependence of the natural and human worlds on their Creator, rather than explaining how exactly these worlds first took shape, then perhaps one ought be just as wary of drawing a cosmological moral from the salvation story as a whole as from the controversial passages in *Genesis*.

What must be disputed, to my mind, is a modern analogue of the first of Galileo's principles,[12] which would, equivalently, reaffirm the presumption that the biblical text was partially intended as a cosmology, so that in cases of apparent conflict between the biblical and current scientific accounts we should evaluate the strength of the scientific account as a means of deciding which of the two "competitors" to accept. Plantinga proposes such a "balancing of likelihoods" methodology: a literalist understanding of God's making of the ancestors of the main natural kinds, he concludes, should be preferred unless and until a far stronger case can be made for the evolutionary alternative.

What I am urging is that it is potentially destructive (as the Galileo case amply shows) to treat the biblical and the scientific as competitors in the realm of cosmological explanation. The cosmological implications (if any) of the Scripture story are to be discovered by *theological* study, not be assessing the credentials of the supposedly competing scientific account. Even if the theory of evolution could be entirely dismissed on scientific grounds, this would not of itself give us any warrant for supposing that the biblical account of origins ought, therefore, be taken literally. It might well be that, in the absence of a plausible evolutionary account the only reasonable alternative for the Christian would be something like special creation. But even this would not warrant the supposition that special creation, in the literal sense, is what we ought properly infer from the biblical texts, whether the accounts of Creation in *Genesis*, or the story of Israel leading up to the coming of Christ. The interpretation of these texts is primarily a hermeneutic problem for the theologian; criticism of supposedly competing scientific theories will rarely be relevant.[13]

Where the theologian is unsure of the best interpretation to give a text, it is not inappropriate, of course, to take into account that some of the possible interpretations may be closed off by the findings of the natural or social sciences. But even in such a case (as Augustine's own practice might easily be made to illustrate) primary weight should be given to the *hermeneutic* issue as to what the disputed text was originally intended to convey and the theological issue of what the tradition has made of it. This way of handling relationships between Scripture and the sciences does not close them off from one another entirely, as we have already seen. But it encourages us to resist the temptation to construe the two as normally belonging to the same order of explanation or historical claim.

As an illustration of how Scripture could "correct current science," Plantinga remarks: "If, for example, current science were to return to the view that the world has no beginning, and in infinitely old, then current science would be wrong."[14] I do not believe that Scripture *does* prescribe that the universe had a beginning in time, in some specific technical sense of the term, "time"; the point of the Creation narratives is the dependence of the world on God's creative act, to my mind, not that it all began at a finite time in the past.[15] A world that has always existed would still require a Creator.

## The Thesis of Common Ancestry

Plantinga dismisses the evidence ordinarily presented in support of what he calls the Thesis of Common Ancestry (TCA) as inconclusive, after a rather cursory review. His conclusion:"It isn't particularly likely, given the Christian faith and the biological evidence, that God created all the flora and fauna by way of some mechanism involving common ancestry." Though my disagreement with him centers especially on the conclusion he draws from Christian faith in regard to the antecedent likelihood of special creation, I am going to spend some time on the scientific issues first. The credentials of a thesis encompassing as much of past and present as TCA does cannot, of course, be dealt with satisfactorily in a few pages.[16] This is particularly true when these credentials are being *denied*, contrary to the firm conviction of the great majority of those professionally engaged in the many scientific fields involved.

Though a full-scale defence of TCA cannot be attempted here, and would in any event be beyond my competence, it may be worthwhile to indicate how in a general way such a defence might proceed.[17] First, an important distinction, one alluded to by Plantinga. TCA is a *historical* claim that the kinds of living things originated somehow from one another. On the other hand, the various theories of evolution are an attempt to *explain* how that could have occurred.[18] The dominant theory of evolution at the present time is the so-called modern synthesis, associated with such figures as Simpson, Dobzhansky, and Mayr. It has its critics: Goldschmidt and Schindewolf a generation ago, for example, Gould and Kimura today. Though all of these have found fault with the Darwinism of the modern synthesis and have proposed alternatives to it, none would for a moment question TCA. Their confidence in TCA does not depend, then, on a similar degree of confidence in the explanatory adequacy of a specifically Darwinian account of the origin of species. Is it, perhaps, that they implicitly reject God's existence, and thus TCA is for them (in Plantinga's phrase) "the only game in town"? I don't think it is quite as simple as this, although the implicit setting aside of a theistic alternative obviously could play a role.

Much of the evidence for TCA functions independently of the *detail* of any specific evolutionary theory. Plantinga mentions three such categories of evidence, so I will confine myself to those. There is the fossil record which has already yielded innumerable sequences of extinct forms, where the development of specific anatomical features can be traced in detail through the rock-layers. Paleontologists have traced the development of eyes in no less than forty *independent* animal lineages ("lineages" being determined by overall morphological similarities).[19] They continue to uncover stage after stage in crucial "linking" forms, such as the therapsids, for example, the forms that relate reptiles with the earliest mammals. In cases like these (and there are a *lot* of them), paleontologists can point to a variety of morphological features that gradually shift over time, retaining a basic likeness (a so-called *Bauplan*) throughout. Gould's objection regarding the rarity of transitional forms (quoted by Plantinga) has to be taken in context. Gould would not deny the morphological continuities of the fossil record; like thousands of other researchers he has given too much of his time to tracing these continuities for him to underrate their significance. What he *would* say (and what many defenders of the modern synthesis would now be disposed to admit) is that species often make their appearance in the record without the prior gradual sequence of modifications one would have expected from the traditional gradualist Darwinian standpoint. But this leaves untouched the implications, overall, of the fossil record for TCA. It *does*, of course, affect the sort of theory that could account for the sequence found in the record.[20]

Instead of scrutinizing the fossil record, we might look to the living forms around us, and there discover all sorts of homologies and peculiar features of geographical distribution, which are best understood in terms of TCA. The arguments here are long familiar and I will not dally on them. But there is a further category of evidence which has taken on a great deal of importance in the last twenty years. This comes from molecular biology.[21] Comparison of the DNA, as well as of the proteins for which DNA codes, between different types of organisms shows that there are striking similarities in chemical composition between them. Cytochrome C, for example, found in all animals, is involved in cell respiration. It contains 104 amino-acids, in a sequence which is invariable for any given species. For humans and rhesus monkeys, the sequence is identical except in one position; for horses and donkeys the sequence also differs in only one position. But for humans and horses, the difference is 12; for monkeys and horses, the difference is 11. If instead of cytochrome C another homologous protein is chosen, similar (though not necessarily identical) results are found. These very numerous resemblances and differences between the macromolecules carrying hereditary information can be explained by supposing a very slow rate of change in the chemical sequences constituting these molecules, and thus a relationship of common descent among the organisms themselves. Thus, the "molecular" differences between any two species become (on this hypothesis) a rough indication of how long since the ancestors of these species diverged; rather more securely, one can infer the relative order of branching between three or more species; one can infer whether A branched from B before C did. What is impressive here is the *coherence* of the results given by examining many different macromolecules in this light.

But much more impressive is that these results conform reasonably well with the findings of both paleontology and comparative anatomy in regard to the ancestral relations between species, the postulated tree of descent that had already been worked out in some detail in these other disciplines. The fit, as one would expect, is not exact in each case in regard to the "closeness" between the species, but it is nevertheless quite good. When a single explanatory hypothesis (TCA) underlies the binding together of three domains so diverse in character, we have the sort of consilience that carries more weight with scientists that does, perhaps, any other virtue of theory.

It should be underlined that specific theories of evolution are not yet involved here. The support given TCA by these diverse types of evidence does not depend on any particular explanatory account of *how* species-change takes place. One could reject natural selection as the primary agent of evolutionary change, for example, and still find this argument for TCA convincing.[22] Of course, a satisfactory explanatory account of how evolutionary change occurred would greatly strengthen the case for TCA. But in the light of the continuing debates about the adequacy of this or that feature of the neo-Darwinian model, it is important to stress that there is a vast body of evidence for common descent that does not depend for its logical force on the further issue of how *exactly* the transitions from one life-form to another came about.

Plantinga raises one objection that bears on TCA directly. Does there not seem to be an "envelope of limited variability" surrounding each species, so that a departure of more than a small amount from the central species-norm leads to reversion or sterility? Would one not expect to find evidence of new species now and then appearing in the present (or perhaps being deliberately produced) if indeed TCA is true? The first and simplest response is to note that in the plant world (in the forest, for example) new species have indeed been observed. And the production of fertile hybrids is an important part of agricultural research. The ability of populations of microorganisms to alter their structures quite basically over relatively short times under the challenge of antibiotics is all too well-known. But defenders of the modern synthesis themselves insist on the extraordinary stability of the genotype, in the animal realm particularly; this stability is essential to the maintenance of species differences, and some progress has been made toward an understanding of its molecular basis in the constellations of genes.

TCA does not require rapid change. The presumption is that the kind of species-changes that would sustain TCA could take thousands of generations to accomplish. The rate of change required (as has been shown in detail in recent studies in population genetics) is far too slow for the sort of direct evidence to accumulate that Plantinga is asking for. But even more fundamentally, there are serious problems with the species-concept itself, the concept underlying this objection. Ought it, for example, be based on morphological differences (of the kind that paleontologists or comparative anatomists can attest to), or ought it be based on interbreeding boundaries (as naturalists have long preferred to maintain)?[23] If we were to find the fossil remains of animals as different as a St. Bernard and a chihuahua in the rock strata, we should assuredly label them different species. But if we adopt the biological species concept according to which "species are groups of interbreeding natural populations that are reproductively isolated from other such groups,"[24] how are we to apply this to populations that are widely separated in space or time? Mayr emphasizes that such application always involves complex and indirect forms of inference.[25] The moral is not that the species-concept is so ambiguous as to be unusable, but only that such notions as species-change are far more difficult to handle than at first sight they seem to be. And more specifically, the claim that an "envelope of limited variability" surrounds each species has no precise empirical foundation.

I suspect that in the end this claim simply begs the question against TCA. It asserts that the sort of change TCA would require does not occur. But this is just the issue, and this is what is challenged by the three kinds of evidence described above, all of them pointing to TCA as the most reasonable explanation. What does Plantinga make of these? He deals with them, to my mind, in a quite unsatisfactory way: "Well, what would prevent [God] from using similar structures?", referring to the argument from homologies; and: "As for the similarity in the biochemistry of all life, this is reasonably probable on the hypothesis of special creation." Any attempt to reconstruct the past on the basis of traces found in the present can, of course, always be met with the objection: but God *could* have disposed matters so as to make it look as though it happened that way, even though it didn't. If TCA is correct, one would *expect* the sort of coherences that molecular biology is now turning up in such abundance. I can see no reason whatever to suppose that the hypothesis of special creation would antecedently have led us to expect this same range of evidence. Recalling the lines of a famous debate long ago about fossils, the God of the Christian tradition is surely not one who would deceive us by strewing around what would amount to misleading clues!

Let me stress once again the criterion of consilience. Evidence from three quite disparate domains supports a single coherent view of the sequence of branchings and extinctions that underlie TCA. If TCA is *false*, if in fact the different kinds of organisms do not share a common ancestry, this consilience goes entirely unexplained. It is all very well to say: "but God *could* have...." This hypothesis leaves the consilience exhibited by TCA an extraordinary coincidence. So it is not as though allowing the theistic alternative into the range of possible explanations alters the balance of probability drastically, as Plantinga supposes. TCA is, of course, an *hypothesis*, as any reconstruction of the past must be. But it remains by far the best-supported response, even for the theist, to the fast-multiplying evidence available to us.

## 5.   Theories of Evolution

What about the objections to the neo-Darwinian theory of evolution, as such, as distinct from TCA? Plantinga outlines a familiar objection to any theory which relies on natural selection as the primary mechanism of evolutionary change. There is no plausible evolutionary pathway (he argues) linking an eyeless organism, say, with an organism possessing the complex structures of the mammalian eye, such that *every* single stage along the way can be *shown* to be adaptively advantageous. This is the oldest of objections to Darwin's theory; it was the primary criticism raised by Mivart in his *Genesis of Species* (1871). Darwin's own first response was to emphasize that his theory did not rely on natural selection alone.[26]

Among the other processes that he proposed, one in particular is still emphasized: change of function, where a structure that originally developed because of the adaptive advantage offered by one particular function takes on (especially under the impact of change of habitat or the like) a new function. Another process whose importance has only recently come to be recognized is genetic drift. In the isolated and often small populations that furnish the likeliest starting-point for the speciation-process, there can be a sort of genetic random sampling error that eventually marks off the smaller population from the parent population. And there can be "hitchhiker" effects of all sorts due to genetic linkage. These processes do not operate independently of natural selection, but they can easily bring about results that the adaptive-advantage-at-every-step model of evolutionary change could not.[27] Defenders of the modern synthesis are as quick as Darwin was to insist that they are *not* limited in their explanatory strategies to the selectionist model only.[28]

Nevertheless, to some critics of the modern synthesis, these concessions are not enough. Gould, for example, has criticized what he calls the "adaptationist program" for its failure to take seriously the many alternatives to trait-by-trait selection on the basis of adaptive advantage. As his own favored alternative he notes the constraints that the integrity of the structure of the organism as a whole sets on possible pathways of change, so that the outcome is explicable rather more by the nature of the constraints than by the application of selectionist norms to individual traits.[29] Kimura has developed a controversial molecular-level theory according to which most changes in gene-frequencies are of no selective advantage, but are neutral. More radical challenges come from those who rely on macro-mutations (saltations) to bridge major discontinuities in the fossil record; theories of this sort, it is generally thought, face intractable problems.[30]

Where does all this leave us? The confidence of the defenders of the modern synthesis is based on the substantial explanatory successes of their model. They have no illusions about having explained everything; in particular, they concede that the processes responsible for the origin of the main phyla are not well understood.[31] Their explanatory model has already been substantially reshaped over the last fifty years, while retaining the original emphasis on the transformative powers of selection operating on individual differences. Undoubtedly, more such reshaping lies ahead. Like any other active scientific theory, the modern synthesis is incomplete, but its exponents argue, in great detail, that there are no *in-principle* barriers to its continued successful extension to the difficult cases. A minority has proposed that a more radical transformation is needed, one which abandons either the gradualism or the heavy reliance on selection that have marked the Darwinian approach.[32] The most extreme view is represented by Michael Denton, who argues that *all* current theories of evolution are in principle inadequate to handle macroevolution, and that we have to await another quite different sort of theory.

Where does the burden of proof lie in a matter of this sort? The claim that principles of a broadly Darwinian sort are capable of explaining the origins of the diversity of the living world rests on the successes of the theory to date. These are very considerable; they span many fields and have shown intricate linkages between those fields. In particular, the theory has shown an extraordinary fertility as it has been extended into new domains; even when it has encountered anomalies, it has shown the capacity to overcome these in creative ways that are clearly not *ad hoc*.[33] This is the sort of thing that impresses those who are actually in touch with the detail of this research. And it gives a *prima facie* case for supposing that the theory can be further extended to contexts not yet successfully treated. But, of course, this cannot in the strong sense be *proved*; it can only be made seem more (or less) plausible.

On the other side is the claim that theories of a Darwinian type are in *principle* incapable of handling certain classes of data: gaps in the fossil record, the origin of complex organs like the eye, the origin of the broad divisions of the living world (the phyla), or the like. Claims of this sort are hard to establish because they cannot anticipate the trajectory that the theory itself may follow as it is reworked in the light of new challenge. (Could the changes of the last century leading up to the modern synthesis have been foreseen?) This is not to say that such claims can *never* be established, or at least shown to be strongly supported. So it is not that the burden of proof falls to one side rather than the other. Rather, it is a matter of weighing up the merits of the case on each side, and making some kind of comparative assessment, informed by parallels from the earlier history of science, and a very detailed knowledge of the history and contemporary situation of the various fields where the evolutionary paradigm is applied.

Plantinga formulates a second sort of objection to theories of evolution in general: they can never tell the *whole* story of the genetic changes involved, the rates of mutation, the links between gene adaptation, and so forth: "Hence we don't really know whether evolution is so much as biologically possible." But first of all, evolutionary explanation begins at the level of the biological individual and the population, not the gene; natural selection operates on adaptations of whose genetic basis we may be (and usually are) entirely unaware. And the explanation is none the less real for that. But, more important, evolutionary explanation is of its nature *historical*, and historical explanation is not like explanation in physics or chemistry. It deals with the singular and the unrepeatable; it is thus *necessarily* incomplete. One must be careful to apply the appropriate criteria when assessing the merits of a particular explanation. An evolutionary explanation can never be better than plausible; the real problem lies in discriminating between different degrees of plausibility. The dangers of settling for a very weak sort of plausibility are real (Gould's "just so" stories). But the dangers of requiring too strong a degree of confirmation before allowing *any* standing to an evolutionary explanation ("Hence we don't really *know* ...") are just as great.

The presumed inadequacy of current theories of evolution is part of what leads Plantinga to propose his own alternative. What *exactly* is it? Is it that God brought to be in a miraculous way *each* of the millions of species that have existed since life first appeared on earth? More than 99.99% of these are now extinct. May one ask why God would have created them? The thesis that Plantinga deems more probable than TCA is simply that "God created mankind, as well as many kinds of plants and animals, separately and specially." Perhaps he means that God just created the phyla (including the ones that have gone extinct?). As we have seen, it is the manner in which the major divisions of the living world came to be that has provided the theory of evolution with its largest challenge. But why not *all* species? How is Plantinga to decide just which thesis *is* more probable than TCA? Presumably by checking to see what evolutionary theory has, in *his* view, been able to explain success-fully. And then whatever is left over, God is more likely to have brought about miraculously.

God of the gaps? It certainly *sounds* like that. Whatever science cannot currently explain, or, more exactly, whatever one can make a case for holding that science could never in principle explain, is to be deemed the "special" work of God. One is reminded of eighteenth-century natural theology. But Plantinga's intent is not apologetic, as that of natural the-ology was. It is not that he is using the supposed gaps in evolutionary explanation to support belief in the existence of a God who could plug the gaps. Rather, he considers it *antecedently* probable that God would intervene in ways like this; his critique of evolutionary theory is intended to locate the spots at which He is most likely to have intervened. When-ever evolutionary theory is unable to explain in a totally convincing way the origins of a particular kind, the hand of God is to be seen at work.

Plantinga claims that the Christian believer "has a freedom not avail-able to the naturalist," because he or she is "free to look at the evidence ... and follow where it leads." This might be correct if he were to hold only that the believer holds open an extra alternative that allows him or her to be more critical of the shortcomings of the scientific theory. But he holds something much stronger than that: there is an antecedent *likeli-hood* of "special" intervention of this kind in cosmic process, and hence unless the scientist has a strong case, the hypothesis of Divine interven-tion has to be allowed the higher likelihood. Recall the two hermeneutic principles sketched by Galileo: the "confrontation" principle suggests that unless a clear case can be made for the scientific theory, the theological alternative should take preference. I am not sure that this *does* in the end allow the Christian believer more freedom than the naturalist. But whatever of that, it certainly ensures conflict; it is likely to maximize the strain between faith and reason.

## 6.  The Integrity of God's Natural World

Plantinga's argument relies first and foremost on the premise that God's "special" intervention in cosmic process is antecedently probable. Here is where he and I really part ways. My view would be that from the theological and philosophical standpoints, such intervention is, if anything, antecedently *improbable*. Plantinga builds his case by recalling that "according to Scripture, [God] has often intervened in the working of his cosmos." And the examples he gives are the miracles of Scripture and the life, death, and resurrection of Jesus Christ. I want to recall here a set of old and valuable distinctions between nature and supernature, between the order of nature and the order of grace, between cosmic history and salvation history. The train of events linking Abraham to Christ is not to be considered an analogue for God's relationship to His creation generally. The Incarnation and what led up to it was unique in its manifestation of God's creative power and His loving concern for His universe. To overcome the consequences of human freedom, a different sort of action on God's part was required, a transformative action culminating in the promise of resurrection for the children of God, something that (despite the immortality claims of the Greek philosophers) lies altogether outside the bounds of nature.

The story of salvation is a story about men and women, about the burden and the promise of being human. It is not about plants and animals; it provides no warrant whatever for supposing that God would have brought the ancestors of the various kinds of plants and animals to be outside the ordinary order of nature. The story of salvation *does* bear on the origin of the first humans. If Plantinga were merely to say that God somehow "leant" into cosmic history at the advent of the human, Scripture would clearly be on his side. How this "leaning" is to be interpreted is, of course, another matter.[34] But his claim is a much stronger one.

To carry the argument a stage further: what would the eloquent texts of *Genesis, Job, Isaiah* and the *Psalms* lead one to expect? What have theologians made of these texts? This is obviously a theme that far transcends the compass of an essay such as this one. I can make a couple of simple points. The Creator whose powers are gradually revealed in these texts is omnipotent and all-wise, far beyond the reach of human reckoning. His Providence extends to all His creatures; they are all part of His single plan, only a fragment of which we know, and that darkly. Would such a Being be likely to "intervene" in His creation in the way that Plantinga describes? (I am uncomfortable with this language of "likelihood" in regard to God's action, as though we were somehow capable of catching the Creator of the galactic universe in the nets of our calculations. But let that be.) If one can use the language of antecedent probability at all here, it surely must point in the opposite direction.

St. Augustine is the most significant guide, perhaps, to the proper theological response to this question. He was the first to weave from biblical texts and his own best understanding of the Church's tradition the full doctrine of creation *ex nihilo*, as Christians understand it today. And in the *De Genesi ad litteram*, his commentary on the very texts in *Genesis* where the writer speaks of the coming to be of the plant and animal world on the fifth and sixth "Days" of Creation, he enunciated the famous theory of the *rationes seminales*, the seed-principles which God brings into being in the first moment of creation, and out of which the kinds of living things will, each in its own time, appear.[35] The "days," said Augustine, must be interpreted metaphorically as indefinite periods of time. And instead of God inserting new kinds of plants and animals ready-made, as it were, into a pre-existent world, He must be thought of as creating in that very first moment the potencies for all the kinds of living things that would come later, including the human body itself:

In the seed, then, there was invisibly present all that would develop in time into a tree. And in this same way we must picture the world, when God made all things together, as having had all things which were made in it and with it when day was made. This includes not only the heavens with sun, moon, and stars ... but also the beings which water and earth contained in potency and in their causes, before they came forth in the course of time.[36]

Augustine, for one, would not have attributed an antecedent probability to God's "intervening" to bring the first kinds of plants and animals abruptly to be, rather than having them develop in the gradual way that seeds do.

But what are we to make of Plantinga's objection that having life coming gradually to be according to the normal regularities of natural process is "semi-deistic," that it attributes too much autonomy to the natural world? He says:

God could have accomplished this creating in a thousand different ways. It was entirely within his power to create life in a way corresponding to the Grand Evolutionary Scenario ... to create matter ... together with laws for its behavior, in such a way that the inevitable[37] outcome of matter's working according to these laws would be first, life's coming into existence three or four billion years ago, and then the various higher forms of life, culminating as we like to think, in humankind. This is a semi-deistic view of God and his workings.

He contrasts this alternative with the one he favors:

Perhaps these laws are *not* such that given enough time, life would automatically emerge. Perhaps he did something different and special in the creation of life. Perhaps he did something different and special in creating the various kinds of animals and plants.

His characterization of the first alternative as 'semi-deistic" is intended to validate the second alternative as the appropriate one for the Christian to choose. But why should the first alternative be regarded as semi-deistic? He allows that it was within God's power to bring about cosmic evolution, but then asserts that to say He *did* in fact fashion the world in this way would be semi-deistic. This is puzzling. It would be semi-deistic in an extended sense perhaps, if we *already* knew that God had intervened in bringing to be some kinds of plants and animals, in which case the "Grand Evolutionary scenario" would attribute a greater degree of autonomy to the natural world than would be warranted. But this is exactly what we *don't* know. And to assume that we *do* know it would beg the question.

The problem may lie in the use of the label, "semi-deistic." A semi-deist, Plantinga remarks, could go so far as to allow that God "starts everything off," and "constantly sustains the world in existence," and even maintain that "any given causal transaction in the universe requires specific divine concurrent activity." All this would, apparently, not be enough to make such a view orthodox from the Christian standpoint. What more could be needed? Defining God's relationship with the natural order in terms of creation, conservation, and *concursus*, has after all been standard among Christian theologians since the Middle Ages. Perhaps what still needs to be made explicit is that God *could* also, if He so chose, relate to His creation in a different way, in the dramatic mode of a grace that overcomes nature and of wonders that draw attention to His covenant with Israel and ultimately to the person of Jesus. The possibility of such an "intrusion" on God's part into human history, of a mode of action that lies *beyond* nature, must not be excluded in advance, must indeed be affirmed. I take it that the denial that such a mode of action *is* possible on the part of the Being who creates and may even also conserve and concur is what constitutes semi-deism, in Plantinga's sense of that term.

But someone who asserts that the evolutionary account of origins is the best-supported one is *not* necessarily a semi-deist in this sense. Some defenders of evolution—notably those who deny the existence of a Creator and are, therefore, not deists of *any* sort—would, of course, exclude special creation in this inprinciple way. But there is no intrinsic connection whatever between the claim that God did, in fact, choose to work through evolutionary means and the far stronger claim that He *could* not have done otherwise. Nor, of course, is there any reason why someone who defends the evolutionary account of origin should go on to deny that God might intervene in the later human story in the way that Christians believe Him to have done.

In sum, then, at least *four* alternatives would have to be taken into account here. There are those who defend the evolutionary account of origins, and, rejecting the existence of God, would (if pressed) say that life could not *possibly* have come to be except through evolution. There may be those who maintain that God created, conserves, and concurs in the activity of the universe, but that He *could* not "intervene" in its history to bring new kinds of animals and plants to be, for example. These (if there are any such) are the semi-deists Plantinga describes. Then there are those who prefer the evolutionary account of origins on the grounds of evidence that this is in fact most probably the way it happened, but are perfectly willing to allow that it was within the Creator's power to speed up the story by special creation of ancestral kinds of plants and animals, even though (in their view) this was not what He did. This is a view that a great many Christians from Darwin's day to our own have defended; it is the view I am proposing here. It is *not* semi-deistic. And finally, there is the option of special creation: that God *did*, in fact, intervene by bringing various kinds of living things suddenly to be.

When Plantinga presents two alternatives only, the second being that God might "perhaps" have intervened as defenders of special creation believe He did, he must be supposing that the other alternative, the "Grand Evolutionary scenario," is one that excludes such a "perhaps," i.e., that excludes, *in principle*, the possibility that God could have intervened in the natural order. What I am challenging is this supposition. The Thesis of Common Ancestry can claim, as we have seen, an impressive body of evidence in its own right. It need not rely on, nor does it entail any in-principle claim about what God could or could not do.[38]

So, finally, how *should* the Christian regard this thesis? Perhaps better, since there are evidently "distinctive threads in the tapestry of Christianity," in Plantinga's evocative metaphor, how might someone respond who sees in the Christian doctrine of creation an affirmation of the integrity of the natural order? TCA implies a cousinship extending across the entire living world, the sort of coherence (as Leibniz once argued) that one might expect in the work of an all-powerful and all-wise Creator. The "seeds," in Augustine's happy metaphor, have been there from the beginning; the universe has in itself the capacity to become what God destined it from the beginning to be as a human abode, and for all we know much else.

When Augustine proposed a developmental cosmology long ago, there was little in the natural science of his day to support such a venture. Now that has changed. What was speculative and not quite coherent has been transformed, thanks to the labors of countless workers in a variety of different scientific fields. TCA allows the Christian to fill out the metaphysics of creation in a way that (I am persuaded) Augustine and Aquinas would have welcomed. No longer need one suppose that God added plants here and animals there. Though He *could* have done so, the evidence is mounting that the resources of His original creation were sufficient for the generation of the successive orders of complexity that make up our world.

Thus, common ancestry gives a meaning to the history of life that it previously lacked. In another perspective, this history now appears as preparation. The uncountable species that flourished and vanished have left a trace of themselves in us. The vast stretches of evolutionary time no longer seem quite so terrifying. Scripture traces the preparation for the coming of Christ back through Abraham to Adam. Is it too fanciful to suggest that natural science now allows us to extend the story indefinitely further back? When Christ took on human nature, the DNA that made him son of Mary may have linked him to a more ancient heritage stretching far beyond Adam to the shallows of unimaginably ancient seas. And so, in the Incarnation, it would not have been just human nature that was joined to the Divine, but in a less direct but no less real sense all those myriad organisms that had unknowingly over the aeons shaped the way for the coming of the human.[39]

Anthropocentric? But of course: the story of the Incarnation *is* anthropocentric. Reconcilable with the evolutionary story as that is told in terms of chance events and blind alleys? I believe so, but to argue it would require another essay. Unique? Quite possibly not: other stories may be unfolding in very different ways in other parts of this capacious universe of ours. Terminal? Not necessarily: we have no idea what lies ahead for humankind. The transformations that made us what we are may not yet be ended. Antecedently probable from a Christian perspective? I will have to leave that to the reader.

## Notes

1. The most obvious difference scarcely needs be stated. Plantinga is one of the most highly respected philosophers in the U.S., justly renowned for the quality of his scholarship and the care and rigor of his arguments. I bracket him here with the creation-science group, incongruous as such an association may seem, only because of the broad similarity of their theses in regard to special creation. I very much fear that this similarity may be sufficient to encourage creation-scientists to co-opt his essay to their own purposes.

2. For a taxonomy of the ways in which faith and knowledge have been related by different Christian thinkers, see James Kellenberger, *Religious Discovery, Faith, and Knowledge,* Englewood Cliffs N.J.: Prentice-Hall, 1972, chap. 10.

3. William B. Provine, review of *Trial and Error: The American Controversy over Creation and Evolution, Academe,* 73 (1), 1987, 50–52.

4. See McMullin, "The Rise and Fall of Physico-theology," section 4 of "Natural Science and Belief in a Creator," in *Physics, Philosophy, and Theology,* ed. R. J. Russell *et al.,* Rome: Vatican Observatory Press, 1988, 63–67.

5. Plantinga is using the term "myth" here in a technical sense, he reminds us, one that should not of itself be made to connote falsity or fiction.

6. Maurice Finocchiaro provides a new translation of the *Letter* in his *The Galileo Affair*, Berkeley: University of California Press, 1989, pp. 87–118.

7. Galileo introduces one further way of dealing with tensions between Scripture and natural science, suggesting that the biblical authors accommodated themselves to their hearers. This does not, in practice, reduce to either of the principles above. The notion of accommodation had already been hinted at by theologians as diverse as Thomas Aquinas and John Calvin. But this is not the place for an exhaustive analysis of the logical complexities of the famous Letter. See my "Galileo as a Theologian," Fremantle Lecture, Oxford, 1983 (unpublished), and Jean Dietz Moss, "Galileo's *Letter to Christina:* Some Rhetorical Considerations," *Renaissance Quarterly, 36,* 1983, 547–576.

8. Finocchiaro, *The Galileo Affair*, p. 94.

9. See my Introduction to *Galileo, Man of Science*, New York: Basic Books, 1967, pp. 33–35.

10. There is an abundant literature on this topic. See, for example, Robert Clifford S.J., "Creation in the Hebrew Bible," in *Physics, Philosophy, and Theology: A Common Quest for Understanding*, ed. R. J. Russell, W. R. Stoeger and G. V. Coyne S.J., Notre Dame: University of Notre Dame Press, 1988, 151–170; Dianne Bergant CSA and Carroll Stuhlmueller CP, "Creation according to the Old Testament," in *Evolution and Creation*, ed. E. McMullin, Notre Dame: University of Notre Dame Press, 1985, 153–175; Bernhard W. Anderson, "The Earth is the Lord's: An Essay on the Biblical Doctrine of Creation," in *Is God a Creationist?* ed. R. M. Frye, New York: Scribner, 1983, 176–196.

11. There is a larger issue here of deciding on the proper approach to Scripture generally. Plantinga characterizes the Reformed Christian as one who takes "Scripture to be a special revelation from God himself." Thus, for example, the story of Abraham, including the details of where he lived and journeyed and how he came to father a son, becomes a matter of history in the modern sense of that term, to be construed (in Plantinga's view) as having the standing of science. There is an implicit literalist presumption here that an Unreformed Christian like myself, someone unsympathetic, that is, to the constraints of the *"sola Scriptura"* maxim, would surely question. But to debate this would lead us far afield indeed.

12. I am reluctant to label it "Augustinian," despite its obvious basis in the text of Augustine's *De Genesi ad litteram*. Augustine himself was not bound by it in practice. He did not require a conclusive demonstration over against the literal reading of the story of the Six Days before abandoning such a reading and espousing a highly metaphorical one. And he stressed the importance of literary norms in general in the understanding of the import of the biblical text. The presumption in favor of literalism that we have seen to be the main source of conflict ("the text is to be interpreted literally unless a contrary reading can be established from an extrinsic source such as natural science") would be much more characteristic of later theology, notably the post-Reformation theology of Galileo's own day.

13. In his *Letter to Christina*, Galileo cited Augustine to warn against the dangers of opposing "unbelievers" (read: experts) on the basis of inadequate scientific knowledge (Finocchiaro, *The Galileo Affair*, p. 112). I suppose that this is what worries me most about the strategy Plantinga urges on his Christian readers. Though he stresses the importance of scholarship and the patient effort to understand, the reader who proceeds on his advice to do battle with defenders of evolution all too easily risks causing the sort of "laughter" that Augustine deplores, because of its negative effects on the credibility of the Christian message generally.

14. As an illustration of how complex the notion of temporal beginnings has become, the Hawking model does *not* imply that the universe is infinitely old (as that phrase would ordinarily be understood) but rather that as we trace time backwards to the Big Bang, the normal concept of time may break down as we approach the initial singularity some fifteen billion years ago. The history of "real time" (as Hawking calls it) would still be finite in the same terms as before, as he explicitly points out (*A Brief History of Time*, New York: Bantam Books, 1988, p. 138).

15. The question of whether or not the time elapsed in cosmic history is finite or infinite depends, in part, on the choice of physical process on which to base the time-scale, particularly on whether it is cyclic or continuous. The question of the finitude or infinity of past time, so much debated by medieval philosophers and theologians, cannot straightforwardly be answered in absolute terms. The notion of time-measurement is far more complex and theory-dependent than earlier discussions allowed. But the theological *point* of the biblical account of creation remains untouched by technical developments such as this. See McMullin, "How Should Cosmology Relate to Theology?" in *The Sciences and Theology in the Twentieth Century*, ed. A. R. Peacocke, Notre Dame: University of Notre Dame Press, 1981, p. 35.

16. As Plantinga himself recognizes: "There isn't the space here for more than the merest hand waving ..." But when the stakes are as high as they become by the end of his essay, one may fairly question whether hand waving is likely to be enough.

17. I would like to acknowledge my debt to the many who in discussion have helped me overcome the bafflement that evolutionary theory induces in the non-expert. In particular, my thanks go to Phil Sloan, Bill Charlesworth, Francisco Ayala, Bob Richards, and John Beatty.

18. Plantinga distinguishes between the Thesis of Common Ancestry and the attempt to *explain* common ancestry by some mechanism or other. But he calls the latter "Darwinism," which might confuse, since non-Darwinian explanations of common ancestry have, of course, also been proposed. And he says that TCA "is what one most naturally thinks of as the Theory of Evolution." What he may mean here is that TCA is theoretical and it implicitly involves evolution in some guise. But it is preferable, I would suggest, not to equate TCA and "the Theory of Evolution" because for most readers "the" theory of evolution is Darwin's.

19. See L. von Salvini-Plawen and E. Mayr. "On the Evolution of Photoreceptors and Eyes," *Evolutionary Biology, 10*, 1977, 207–263.

20. In his most recent discussion of the relation between microevolution and macroevolution, Mayr writes: "Almost every careful analysis of fossil sequences has revealed that a multiplication of species does not take place through a gradual splitting of single lineages into two and their subsequent divergence but rather through the sudden appearance of a new species. Early paleontologists interpreted this as evidence for instantaneous sympatric speciation [speciation over a single area], but it is now rather generally recognized that the new species had originated somewhere in a peripheral isolate and had subsequently spread to the area where it is suddenly found in the fossil record. The parental species which had budded off the neospecies showed virtually no change during this period. The punctuation is thus caused by a localized event in an isolated founder population, while the main species displays no significant change" (*Toward a New Philosophy of Biology*, Cambridge, Mass.: Harvard University Press, 1988, p. 415). This theory of allopatric speciation (speciation involving a second—in this case a geographically isolated but adjoining—territory) allows Mayr to modify the gradualism of the original Darwinian proposal, while retaining the basic Darwinian mode of explanation and avoiding the (to him) objectionable "punctual" events of the Gould-Eldredge scenario. But the debate is by no means closed.

21. For a brief review of this evidence, see Francisco Ayala, "The Theory of Evolution: Recent Successes," in *Evolution and Creation*, ed. E. McMullin, 59–90.

22. In this regard, the position adopted by Michael Denton, perhaps the most sweeping critic of evolutionary theory now writing, is quite puzzling. On the one hand, he finds the sort of consilience described above altogether remarkable: "It became increasingly apparent as more and more sequences accumulated that the differences between organisms at a molecular level corresponded to a large extent with their differences at a morphological level; and that all the classes traditionally identified by morphological criteria could also be detected by comparing their protein sequences.... The divisions turned out to be more mathematically perfect than even the most die-hard typologists would have predicted" (*Evolution: A Theory in Crisis*, Bethesda, Md.: Adler and Adler, 1986, pp. 276, 278). But the distances between the molecular sequences characteristic of different species can only be explained (he argues) by postulating a remarkably uniform "molecular clock" marking the rate of change in the constituents of particular kinds of molecules (and varying from one kind to another), and such a "clock" (he maintains) is impossible to understand on neo-Darwinian principles. What would seem, at best, to follow from this is that neo-Darwinian theory cannot explain the uniformity of the postulated "clock." But he assumes that he has also refuted TCA, while providing no hint himself as to how the correspondences he finds so remarkable might be explained by something *other* than common ancestry.

23. These are only two of the many possibilities. There is an enormous literature on this topic. See, for example, the papers gathered in Section VII of *Conceptual Issues in Evolutionary Biology*, ed. Elliott Sober, Cambridge, Mass.: MIT Press, 1984, and E. Mayr, *Animal Species and Evolution*, Cambridge, Mass.: Harvard University Press, 1963, 400–423.

24. Mayr, *Toward A New Philosophy of Biology* (TNPB), p. 318.

25. A further problem is suggested by the notion of a "natural" population. Reproductive isolation in the animal world is due, in the first instance, to *behavioral* barriers, which are the main isolating mechanisms (Mayr, *TNPB*, p. 320). Under artificial circumstances, such barriers can be overcome, but this will not necessarily give rise to new biological species. Likewise, deliberate interbreeding to produce new varieties of domestic dog, for example, will not produce a natural population with its own behavioral barriers to outbreeding.

26. Indeed, he showed some uncharacteristic indignation in his comment in the last edition of the *Origin of Species* (1872): "As my conclusions have lately been much misrepresented, and it has been stated that I attribute the modification of species exclusively to natural selection, I may be permitted to remark that in the first edition of that work, and subsequently, I placed in a most conspicuous position—namely, at the close of the Introduction—the following words: 'I am convinced that natural selection has been the main but not exclusive means of modification.' This has been of no avail. Great is the power of misrepresentation" (p. 395).

27. For a useful review, from the point of view of the modern synthesis, see E. Mayr, "The Emergence of Evolutionary Novelties," in *Evolution after Darwin*, vol. 1: *The Evolution of Life*, ed. Sol Tax, Chicago: University of Chicago Press, 1960, 349–380.

28. Mayr, for instance, repudiates what he calls "selectionist extremism," and says that "much of the phenotype is a byproduct of the evolutionary past, tolerated by natural selection but not necessarily produced under current conditions.... The mere fact of the vast reproductive surplus in each generation, together with the genetic uniqueness of each individual in sexually reproducing species, makes the importance of selection inescapable. This conclusion, however, does not in the least exclude the probability that random events also affect changes of survival and of the successful reproduction of an individual. The modern theory thus permits the inclusion of random events among the causes of evolutionary change. Such a pluralistic approach is surely more realistic than any one-sided extremism" (TNPB, pp. 136, 140). Still, he also wants to say that the modern synthesis for which he is perhaps the leading spokesman "was a reaffirmation of the Darwinian formulation that all *adaptive* evolutionary change is due to the directing force of natural selection on abundantly available variation" (TNPB, p. 527; emphasis mine).

29. See S. J. Gould and R. Lewontin, "The Spandrels of San Marco and the Panglossian Paradigm: A Critique of the Adaptationist Programme," *Proc. Royal Society London*, B205, 1979, 581–598.

30. One such problem is that a mutation affecting the phenotype in a major way would require co-ordinated change in hundreds of genes; another is that a macromutation in a *single* individual would not be enough, in a sexually reproducing species, to establish a new kind right away. The role of mutations in evolutionary change is much less dramatic than is often conveyed in popular accounts; they serve mainly to augment the stock of variations in a population upon which

recombination can work. (Recombination is the blending of fraternal and maternal DNA in each new biological individual in a sexually reproducing species; it is responsible for the fact that each such individual is different from all others.)

31. In the early stages of life's development on earth, sixty or seventy different phyla (morphological types) developed, most of which became extinct. Not a single new phylum has originated since the Cambrian period, more than four hundred million years ago. It would seem that the genetic structures of this early period were not as fixed as they later became. Selection may thus have had fewer constraints then than later on when highly cohesive genotypes developed; the rate of species-change might thus have been quite rapid, lowering the chances of an adequate fossil record of the changes.

32. The differences between the "punctuated equilibrium" model of Gould and Eldredge and the standard one of the modern synthesis are not nearly as great as was originally claimed. In particular, Gould's original assertion that only a "non-Darwinian" theory could handle the evidence from the fossil record was quite clearly based on a very narrow construal of what ought to count as "Darwinian." Mayr has to my mind convincingly shown that Gould's own model is compatible with Darwinian principles (TNPB, chapter 26).

33. Denton's comparison of the modern synthesis to late Ptolemaic astronomy with its profusion of epicycles, and his conclusion that it is a paradigm in crisis (*Evolution*, chapter 15) cannot, I think, be sustained. The crucial disagreement between us would be as to what constitutes an *ad hoc* modification (what he, oddly, calls a "tautology").

34. God fashioned man from the dust of the earth and breathed into his nostrils the breath of life" (*Genesis* 2, 7). The "fashioning" here could be that of a billion years of evolutionary preparation of that "dust" to form a being that for the first time could freely affirm or freely deny his Maker. Pius XII in his encyclical *Humani Generis* (1951) allowed that such an evolutionary origin of the human body was an acceptable reading of the *Genesis* text. But he added that the human *soul* could not be so understood; it had to be specially "infused" by God, presumably not just in the case of the first humans, but all humans since. Many theologians today would find the Platonic-sounding dualism underlying this assertion troubling. The uniqueness of God's covenant with men and women and of the promise of resurrection does not require that there be a naturally immortal soul, distinct in its genesis and history from its "attendant" body. But it is unnecessary to develop this issue here, since Plantinga's challenge extends to the evolutionary account of the plant and animal worlds, and not simply of the human.

35. For a full discussion, see "Augustine's 'seed principles,'" section 4 of my Introduction to *Evolution and Creation*, 11–16.

36. *The Literal Meaning of Genesis*, 2 vols., translated by J. H. Taylor, New York: Newman, 1982, V, 23; p. 175 (translation slightly modified).

37. "Inevitable" is a word that defenders of evolution, whether theists or not, would be uneasy with. It suggests that the evolutionary process is at least in a general way deterministic or predictable. But this is just what nearly all theorists of evolution would deny.

38. There is one further perspective on this matter of semi-deism that I have deliberately set aside above. The occasionalists of the fourteenth century maintained that God is the *only* cause, strictly speaking, of what happens in the world. What appears to be causal action within the world is for them no more than temporal succession. Things do not have natures that specify their actions; rather, the fact that they act according to certain norms must be directly attributed to God's intentions. There is no reason why God should not, for example, suddenly make new kinds of plants and animals appear, if He so wishes; since there is no order of *nature*, God is committed only to the reasonable stability of (more or less) regular succession on which human life depends. The issue that separated the nominalists from the Aristotelian defenders of real causation in nature is brought out very well in the essay by Alfred Freddoso cited by Plantinga: "Medieval Aristotelianism and the Case against Secondary Causation in Nature," in *Divine and Human Action*, ed. T. V. Morris, Ithaca: Cornell University Press, 1988.

In this perspective, the issue of "special creation" comes to be posed in a quite different way. Any view which affirms the sufficiency of the natural order for bringing about the origins of life might be dubbed by the occasionalist as "semi-deist." When I read the paragraph in which Plantinga says that someone who maintains that God creates, conserves, and concurs in the activity of, the universe is "semi-deistic," my first reaction was to assume that this committed him to occasionalism, since it is *only* from the occasionalist perspective that this view of God's relationship with the natural order would be classed as "semi-deist." But Plantinga is quite evidently *not* an occasionalist; his treatment of natural science implies that he believes in the operation of secondary causation in nature. Thus, I have assumed in the discussion above that Plantinga must have had something else in mind, namely, the openness of creation to a supernatural order of grace and miracle. Incidentally, the occasionalist *would* be likely to believe that special creation is antecedently more probable, and (in Berkeley's version, at least) might tend to question a theory, like the theory of evolution, which depends on the reality of such causes as genetic mutation.

39. Though the alert reader will have caught echoes of the theology (not the biology; see section 5 above) of Teilhard de Chardin, the affinities with the Christology of Karl Rahner are more immediate perhaps. See, for example, his "Christology within an Evolutionary View of the World," *Theological Investigations*, Baltimore: Helicon, vol. 5, pp. 157–192.

# 9

# Evolution, Neutrality, and Antecedent Probability: A Reply to McMullin and Van Till

Alvin Plantinga

First, I'd like to thank Professors Van Till, Pun, and McMullin for their careful and thoughtful replies. There is a deep level of agreement among all four of us; as is customary with replies and replies to replies, however, I shall concentrate on our areas of disagreement. In the cases of Van Till and McMullin, this may give an impression of deeper disagreement than actually exists. In the case of Pun it leaves me with little to say except Yea and Amen; I find no serious disagreement between us.[1]

## I.  Ad Van Till

When my friend Howard Van Till first heard this lecture, he said he found himself in 98% agreement with it. Sadly enough, additional disagreement seems to have reared its ugly head.

### A.   Misunderstandings

First, some cases of "failure of communication," as they say, due no doubt as much to expository inadequacy on my part as to hermeneutical inadequacy on his. Van Till thinks "it was most inappropriate for Plantinga to employ the conflict metaphor as frequently as he did in his paper." But here there is in part a misunderstanding. I did speak of faith and reason as sometimes apparently conflicting, in the sense that what seems initially to be among the deliverances of the faith sometimes appears to conflict with what seems initially to be among the deliverances of reason; and of course this certainly happens. I employed the battle or

Originally published in *Christian Scholar's Review* (1991, vol. 21, no. 1, pp. 80–109). © 1991 by *Christian Scholar's Review*. Reprinted by permission.

conflict metaphor much more frequently, however, in connection with something quite different: the conflict between competing and fundamentally religious ways of thinking about ourselves, the world, and God. Here I was following Kuyper and Augustine (and many others) and here the conflict metaphor is perfectly appropriate. Indeed, as I see it, here it isn't really a metaphor at all; there really is a conflict between these competing world views, and it is of the first importance to be aware of it.

Second, Van Till seems puzzled by my use of the terms "faith" and "deliverances of the faith":

Perhaps my complaint is primarily with Plantinga's use of the term *faith*. In most instances in this paper the term does not refer to one's personal commitment to act in the warranted confidence that the object of one's faith is trustworthy ... rather, Plantinga employs the term principally as an abbreviated version of "a deliverance of the faith" ... I take this to be a reference to some specific belief concerning what the Scriptures require a Christian to affirm, a belief held mostly for the reason that it constitutes an element in the received Christian tradition (e.g., that the Bible teaches the concept of special creation).

But I wasn't using the term "faith" in *that* way; I meant to use it to denote or refer to the essential elements of Christian belief. I believe that this is perfectly standard use of the term. (Perhaps its origin is to be found in *Jude* v. 3 [King James translation]: "Beloved, when I gave all diligence to write unto you of the common salvation, it was needful for me to write unto you, and exhort you that ye should earnestly contend for the faith which was once delivered to the saints.") So taken, the deliverances of the faith would *of course* be beliefs or propositions (not personal commitments): such propositions, for example, as that God has created the heavens and the earth, that Jesus was in fact the divine son of God, and that God was in Christ reconciling the world to himself. The deliverances of the faith would be what the Bible teaches; they would not be mere "received traditions about what the Bible teaches"; nor would an element of the deliverances of the faith be held "mostly for the reason that it constitutes an element in the received Christian tradition." It would be accepted, instead, because it is taken to be what the Lord teaches. There is misunderstanding when Van Till says, "In places where we might have expected Plantinga to pose penetrating questions that would challenge (in the constructive sense) the accuracy of the relevant 'deliverances of the faith'...." But you *wouldn't* expect a Christian to challenge the deliverances of the faith (though she might challenge one or another understanding or construal of them).

Third, Van Till thinks he detects a certain *imbalance* in my paper: I seem "well aware," he says, of the need for paying careful attention to matters of epistemology and hermeneutics in the arena of the natural sciences, but "Where is the evidence for a comparable level of concern for dealing with the equally difficult and relevant issues of epistemology and hermeneutics in the arena of biblical exegesis?" Here the answer is twofold. First, academics, other intellectuals, the readers of this journal and the audience of my original lecture all get told about a dozen times a day that there are epistemological and hermeneutical difficulties in determining what the Bible teaches; this hardly needs further emphasis. But secondly, I explicitly set aside questions of the proper understanding of the early chapters of *Genesis*, just *because* this is a difficult area, an area where I am not sure where the truth lies. I do believe that the Lord intends to teach us here not only that the world depends upon him for its existence, but also (at least) that the world has not existed for an infinite stretch of time, and that there was an original pair of human beings whose sin brought calamity upon the human race (a calamity the remedy for which is the life and death and resurrection of Christ). I also think it likely that he intends to teach us that human beings were created in a special way and in an act of special creation; but I could be persuaded otherwise. Nothing in my paper hinges upon these exegetical beliefs, however, or, as far as I can see, upon any other exegetical beliefs about which there is sensible controversy.

Still another misunderstanding: I spoke of what the Bible "initially seems" to tell us about the origin and development of life, and about what the Bible "taken at face value" seems to teach. Van Till asks: "But how well does this hermeneutic of 'initially seems' and 'taken at face value' actually hold up?" Here there is serious misunderstanding: of course I don't propose this as a *hermeneutic*. I don't mean for a moment to suggest that what the Bible *initially seems* to teach must be what it *really does* teach. Of course not. Indeed, I said that while the Bible initially seems to teach that the earth is very young, this is *not* what the Lord intends to teach us in early *Genesis*. Van Till goes on to say that "we need far more than a naive biblical hermeneutic or a simple 'folk exegesis.'" That is hard to dispute, but I can't see why Van Till felt obliged to say it.

## B.  Disagreements

**1.  Is Science Religiously Neutral?**  I come now to two matters, not of misunderstanding, but of disagreement. First, there is the question whether science is religiously neutral. I said I thought there was a three-way conflict between the *Civitas Dei*, Perennial Naturalism, and Creative anti-Realism. And I added that "it would be excessively naive to think that contemporary science is religiously and theologically neutral, standing serenely above this battle and wholly irrelevant to it." Now Van Till suggests that it is "profoundly misguided" "to expect the Scriptures to provide us with the kind of statements that would be directly relevant to the evaluation of contemporary scientific theory on the world's formative history." This seems to me mistaken. It seems entirely possible that Scripture should teach us something running wholly contrary to a given scientific theory, and even to a scientific theory on the world's formative history; and the fact that Scripture does not directly involve that list of concepts given by Van Till (galactic red shift, thermonuclear fission, etc.) seems irrelevant to this question. For if a scientific theory entails, for example, that the universe has existed for an infinite stretch of time (i.e., if it entails that for any number $n$, the universe has existed for more than $n$ years), then that theory would be in conflict, so I believe, with the clear teaching of Scripture. The Bible clearly teaches (so I believe), that every concrete object distinct from God was created by God, and was created by God in such a way that it has not existed for an infinite stretch of time. That a scientific theory should say something inconsistent with this is not merely a possibility; until fairly recently, that is precisely what the accepted scientific theories did.

Secondly: Van Till speaks here only of "contemporary scientific theory on the world's formative history"; but of course the question whether science is religiously neutral extends much further than that. Consider the human sciences: economics, sociology, psychology, sociobiology, and the like. Here too it seems wholly obvious that a scientific theory belonging to one of those areas *could* assert something inconsistent with biblical teaching; and here too some scientific theories *do* precisely that.

Consider, for example, Herbert Simon's recent article, "A Mechanism for Social Selection and Successful Altruism."[2] According to Simon, there is a problem with altruistic behavior, such as that displayed by Mother Teresa, or The Little Sisters of the Poor. Why do they do the sorts of things they do? The rational thing to do, says Simon, is to act or try to act in such a way as to increase one's personal fitness, i.e., to act so as to increase the probability that one's genes will be widely propagated, thus doing well in the evolutionary derby.[3] Altruistic behavior, such as that of Mother Teresa and The Little Sisters, does no such thing; so what is the explanation of their behaving as they do? Simon proposes two mechanisms: "bounded rationality," and "docility":

Docile persons tend to learn and believe what they perceive others in the society want them to learn and believe. Thus the content of what is learned will not be fully screened for its contribution to personal fitness. (p. 1666)

Because of bounded rationality, the docile individual will often be unable to distinguish socially prescribed behavior that contributes to fitness from altruistic behavior. In fact, docility will reduce the inclination to evaluate independently the contributions of behavior to fitness.... By virtue of bounded rationality, the docile person cannot acquire the personally advantageous learning that provides the increment, $d$, of fitness without acquiring also the altruistic behaviors that cost the decrement, $c$. (p. 1667)

So the idea is that a Mother Teresa displays "bounded rationality"; she adopts those culturally transmitted altruistic behaviors without making an independent evaluation of their contribution to her personal fitness. If she *did* make such an independent evaluation (and was rational enough to do it properly) she would see that this sort of behavior did not contribute to her personal fitness, would stop engaging in it, and would instead get to work on her expected number of progeny.

But isn't this in clear conflict with scriptural teachings about what it is rational for human beings to do? Behaving like Mother Teresa is not at all a manifestation of "bounded rationality"—as though if she thought about the matter with greater clarity and penetration, she would instead act so as to increase her personal fitness. Behaving as she does is instead a manifestation of a Christ-like spirit; she is reflecting in her limited human way the splendid glory of Christ's sacrificial action in the Atonement. (No doubt she is also laying up treasure in heaven.) From a Christian perspective, there is no sense of "rational" in which there is anything at all a human being can do, that is *more* rational than what she does. Of course we might be tempted to claim that Simon's article really isn't science; but can we sensibly make that claim in these post-Kuhnian days? If the scientists call it science and get grants from the National Science Foundation for doing it, can we sensibly claim that it really isn't science?

And obviously the connection between the explicit tenets of Christianity or the explicit teachings of Scripture, on the one hand, and a given scientific theory, on the other, can be vastly more subtle. Perhaps, for example, a scientific theory doesn't explicitly *contradict* the deliverances of the faith, but offers evidence of some sort against it, or is unlikely to be true if the faith is, or is such that its conjunction with some other epistemically probable proposition meets one or both of those conditions.

## 2. Is the Grand Evolutionary Story (GES) Religiously Neutral?
Coming to the second area of disagreement (a special case of the first), I said I thought the Grand Evolutionary Story wasn't religiously neutral:

> Now it would be excessively naive to think that contemporary science is religiously and theologically neutral, standing serenely above this battle and wholly irrelevant to it. Perhaps *parts* of science are like that: mathematics, for example, and perhaps physics, or parts of physics—although even in these areas there are connections. Other parts are obviously and deeply involved in this battle: and the closer the science in question is to what is distinctively human, the deeper the involvement.

Now I *think* Van Till and I disagree here, but I'm not sure; and this is again because of misunderstanding. Van Till apparently believes what I meant to say when I said GES was not neutral, was only that scientific theories (for example, GES) could be incorporated into various different world views (for example, naturalism). To put it in his words, what I meant was that science did not have "mythological immunity": science is not "immune from employment in the mythology of all religious cultures, including atheistic ones." But of course this isn't at all what I meant. To say that science was not neutral in *that* sense would be to make a statement weak *in excelsis*: a scientific theory would be neutral in that sense (if I understand Van Till) only if there were no religious ways of looking at the world *at all* into which it could be coherently incorporated. Only inconsistent scientific theories, I suppose, could qualify for that distinction.

Rather, I meant to point out a much more specific way in which, as I see it, the Grand Evolutionary Story is not religiously neutral. GES plays an important role in the conflict between Christian theism and naturalism (taken as a mythology, a deep account of ourselves and the world around us). This role is that of providing an answer to a question that is both insistent and monumentally difficult from a naturalistic perspective: how did all this astounding variety of life with its millions of species get here? Their ancestors can't have just popped into existence; but neither, from a naturalistic perspective, could they have been created by God; so where does all this life come from, and how did it get here? Evolution gives an answer the naturalist can accept, and it gives the only such answer anyone can presently think of. So it isn't just that evolution can be incorporated into naturalism; that is of course so, but it can also be incorporated into theism and most anything else. It is rather that evolution serves to answer what would otherwise be a crushing objection to naturalism. It functions as a "defeater-defeater" for naturalism: it offers a way to answer and hence defeat what would otherwise be a crushing objection or defeater for naturalism. It is therefore crucially important to naturalism; and this, I said, partly accounts for the insistence that this theory is no mere theory but a rock-hard certainty, and the venom, the *odium theologicum* with which dissent is greeted. Another example of that *odium theologicum*: according to the January, 1991, issue of *First Things*, the *New York Times* reported recently that

*Scientific American* denied a job to the gifted science writer, Forrest M. Mims, III. Mr Mims had been doing a column for the magazine, titled "Amateur Scientist." But then his awful secret was discovered. According to Armaund Schwab, who was managing editor when the decision was made, Mr. Mims "was a nonbeliever in evolution." ... Ever vigilant against extremisms, editor Jonathan Piel determined that hiring Mims would compromise the magazine's integrity (p. 64).

One can see why. A nonbeliever in evolution at *Scientific American* would be like a Unitarian minister who was a closet follower of Jerry Falwell or a seventeenth-century Scottish Calvinist who secretly accepted the authority of the Pope: at best appalling.

So I say that neither GES nor TCA (the Theory of universal Common Ancestry) is neutral with respect to the issue between Christian theism and naturalism. Now Van Till seems to think that the only question of interest, with respect to this neutrality issue, is the question whether these theories are *logically consistent* with Christian theism (as well as naturalism). This is a reasonable place to start; and the answer (roughly speaking anyway) is that of course evolution is indeed logically consistent both with Christian theism and with naturalism. But this doesn't anywhere nearly suffice to show that GES and TCA are neutral with respect to these two world views. You might as well argue that the evidence for relativity theory is neutral with respect to that theory; after all, it is consistent both with that theory and with its denial. You might as well argue that the evidence for the universe's being very old is neutral with respect to the dispute between young earthers and old earthers; after all that evidence is consistent with both views. This question about logical consistency is a reasonable place to start; but it is not a reasonable place to end. Another good question to ask is whether GES or TCA *offers evidence* for or against Christian theism and for or against naturalism; another question to ask is whether its conjunction with things we all believe, offers such evidence; another question is whether GES makes it easier to accept naturalism; and another question is whether the Grand Evolutionary Story is equally probable, with respect to Christian theism and naturalism. Here I wanted to argue that TCA does indeed make it easier to believe naturalism (by offering that defeater-defeater) and that it is indeed much more probable with respect to naturalism that with respect to theism; this is one source of those claims of certainty for it. But of course a Christian or other believer in God won't have *that* reason for thinking it certain; so unless there is some *other* reason, a reason a *Christian* might have for that conclusion, a Christian ought to *reject* that claim of certainty and all the sociological trappings that go along with it.

Accordingly Dawkins is right or partly right: evolution does something to make it easier to be an intellectually fulfilled atheist. No doubt it is still extremely difficult (and in the long run impossible); but GES answers what would otherwise be an unanswerable objection. Van Till, indeed, claims that this answer "encourage(s) a mischievous misconstrual of the issue. The central problem ... is a failure to distinguish authentically religious questions from questions accessible to modern empirical science— the common error of treating *creation* and *evolution* as if they were in essence alternative answers to the same question." But in this context they *are* alternative answers to the same question. God created all of life; we ask how he did it: did he do it by way of TCA or even GES, the Grand Evolutionary Story? We may answer either "yes" or "no"; and of course these are alternative answers to the same question.

Finally, there is a theological issue that separates us; but since the same theological issue separates McMullin and me, it will be convenient to treat what Van Till and McMullin say on this head together. In conclusion, however, I wish to say that I entirely applaud Van Till's conclusion: that faith and reason should *cooperate*. Indeed they should. This is the heart of the suggestion I meant to make about Christian scholarship: in understanding the issues involved in sociology, or psychology, economics, or biology or whatever, we should use *everything* we know—what we know by faith as well as what we know in other ways. Faith and reason must indeed cooperate; this is precisely why we need distinctively Christian scholarship in these areas. We must approach a topic like evolution from the perspective of faith—of the deliverances of the faith—as well as that of current science; what faith teaches here is of crucial importance and must not be silenced.

## II.   Ad McMullin

### A.   Misunderstandings
As with Van Till, so with my friend and colleague Ernan McMullin: before turning to real disagreement I want to try to clear up some misunderstandings (which, again, are not to be laid to his account rather than to mine).

### 1.   God of the Gaps?   I begin with a serious misunderstanding: McMullin suggests that my view is an example of "God of the gaps" thought:

God of the gaps? It certainly *sounds like* that [McMullin's emphasis]. Whatever science cannot currently explain, or, more exactly, whatever one can make a case for holding science could never in principle explain, is to be deemed the "special" work of God. One is reminded of 18th century natural theology.

I say it doesn't sound like that at all. God of the gaps thought, as McMullin recognizes, is essentially an *apologetic* enterprise. One who takes part in this project argues for the existence of God by pointing to phenomena science can't currently explain, suggesting that the only explanation is to be found in the activity of a divine being. From a theistic perspective, of course, this leaves a great deal to be desired. First, this procedure suggests that God is a gap plugger, that his activity in the natural world is limited to plugging gaps in a few areas of the natural world, while in the rest of nature everything goes on entirely independent of him and his activity. But the theist does not, of course, think of God as a mere gap plugger; God is crucially active in every transaction in nature, from the smallest most insignificant event to the largest cataclysmic event. God was active in the Big Bang; he is equally active in the sparrow's fall. According to traditional theism, this activity is at least of two sorts. First, theists have agreed that in any natural transaction, God *conserves the transactors in existence*; were he to withdraw this conserving activity the created universe would vanish like a computer image when you pull the plug. Furthermore, many theistic thinkers have added that every causal transaction on the part of creatures requires a concurrent causal act on the part of God, a kind of concurrent ratifying activity. A second danger of God of the gaps thought: if science progresses, the area of God's activity gets progressively diminished. And third, this procedure suggests that the correct way to believe in God is as in a scientific or explanatory *hypothesis*: there are some things such that the best explanation for them is the work of a divine agent.

But nothing could be further from my intent. What I said had no apologetic intent at all; I certainly wasn't proposing anything like a theistic argument. Furthermore, I don't for a moment think that belief in God is properly taken as an explanatory or scientific hypothesis: certainly not. It is no more to be taken as such a hypothesis than our beliefs that there are other persons, and a past, and an external world.[4] Instead, I was thinking of the matter as follows: a Christian (naturally) believes that there is such a person as God, and believes that God has created and sustains the world. *Starting* from this position (taking it for granted and not trying to argue *to* it), we recognize that there are many ways in which God could have created the living things he has in fact created; how, in fact, did he do it? Did he create matter, with its nature and it ways of working, in such a way that he could foresee that the result of its working in those ways would eventually be life, and then the various kinds of plants and animals, and then finally humankind? Or did he do something special in the creation of life? And did he do something special in the creation of his image bearers, human beings? And did he perhaps do something special in the creation of some other kinds of creatures? Did it all happen just by way of the working of the laws of physics, or was there further divine activity (activity not restricted to the upholding of matter in existence and concurring in the causal transactions expressing its nature)? That's the question, and the way to try to answer it, so it seems to me, is to ask two others: first what is the antecedent probability of his doing it the one way rather than the other? And second what does the evidence at our disposal suggest? Can we see how it could or would have happened just by the workings of the laws expressing the behavior and activity of matter? (I shall argue below that the second sort of consideration is more important than the first.) Starting from the belief in God, we must look at the evidence and consider the probabilities as best we can. But of course none of this has anything at all to do with "God of the gaps" apologetics, which in this context is nothing but a red herring.

**2. Does It Look As If TCA Is True?**    A second misunderstanding: I considered the evidence Stephen Gould gives for TCA (the Theory of Common Ancestry, according to which any two living things can trace their line of descent back to a common progenitor). Included in this evidence is that offered by homologies, and by the similarity in the biochemistry of all life. Since this is evidence that is supposed to support TCA as opposed to the hypothesis that God did something different and special in the creation of some of the various kinds of creatures, I pointed out that these two kinds of phenomena (homologies and the biochemical similarities) are also at least reasonably probable on the view that God created some creatures specially. McMullin says:

Plantinga ... deals with them, to my mind, in a quite unsatisfactory way.... Any attempt to reconstruct the past on the basis of traces found in the present can, of course, always be met with the objection: but God *could* have disposed matter so as to make it look as though it happened that way, even though it didn't.

But this, again, is a misunderstanding. My claim was not that it *looks* as if TCA is true, but God could nonetheless have done things differently (so he probably *did* do things differently?)—a sort of contemporary reflection of Philip Gosse's proposal that God put fossils into the rocks to deceive agnostic scientists. Not at all. That would be like conceding that it looks as if it is more than five miles from South Bend to Chicago, but claiming that it really isn't, on the grounds that God could bring it about that things would look like that even if they aren't that way; clearly enough that would not be the method of true philosophy. But it also isn't what I did. My claim is that from the perspective of theism, it *doesn't* look as if TCA is true—too many huge gaps in the fossil record; too many questions about whether TCA is even biologically possible, or whether, if it is, there has been enough time; too many epicycles needed to deal with some of the recent discoveries in molecular biology; and so on. A main issue between McMullin and me is just whether it *does* look (from a theistic perspective and given the empirical evidence) as if TCA is true.

**3.  An Alternative Explanation?**   Another misunderstanding: McMullin apparently thinks I mean to be proposing an *explanation* of the various forms of contemporary life, an explanation at the same level as evolution: "Like creation scientists, he maintains that the best explanation of the origin of many 'kinds of plants and animals' is an interruption of the ordinary course of natural process, a moment when. . . ." And later on in his paper he says:

The presumed inadequacy of current theories of evolution is part of what leads Plantinga to propose his own alternative. What *exactly* is it? Is it that God brought to be in a miraculous way *each* of the millions of species that have existed since life first appeared on earth? ... Perhaps he means that God just created the phyla.... But why not *all* species? How is Plantinga to decide just which thesis is more probable than TCA? Presumably by checking to see what evolutionary theory has, in *his* view (McMullin's emphasis) been able to explain successfully. And then whatever is left over, God is more likely to have brought about miraculously.

Now as a matter of fact I think *many* hypotheses are more likely than either GES or TCA. So far as GES goes, I think it overwhelmingly likely that God did something special in creating life. And as for TCA, that God did something special in creating human beings, for example, everything else going much as according to TCA, seems to me more probable than TCA. That God did something special in creating initial forms of life, then something special in creating some other forms of life, then something special in creating human beings is also more probable. But in fact McMullin's question is mistaken. Speith, Ayala, Gould, Denton and others claim that TCA is *certain*, as certain as that the earth revolves around the sun. My primary claim, as you recall, was that this is at best foolish exaggeration, due in part to the religious role evolution plays. Secondly, what I claimed is that from a theistic point of view TCA is unlikely, not more likely than not, less likely than its denial. But of course that does not commit me to finding some *other* explanation (some other explanation on the same logical level as TCA), and claiming that *that* one *is* probable with respect to the evidence—not unless we mistakenly take the denial of TCA as *itself* an explanation or explanatory hypothesis. What I say is that from a theistic or Christian point of view, TCA is unlikely, somewhat less likely than its denial. That is all I am claiming; I am not proposing an alternative explanation (unless, of course, the *denial* of TCA is an alternative, in which case my answer to the question, "What *exactly* is [his alternative?]" is "The denial of TCA"). In order to claim quite properly that an explanation is improbable, you are not obliged to be able to point to a better alternative. We might very well know that a given theory is improbable, even if we don't know of a more probable alternative to it (again, unless we take its denial to be an alternative to it). I may think is unlikely that Oswald shot Kennedy, even if I don't have a good idea as to who did.

"How is Plantinga to decide just which thesis is more probable than TCA?" But of course I don't have to be able to do that in order to hold, quite properly, that TCA itself is unlikely on the evidence; all I need to know is that the *disjunction* of these theses is more likely than TCA. I may think it unlikely, on the evidence, that I will be the next Pope; that doesn't commit me to having a candidate I think is likely to win. Of course I am committed to thinking that *there* is a theory that is both true and incompatible with TCA; but thinking TCA unlikely doesn't require that I know which theory that is. You and I find a watch in the woods behind your house; I propose the hypothesis that it was dropped there by Saddam Hussein; you are entirely within your rights in pointing out that this hypothesis is unlikely, on our evidence, even if you are not prepared to propose as probable some hypothesis as to who *did* drop it there.

**4. Creation Science**    McMullin says he is willing to apply the term "fanatic" to certain creation scientists (Henry Morris, for example) despite the fact that the term carries moral as well as epistemic disapproval. And here again he seems to be misunderstanding me, at least if he thinks he is disagreeing with what I said. What I said is that one could reject the current deliverances of science and hold that the earth is young without being ignorant, immoral, fanatical, deranged or whatever. Of course it is quite consistent with that to hold that it is *also* possible to reject the current teachings of science, and *be* ignorant or fanatical or deranged or whatever—just as it is possible to *endorse* contemporary evolution and display those unhappy properties. (In fact, I gave an example of an argument found in creation science literature that seems to me either ignorant or badly confused.) So here we have no disagreement.

But I do feel obliged to add something here. Creation scientists are wrong (so I think), but some of them are nonetheless admirable. Their aim is to be faithful to the Christian faith and to the Lord; they do their best to do so, often at considerable personal cost. (They don't, after all, *enjoy* being called fundamentalist ignoramuses; nor do they take delight in the rest of the ridicule and disapprobation heaped upon them by the scientific establishment.) I happen to think they are mistaken; but their errors, to my mind, are enormously less important than the errors of many of those—the Dawkins and Provines and Sagans of this world, for example—who load abuse upon them. It is vastly more important to be clear that the Lord created the heavens, the earth, and all that they contain, than to know that he didn't do it 10,000 years ago.[5] I disagree with the creation scientists, and, like most other academics, I don't relish the scorn and obloquy that goes with being associated with them; but at a deep level I fell much closer to them, both spiritually and intellectually, than to their cultured despisers. Christians who disagree with them should treat them as Christian brothers and sisters who, perhaps through an excess of zeal, err on a point of some importance; but Christians should *not* treat them as intellectual pariahs, or join in the cultural chorus expressing scorn, contempt, and disdain for them. Creation scientists reject wholesale large areas of well confirmed theory; the other side insists that on the empirical evidence alone, evolution is certain; from the point of view of respect for the evidence is there much to choose from between these two excesses?

**5.  Semi-deism**    A final locus of misunderstanding has to do with my use of the term "semi-deistic." Here perhaps my expository inadequacy is particularly evident; perhaps I should not have used that term at all.

I referred to a certain as "semi-deistic"; I used that term to highlight a claim that view shares with deism. According to deism, God originally created the world and started it off; then he sat back to watch, not acting upon his creation in any way at all. So the deist of course rejects miracles (though he concedes that surprising things can happen) and he also rejects divine conservation and concurrence. The *semi*-deist I had in mind concurs with the deist in supposing that once God has created things and started them off, he does nothing more special. (Semi-deism, of course, is topic-specific: you might be a semi-deist about the creation of the various forms of life but not about salvation history.) But I did not mean for a moment to suggest that the view I called "semi-deist" was unorthodox; when McMullin says, "His characterization of the first alternative as 'semi-deistic' is intended to validate the second alternative as the appropriate one for the Christian to choose" he errs, as when he says (on the same page) "All this ['starting everything off,' conservation, concurrence] would, apparently, not be enough to make such a view orthodox from the Christian standpoint." The label is meant to be a description of the view in question, not a subrosa way of impugning its orthodoxy. McMullin goes on to add that I must mean by "semi-deist" one who thinks God *could not* do anything other than go along with the laws of physics and chemistry, could not do something different and special in the creation of life, or human life, or other plant or animal life. But that's a misunderstanding; I meant the term to refer to those who say God *doesn't* do this, whether or not they go on to add that he *couldn't*.[6]

**6. Galileo, Scripture, and Scripture Scholarship**   According to "Galileo's Principle," "in cases of apparent conflict the literal interpretation of Scripture is to be maintained, unless the opposing scientific claim can be *demonstrated*" (McMullin's emphasis). He goes on:

What must be disputed, to my mind, is a modern analogue of the first of Galileo's principles, which would, equivalently, reaffirm the presumption that the Biblical text was partially intended as a cosmology, so that in cases of apparent conflict between the Biblical and current scientific account we should evaluate the strength of the scientific account as a means of deciding which of the two "competitors" to accept. Plantinga proposes such a "balancing of likelihoods" methodology: a literalist understanding of God's making the ancestors of the main natural kinds, he concludes, should be preferred unless and until a far stronger case can be made for the evolutionary alternative.

Galileo's principle seems clearly false.[7] It seems to imply, for example, that we should suppose it is possible to shout back and forth between hell and heaven, as in the parable of the rich man and Lazarus, unless we have scientific demonstration that this cannot be done; and that is not at all plausible. As it stands, furthermore, Galileo's principle will have little application: there is hardly ever a knock-down, drag-out *demonstration* of anything of much scientific interest.[8] But even if there were, it doesn't seem at all obvious that the face value of scriptural teaching must always be preferred to any deliverance of science that isn't backed up by something as strong as demonstration. Couldn't it happen that Scripture, taken at face value, seems to teach some proposition *P*, where it isn't at all obvious that *P* is what in fact the Lord intends to teach us, and where there is massive evidence from other sources that *P* is false? In such a case, shouldn't we reject *P* even if that massive evidence falls short of demonstration?

Indeed, a good example is offered by the very case under consideration: perhaps you think, with many Christians over the ages, that it initially *looks*, at any rate, as if early *Genesis* means to teach that the earth is young. Suppose you are also aware of Augustine's idea that early *Genesis* really *isn't* intended to teach us this, but you think the face value understanding slightly more likely to be correct than Augustine's. Then suppose (as in fact has happened to the Christian community) you encounter powerful and massive evidence that the earth is in fact very old; but though the evidence is powerful and massive, it stops short of being a full *demonstration*. I should think the sensible course, under these conditions, would be to move to the Augustinian understanding. Of course there won't be anything like an *algorithm* here. It would be nice if there were always some way to say just how strong the warrant (on a scale from 0 to 1, perhaps) is for the claim that the face value understanding *P* of a given passage is in fact the correct understanding, how strong the scientific evidence against *P* is, and what level of evidence of the latter kind defeats what level of warrant of the former. But there isn't.

Now McMullin attacks Galileo's Principle by arguing that it has a "disastrous consequence: it sets theologians evaluating the validity of the arguments of the natural philosophers, and natural philosophers defending themselves by composing theological tracts." If I understand him, I disagree with him here. Where there is apparent conflict between

Scripture and science, we must try the best way we can to see how to resolve it; given present academic arrangements, this will inevitably result in *someone's* making pronouncements that are outside her field. (Thus, in the present context, Van Till makes pronouncements on theology and philosophy, and McMullin and I do the same for biological and theological matters.) This could be avoided only if there were professionals, experts, who were expert in the relevant science, and also in philosophy and philosophy of science, and also in theology. None of us, as I said in my paper, fills a bill like that. So if McMullin means to suggest that philosophers should stick to their philosophy, theologians to their theology, and scientists to their science, then no one could address apparent conflicts of the sort that occasioned my paper. But we, the Christian community, *need* answers to these questions; we need to know how to think about these matters from a Christian perspective; we need to know whether thinking about them from that perspective will make a difference; we need to know how much of current culture—current scientific culture included—is to be seen as deliverance of reason, and how much comes from broadly speaking religious sources. If no one seriously addresses these questions, the answers we accept will be at best superficial and at worst calamitous for the intellectual and spiritual health of the Christian community. *That*, to my mind, is what would be disastrous.

In the passage I just quoted, McMullin apparently rejects my suggestion that what one must do, in cases of apparent conflict, is try to see how strong the case is for supposing that God teaches $P$ in the Scripture under consideration, how strong the evidence from reason and science for the denial of $P$ is, and then try to come to some resolution. Perhaps in some cases where the scientific evidence is very strong and the evidence for $P$'s being what God intends to teach weak, we should move to another understanding of $P$; in cases where the evidence from science and reason is weak and the evidence for $P$'s being what the Lord intends to teach strong, we should reject the bit of science in question.

On the next page, however, he says something that leads me to wonder whether I have properly understood him:

When the theologian is unsure of the best interpretation to give to a text, it is not inappropriate, of course, to take into account that some of the possible interpretations may be closed off by the findings of the natural or social sciences. But even in such a case, primary weight should be given to the *hermeneutic* issue as to what the disputed text was originally intended to convey....

I find this puzzling for two reasons: first, the first sentence seems to enjoin what is a special case of the procedure I was suggesting, and opens the believer or more exactly the believing community to precisely the sort of balancing of likelihoods McMullin describes as disastrous.

And second (with respect to the second sentence), what I was proposing *is*, of course, a hermeneutical principle: it makes a suggestion as to how to interpret Scripture in circumstances of apparent conflict. As we have just seen, it implies, for example, that if there is powerful evidence from reason for the claim that not-*P*, and some but much less powerful warrant from other considerations for the claim that God intends to teach us *P*, then we should take it that God does not intend to teach us *P*. But I suspect that's not the sort of hermeneutics McMullin has in mind. I think *he* thinks what is decisive here is what the *human author(s)* of the text in question had in mind. If that *is* what he means, I am obliged to disagree with him. In order to understand Scripture, we must know who its author and audience is. As to the latter, it is the Christian church over the ages; as to the former, as Aquinas and Calvin agree, the principle and primary author of Scripture is the Lord. (Of course this doesn't imply any kind of crude dictation theory.) What we really need to know, therefore, is what *he* intends to teach in the text in question. This may very well be what the human author had in mind in writing that text; but of course it needn't be. It might be that the Lord proposes to teach us (coming where we do in the whole history of his interactions with his children) something that hadn't occurred to the person or persons actually composing the text in question. I would concur with those Christians, for example, who see various Old Testament passages (*Isaiah* and elsewhere) as really referring to Christ, the second person of the Trinity, and making assertions about him; it is unlikely, however, that the original author intended to make assertions about the second person of the Trinity. What the original authors had in mind will ordinarily be of importance, but it will not necessarily settle the issue as to how to understand the text in question.

I was surprised to read the last sentence of the above quotation ("a literalist understanding of God's making the ancestors of the main natural kinds"), since I explicitly and in more than one place said I proposed to set aside the evidence, whatever exactly it is, from early *Genesis*. McMullin apparently finds it hard to credit my explicit claims here:

"Despite his silence in regard to *Genesis*, I do not think that a linkage between his argument and the more traditional *Genesis*-based argument can be denied." But let me assure him (and you) that I really *am* leaving out of account questions about the right understanding of early *Genesis*, because I am unsure about the answers to some of those questions. Because I am unsure of these matters, I am not resting any part of my argument on them. I appeal instead only to (1) the theistic claim that God constantly supports and sustains all his creatures in existence,[9] (2) the Christian claims that God did indeed create the heavens and the earth, and that he did many things in a special way in connection with salvation history and (3) the empirical evidence.

McMullin speculates that perhaps my reason for thus keeping my council about early *Genesis* is a prudent desire to avoid confrontation with the theologians: "He already has the scientists on his hands. Why open a second front and take on the theologians too?" He claims that "the great majority of contemporary Scripture scholars agree" that to interpret early *Genesis* "literally or quasi-literally is to misunderstand the point that the writers of those narratives were trying to make." These theologians rise as one man (or woman), to assure us, says McMullin, that the writer(s) of the early chapters of *Genesis* meant to tell us no more than that the world was indeed created by and is dependent upon God. They do not mean to tell us anything at all about *how* God created—whether he did it in seven 24-hour days, whether he created humankind separately, whether there was an original human pair in the garden of Eden—they mean to tell us only that the world depends upon God: "... and if the Biblical theologians are right in holding that the cosmological references in the Old Testament ought to be understood as conveying fundamental truths about the dependence of the natural and human worlds on their Creator, rather than explaining how exactly these worlds first took shape, then perhaps we ought to be just as wary...."

Now strictly speaking, this is doubly irrelevant to the questions under consideration. For first, I didn't rest any part of my case on the correct understanding of the first chapters of *Genesis*; and second, as I argued above, the question what the early authors had in mind is not the primary question. (Furthermore, as a glance at contemporary Scripture scholarship will attest, questions as to what they *did* have in mind often become enormously speculative.) Still, I cannot forebear noting that

McMullin's appeal to contemporary Scripture scholarship is extraordinarily selective, not to say tendentious (or perhaps it is only that he and I have not met the same theologians). There are indeed theologians who deny that the (human) writer(s) of *Genesis* meant to say more than that the world depends upon God; but there are many more who think that the original (human) authors had a great deal more in mind.

First, of course, there are whole coveys or phalanxes of conservative critics—e.g., E. J. Young and G. C. Aalders—who think that the writer(s) of *Genesis* meant to teach much more than that creation depends upon the Lord. (There is also, of course, Thomas Aquinas, who took early *Genesis* to teach that God created the world in six 24 hour days.) But the same goes for their more liberal colleagues. Thus for example, Julius Wellhausen,[10] speaking of the author of *Genesis*,

He undoubtedly wants to depict faithfully the factual course of events in the coming-to-be of the world, he wants to give a cosmogonic theory. Anyone who denies that is confusing the value of the story for us with the intention of the author.

Wellhausen's last point, here, deserves careful attention: many who claim that the author(s) of *Genesis* did not mean to say anything about "the factual course of events" seem to be motivated more by what they think is the *correct* view of the matter than by what it is likely the historical authors had in mind. This is a common human phenomenon and we can easily think of many other examples (for example, the way the medievals often found "the Philosopher" holding just what they thought the truth of the matter). Nevertheless it is hard to see how it makes for accurate Scripture scholarship.

We may add the voice of Herman Gunkel, here, who says, "People should never have denied that *Genesis* 1 wants to recount how the coming-to-be of the world actually happened."[11] Still further, James Barr, Regius Professor of Hebrew in the University of Oxford until he joined the brain-drain to the US, and an Old Testament scholar than whom there is none more distinguished, writes as follows:

To take a well-known instance, most conservative evangelical opinion today does not pursue a literal interpretation of the creation story in *Genesis*. A literal interpretation would hold that the world was created in six days, these days being the first of the series which we still experience as days and nights.[12]

After buttressing this claim that most evangelicals (he also calls them "fundamentalists") indeed do not pursue a literal interpretation, he writes

In fact the only natural exegesis is a literal one, in the sense that this is what the author meant.

Elsewhere he goes much further:

... so far as I know there is no professor of Hebrew or Old Testament at any world-class university who does not believe that the writer(s) of *Genesis* 1–11 intended to convey to their readers the ideas that: (a) creation took place in a series of six days which were the same as the days of 24 hours we now experience; (b) the figures contained in the *Genesis* genealogies provide by simple addition a chronology from the beginning of the world up to the later stages of the Biblical story, and (c) Noah's flood was understood to be worldwide, and to have extinguished all human and land animal life except for those in the ark.[13]

Here we must bear in mind the polemical context in which Barr is writing: he means to discredit the "fundamentalists" or "evangelicals" by showing that they profess to take Scripture at its literal word, but in this case clearly do not do so, since it is obvious (at any rate to those professors at the world class universities) that the writer(s) of *Genesis* meant to assert the three things Barr mentions. According to Albert Wolters, this second quotation certainly contains exaggeration, since "the highly respected German commentaries on *Genesis* by G. von Rad and K. Westermann (who both taught at major German universities) would not fit his description" (personal communication). Even allowing for a bit of exaggeration, however, the picture presented by Barr, the distinguished Old Testament scholar, is enormously different from that presented by McMullin. McMullin's claims about "the great majority of contemporary theologians" are dubious at best. But, as I say, this issue is really peripheral to what concerns us.

## B.   Disagreements

**1.   Science and "Science."**   I say the Christian community ought to think about the subjects of the various sciences—the so-called natural sciences, such as physics and chemistry and biology, and also the human sciences, such as psychology, sociology, economics—from an explicitly theistic or Christian point of view; and I suggested calling the result

"Unnatural Science," or "Creation Science," or "Theistic Science." As far as I can see, McMullin agrees that the Christian community should pursue this sort of study, but he objects to calling the fruit of such study "science":

I do not think, however, that theistic science should be described *as* science. It lacks the universality of science, as that term has been understood in the Western tradition. It also lacks the sort of warrant that has gradually come to characterize natural science, one that points to systematic observation, generalization, and the testing of explanatory hypotheses. It appeals to a specifically Christian belief, one that lays no claim to assent from a Hindu or an agnostic.

Now in a way, perhaps, it doesn't matter what we *call* this enterprise that (as McMullin and I agree) ought to be undertaken by the Christian community—and no doubt we would agree further that it is the scientists of the Christian community, the practitioners of the discipline in question, who ought to undertake it. But there *is* something (research grants, for example) in a name, and I disagree with McMullin's reasons for denying the name of science to this enterprise. He makes two points. First, theistic science lacks the *universality* of science properly so-called; it couldn't be practiced by an agnostic or a Hindu. And second, it lacks the sort of warrant that "points to systematic observation, generalization, and the testing of explanatory hypotheses." As for the second point, here again there is misunderstanding. The way to try to understand, from a theistic perspective, how God created plants and animals and human beings is to take account of all that you know: what you know by faith, what you know as a Christian, as well as what you know in other ways. In the case at hand, what would be relevant would be what Scripture teaches or suggests on the matter, together with the antecedent probability from a theistic perspective, together with the "empirical evidence": the fossil record, the molecular evidence, homologies, and the like. Clearly this involves precisely the sort of systematic observation, generalization and testing of explanatory hypotheses that McMullin cites as the hallmark of science. It may involve *more*; but it certainly involves this much. To establish his point along these lines, McMullin would have to argue something else: that science (properly so-called) somehow *couldn't* involve the other matters, the looking to see what (if anything) Scripture says on the matter and the consideration of the antecedent probability of a theory on theism. And I haven't the faintest idea how that could be argued. Where is it laid down that anything that does that is not science?

The answer, McMullin thinks, lies in that methodological naturalism he thinks essential to science (or perhaps natural science). Speaking of methodological naturalism, he writes, "Scientists *have* to proceed in this way; the methodology of natural science gives no purchase on the claim that particular event or type of event is to be explained by invoking God's creation directly." But where does this embargo come from? It is ordinarily supported only by bad arguments of the type "God is not part of the universe; in science we can only refer to parts of the universe; therefore. . . ."; or even "To refer to God in science is to treat God as an object, which is idolatry; therefore. . . ." Why believe that scientists have to proceed the way McMullin says they have to? You might as well argue that science can't start from the assumption that inductive procedures will continue to be successful (or that there has really been a past), on the grounds that scientific method isn't competent to decide these matters.

Consider the question how life originated: as a theist I believe that God created it in one way or another, and now the question is: how did he do it? Did he do it by way of the ordinary workings of the laws of physics and chemistry (the ordinary behavior of matter, so far as we understand it) or did he do something special? If after considerable study, we can't see how it could possibly have happened by way of the ordinary workings of matter, the natural thing to think, from that perspective, is that God did something different and special here. (Such a conclusion, of course, would not be written in stone; the inquiry wouldn't be finally closed at any point.) And why couldn't one conclude this precisely as a scientist? Where is it written that such a conclusion can't be part of science? Where is it written that science inevitably involves that methodological naturalism?

We must note that nothing is settled by a definition: you can define the word "science" as you please, but no questions are settled and no perplexities answered just by offering a definition. In particular, it isn't settled whether a proper scientific understanding of nature may include a reference to God. You may insist that the answer is just given by the very definition of "science": "Science," you say, means something like "empirical inquiry in which one never refers to God." But this doesn't answer the original question; it just means that you now have to use different words to *ask* the question.

And now consider the first point, "the universality of science, as that term has been understood in the Western tradition." Here we must remind ourselves that in the Western tradition, *theology*, of course, was long thought to be a science. (And not just any old science, but the *queen* of the sciences.) Until fairly recently, hardly anyone would have doubted that theology is a science—even though, of course, it is based upon principles laying no claim to assent from a Hindu or an agnostic. McMullin might be closer to the truth, here, if he appealed, not to the whole Western tradition, but to the Western tradition since the mid-nineteenth century or so; and if he spoke, not of "science" *simpliciter*, but of "natural science." Then the point would be just that as a matter of fact "natural science" hasn't been applied, in that stretch of our tradition, to what I called "Theistic Science"; that might be right, but it is hardly determinative.

On the other side, it is also clear that science, if it is practiced in such a way as to honor the methodological naturalism McMullin urges, is by no means always universal. Think again of the piece by Simon I mentioned above, according to which (1) the rational thing to do is to act so as to increase your personal fitness, i.e., so as to maximize the probability that your genes will be widely disseminated, and (2) the (or an) explanation of the behavior of someone like Mother Teresa or The Little Sisters of the Poor is "bounded rationality" (i.e., not to put too fine a point upon it, stupidity) together with docility. I should think no Christian could even for a moment take this seriously as an explanation of their behavior. The explanation is something wholly different, something much more along the lines I mentioned above. Of course someone might insist that Simon's piece and perhaps even sociobiology more generally really isn't science. But why not? Perhaps it isn't *good* science; but on what grounds are we to declare that it isn't science at all? It is published in scientific journals and written in that stiff, impersonal scientific style; people get grants from the National Science Foundation to pursue it; it involves experiments, mathematical models, and the attention, customary in science, to the fit between model and data; it certainly has all the earmarks and trappings of science. It if looks like a duck, walks like a duck, and quacks like a duck, it's best thought of as a duck. We know better, in this post Kuhnian and post-positivistic age, than to try to give precise criteria for distinguishing science from non-science; we know about the notorious problem of demarcation.[14]

**2. Antecedent Probabilities.** The last two areas of disagreement I want to mention both have to do with the probability of TCA from a theistic perspective. The first has to do, we might say, with its *antecedent* likelihood—its likelihood on Christian theism prior to the consideration of the empirical evidence such as the fossil record, the homologies, the molecular evidence, and the like of that; and the second has to do with the evidential force of that empirical evidence. Both Van Till and McMullin argue that the antecedent probability of TCA, on Christian theism, is much greater than the antecedent probability of God's doing something special in creating life, or humankind, or other forms of life. I shall look first and briefly at what Van Till has to say.

Van Till argues that "the world created by the God who reveals himself in Scripture is a world characterized by ... functional integrity." What is functional integrity?

By this term I mean to denote a created world that has no functional deficiencies, no gaps in its economy of the sort that would require God to act immediately, temporarily assuming the role of creature to perform functions within the economy of the creation that other creatures have not been equipped to perform.

The suggestion seems to be that God does nothing *immediately* in creation; everything he does, he does mediately, indirectly, by way of having some *creature* do the immediate acting. And the suggestion seems further to be that if God *did* act immediately in creation, he would be assuming the role of a creature; the divine role is to act only mediately and indirectly. Van Till attempts to buttress his case by appealing to a passage from the theologian John Stek, who writes "[creation] contains no gaps that have to be filled with continuous or sporadic immediate operations of divine power; God is not himself a component within the internal economy of his creaturely realm."[15]

Here I wonder whether Van Till and Stek really mean what their words seem to say. For (as I have already noted) of course theists have always held that God acts immediately upon his creation by way of conserving it in existence; this divine activity is absolutely essential to the existence of creation. (Many theists have added that every creaturely causal transaction also requires an additional immediate divine concurrent action.) The view that God does nothing *immediately*, in or with respect to his creation, is neither Christian nor even theistic. Furthermore, it is of doubtful coherence: how could God do anything mediately unless he did *something* immediately?

But perhaps Van Till doesn't really mean to say that God does nothing immediately; perhaps the idea is that God does nothing immediately, in creation, in addition to the conservation and perhaps concurrence necessary for creaturely existence and causal activity. Instead, he creates the world with all its potentialities, and then restricts his activity to conservation and coherence. But why should we believe this? What is the reason for this assumption? Is there anything in Scripture to suggest it? I should think not; Scripture seems to be full of accounts of divine activity (miracles, for example) that go far beyond conservation and concurrence. And towering above all, of course, there is the *Incarnation*, which can hardly be described as an act merely of conservation or concurrence.

Is it perhaps Van Till's idea that God limits his activity to conservation and concurrence, except for the events of salvation history? But what is the reason for believing that? Scripture does not so much as slyly suggest it; the fact that God does in fact go beyond conservation and concurrence in his salvific activity suggests that at any rate he is not averse to so doing. Van Till produces no arguments for this assumption; so what is the backing for it? Why should we believe it? In considering our topic—how God created life and whether he did so by way of Theory of Common Ancestry or even the Grand Evolutionary Story—I suggest that we should pay more attention to the empirical evidence, rather than rely upon unsubstantiated and apparently *a priori* theological assumption. Here, perhaps, is a place where theological assumption ought not intrude into empirical science.

So far then, we have been given no reason at all for supposing that there is a substantial antecedent probability of God's creating by way of TCA, or that antecedent probability of TCA is greater than that of its denial. I turn now to McMullin's suggestions on this topic. He makes initial heavy weather over the very asking of the question what God is likely to do; if I understand him, however, he goes on to claim that in fact it is unlikely, indeed *very* unlikely, that God would do something special and different, create something special and different in bringing it about that there are human beings, or certain kinds of plants and animals, or even, presumably, life itself:

To carry the argument a stage further: what would the eloquent texts of *Genesis, Job, Isaiah* and the *Psalms*, lead one to expect? What have theologians made of these texts? This is obviously a theme that far transcends the compass of an essay such as this one. I can make a couple of simple points. The Creator whose powers are gradually revealed in these texts is omnipotent and all-wise, far beyond the reach of human reckoning. His Providence extends to all His creatures; they are all part of His single plan, only a fragment of which we know, and that darkly. Would such a Being be likely to "intervene" in His creation in the way Plantinga describes? (I am uncomfortable with this language of "likelihood" in regard to God's actions, as though we were somehow capable of catching the Creator of the galactic universe in the nets of our calculations. But let that be.) If one can use the language of antecedent probability at all here, it surely must point in the opposite direction.[16]

A couple of comments on this passage: first, I am not suggesting, of course, that there is some way to *calculate* the probability that God would do this or that; at best, on a topic like that, we have little more than crude guesses. And I agree that any ideas we might have about the antecedent likelihood that God would do things a certain way should be at best extremely tentative. What I said was only that I thought it a bit more probable that God would do something different and special in the creation of life, and human beings, and perhaps some other forms of life. But any such views, surely, should be tentative and held with appropriate diffidence. It certainly befits no one to be at all cocksure here.

This said, however, I fail to see any force in the considerations McMullin puts forward. God is indeed omnipotent and all wise; his providence does indeed extend to all creatures; and we know only a fragment of his total plan. These things are all true; but how do they bear on the question whether God would or would not, for example, create in stages: first creating inanimate material, say, then later doing something special in creating life, perhaps, and then still later in creating human life? (It is part of the major theistic religions to think that God has created humankind *in his own image*; might he not have thought it appropriate to create human life in a special way, by way of an act of special creation?) We know, after all, that God is not averse to acting in special ways, as the many miracles recorded in the Bible attest.

True, McMullin insists that what God does with respect to salvation tells us nothing about how God would create: "The story of salvation is a story about men and women, about the burden of being human. It is not about plants and animals; it provides no warrant whatever for supposing that God would have brought the ancestors of the various kinds of plants and animals to be outside the ordinary order of nature." There is of course some question as to what the *ordinary* order of nature is. According to McMullin, Pope Leo XII taught that God creates specially a new human soul or a new person whenever a human being is conceived; if so, then the order of nature regularly and ordinarily involves very many acts of special divine creation.

What we know, here, is this. First, God is in constant, close, intimate causal contact with his creation: he continuously upholds it, and perhaps also acts concurrently with every creaturely causal transaction. We also know that he has often done things in a special way, a way in which he doesn't ordinarily do them, so that, for example, water turns into wine, or human beings emerge from a fiery furnace unhurt, or a human being rises from the dead; he is apparently not averse to working in his creation in a special way. Accordingly, there is no particular antecedent probability in favor of the idea that he wouldn't do anything different or special in the creation of life, say, or in the creation of special kinds of life. So it is hard to see how there is any antecedent probability in favor of GES. In thinking about the antecedent probability of TCA, furthermore, we must also note that we have every reason to doubt that life arose simply by the workings of the laws of physics. According to Francis Crick, life must be regarded as the next thing to miracle; according to Harold P. Kein of Santa Clara University, chairman of a National Academy of Sciences committee that recently reviewed origin-of-life research, "The simplest becterium is so damn complicated from the point of view of a chemist that it is almost impossible to imagine how it happened." It therefore looks as if God did something special in the creation of life. (Of course things may change; that is how things look *now*.) And if he did something special in creating life, what would prevent him from doing something special at other points, in creating human life, for example, or other forms of life? These things taken together suggest that the Lord might very well have done something different in creating life, something different in creating human life, and perhaps something different in creating other forms of life. It would seem to be entirely in character. I am therefore inclined to maintain my suggestion that the antecedent probability, from a theistic point of view, is somewhat against the idea that all the kinds of plants and animals, as well as humankind, would arise just by the workings of the laws of physics and chemistry.

Of course these are deep and difficult waters; as we read in *Ecclesiastes*, "... I saw all the work of God, that man cannot find out the work that is done under the sun. However much man may toil in seeking, he will not find it out; even though a wise man claims to know, he cannot find it out." Perhaps the most reasonable attitude, here, is one of agnosticism: one just doesn't know what these antecedent probabilities are. What seems to me unreasonable, however, is to be confident that that antecedent probability favors TCA or GES. It seems to me unreasonable to suppose that there is an antecedent probability favoring what I called semi-deism. And if I am right, then we must rely most heavily, here, on the empirical evidence. Our view should depend heavily upon our judgment as to the strength of that evidence for TCA. Does evidence suggest that the Lord did it by way of TCA? Or does it suggest something else? Here we should rely upon the empirical evidence much more heavily than upon *a priori* theological assumption. And when we look at the matter this way, then it looks as if the evidence is at best ambiguous. God *could* have done it by way of TCA; but, as it seems to me, it is somewhat more probable that he did not. For if he had done it that way, then we should expect much stronger evidence than we actually have. We shouldn't expect those enormous gaps in the fossil record; we shouldn't expect those epicycles; we shouldn't expect the problems with the molecular evidence. The actual empirical evidence must be allowed to speak more loudly than speculative theological assumption.

**3.   The Empirical Evidence**   We now turn to that evidence. Here I don't mean to use the word "empirical" to mark any philosophically important property; in this context (but for intolerable prolixity) I could just as well enumerate the sort of evidence of which I mean to speak: the fossil record, homologies, geographical evidence, the molecular evidence, and so on. And here I must first join my critics in pointing out an error in my paper. I represent myself as arguing against TCA, the Theory of Common Ancestry; as a matter of fact, however, I am questioning the hypothesis that eyes, wings, brains and the like have developed according to the mechanisms suggested by contemporary evolutionary theory.

These two hypotheses are of course intimately connected; in particular, it is hard to imagine (given naturalism) how the former could be true unless some version of the latter were. Nevertheless, of course, they are distinct hypotheses.[17]

With respect to the empirical evidence, I have three points. First, I said that the evidence for TCA was defective, in many ways, so that it was foolish exaggeration to declare the latter certain. Here McMullin reminds me that the evidence for TCA is *necessarily* incomplete, since TCA is an historical hypothesis: "... evolutionary explanation is of its nature *historical* and historical explanation is not like explanation in physics or chemistry. It deals with the singular and the unrepeatable; it is thus *necessarily* incomplete." This is true, but it is also part of my point; it is (partly) for this reason that it is absurd to claim that TCA is certain; those strident declarations of certainty must come from some source other than a cool, reasoned, dispassionate look at the evidence. Furthermore, McMullin is right in pointing out that one shouldn't apply inappropriate criteria when assessing the merits of a particular explanation; but an explanation for which the evidence is necessarily weak is still an explanation for which the evidence is weak.

Second, the fossil evidence. As McMullin says, the fossil record contains many sequences of extinct forms (e.g., trilobites) "where the development of specific anatomical features can be traced in detail through the rock layers." This is indeed so, but does not bear on the main problems for TCA with the fossil record, which have to do with the lack, in that record, of sequences of intermediary forms between the really major taxa. Consider, for example, the Cambrian explosion. There is fossilized record of unicellular life going all the way back, so they tell us, to 3 or 3.5 billion years ago—only a billion years or so after the formation of the earth itself. There is no fossil record of the development of multicellular life until about 570 million years ago, 2.5 to 3 billion years after the appearance of unicellular life. Then there is a veritable explosion of invertebrate life, a riot of shapes and anatomical designs, with ancestors of the major contemporary forms represented, together with a lot of forms wholly alien in the contemporary context.[18] Now of course this enormous gap in the fossil record is logically compatible with TCA, but, given that God might very well have created specially, TCA seems at present unlikely. It requires epicyclic explanations of this and the other major gaps.

It is in areas like this that the problems lie; and finding sequences of intermediary forms between varieties of trilobites, e.g., does nothing to narrow such gaps. McMullin goes on: "They continue to uncover stage after stage in crucial "linking" forms, such as the therapsids, for example, the forms that related reptiles with the earliest mammals." This is misleading. It suggests that "they" have discovered and continue to discover many linking forms that link, say, reptiles with birds or reptiles with mammals. But so far as I know, this is not so. So far as I know, for example, therapsids are the only candidates for a link between reptiles and mammals (and they have been known for some time). Although there is some controversy about the therapsids, perhaps they really could be thought of as linking forms between reptiles and mammals; but if TCA were true, one would expect vastly many more such forms. (Furthermore, Archaeopteryx, which was formerly the only serious candidate for a similar post linking reptiles and birds, has been demoted by the discovery of modern birds antedating it.) The fossil record fits versions of special creation considerably better than it fits TCA: it suggests the independent appearance of the major bauplans to which McMullin refers, with substantial evolution proceeding out from these *Ur* forms. The enormous gaps between the major forms would be much better accommodated on such a view than on TCA.

Finally, I referred briefly to the puzzles involving the *molecular* evidence in my paper; I'm happy to be given a chance to explain myself a bit. McMullin paints a rosy picture here: consilience piled upon serendipitous consilience. But I'd be inclined to see much of this thinking as more like a pious hope that has blossomed nicely into a foregone conclusion (as Quine says in another connection). As far as I can see, things here are not anywhere nearly as rosy as McMullin suggests. As a matter of fact (so it seems to me) the molecular evidence too fits better with some version of special creation than it does with TCA. McMullin summarizes the molecular evidence thus:

Comparison of the DNA, as well as of the proteins for which DNA codes, between different types or organisms show that there are striking similarities in chemical composition between them. Cytochrome C, for example, found in all animals, is involved in cell respiration. It contains 104 amino acids in a sequence which is invariable for any given species. For humans and rhesus monkeys, the sequence is identical except in one position; for horses and donkeys the sequence also differs in only one position. But for humans and horses, the difference is 13; for monkeys and horses the difference is 11. If instead of cytochrome C another homologous protein is chosen, similar (though not necessarily identical) results are found. These very numerous resemblances and differences between the macromolecules carrying hereditary information can be explained by supposing a very slow rate of change in the chemical sequences constituting these molecules, and thus a relationship of common descent among the organisms themselves.

Cytochrome C is to be found in all living things (plants as well as animals); and the striking thing about it is that in general[19] morphological differences among species are reflected in differences in their cytochrome C; that is, the greater the morphological difference between a pair of species, the greater the difference in their cytochrome C. Thus the cytochrome C difference between a horse and a dog is 6%; between a horse and a duck 10%; between a horse and a tuna fish, silkworm moth, sunflower, yeast, and bacterium it is respectively 18%, 27%, 41%, 42%, and 64%. (A similar pattern is found for hemoglobin, although nearby species differ more in their hemoglobin than they do in their cytochrome C.) McMullin goes on:

Thus, the "molecular" differences between any two species become (on this hypothesis) a rough indication of how long since the ancestors or these species diverged; rather more securely, one can infer the relative order of branching between three or more species; one can infer whether A branched from B before C did. What is impressive here is the *coherence* [M's emphasis] of the results given by examining many different macromolecules in this light.

A bit later he adds: "If TCA is correct, one would *expect* the sort of coherences that molecular biology is now turning up in such abundance."

Now my reading of these matters is different. (Of course the reader must remember that neither of us is expert here.) First, in terms of the overall picture, there is indeed that similarity between the morphological and molecular differences: morphologically similar contemporary forms of life tend to be molecularly similar, and morphologically different contemporary forms of life tend to be molecularly different. This initially fits well with TCA, but of course equally well with various versions of special creation. (It fits particularly well with those that involve typology, the idea that God created ancestors of the main types of animal and plant life, with subsequent evolution.) So far, this evidence seems equal with respect to TCA and its denial.

But second and much more important. There are a number of representatives of very ancient lines: cockroaches, cyclostomes or lampreys, lungfish, some kinds of reptiles, and some of the forms to be found in the Cambrian explosion. These lines vary at the molecular level to a great degree; lampreys are molecularly just as far from the lungfish as they are from modern fishes. If TCA is correct, this means that in the past these lines must have been much closer to each other, in terms of their molecules such as Cytochrome C, hemoglobin, and so on, than they are now. If so, then, molecular evolution must continue on when morphological evolution has stopped. Present day lampreys are morphologically very similar to their ancestors of 350 million years ago, which is more than half way back to the beginnings of multicellular life. But while they are *morphologically* similar to those ancestors, they are *molecularly* very different from them—on TCA. The same goes even more emphatically for those contemporary forms of life that have representatives all the way back in the Cambrian.

This same conclusion can be reached by a different route. Consider the fact that bacterial Cytochrome C displays almost the same degree of difference from all the major groups. Lampreys and lungfish, for example, display the same degree of difference from bacteria (some 65%) as do human beings, pigs, chickens, rattlesnakes, carp, fruit flies, and wheat. There is no trace, at this level, of the traditional sequence: invertebrate, fish, amphibian, reptile, and mammal. All these forms, so far as their proteins go, are equally distant from bacteria; none can be thought of as molecularly more primitive than others. The so-called "higher" forms of life are exactly as distant, from bacteria, as the lower forms. (The same pattern is repeated *within* each of the major groups.) Lampreys and lungfish, however, have remained morphologically unchanged for hundreds of millions of years; but the line culminating in human beings, for example, has (again, according to the theory) undergone a number of enormous changes. This sort of evidence can be understood, on TCA, only if we suppose that molecular evolution is "decoupled" from morphological evolution, so that in a line like that of lungfish and lampreys, while their remote ancestors were *morphologically* very similar to present representatives of those lines, those remote ancestors must have been very different from these present representatives at the *molecular* level. Molecular evolution must continue on after morphological evolution ceases; there must be something like a "molecular clock" that goes on ticking, in these forms, through the hundreds of millions of years when morphological evolution does not occur. Given the typological pattern of molecular similarities and differences among present organisms, TCA requires some kind of molecular clock: it requires, as McMullin says, "a very slow rate of change in the chemical sequences constituting these molecules" in all lines,[20] whether or not they have undergone much morphological change, even if they have undergone no significant morphological change for hundreds of millions of years.

Is this "what one would *expect* [his emphasis] if TCA is correct?" I don't think so. It certainly wasn't what E. Zuckerkandl, one of the pioneers in this area and one of the first to endorse the idea of a molecular clock, expected. Writing in 1963 (when the idea of a molecular clock was first being broached), he says:

Contemporary organisms that look much like ancient ancestral organisms probably contain a majority of polypeptide chains that resemble quite closely those of the ancient organisms. In other words, certain animals said to be "living fossils," such as the cockroach, the horseshoe crab, the shark and, among mammals, the lemur, probably manufacture a great many polypeptide molecules that differ only slightly from those manufactured by their ancestors millions of years ago.[21]

This, it seems to me, is more like what one would expect; but things have turned out very differently. This decoupling of molecular and morphological evolution is not what one would have expected, if TCA is correct, and it constitutes still another epicycle for that theory.

The rate of molecular change is supposed to be much the same, for a given molecule, in all the various lines. But here there is another problem. One would expect the rate of change of this sort of variation to be a function (among other things) of generational length: forms of life with more generations per unit time should change more rapidly than those with fewer generations per unit time. But of course in some lines generations are vastly longer than in others. Thus some plants don't reproduce until they are 80 years old or so; elephants have a generational length of some 14 years; mice of some 2 months or so, and the generational length of yeast can be measured in minutes. If so, how can it be that the rate of change in the homologous molecules of these forms of life is approximately the same? Michael Denton (who cannot be suspected of theological special pleading, since he is not a theist) may be exaggerating when he writes, "There is simply no way of explaining how a uniform rate of evolution could have occurred in any family of homologous proteins by either chance or selection; and, even if we could advance an explanation for one particular protein family, we would still be left with the mystifying problem of explaining why other protein families should have evolved at different rates." Human ingenuity is nearly limitless; almost anything can be explained in one way or another; but it looks as if any such explanation will exact an epistemic penalty in the form of another epicycle or two.

Finally, the whole idea of a molecular clock has recently come under heavy fire. In the 1990 volume of *Evolutionary Biology*, Sigfried Scherer, after what appears to the layman (this layman, anyway) to be an impressive and thorough study of the molecular clock hypothesis, summarizes his findings as follows:

The protein molecular clock hypothesis (i.e., linearity of amino acid replacements compared with geologic time) has been tested empirically using ten different proteins, altogether representing more than 500 individual sequences from plants, animals and prokaryotes [i.e., bacteria]. In no case a linearity within reasonable limits of confidence could be found as would be expected based on the clock concept: a reliable molecular clock with respect to protein sequences seems not to exist. This hold also for proteins such as cytochrome C or fibrinopeptides which usually have been considered as being reliable molecular clocks. It is shown that the relative rate test of the molecular clock is inconclusive. Thus, the prediction of divergence times based on protein structure is prone to error. Different approaches accounting for a nonconstancy of the rate of molecular evolution, questioning the molecular clock concept for theoretical reasons, are reviewed. It is concluded that the protein molecular clock hypothesis should be rejected.[22]

The picture McMullin paints of the molecular evidence as wholly coherent with TCA and offering one new confirmation of it after another, a picture of consilience unrelieved by the slightest suggestion of disharmony—this picture seems to me to be very much at odds with the facts of the matter. So far as I can see, the molecular evidence does not confirm TCA; if anything it disconfirms it, by requiring still more epicycles. On the other hand, the typological structure of the molecular evidence fits very well with various typological views as to how God might have created some forms of life specially.

By way of conclusion: after reflecting on the replies offered by Van Till and McMullin, I remain confident that TCA is relatively unlikely given a Christian or theistic perspective and the empirical evidence. But this isn't where I want to place central emphasis. I would prefer to leave that question to Christian biologists, to people who know the evidence much better than I do, but are also keenly aware of the philosophical and theological issues involved. (People like Jitse Van der Meer of Redeemer College.) I am most interested, first, in emphasizing the fact that there is the sort of battle I mentioned in the paper, and that current science, crucially including current evolutionary theory, is intimately involved in it, thus partly accounting for the striking and unusual role played by evolution in contemporary society. Second, I wish to emphasize the fact that there is no antecedent probability that God would create either by way of TCA or by way of GES; third, I want to argue that at the moment, anyway, it looks as if GES is very unlikely, given theism and the empirical evidence; this is because of the difficulties in seeing how life could have arisen from nonliving matter just by way of the workings of the laws of physics and chemistry. Last and least, I think TCA is no more likely than its denial on theism together with the empirical evidence.

McMullin finds these things worrisome; what worries him, he says, is the "implication that Christian intellectuals should ally themselves with critics of evolution, despite the almost universal support it has among experts in the relevant fields of natural science." I think he finds this worrisome because he is afraid it will subject Christianity to the "laughter" that Galileo, following Augustine, worried about: "I suppose this is what worries me most about the strategy Plantinga urges on his Christian readers.... the reader who proceeds on his advice to do battle with defenders of evolution all too easily risks causing the sort of 'laughter' that Augustine deplores, because of its negative effects on the credibility of the Christian message generally." But first, the expert consensus doesn't have quite the monolithic solidarity we might think. It is rather solid in the United States, but (for reasons that are no doubt sociological but remain obscure) less so in Europe. Pierre Grassé's doubts on the matter have been long known, for example; and according to Paul Lemoine (professor at the Museum of Paris), writing in *Encyclopédie française* (1965):

The result of this exposé is that the theory of evolution is impossible. Basically, despite appearances, no one believes it anymore, and one says—without attaching any other importance to it—"evolution" in order to signify a "series of events in time"; or "more evolved" or "less evolved" in the sense of "more perfected," "less perfected," because such is the language of convention, accepted and almost obligatory in the scientific world. Evolution is a sort of dogma which the priests do not believe in any more, but which they keep up for the sake of their flocks.

It is necessary to have the courage to say this in order that the next generation may direct their research in another way.

And second, Augustine was speaking of stupid, or silly, or wildly uninformed ("delirious") mistakes: "Now it is very scandalous, as well as harmful and to be avoided at all costs, that any infidel should hear a Christian speak about these things as if he were doing so in accordance with Christian Scripture and should see him err so deliriously as to be forced into laughter."[23] That kind of error is, of course, deplorable; but opposition to current orthodoxy with respect to evolution does not automatically constitute delirious error. A Christian intellectual must faithfully speak the truth as she sees it; the results can safely be left to God. Christians who oppose this current orthodoxy may no doubt sometimes have to put up with a bit of contempt or abuse from some quarters; but is that a matter of great concern? We Christians are called to speak the truth as best we can; if we do so, we run the risk of being despised by the cultured despisers; but is that really worth worrying about? I doubt it.

## Notes

1. Although the paper by Scherer cited below suggests that Pun may be unduly optimistic about the molecular clock.

2. *Science* (vol 250 [December, 1990]) pp. 1665 ff. Simon won a Nobel Prize in economics, but also works in computer studies and psychology; he is currently professor of computer studies and psychology in the Department of Psychology at Carnegie Mellon.

3. More simply, says Simon, "Fitness simply means expected number of progeny" (p. 1665). *Why* he thinks this is the rational course he doesn't explain; couldn't I sensibly say that while having lots of progeny might be best for my genes, *I'm* interested in *my* welfare, not theirs?

4. See my "Is Theism Really a Miracle?" *Faith and Philosophy*, April, 1986, pp. 131 ff.

5. "W. J. Bryant may have been a fool in many respects, but he had a more accurate picture of the cosmos than Carl Sagan (who if we may trust the Fourteenth Psalm, is also a fool)." Peter van Inwagen, "Genesis and Evolution" (The Kraemer Lecture at the University of Arkansas, March 31, 1989).

6. In rejecting the semi-deist view, I am not claiming that God must have created "by bringing various kinds of living things suddenly to be" as McMullin suggests; it could be by a process as slow and gradual as you please. Claiming that TCA is improbable does not commit me to God's creating horses by, for example, saying "Let there be a horse" and there immediately springing into being a horse. Perhaps God did it that way and perhaps not; but there are a thousand other ways he could have done it, and I make no suggestions as to which way he *did* do it. Augustine's seeds, for example, would do very nicely. (Augustine, by the way, did not accept TCA; he didn't believe, e.g., that mammals evolved from reptiles or amphibia from fish. Rather, what he held was that mammals arose from mammal seeds, reptiles from reptile seeds, amphibia from amphibia seeds, and so on.)

7. Perhaps what Galileo really meant to say has less to do with the *literal* interpretation of Scripture, than with something else, something that isn't easy to state exactly, something like what the Scriptures seem initially, taken at face value, to teach, or what it seems initially (prior to much reflection) that the Lord intends to teach in the Scripture in question. In *I Kings* 7 v. 23 we read: "Then he made the molten sea; it was round, ten cubits from brim to brim, and five cubits high, and a line of thirty cubits measured its circumference." Taken at face value, this seems to suggest that the value of pi is 3. One supposes that Galileo would not hold that here the Lord intends to teach us that current mathematics is wrong about pi; no doubt he would agree that we shouldn't take this bit of Scripture at face value. This is not, however, because the bit of writing in question is to be understood *figuratively*.

8. Consider even such an assured result as that the earth revolves around the sun and rotates on its axis. According to the usual interpretations of current relativity theory, there is no privileged frame of reference, no frame at absolute rest. But if that is true, then it isn't even clear what it means to say that in fact, contrary to Copernicus, the earth revolves around the sun rather than *vice versa*. That's true in some frames of reference, but not in others; and in principle (apart from matters of convenience, and the like) there is no more to be said for the former than for the latter.

It might be replied that at least Copernicus is controverted in that *he* held that there *is* a frame at absolute rest, which we now know is false. But the usual interpretations of relativity theory are not themselves supported by knock-down drag-out arguments. One can also interpret relativity theory as nothing more than a recipe for translation from one frame of reference to another; so taken it makes no pronouncements on the question whether there is a frame at absolute rest. So taken, the claim that there is such a frame is quite consistent with it; perhaps the frame at absolute rest is given by the way God sees things. (And hence it could be, so far as knock-down drag-out demonstration goes, that the earth is the center of the universe after all!)

9. I also think reasonably plausible the traditional theistic affirmation of divine *concurrence* in the causal activity of his creatures.

10. *Prolegomena zur Geschichte Israels* (Berlin and Leipzig: 6th edition, 1905; republished in 1927) p. 296 (translation by Albert Wolters). Wellhausen is widely recognized as one of the giants of critical biblical scholarship. According to Albert Wolters, his *Prolegomena* has approximately the same status in biblical studies as Kant's first Critique in philosophy. My gratitude to Wolters for the references to Wellhausen, Gunkel, Young, Aalders, and Barr.

11. *Genesis* in Nowack's *Hand kommentar zum Alten Testament* (translation by Albert Wolters).

12. *Fundamentalism* (London: SCM Press, 2nd edition, 1981), p. 40.

13. Personal letter to David C. K. Watson, (April 23, 1984), published in the *Newsletter* of the Creation Science Association of Ontario, vol. 3, no. 4, 1990/91.

14. Here there is another important point, one familiar from discussions of public education; if you hew to the methodological naturalism McMullin favors, you are very likely to wind up with the very sort of science you would aim at if your naturalism weren't merely *methodological* but (as you see it) the sober metaphysical truth of the matter. I therefore think "provisional atheism" is a good name for methodological naturalism, not because anyone who employs it is properly called an atheist (of course not), but because the results of using it in a given area will often be the very same as the results of assuming atheism or taking it for granted and arguing from there.

15. *Portraits of Creation*, ed. Van Till, Snow, Stek and Young (Grand Rapids: Eerdmans, 1990), p. 254.

16. McMullin's word "intervene' isn't a good one in the theistic context; God constantly supports every created thing in existence by way of activity without which the creature would disappear like your breath on a frosty morning; furthermore, according to the bulk of the theistic community, no creaturely action is so much as possible without his further concurrent activity. So he necessarily and constantly takes a hand in the operation of his creatures; he necessarily and constantly "intervenes" in his creation; were he to leave it to its own devices, even for an instant, it would vanish like a dream upon awakening. Creating something new—life, for example, or human life—isn't really in any sensible sense an "intervention" for him.

17. I am grateful to William Hasker and Richard Swinburne for calling this point to my attention.

18. " 'The real shocker for me is the worm that looks like it has kneecaps,' said Dr. Ellis L. Yochelson, a paleontologist at the Smithsonian Institution. He was referring to an animal known as Microdictyon." ("Spectacular Fossils Record Early Riot of Creation" *New York Times*, 4/23/91.) (I am indebted to Phillip Johnson for this reference.) Until recently, the main source of fossils for the Cambrian explosion was the Burgess shale (in the Canadian Rockies) dating back, so they say, to 530 million years ago. In 1984, however, a sediment dating back to 570 million years was discovered in Yunnan, China, containing a vast

wealth of fossils of invertebrate life—sponges, arthropods, worms, crustaceans, and the like. (Reported in the article just referred to.)

19. Though not in every instance: cytochrome C differs in only one place between humans and chimpanzees, but in three places between very similar species of fruit flies.

20. Another result may at first be surprising. Cytochrome C is still evolving slowly, and is doing so at a rate that is approximately constant for all species, when the rate is averaged over geological time periods." "The Structure and History of an Ancient Protein," by Richard E. Dickerson, *Scientific American*, April, 1972, p. 8.

21. "The Evolution of Hemoglobin," *Scientific American*, 213 (5), p. 111.

22. I am indebted to Jitse Van der Meer for this reference, as for other good counsel and advice.

23. *De Genesi ad Litteram Libri Duodecimi*, chapter 19.

# IV
## Intelligent Design's Scientific Claims

In addition to their philosophical attacks on scientific method, IDCs follow the pattern of classic creationists in launching a barrage of negative arguments against evolution, citing things such as gaps in the fossils record that scientists supposedly cannot explain. *Of Pandas and People*, the ID "supplemental textbook" written for high schools and junior high schools, proposes to let students decide for themselves between "the two theories" of Darwinian evolution and intelligent design. However, members of the intelligent design group have yet to publish in professional scientific journals any research supporting intelligent design as a better explanation of the origin of species. This section therefore deals with the scientific claims they regularly make in the books and articles that they have published.

The first article, "Molecular Machines: Experimental Support for the Design Inference," is by Lehigh University professor Michael Behe. In this article he presents the key ideas and argument that he wrote about in *Darwin's Black Box: The Biochemical Challenge to Evolution* (Free Press, 1996), which is the most influential ID book next to Johnson's *Darwin on Trial*. The article is reprinted from *Cosmic Pursuit*, a magazine that is edited by fellow IDC Fred Heeren. A biochemist, Behe argues that biomolecular systems function like tiny, intricate machines and that they exhibit what he calls "irreducible complexity"—the function performed depends on all the components of the system such that it would be lost were any one of the components removed. Behe claims that irreducible complexity proves that such molecules could not have evolved gradually as Darwinism holds, and he claims that it is more reasonable to believe that they were intelligently designed.

Philip Kitcher, professor of philosophy at Columbia University, provides the first critical assessment of IDC's scientific claims. Kitcher's 1982 book *Abusing Science: The Case against Creationism*, remains one of the best critiques of young-earth creation-science as it stood at the critical period at the beginning of the 1980s. In this article, focusing his attention on the arguments of Michael Behe and Philip Johnson, he looks at how creationism has been "born again" in the intelligent design neocreationist movement. Kitcher writes that Behe's argument about irreducible complexity amounts to no more than "an attempt to parlay ignorance of molecular details into an impossibility proof," and that the quantitative argument he uses to try to back it up "smacks more of numerology than of science." Kitcher then turns to the series of arguments that Philip Johnson made in *Darwin on Trial* to try to cast doubt upon the evidential basis of evolutionary theory. Analyzing both the substance and rhetoric of Johnson's arguments, he concludes that Johnson "distorts the positions he opposes, shifts standards of evidence, quotes people out of context, and uses ambiguity to cover his argumentative gaps."

In the third article of this section, biologists Matthew Brauer and Daniel Brumbaugh also take on Johnson and Behe and find many additional problems with their criticisms of evolution. They also examine Michael Denton's anti-evolutionary arguments. Denton is an important figure in that Johnson, Behe, and other IDCs cite his book—*Evolution: A Theory in Crisis* (1985)—as revealing huge, unaddressed problems in evolutionary theory that provided early inspiration for the design movement. Brauer and Brumbaugh focus on what Denton cites as perhaps the most dramatic example of discontinuities in nature that call Darwinism into question—supposed biochemical discontinuities that he says are an "echo of typology." They show how Denton's key criticism here rests on an "absurd caricature of evolutionary biology" and "demonstrates a fundamental lack of understanding of ancestor-descendant relationships." Quite in contrast to Denton's claim that evolution is a "theory in crisis" or Behe's suggestions of the same, evolutionary biology is a healthy, robust field of inquiry. To illustrate its continual advance, Brauer and Brumbaugh briefly describe two new lines of evidence that are beginning to extend our knowledge of evolution history. (1) SINES and LINES, recently discovered "molecular fossils" in the genome of organisms, provide a fresh, independent source of evidence about evolutionary relationships. (2) A combination of theoretical and experimental work on visual pigment evolution opens a new way for testing hypotheses that distinguish sequence similarity that is the result of shared ancestry versus functional convergence. They conclude that while "there is there is abundant need for spiritual problem-solving in the world," it does not make sense to "let tired, erroneous, and dogmatic creationist assertions distract science from the successful methodologies and logic used to solve scientific problems."

# 10

## Molecular Machines: Experimental Support for the Design Inference

Michael J. Behe

### Darwinism's Prosperity

Within a short time after Charles Darwin published *The Origin of Species* the explanatory power of the theory of evolution was recognized by the great majority of biologists. The hypothesis readily resolved the problems of homologous resemblance, rudimentary organs, species abundance, extinction, and biogeography. The rival theory of the time, which posited creation of species by a supernatural being, appeared to most reasonable minds to be much less plausible, since it would have a putative Creator attending to details that seemed to be beneath His dignity.

As time went on the theory of evolution obliterated the rival theory of creation, and virtually all working scientists studied the biological world from a Darwinian perspective. Most educated people now lived in a world where the wonder and diversity of the biological kingdom were produced by the simple, elegant principle of natural selection.

However, in science a successful theory is not necessarily a correct theory. In the course of history there have also been other theories which achieved the triumph that Darwinism achieved, which brought many experimental and observational facts into a coherent framework, and which appealed to people's intuitions about how the world should work. Those theories also promised to explain much of the universe with a few simple principles. But, by and large, those other theories are now dead.

A good example of this is the replacement of Newton's mechanical view of the universe by Einstein's relativistic universe. Although Newton's model accounted for the results of many experiments in his time, it failed to explain aspects of gravitation. Einstein solved that problem and others by completely rethinking the structure of the universe.

Similarly, Darwin's theory of evolution prospered by explaining much of the data of his time and the first half of the 20th century, by my article will show that Darwinism has been unable to account for phenomena uncovered by the efforts of modern biochemistry during the second half of this century. I will do this by emphasizing the fact that life at its most fundamental level is irreducibly complex and that such complexity is incompatible with undirected evolution.

**A Series of Eyes**

How do we see?

In the 19th century the anatomy of the eye was known in great detail and the sophisticated mechanisms it employs to deliver an accurate picture of the outside world astounded everyone who was familiar with them. Scientists of the 19th century correctly observed that if a person were so unfortunate as to be missing one of the eye's many integrated features, such as the lens, or iris, or ocular muscles, the inevitable result would be a severe loss of vision or outright blindness. Thus it was concluded that the eye could only function if it were nearly intact.

As Charles Darwin was considering possible objections to his theory of evolution by natural selection in *The Origin of Species* he discussed the problem of the eye in a section of the book appropriately entitled "Organs of extreme perfection and complication." He realized that if in one generation an organ of the complexity of the eye suddenly appeared, the event would be tantamount to a miracle. Somehow, for Darwinian evolution to be believable, the difficulty that the public had in envisioning the gradual formation of complex organs had to be removed.

Darwin succeeded brilliantly, not by actually describing a real pathway that evolution might have used in constructing the eye, but rather by pointing to a variety of animals that were known to have eyes of various constructions, ranging from a simple light sensitive spot to the complex vertebrate camera eye, and suggesting that the evolution of the human eye might have involved similar organs as intermediates.

But the question remains, how do we see? Although Darwin was able to persuade much of the world that a modern eye could be produced gradually from a much simpler structure, he did not even attempt to explain how the simple light sensitive spot that was his starting point actually worked. When discussing the eye Darwin dismissed the question of its ultimate mechanism by stating: "How a nerve comes to be sensitive to light hardly concerns us more than how life itself originated."

He had an excellent reason for declining to answer the question: 19th century science had not progressed to the point where the matter could even be approached. The question of how the eye works—that is, what happens when a photon of light first impinges on the retina—simply could not be answered at that time. As a matter of fact, no question about the underlying mechanism of life could be answered at that time. How do animal muscles cause movement? How does photosynthesis work? How is energy extracted from food? How does the body fight infection? All such questions were unanswerable.

## The Calvin and Hobbes Approach

Now, it appears to be a characteristic of the human mind that when it lacks understanding of a process, then it seems easy to imagine simple steps leading from nonfunction to function. A happy example of this is seen in the popular comic strip *Calvin and Hobbes*. Little boy Calvin is always having adventures in the company of his tiger Hobbes by jumping in a box and traveling back in time, or grabbing a toy ray gun and "transmogrifying" himself into various animal shapes, or again using a box as a duplicator and making copies of himself to deal with worldly powers such as his mom and his teachers. A small child such as Calvin finds it easy to imagine that a box just might be able to fly like an airplane (or something), because Calvin doesn't know how airplanes work.

A good example from the biological world of complex changes appearing to be simple is the belief in spontaneous generation. One of the chief proponents of the theory of spontaneous generation during the middle of the 19th century was Ernst Haeckel, a great admirer of Darwin and an eager popularizer of Darwin's theory. From the limited view of cells that 19th century microscopes provided, Haeckel believed that a cell was a "simple little lump of albuminous combination of carbon," not much different from a piece of microscopic Jell-O®. Thus it seemed to Haeckel that such simple life could easily be produced from inanimate material.

In 1859, the year of the publication of *The Origin of Species*, an exploratory vessel, the H.M.S. Cyclops, dredged up some curious-looking mud from the sea bottom. Eventually Haeckel came to observe the mud and thought that it closely resembled some cells he had seen under a microscope. Excitedly he brought this to the attention of no less a personage than Thomas Henry Huxley, Darwin's great friend and defender, who observed the mud for himself. Huxley, too, became convinced that it was Urschleim (that is, protoplasm), the progenitor of life itself, and Huxley named the mud *Bathybius haeckelii* after the eminent proponent of abiogenesis.

The mud failed to grow. In later years, with the development of new biochemical techniques and improved microscopes, the complexity of the cell was revealed. The "simple lumps" were shown to contain thousands of different types of organic molecules, proteins, and nucleic acids, many discrete subcellular structures, specialized compartments for specialized processes, and an extremely complicated architecture. Looking back from the perspective of our time, the episode of *Bathybius haeckelii* seems silly or downright embarrassing, but it shouldn't. Haeckel and Huxley were behaving naturally, like Calvin: since they were unaware of the complexity of cells, they found it easy to believe that cells could originate from simple mud.

Throughout history there have been many other examples, similar to that of Haeckel, Huxley, and the cell, where a key piece of a particular scientific puzzle was beyond the understanding of the age. In science there is even a whimsical term for a machine or structure or process that does something, but the actual mechanism by which it accomplishes its task is unknown: it is called a "black box." In Darwin's time all of biology was a black box: not only the cell, or the eye, or digestion, or immunity, but every biological structure and function because, ultimately, no one could explain how biological processes occurred.

Biology has progressed tremendously due to the model that Darwin put forth. But the black boxes Darwin accepted are now being opened, and our view of the world is again being shaken.

Take our modern understanding of proteins, for example.

## Proteins

In order to understand the molecular basis of life it is necessary to understand how things called "proteins" work. Proteins are the machinery of living tissue that build the structures and carry out the chemical reactions necessary for life. For example, the first of many steps necessary for the conversion of sugar to biologically-usable forms of energy is carried out by a protein called hexokinase. Skin is made in large measure of a protein called collagen. When light impinges on your retina it interacts first with a protein called rhodopsin. A typical cell contains thousands and thousands of different types of proteins to perform the many tasks necessary for life, much like a carpenter's workshop might contain many different kinds of tools for various carpentry tasks.

What do these versatile tools look like? The basic structure of proteins is quite simple: they are formed by hooking together in a chain discrete subunits called amino acids. Although the protein chain can consist of anywhere from about 50 to about 1,000 amino acid links, each position can only contain one of 20 different amino acids. In this they are much like words: words can come in various lengths but they are made up from a discrete set of 26 letters.

Now, a protein in a cell does not float around like a floppy chain; rather, it folds up into a very precise structure which can be quite different for different types of proteins. Two different amino acid sequences—two different proteins—can be folded to structures as specific and different from each other as a three-eighths inch wrench and a jigsaw. And like the household tools, if the shape of the proteins is significantly warped then they fail to do their jobs.

## The Eyesight of Man

In general, biological processes on the molecular level are performed by networks of proteins, each member of which carries out a particular task in a chain.

Let us return to the question, how do we see? Although to Darwin the primary event of vision was a black box, through the efforts of many biochemists an answer to the question of sight is at hand. The answer involves a long chain of steps that begin when light strikes the retina and a photon is absorbed by an organic molecule called 11-cis-retinal, causing it to rearrange itself within picoseconds. This causes a corresponding change to the protein, rhodopsin, which is tightly bound to it, so that it can react with another protein called transducin, which in turn causes a molecule called GDP to be exchanged with a molecule called GTP.

To make a long story short, this exchange begins a long series of further bindings between still more specialized molecular machinery, and scientists now understand a great deal about the system of gateways, pumps, ion channels, critical concentrations, and attenuated signals that result in a current to finally be transmitted down the optic nerve to the brain, interpreted as vision. Biochemists also understand the many chemical reactions involved in restoring all these changed or depleted parts to make a new cycle possible.

## To Explain Life

Although space doesn't permit me to give the details of the biochemistry of vision here, I have given the steps in my talks. Biochemists know what it means to "explain" vision. They know the level of explanation that biological science eventually must aim for. In order to say that some function is understood, every relevant step in the process must be elucidated. The relevant steps in biological processes occur ultimately at the molecular level, so a satisfactory explanation of a biological phenomenon such as sight, or digestion, or immunity, must include a molecular explanation.

It is no longer sufficient, now that the black box of vision has been opened, for an "evolutionary explanation" of that power to invoke only the anatomical structures of whole eyes, as Darwin did in the 19th century and as most popularizers of evolution continue to do today. Anatomy is, quite simply, irrelevant. So is the fossil record. It does not matter whether or not the fossil record is consistent with evolutionary theory, any more than it mattered in physics that Newton's theory was consistent with everyday experience. The fossil record has nothing to tell us about, say, whether or how the interactions of 11-cis-retinal with rhodopsin, transducin, and phosphodiesterase could have developed, step by step.

"How a nerve comes to be sensitive to light hardly concerns us more than how life itself originated," said Darwin in the 19th century. But both phenomena have attracted the interest of modern biochemistry in the past few decades. The story of the slow paralysis of research on life's origin is quite interesting, but space precludes its retelling here. Suffice it to say that at present the field of origin-of-life studies has dissolved into a cacophony of conflicting models, each unconvincing, seriously incomplete, and incompatible with competing models. In private even most evolutionary biologists will admit that science has no explanation for the beginning of life.

The same problems which beset origin-of-life research also bedevil efforts to show how virtually any complex biochemical system came about. Biochemistry has revealed a molecular world which stoutly resists explanation by the same theory that has long been applied at the level of the whole organism. Neither of Darwin's black boxes—the origin of life or the origin of vision (or other complex biochemical systems)—has been accounted for by his theory.

**Irreducible Complexity**

In *The Origin of Species* Darwin stated:

If it could be demonstrated that any complex organ existed which could not possibly have been formed by numerous, successive, slight modifications, my theory would absolutely break down.

A system which meets Darwin's criterion is one which exhibits irreducible complexity. By irreducible complexity I mean a single system which is composed of several interacting parts that contribute to the basic function, and where the removal of any one of the parts causes the system to effectively cease functioning. An irreducibly complex system cannot be produced directly by slight, successive modification of a precursor system, since any precursor to an irreducibly complex system is by definition nonfunctional.

Since natural selection requires a function to select, an irreducibly complex biological system, if there is such a thing, would have to arise as an integrated unit for natural selection to have anything to act on. It is almost universally conceded that such a sudden event would be irreconcilable with the gradualism Darwin envisioned. At this point, however, "irreducibly complex" is just a term, whose power resides mostly in its definition. We must now ask if any real thing is in fact irreducibly

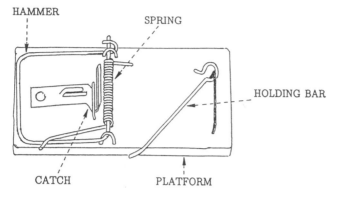

**Figure 10.1**
A household mousetrap. The working parts of the trap are labeled. If any of the parts is missing, the trap does not function.

complex, and, if so, then are any irreducibly complex things also biological systems?

Consider the humble mousetrap (Figure 10.1). The mousetraps that my family uses in our home to deal with unwelcome rodents consist of a number of parts. There are: 1) a flat wooden platform to act as a base; 2) a metal hammer, which does the actual job of crushing the little mouse; 3) a wire spring with extended ends to press against the platform and the hammer when the trap is charged; 4) a sensitive catch which releases when slight pressure is applied; and 5) a metal bar which holds the hammer back when the trap is charged and connects to the catch. There are also assorted staples and screws to hold the system together.

If any one of the components of the mousetrap (the base, hammer, spring, catch, or holding bar) is removed, then the trap does not function. In other words, the simple little mousetrap has no ability to trap a mouse until several separate parts are all assembled.

Because the mousetrap is necessarily composed of several parts, it is irreducibly complex. Thus, irreducibly complex systems exist.

**Molecular Machines**

Now, are any biochemical systems irreducibly complex? Yes, it turns out that many are.

Earlier we discussed proteins. In many biological structures proteins are simply components of larger molecular machines. Like the picture tube, wires, metal bolts and screws that comprise a television set, many proteins are part of structures that only function when virtually all of the components have been assembled.

A good example of this is a cilium. Cilia are hairlike organelles on the surfaces of many animal and lower plant cells that serve to move fluid over the cell's surface or to "row" single cells through a fluid. In humans, for example, epithelial cells lining the respiratory tract each have about 200 cilia that beat in synchrony to sweep mucus towards the throat for elimination.

A cilium consists of a membrane-coated bundle of fibers called an axoneme. An axoneme contains a ring of 9 double microtubules surrounding two central single microtubules. Each outer doublet consists of a ring of 13 filaments (subfiber A) fused to an assembly of 10 filaments (subfiber B). The filaments of the microtubules are composed of two proteins called alpha and beta tubulin. The 11 microtubules forming an axoneme are held together by three types of connectors: subfibers A are joined to the central microtubules by radial spokes; adjacent outer doublets are joined by linkers that consist of a highly elastic protein called nexin; and the central microtubules are joined by a connecting bridge. Finally, every subfiber A bears two arms, an inner arm and an outer arm, both containing the protein dynein.

But how does a cilium work? Experiments have indicated that ciliary motion results from the chemically-powered "walking" of the dynein arms on one microtubule up the neighboring subfiber B of a second microtubule so that the two microtubules slide past each other (Figure 10.2). However, the protein cross-links between microtubules in an intact cilium prevent neighboring microtubules from sliding past each other by more than a short distance. These cross-links, therefore, convert the dynein-induced sliding motion to a bending motion of the entire axoneme.

Now, let us sit back, review the workings of the cilium, and consider what it implies. Cilia are composed of at least a half dozen proteins: alpha-tubulin, beta-tubulin, dynein, nexin, spoke protein, and a central bridge protein. These combine to perform one task, ciliary motion, and all of these proteins must be present for the cilium to function. If the

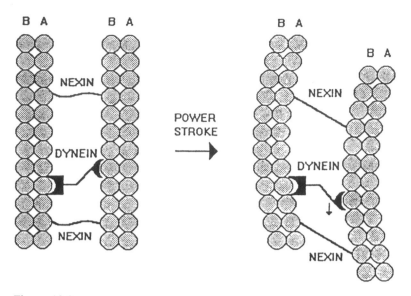

**Figure 10.2**
Schematic drawing of part of a cilium. The power stroke of the motor protein dynein, attached to one microtublule, against subfiber B of a neighboring microtubule causes the fibers to slide past each other. The flexible linker protein, nexin, converts the sliding motion to a bending motion.

tubulins are absent, then there are no filaments to slide; if the dynein is missing, then the cilium remains rigid and motionless; if nexin or the other connecting proteins are missing, then the axoneme falls apart when the filaments slide.

What we see in the cilium, then, is not just profound complexity, but it is also irreducible complexity on the molecular scale. Recall that by "irreducible complexity" we mean an apparatus that requires several distinct components for the whole to work. My mousetrap must have a base, hammer, spring, catch, and holding bar, all working together, in order to function. Similarly, the cilium, as it is constituted, must have the sliding filaments, connecting proteins, and motor proteins for function to occur. In the absence of any one of those components, the apparatus is useless.

The components of cilia are single molecules. This means that there are no more black boxes to invoke; the complexity of the cilium is final, fundamental. And just as scientists, when they began to learn the complexities of the cell, realized how silly it was to think that life arose spontaneously in a single step or a few steps from ocean mud, so too we now realize that the complex cilium can not be reached in a single step or a few steps.

But since the complexity of the cilium is irreducible, then it can not have functional precursors. Since the irreducibly complex cilium can not have functional precursors it can not be produced by natural selection, which requires a continuum of function to work. Natural selection is powerless when there is no function to select. We can go further and say that, if the cilium can not be produced by natural selection, then the cilium was designed.

### A Non-Mechanical Example

A non-mechanical example of irreducible complexity can be seen in the system that targets proteins for delivery to subcellular compartments. In order to find their way to the compartments where they are needed to perform specialized tasks, certain proteins contain a special amino acid sequence near the beginning called a "signal sequence."

As the proteins are being synthesized by ribosomes, a complex molecular assemblage called the signal recognition particle or SRP, binds to the signal sequence. This causes synthesis of the protein to halt temporarily. During the pause in protein synthesis the SRP is bound by the transmembrane SRP receptor, which causes protein synthesis to resume and which allows passage of the protein into the interior of the endoplasmic reticulum (ER). As the protein passes into the ER the signal sequence is cut off.

For many proteins the ER is just a way station on their travels to their final destinations (Figure 10.3). Proteins which will end up in a lysosome are enzymatically "tagged" with a carbohydrate residue called mannose-6-phosphate while still in the ER. An area of the ER membrane then begins to concentrate several proteins; one protein, clathrin, forms a sort of geodesic dome called a coated vesicle which buds off from the ER. In the dome there is also a receptor protein which binds to both the clathrin and to the mannose-6-phosphate group of the protein which is being transported. The coated vesicle then leaves the ER, travels through the cytoplasm, and binds to the lysosome through another specific receptor protein. Finally, in a maneuver involving several more proteins, the vesicle fuses with the lysosome and the protein arrives at its destination.

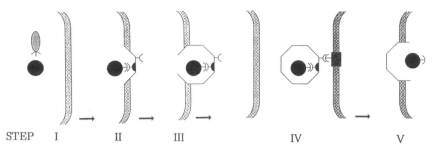

STEP    I        II        III                    IV              V

**Figure 10.3**
Transport of a protein from the ER to the lysosome. Step I: A specific enzyme (gray oval) places a marker on the protein (black sphere). This takes place within the ER, which is delimited by a barrier membrane (cross-hatched bar with ends curving to the left). Step II: The marker is specifically recognized by a receptor protein and the clathrin vesicle (hexagonal shape) begins to form. Step III: The clathrin vesicle is completed and buds off from the ER membrane. Step IV: The clathrin vesicle crosses the cytoplasm and attaches through another specific marker to a receptor protein (dark gray box) on the lysosomal membrane and releases its cargo.

During its travels our protein interacted with dozens of macromolecules to achieve one purpose: its arrival in the lysosome. Virtually all components of the transport system are necessary for the system to operate, and therefore the system is irreducible. And since all of the components of the system are comprised of single or several molecules, there are no black boxes to invoke. The consequences of even a single gap in the transport chain can be seen in the hereditary defect known as I-cell disease. It results from a deficiency of the enzyme that places the mannose-6-phosphate on proteins to be targeted to the lysosomes. I-cell disease is characterized by progressive retardation, skeletal deformities, and early death.

## The Study of "Molecular Evolution"

Other examples of irreducible complexity abound, including aspects of protein transport, blood clotting, closed circular DNA, electron transport, the bacterial flagellum, telomeres, photosynthesis, transcription regulation, and much more. Examples of irreducible complexity can be found on virtually every page of a biochemistry textbook. But if these things cannot be explained by Darwinian evolution, how has the scientific community regarded these phenomena of the past forty years?

A good place to look for an answer to that question is in the *Journal of Molecular Evolution. JME* is a journal that was begun specifically to deal with the topic of how evolution occurs on the molecular level. It has high scientific standards, and is edited by prominent figures in the field. In a recent issue of *JME* there were published eleven articles; of these, all eleven were concerned simply with the analysis of protein or DNA sequences. None of the papers discussed detailed models for intermediates in the development of complex biomolecular structures.

In the past ten years *JME* has published 886 papers. Of these, 95 discussed the chemical synthesis of molecules thought to be necessary for the origin of life, 44 proposed mathematical models to improve sequence analysis, 20 concerned the evolutionary implications of current structures and 719 were analyses of protein or polynucleotide sequences. However, there weren't any papers discussing detailed models for intermediates in the development of complex biomolecular structures. This is not a peculiarity of *JME*. No papers are to be found that discuss detailed models for intermediates in the development of complex biomolecular structures in the *Proceedings of the National Academy of Science, Nature, Science,* the *Journal of Molecular Biology* or, to my knowledge, any journal whatsoever.

Sequence comparisons overwhelmingly dominate the literature of molecular evolution. But sequence comparisons simply can't account for the development of complex biochemical systems any more than Darwin's comparison of simple and complex eyes told him how vision worked. Thus in this area science is mute.

## Detection of Design

What's going on? Imagine a room in which a body lies crushed, flat as a pancake. A dozen detectives crawl around, examining the floor with magnifying glasses for any clue to the identity of the perpetrator. In the middle of the room next to the body stands a large, gray elephant. The detectives carefully avoid bumping into the pachyderm's legs as they crawl, and never even glance at it. Over time the detectives get frustrated with their lack of progress but resolutely press on, looking even more closely at the floor. You see, textbooks say detectives must "get their man," so they never consider elephants.

There is an elephant in the roomful of scientists who are trying to explain the development of life. The elephant is labeled "intelligent design." To a person who does not feel obliged to restrict his search to unintelligent causes, the straightforward conclusion is that many biochemical systems were designed. They were designed *not* by the laws of nature, not by chance and necessity. Rather, they were *planned*. The designer knew what the systems would look like when they were completed; the designer took steps to bring the systems about. Life on earth at its most fundamental level, in its most critical components, is the product of intelligent activity.

The conclusion of intelligent design flows naturally from the data itself—not from sacred books or sectarian beliefs. Inferring that biochemical systems were designed by an intelligent agent is a humdrum process that requires no new principles of logic or science. It comes simply from the hard work that biochemistry has done over the past forty years, combined with consideration of the way in which we reach conclusions of design every day.

What is "design"? Design is simply the *purposeful arrangement of parts*. The scientific question is how we detect design. This can be done in various ways, but design can most easily be inferred for mechanical objects.

Systems made entirely from natural components can also evince design. For example, suppose you are walking with a friend in the woods. All of a sudden your friend is pulled high in the air and left dangling by his foot from a vine attached to a tree branch.

After cutting him down you reconstruct the trap. You see that the vine was wrapped around the tree branch, and the end pulled tightly down to the ground. It was securely anchored to the ground by a forked branch. The branch was attached to another vine—hidden by leaves—so that, when the trigger-vine was disturbed, it would pull down the forked stick, releasing the spring-vine. The end of the vine formed a loop with a slip-knot to grab an appendage and snap it up into the air. Even though the trap was made completely of natural materials you would quickly conclude that it was the product of intelligent design.

Intelligent design is a good explanation for a number of biochemical systems, but I should insert a word of caution. Intelligent design theory has to be seen in context: it does not try to explain everything. We live in a complex world where lots of different things can happen. When deciding how various rocks came to be shaped the way they are a geologist might consider a whole range of factors: rain, wind, the movement of glaciers, the activity of moss and lichens, volcanic action, nuclear explosions, asteroid impact, or the hand of a sculptor. The shape of one rock might have been determined primarily by one mechanism, the shape of another rock by another mechanism.

Similarly, evolutionary biologists have recognized that a number of factors might have affected the development of life: common descent, natural selection, migration, population size, founder effects (effects that may be due to the limited number of organisms that begin a new species), genetic drift (spread of "neutral," nonselective mutations), gene flow (the incorporation of genes into a population from a separate population), linkage (occurrence of two genes on the same chromosome), and much more. The fact that some biochemical systems were designed by an intelligent agent does not mean that any of the other factors are not operative, common, or important.

## Conclusion

It is often said that science must avoid any conclusions which smack of the supernatural. But this seems to me to be both bad logic and bad science. Science is not a game in which arbitrary rules are used to decide what explanations are to be permitted. Rather, it is an effort to make true statements about physical reality. It was only about sixty years ago that the expansion of the universe was first observed. This fact immediately suggested a singular event—that at some time in the distant past the universe began expanding from an extremely small size.

To many people this inference was loaded with overtones of a supernatural event—the creation, the beginning of the universe. The prominent physicist A. S. Eddington probably spoke for many physicists in voicing his disgust with such a notion:

Philosophically, the notion of an abrupt beginning to the present order of Nature is repugnant to me, as I think it must be to most; and even those who would welcome a proof of the intervention of a Creator will probably consider that a single winding-up at some remote epoch is not really the kind of relation between God and his world that brings satisfaction to the mind.

Nonetheless, the big bang hypothesis was embraced by physics and over the years has proven to be a very fruitful paradigm. The point here is that physics followed the data where it seemed to lead, even though some thought the model gave aid and comfort to religion. In the present day, as biochemistry multiplies examples of fantastically complex molecular systems, systems which discourage even an attempt to explain how they may have arisen, we should take a lesson from physics. The conclusion of design flows naturally from the data; we should not shrink from it; we should embrace it and build on it.

In concluding, it is important to realize that we are not inferring design from what we do not know, but from what we do know. We are not inferring design to account for a black box, but to account for an open box. A man from a primitive culture who sees an automobile might guess that it was powered by the wind or by an antelope hidden under the car, but when he opens up the hood and sees the engine he immediately realizes that it was designed. In the same way biochemistry has opened up the cell to examine what makes it run and we see that it, too, was designed.

It was a shock to the people of the 19th century when they discovered, from observations science had made, that many features of the biological world could be ascribed to the elegant principle of natural selection. It is a shock to us in the twentieth century to discover, from observations science has made, that the fundamental mechanisms of life cannot be ascribed to natural selection, and therefore were designed. But we must deal with our shock as best we can and go on. The theory of undirected evolution is already dead, but the work of science continues.

# 11

## Born-Again Creationism

Philip Kitcher

### 1. The Creationist Reformation

In the beginning, creationists believed that the world was young. But creation "science" was without form and void. A deluge of objections drowned the idea that major kinds of plants and animals had been fashioned a few thousand years ago and been hardly modified since. Then the spirit of piety brooded on the waters and brought forth something new. "Let there be design!" exclaimed the reformers—and lo! there was born-again creationism.

Out in Santee, California, about twenty miles from where I used to live, the old movement, dedicated to the possibility of interpreting *Genesis* literally, continues to ply its wares. Its spokesmen still peddle the familiar fallacies, their misunderstandings of the second law of thermodynamics, their curious views about radiometric dating with apparently revolutionary implications for microphysics, the plundering of debates in evolutionary theory for lines that can be usefully separated from their context, and so forth. But the most prominent creationists on the current intellectual scene are a new species, much smoother and more savvy. Not for them the commitment to a literal interpretation of *Genesis* with all the attendant difficulties. Some of them even veer close to accepting the so-called fact of evolution, the claim, adopted by most scientists within a dozen years of the publication of Darwin's *Origin*, that living things are related and that the history of life has been a process of descent with modification. The sticking point for the born-again creationists, as it was for many late-nineteenth-century thinkers, is the mechanism of evolutionary change. They want to argue that natural selection is inadequate, indeed that no natural process could have produced the diversity of organisms, and thus that there must be some designing agent, who didn't just start the process but who has intervened throughout the history of life.

From the viewpoint of religious fundamentalists the creationist Reformation is something of a cop-out. Yet for many believers, the new movement delivers everything they want—particularly the vision of a personal God who supervises the history of life and nudges it to fulfill His purposes—and even militant evangelicals may come to appreciate the virtues of discretion. Moreover the high priests of the Reformation are clad in academic respectability, Professors of Law at University of California-Berkeley and of Biochemistry at Lehigh, and two of the movement's main cheerleaders are highly respected philosophers who teach at Notre Dame. Creationism is no longer hick, but *chic*.

## 2.   Why Literalism Failed

In understanding the motivations for, and the shortcomings of, born-again creationism, it's helpful to begin by seeing why the movement had to retreat. The early days of the old-style "creation-science" campaign were highly successfully. Duane Gish, debating champion for the original movement, crafted a brilliant strategy. He threw together a smorgasbord of apparent problems for evolutionary biology, displayed them very quickly before his audiences, and challenged his opponents to respond. At first, the biologists who debated him laboriously offered details to show that one or two of the problems Gish had raised could be solved, but then their time would run out and the audience would leave thinking that most of the objections were unanswerable. In the middle 1980s, however, two important changes took place: first, defenders of evolutionary theory began to take the same care in formulating answers as Gish had given to posing the problems, and there were quick, and elegant, ways of responding to the commonly reiterated challenges; second, and more important, debaters began to fight back, asking how the observable features of the distribution and characteristics of plants and organisms, both those alive and those fossilized, could be rendered compatible with a literal interpretation of *Genesis*.

Suppose that the earth really was created about ten thousand years ago, with the major kinds fashioned then, and diversifying only a little since. How are we to account for the distributions of isotopes in the earth's crust? How are we to explain the regular, worldwide, ordering of the fossils? The only creationist response to the latter question has been to invoke the Noachian deluge: the order is as it is because of the relative positions of the organisms at the time the flood struck. Take this suggestion seriously, and you face some obvious puzzles: sharks and dolphins are found at the same depths, but, of course, the sharks occur much, much lower in the fossil record; pine trees, fir trees, and deciduous trees are mixed in forests around the globe, and yet the deciduous trees are latecomers in the worldwide fossil record. Maybe we should suppose that the oaks and beeches saw the waters rising and outran their evergreen rivals?

Far from being a solution to creationism's problems, the Flood is a real disaster. Consider biogeography. The ark lands on Ararat, say eight thousand years ago, and out pop the animals (let's be kind and forget the plants). We now have eight thousand years for the marsupials to find their way to Australia, crossing several large bodies of water in the process. Perhaps you can imagine a few energetic kangaroos making it—but the wombats? Moreover, creationists think that while the animals were sorting themselves out, there was diversification of species within the "basic kinds"; jackals, coyotes, foxes, and dogs descend, so the story goes, from a common "dog kind." Now despite all the sarcasm that they have lavished on orthodox evolutionary theory's allegedly high rates of speciation, a simple calculation shows that the rates of speciation "creation-science" would require to manage the supposed amount of species diversification are truly breathtaking, orders of magnitude greater than any that have been dreamed of in evolutionary theory. Finally, to touch on just one more problem, creationists have to account for the survival of thousands of parasites that are specific to our species. During the days on the ark, these would have had to be carried by less than ten people. One can only speculate about the degree of ill-health that Noah and his crew must have suffered.

A major difficulty for old-style creationism has always been the fact that very similar anatomical structures are co-opted to different ends in species whose ways of life diverge radically. Moles, bats, whales, and dogs have forelimbs based on the same bone architecture that has to be adapted to their methods of locomotion. Not only is it highly implausible that the common blueprint reflects an especially bright idea from a designer who saw the best ways to fashion a burrowing tool, a wing, a flipper, and a leg, but the obvious explanation is that shared bone structure reflects shared ancestry. That explanation has only been deepened as studies of chromosome banding patterns have revealed common patterns among species evolutionists take to be related, as comparisons of proteins have exposed common sequences of amino acids, and, most recently, as genomic sequencing has shown the affinities in the ordering of bases in the DNA of organisms. Two points are especially noteworthy. First, like the anatomical residues of previously functional structures (such as the rudimentary pelvis found in whales), parts of our junk DNA have an uncanny resemblance to truncated, or mutilated, versions of genes found in other mammals, other vertebrates, or other animals. Second, the genetic kinship even among distantly related organisms is so great that a human sequence was identified as implicated in colon cancer by recognizing its similarity to a gene coding for a DNA repair enzyme in yeast. The evidence for common ancestry is so overwhelming that even the born-again creationist, Michael Behe is moved to admit that it is "fairly convincing" and that he has "no particular reason to doubt it" (DBB 5).[1] (Notice that Behe doesn't quite commit himself here—in fact, to use an example from Richard Dawkins that Behe, and others, have discussed, there's an obvious line to describe Behe's phraseology: METHINKS IT IS A WEASEL.)

Imagine creationists becoming aware, at some level, of this little piece of history, and retreating to the bunker in which they plot strategy. What would they come up with? First, the familiar idea that the best defense is a good offense: they need to return to the tried-and-true, give-'em-hell, Duane Gish fire and brimstone attack on evolutionary theory. Second, they need to expose less to counterattack, and that means giving up on the disastrous "creation model" with all the absurdities that *Genesis*-as-literal-truth brings in its train; better to make biology safe for the central tenets of religion by talking about a design model so softly focused that nobody can raise nasty questions about parasites on the ark or the wombats' dash for the Antipodes. Third, they should do something to mute the evolutionists' most successful arguments, those that draw on the vast number of cross-species comparisons at all levels to establish common descent; this last is a matter of some delicacy, since too blatant a commitment to descent with modification might seem incompatible with creative design. So the best tactic here is a carefully choreographed waltz—advance a little toward accepting the "fact of evolution" here, back away there; as we shall see, some protagonists have an exquisite mastery of the steps.

Surprise, surprise. Born-again creationism has arrived at just this strategy. I'm going to look at the two most influential versions.

## 3.   The Hedgehog and the Fox

Isaiah Berlin's famous division that contrasts hedgehogs (people with one big idea) and foxes (people with lots of little ideas) applies not only to thinkers but to creationists as well. The two most prominent figures on the neo-creo scene are Michael Behe (a hedgehog) and Phillip Johnson (a fox), both of whom receive plaudits from such distinguished philosophers as Alvin Plantinga and Peter van Inwagen. (Since Plantinga and van Inwagen have displayed considerable skill in articulating and analyzing philosophical arguments, the only charitable interpretation of their fulsome blurbs is that a combination of *Schwärmerei* for creationist doctrine and profound ignorance of relevant bits of biology has induced them to put their brains in cold storage.) Johnson, a lawyer by training, is a far more subtle rhetorician than Gish, and he moves from topic to topic smoothly, discreetly making up the rules of evidence to suit his case as he goes. Many of his attack strategies refine those of country-bumpkin creationism, although, like the White Knight in Alice, he has a few masterpieces of his own invention.

Behe, by contrast, mounts his case for born-again creationism by taking one large problem, and posing it again and again. The problem isn't particularly new: it's the old issue of "complex organs" that Darwin tried to confront in the *Origin*. Behe gives it a new twist by drawing on his background as a biochemist, and describing the minute details of mechanisms in organisms so as to make it seem impossible that they could ever have emerged from a stepwise natural process.

## 4. Behe's Big Idea

Here's the general form of the problem. Given our increased knowledge of the molecular structures in cells and the chemical reactions that go on within and among cells, it's possible to describe structures and processes in exceptionally fine detail. Many structures have large numbers of constituent molecules and the precise details of their fit together are essential for them to fulfill their functions. Similarly, many biochemical pathways require numerous enzymes to interact with one another, in appropriate relative concentrations, so that some important process can occur. Faced with either of these situations, you can pose an obvious question: how could organisms with the pertinent structures or processes have evolved from organisms that lacked them? That question is an explicit invitation to describe an ancestral sequence of organisms that culminated in one with the structures or processes at the end, where each change in the sequence is supposed to carry some selective advantage. If you now pose the question many times over, canvass various possibilities, and conclude that not only has no evolutionist proposed any satisfactory sequences, but that there are systematic reasons for thinking that the structure or process could not have been built up gradually, you have an attack strategy that appears very convincing.

That, in outline, is Behe's big idea. Here's a typical passage, summarizing his quite lucid and accessible description of the structures of cilia and flagella:

... as biochemists have begun to examine apparently simple structures like cilia and flagella, they have discovered staggering complexity, with dozens or even hundreds of precisely tailored parts. It is very likely that many of the parts we have not considered here are required for any cilium to function in a cell. As the number of required parts increases, the difficulty of gradually putting the system together skyrockets, and the likelihood of indirect scenarios plummets. Darwin looks more and more forlorn. (DBB 73)

This sounds like a completely recalcitrant problem for evolutionists, but it's worth asking just why precisely Darwin should look more and more forlorn.

Notice first that lots of sciences face all sorts of unresolved questions. To take an example close to hand, Behe's own discussions of cilia frankly acknowledge that there's a lot still to learn about molecular structure and its contributions to function. So the fact that evolutionary biologists haven't yet come up with a sequence of organisms culminating in bacteria with flagella or cilia might be regarded as signaling a need for further research on the important open problem of how such bacteria evolved. Not so! declares Behe. We have here "irreducible complexity," and it's just impossible to imagine a sequence of organisms adding component molecules to build the structures up gradually.

What does this mean? Is Behe supposing that his examples point to a failure of natural selection as a mechanism for evolution? If so, then perhaps he believes that there was a sequence of organisms that ended up with a bacterium with a flagellum (say), but that the intermediates in this sequence added molecules to no immediate purpose, presumably being at a selective disadvantage because of this. (Maybe the Good Lord tempers the wind to the shorn bacterium.) Or does he just dispense with intermediates entirely, thinking that the Creator simply introduced all the right molecules *de novo*? In that case, despite his claims, he really does doubt common descent. Behe's actual position is impossible to discern because he has learned Duane Gish's lesson (Always attack! Never explain!). I'll return at the very end to the cloudiness of Behe's account of the history of life.

Clearly, Behe thinks that Darwinian evolutionary theory requires some sequence of precursors for bacteria with flagella and that no appropriate sequence could exist. But why does he believe this? Here's a simple-minded version of the argument. Assume that the flagellum needs 137 proteins. Then Darwinians are required to produce a sequence of 138 organisms, the first having none of the proteins and each one having one more protein than its predecessor. Now, we're supposed to be moved by the plight of organisms numbers 2 to 137, each of which contains proteins that can't serve any function, and is therefore, presumably, a target of selection. Only number 1, the ancestor, and number 138, in which all the protein constituents come together to form the flagellum, have just what it takes to function. The intermediates would wither in the struggle for existence. Hence evolution under natural selection couldn't have brought the bacterium from there to here.[2]

But this story is just plain silly, and Darwinians ought to disavow any commitment to it. After all, it's a common theme of evolutionary biology that constituents of a cell, a tissue, or an organism, are put to new uses because of some modification of the genotype. So maybe the immediate precursor of the proud possessor of the flagellum is a bacterium in which all the protein constituents were already present, but in which some other feature of the cell chemistry interferes with the reaction that builds the flagellum. A genetic change removes the interference (maybe a protein assumes a slightly different configuration, binding to something that would have bound to one of the constituents of the flagellum, preventing the assembly). "But, Professor Kitcher [creos always try to be polite], do you have any evidence for this scenario?" Of course not. That is to shift the question. We were offered a proof of the impossibility of a particular sequence, and when one tries to show that the proof is invalid by inventing possible instances, it's not pertinent to ask for reasons to think that those instances exist. If they genuinely reveal that what was was declared to be impossible isn't, then we no longer have a claim that the Darwinian sequence couldn't have occurred, but simply an open problem of the kind that spurs scientists in any field to engage in research.

Behe has made it look as though there's something more here by inviting us to think about the sequence of precursors in a very particular way. He doesn't actually say that proteins have to be added one at a time—he surely knows very well that that would provoke the reaction I've offered—but his defense of the idea that there just couldn't be a sequence of organisms leading up to bacteria with flagella insinuates, again and again, that the problem is that the alleged intermediates would have to have lots of the components lying around like so many monkey-wrenches in the intracellular works. This strategy is hardly unprecedented. Country-bumpkin creos offered a cruder version when they dictated to evolutionists what fossil intermediates would have to be like: the transitional forms on the way to birds would have to have had half-scales and half-feathers, halfway wings—or so we are told.[3] Behe has made up his own ideas about what transitional organisms must have been like, and then argued that such organisms couldn't have existed.

In fact, we don't need to compare my guesswork with his. What Darwinism is committed to (at most) is the idea that modifications of DNA sequence (insertions, deletions, base changes, translocations) could yield a sequence of organisms culminating in a bacterium with a flagellum, with selective advantages for the later member of each adjacent pair. To work out what the members of this sequence of organisms might have been like, our ideas should be educated by the details of how the flagellum is actually assembled and the loci in the bacterial genome that are involved. Until we know these things, it's quite likely that any efforts to describe precursors or intermediates will be whistling in the dark. Behe's examples cunningly exploit our ability to give a molecular analysis of the end product and our ignorance of the molecular details of how it is produced.

Throughout his book, Behe repeats the same story. He describes, often charmingly, the complexities of molecular structures and processes. There would be nothing to complain of if he stopped here and said: "Here are some interesting problems for molecularly minded evolutionists to work on, and, in a few decades time, perhaps, in light of increased knowledge of how development works at the molecular level, we may be able to see what the precursors were like." But he doesn't. He tries to argue that the precursors and intermediates required by Darwinian evolutionary theory couldn't have existed. This strategy has to fail because Behe himself is just as ignorant about the molecular basis of development as his Darwinian opponents. Hence he hasn't a clue what kinds of precursors and intermediates the Darwinian account is actually committed to—so it's impossible to demonstrate that the commitment can't be honored. However, again and again, Behe disguises his ignorance by suggesting to the reader that the Darwinian story must take a very particular form—that it has to consist in something like the simple addition of components, for example—and on that basis he can manufacture the illusion of giving an impossibility proof.

Although this is the main rhetorical trick of the book, there are some important subsidiary bits of legerdemain. Like pre-Reformation creationists, Behe loves to flash probability calculations, offering spurious precision to his criticisms. Here's his attack on a scenario for the evolution of a blood-clotting mechanism, tentatively proposed by Russell Doolittle:

… let's do our own quick calculation. Consider that animals with blood-clotting cascades have roughly 10,000 genes, each of which is divided into an average of three pieces. This gives a total of about 30,000 gene pieces. TPA [Tissue Plasminogen Activator] has four different types of domains. By "variously shuffling," the odds of getting those four domains together is 30,000 to the fourth power, which is approximately one-tenth to the eighteenth power. Now, if the Irish Sweepstakes had odds of winning of one-tenth to the eighteenth power, and if a million people played the lottery each year, it would take an average of about a thousand billion years before *anyone* (not just a particular person) won the lottery.... Doolittle apparently needs to shuffle and deal himself a number of perfect bridge hands to win the game. (DBB 94)

This sounds quite powerful, and Behe drives home the point by noting that Doolittle provides no quantitative estimates, adding that "without numbers, there is no science" (DBB 95)—presumably to emphasize that born-again creationists are better scientists than the distinguished figures they attack. But consider a humdrum phenomenon suggested by Behe's analogy to bridge. Imagine that you take a standard deck of cards and deal yourself thirteen. What's the probability that you got exactly those cards in exactly that order? The answer is 1 in $4 \times 10^{21}$. Suppose you repeat this process ten times. You'll now have received ten standard bridge hands, ten sets of thirteen cards, each one delivered in a particular order. The chance of getting just those cards in just that order is 1 in $4^{10} \times 10^{210}$. This is approximately 1 in $10^{222}$. Notice that the denominator is far larger than that of Behe's trifling $10^{18}$. So it must be *really* improbable that you (or anyone else) would ever receive just those cards in just that order in the entire history of the universe. But, whatever the cards were, you did.

What my analogy shows is that, if you describe events that actually occur from a particular perspective, you can make them look improbable. Thus, given a description of the steps in Doolittle's scenario for the evolution of TPA, the fact that you can make the probability look small doesn't mean that that isn't (or couldn't) have been the way things happened. One possibility is that the evolution of blood-clotting was genuinely improbable. But there are others.

Return to your experiment with the deck of cards. Let's suppose that all the hands you were dealt were pretty mundane—fairly evenly distributed among the suits, with a scattering of high cards in each. If you calculated the probability of receiving ten mundane hands in succession, it would of course be much higher than the priority of being dealt those very particular mundane hands with the cards arriving in just that sequence (although it wouldn't be as large as you might expect). There might be an analogue for blood-clotting, depending on how many candidates there are among the 3,000 "gene pieces" to which Behe alludes that would yield a protein product able to play the necessary role. Suppose that there are a hundred acceptable candidates for each position. That means that the chance of success on any particular draw is $(1/30)^4$, which is about 1 in 2.5 million. Now, if there were 10, 000 tries per year, it would take, on average, two or three centuries to arrive at the right combination, a flicker of an instant in evolutionary time.

Of course, neither Behe nor I knows how tolerant the blood-clotting system is, how many different molecular ways it allows to get the job done. Thus we can't say if the right way to look at the problem is to think of the situation as the analogue to being dealt a very particular sequence of cards in a very particular order, or whether the right comparison is with cases in which a more general type of sequence occurs. But these two suggestions don't exhaust the relevant cases.

Suppose you knew the exact order of cards in the deck prior to each deal. Then the probability that the particular sequence would occur would be extremely high (barring fumbling or sleight of hand, the probability would be 1). The sequence only *looks* improbable because we don't know the order. Perhaps that's true for the Doolittle shuffling process as well. Given the initial distribution of pieces of DNA, plus the details of the biochemical milieu, principles of chemical recombination might actually make it very probable that the cascade Doolittle hypothesizes would ensue. Once again, nobody knows whether this is so. Behe simply assumes that it isn't.

Let me sum up. There are two questions to pose: What is the probability that the Doolittle sequence would occur? What is the significance of a low value for that probability? The answer to the first question is that we haven't a clue: it might be close to 1, it might be small but significant enough to make it likely that the sequence would occur in a flicker of evolutionary time, or it might be truly tiny (as Behe suggests). The answer to the second question is that genuinely improbable things sometimes happen, and one shouldn't confuse improbability with impossibility. Once these points are recognized, it's clear that, for all its rhetorical force, Behe's appeal to numbers smacks more of numerology than of science. As with his main line of argument, it turns out to be an attempt to parlay ignorance of molecular details into an impossibility proof.

I postpone until the very end another fundamental difficulty with Behe's argument for design, to wit his fuzzy faith that appeal to a creator will make all these "difficulties" evaporate. As we shall see, both he and Johnson try to hide any positive views. With good reason.

## 5.   Johnson's Kangaroo Court

*Darwin on Trial* is a bravura performance by a formidable prosecutor, able to assemble nuggets of evidence and to present them in the most damning fashion. The defense lawyer isn't even court-appointed, and is simply absent or asleep. I'm going to argue that, when the defense actually shows up, Johnson's apparently devastating attacks turn out to be slick versions of old sophisms.

Unlike Behe, who officially admits the universal relatedness of organisms, Johnson takes some trouble to blur the distinction between the process of descent with modification (the "fact of evolution"), and the mechanism that drives the process. Here are some typical passages:

The arguments among the experts are said to be matters of detail, such as the precise timescale and mechanism of evolutionary transformations. These disagreements are signs not of crisis but of healthy creative ferment within the field, and in any case there is no room for doubt about something called the "fact" of evolution.

But consider Colin Patterson's point that a fact of evolution is vacuous unless it comes with a supporting theory. Absent an explanation of how fundamental transitions can occur, the bare statement that "humans evolved from fish" is not impressive. What makes the fish story impressive, and credible, is that scientists think they know how a fish can be changed into a human without miraculous intervention. (DOT 12)

We observe directly that apples fall when dropped, but we do not observe a common ancestor for modern apes and humans. What we *do* observe is that apes and humans are physically and biochemically more like each other than they are like rabbits, snakes, or trees. The ape-like common ancestor is a hypothesis in a *theory*, which purports to explain how these greater and lesser similarities came about. The theory is plausible, especially to a philosophical materialist, but it may nonetheless be false. The true explanation for natural relationships may be something much more mysterious. (DOT 67)

Paleontologists now report that a *Basilosaurus* skeleton recently discovered in Egypt has appendages which appear to be vestigial hind legs and feet. The function these could have served is obscure. They are too small even to have been much assistance in swimming, and could not conceivably have supported the huge body on land. (DOT 84)

Here, and in other places, Johnson confuses the question of whether the history of life shows a process of descent with modification with problems about evolutionary mechanisms, as well as cleverly raising the standards of evidence appropriate for calling something a "fact."

Contemporary evolutionary theorists, notably Stephen Jay Gould, have wanted to distinguish the "fact" of evolution (the universal relatedness of life, the process of descent with modification) from theories about the mechanisms of evolutionary change, precisely because creationists have exploited debate on the latter issue to cast doubt on the former. The distinction was already clear in the late nineteenth century, when the claim that organisms are related by descent with modification became virtually universally accepted, even though naturalists continued to debate Darwin's preferred account of the causes of evolutionary change. Johnson wants to turn the clock back. His first sally charges that facts are "vacuous" unless they come with supporting theories—and, of course, there's an appeal to authority thrown in. The word choice is interesting. Does Johnson think that the claim that organisms are related isn't true? Or that it's equivalent to some elementary logical truth (such as "All fish are fish")? The latter is completely implausible. Of course, Johnson would like to say that the claim of descent with modification is incorrect, but, since he can't defend that, he insinuates it by using a negative term.

In fact, scientific claims are often made without "supporting theories." Consider Kepler's laws about planetary orbits. Prior to the articulation of Newtonian theory, were these "vacuous"? Were chemists' proposals about chemical composition "vacuous" before we had detailed accounts of molecules and valences? Or Mendelian claims about hereditary factors in the absence of knowledge that genes are made of DNA? The point derived from Patterson seems straightforwardly false, since it often seems a scientific advance to establish *that* something is the case without being able to say why or how it is so. But Johnson cleverly buttresses his argument by misformulating the claim of common descent—instead of "Humans evolved from fishes," we should have "All living things in the history of life on earth are related through a process of descent with modification."

The next step is to offer an appraisal. Johnson opines that the doctrine he dislikes isn't "impressive." Again, it's not obvious that the ability to wow Johnson or his creo friends is the appropriate criterion—shouldn't we be concerned with whether or not the doctrine is *true*? But, of course, it's been made to *seem* less impressive because of the pathetically reduced formulation. Only as an afterthought does Johnson link the irrelevant "impressiveness" to the pertinent criterion, credibility, and then he garbles the relations of evidence. What makes claims about common descent credible is a wide variety of evidence drawn from comparative anatomy and physiology, comparative embryology, and biogeography—the kind of evidence that clinched the case in the post-Darwinian decades, and that has been extended ever since (most notably in recent biochemical studies)—not any embedding in a theory about the causes of evolutionary change. Precisely the point made by defenders of evolutionary theory like Gould is that we have overwhelming evidence for common descent even though we may debate the mechanisms of evolutionary change.

On to the second version. Here Johnson starts by pouncing on an analogy used by Gould, the comparison of relations of descent to the falling of apples. Of course, it's perfectly correct to point out that we don't *observe* all the intermediates hypothesized by the claim that all organisms are related through descent with modification. However, the fundamental point was to differentiate between parts of science that are so firmly accepted that they are classified as "facts" from parts of science that are more controversial. It's not obvious to me that the fact/theory terminology is the best way of marking this distinction—so Johnson may be justified in criticizing the rhetoric of his opponents. But that's just a preliminary point, and the main issue is whether the evidence for claims of descent is much stronger than that for causal explanations of the processes of modification.

Of course, there are plenty of parts of science that are not directly established by observation in the way that statements about falling apples are, but which nevertheless are counted as so firmly in place that scientists see themselves as building on them, rather than disputing them. Consider the claim that water molecules consist of two hydrogen atoms and one oxygen atom, or the identification of DNA as the molecular basis of heredity. Gould's characterization of common descent as a "fact" is meant to assimilate the thesis of universal relatedness to these scientific claims, to point out that its status is equally secure. If Johnson means to dispute this point, it's useless to note that one can't observe hypothetical intermediates—observation is just as inept to confirm molecular composition as it is to disclose ancestral organisms. What must be done is to show that the evidence in favor of common descent is far flimsier than evolutionary theorists have taken it to be, that it is nowhere near as strong as the support that has been garnered for the proposal that water is $H_2O$. To do that he has to explain why all those anatomical and physiological similarities, ranging from matters of gross morphology all the way down to molecular minutiae and including the apparently useless and nonfunctional residues of past structures, have been misinterpreted or overinterpreted by the defenders of evolution.

Johnson's half-hearted attempts to do just that are typified by the third passage I've quoted. As he notes, *Basilosaurus* is a sea-dwelling mammal related to whales, and it appears to retain rudimentary limbs. Evolutionary theorists account for the presence of these limbs by supposing that *Basilosaurus* is a modified descendant of some land-dwelling mammal, in whom the limbs were functional. The genetic changes that have taken place along the lineage have modified the body considerably, but the developmental program continues to produce vestigial versions of the structures present in the ancestors: they proclaim the animal's relatedness to land-dwelling forebears.

What neo-creos have to do at this point is explain that the vestiges don't signal any relationship to other mammals. So why are they there? What's the nonevolutionary explanation? Johnson doesn't tell us. Instead, he changes the subject, pointing out that the limbs aren't functional. But that wasn't the point at issue—indeed, the nonfunctionality was an indication that the limbs had been carried over from ancestral forms! Johnson has let the argument evolve from a dispute about descent with modification to a debate about the *causes* of evolutionary change, and he irrelevantly chides his opponents for not being able to tell a Darwinian selectionist story for these particular features of *Basilosaurus* and its immediate ancestors (DOT 85).

In the end, then, Johnson's attempt to dispute the "fact of evolution" is an exercise in evasion. When the rhetorical tricks are unmasked, it's clear that he's failed to answer the big question: if organisms aren't related by descent with modification, what's the explanation for all the detailed similarities we find among living things? Yes, indeed, the true explanation for observed relationships might be "more mysterious"—as might the true explanation for the data from which chemists justify their views about the composition of water or of genes—but the mysteries are, apparently, to remain the strict property of Johnson and his cronies.

As a lawyer, Johnson has an excellent understanding of ways in which burdens of proof can be shifted, and standards of evidence raised. Here are some samples of his skill:

The question I want to investigate is whether Darwinism is based upon a fair assessment of the scientific evidence, or whether it is another kind of fundamentalism.

Do we really know for certain that there exists some natural process by which human beings and all other living beings could have evolved from microbial ancestors, and eventually from non-living matter? (DOT 14)

*Archaeopteryx* is on the whole a point for the Darwinists, but how important is it? Persons who come to the fossil evidence as convinced Darwinists will see a stunning confirmation, but skeptics will see only a lonely exception to a consistent pattern of fossil disconfirmation. If we are testing Darwinism rather than merely looking for a confirming example or two, then a single good candidate for ancestor status is not enough to save a theory that posits a worldwide history of continual evolutionary transformation. (DOT 79)

The first passage frames the issues so as to impose unnecessarily stringent requirements on defenders of evolutionary theory. We start with two options: either the acceptance of evolutionary theory rests on "a fair assessment of the evidence" or it's a "kind of fundamentalism." Strictly speaking, that doesn't exhaust the possibilities, but let that pass. In the very next sentence Johnson transmutes the first option, reformulating it as the requirement that we know *for certain* that some natural process produced people out of microbes. Now, of course, this is focused directly on the issue of the mechanism of evolutionary change, and explicitly demands knowledge of the mechanisms that have operated over the entire sweep of evolutionary history, but the most glaring distortion occurs in the talk of certainty. In effect, the choices have been reduced to either knowing all the details with certainty or being a fundamentalist, so that no space is left for the thoughtful evolutionary theorist who wants to say "The evidence for the universal relatedness of life is compelling. Further, we know of a number of natural processes that have produced evolutionary change. We can't always say for sure which of these has been operative at which stage of the history of life, nor do we know that our inventory of possible mechanisms is complete, but, on the evidence we have, there's no reason to think that any supernatural process was needed in the evolution of organisms." That type of response is analogous to that of the chemist who declares "The evidence for our views about the kinds of bonds that occur between molecules in a vast number of substances is compelling. Further, we know in principle how the distributions of electrons in bonds result from basic principles of quantum mechanics. But we don't know how to solve the Schrödinger equation for any complex molecule, and it may be that our understanding of the microphysics is limited in various respects. Given the evidence we have, however, there's no reason to think that supernatural processes are needed to keep the constituents of large molecules together." In chemistry, as in evolutionary biology, there are open problems, and, while some parts of the science are quite firmly established (on the basis of compelling evidence), the idea that we should claim certainty überhaupt is as absurd as the thought that, if we can't do so, we've relapsed into fundamentalism.

The second passage occurs in the middle of a discussion of the fossil record (a discussion I'll treat from a different perspective shortly). After clouding the issues about the reptile-bird transition—mainly by claiming that evolutionists ought to produce fine-grained transitional sequences linking ancestral organisms to all the different species of birds—Johnson concedes, grudgingly, that the existence of *Archaeopteryx* is "on the whole a point for the Darwinists." Indeed, since the explicit challenge was to find transitional forms linking major groups, it's hard to see how the production of an intermediate, such as *Archaeopteryx*, could fail to meet the challenge: Johnson's strategy is like that of the child who bets his friend that she can't juggle three balls for a minute and then, when she does it successfully, welshes on the bet on the grounds that she didn't do it with her eyes shut. But, after magnanimously conceding that *Archaeopteryx* is confirming evidence for the view that reptiles and birds are related by descent, he pooh-poohs the significance of this by suggesting that it's a "lonely exception to a consistent pattern of fossil disconfirmation." Let's formulate Johnson's implicit requirement explicitly: it's the demand that the fossil record would confirm evolutionary theory only if we could discover intermediate forms for every major transition (with Johnson reserving the right to decide which transitions count as "major" and also to demand the fineness of grain of the intermediate sequences). This is as arrogant as a counterdemand to be shown the fingerprint of the Creator in specified domains of the living world. As Darwin well knew, and as our improved understanding of the physics and chemistry of fossilization has shown us ever more clearly, the chances that any given species will be represented in the fossil record is extremely low. Our estimates of those chances are not, as Johnson likes to insinuate, specially cooked to favor evolutionary theory; they are based on independent parts of science. Given those estimates, we'd expect that for many major transitions the hypothesized intermediates would not be found in the fossil record, but, when the transitional fossils do occur, they provide striking confirmation of the claim of descent with modification because, if that claim were not true, the existence of such fossils would be highly improbable.

To see this more clearly, consider an analogy. In building the case against the notorious Moriarty, specifically in order to justify the conclusion that Moriarty visited the scene of the crime, the prosecution appeals to the fact that he was observed, just before the crime was committed, halfway between his lair and the crime scene. The defense responds that there has been no evidence of Moriarty's footprints on the pavement throughout the hundred-yard walk, that nobody saw Moriarty within ten yards of the crime scene, and so forth. The defense lawyer is a studious disciple of Phillip Johnson.

In fact, Johnson is sufficiently uneasy about the fossil evidence to go to considerable lengths to respond to examples on which evolutionary theorists (rightly) place special emphasis. He cites the reptile-mammal transition as the "crown jewel of the fossil evidence for Darwinism" (DOT 75). He continues with one of his most accurate condensations of the biology:

At the boundary, fossil reptiles and mammals are difficult to tell apart. The usual criterion is that a fossil is considered reptile if its jaw contains several bones, of which one, the articular, connects to the quadrate bone of the skull. If the lower jaw consists of a single dentary bone, connecting to the squamosal bone of the skull, the fossil is classified as a mammal. (DOT 75)

It might initially appear very difficult for an animal to be "intermediate" between reptiles and mammals, given this criterion in terms of jaw morphology. However, there's a very rich set of fossils showing reduction of the reptilian features and development of the mammalian traits; particularly remarkable are fossils, most famously *Diarthrognathus*, in which both types of jaw-joint are present.

After quoting Stephen Jay Gould's description of the advanced mammal-like reptiles, distinguished by the reduction in the quadrate and articular, Johnson comments:

We may concede Gould's narrow point, but his more general claim that the mammal-reptile transition is thereby established is another matter. Creatures have existed with a skull bone structure intermediate between that of reptiles and mammals, and so the transition with respect to this feature is possible. On the other hand, there are many important features by which mammals differ from reptiles besides the jaw and ear bones, including the all-important reproductive systems. (DOT 76)

Well, when you can't argue the facts, argue the law. The existence of *Diarthrognathus* and friends shows that transitional forms with respect to jaw morphology *actually appeared* (not just [sniff!] that they were "possible"). So Johnson has to contend that these are irrelevant to the case. He reaches into the creationist bag of debating tricks for a well-known tactic, that of specifying just the intermediate forms that would satisfy him—he wants to see transitions in the "reproductive system." Clever indeed! For what he wants are the soft bits, the parts that don't have a prayer of being represented in the fossil record.

Now in fact, as Johnson ought to know, there isn't a single mammalian reproductive system—there are monotremes (egg-laying mammals), marsupials, and, most familiar, the placental mammals. So what is actually happening is that the question is being shifted. Instead of asking for an account of the reptile-mammal transition, Johnson is making a much more sweeping demand for a fine-grained sequence of transitional forms *within* Mammalia to show the gradual emergence of the placental mammals. What Gould actually claimed to be able to do was to show how a feature shared by all mammals (monotremes and marsupials as well as placentals), the structure of the jaw joint, emerges in the fossil record in a fine-grained transition from the structure found in living and extinct reptiles. However much Johnson might like to invoke the character of the reproductive system as a way of separating the mammals from the reptiles, the criterion of jaw morphology is taxonomically fundamental: mammals are animals that have one jaw structure, reptiles are animals that have a different jaw structure, and the mammal-like reptiles are those in which the bones involved in the reptilian jaw are being reduced, the most advanced of them being double-jointed. These criteria aren't pressed into service to save evolutionary theory. They are demanded by the diversity of the mammals with respect to other features. Johnson's revisionary taxonomy would sweep away some of the Antipodean mammals.

There are signs that Johnson recognizes that all is not well with his first line of argument, for he follows it up with an alternative. After noting that the fossil record for the reptile-mammal transition is so rich that a prominent evolutionary biologist (Douglas Futuyma) suggests that it's impossible to tell which species were the ancestors of modern mammals, Johnson continues:

> But large numbers of eligible candidates are a plus only to the extent that they can be placed in a single line of descent that could conceivably lead from a particular reptile species to a particular early mammal species. The presence of similiarities in many different species that are outside of any possible ancestral line only draws attention to the fact that skeletal similarities do not necessarily imply ancestry. The notion that mammals-in-general evolved from reptiles-in-general through a broad clump of diverse therapsid lines is not Darwinism. Darwinian transformation requires a single line of ancestral descent. (DOT 76)

The claim of common descent is, apparently, to be defeated by an *embarras de richesse*.

Plainly, Johnson hasn't been reading contemporary evolutionary theory very carefully, for he seems to have overlooked the modern emphasis on a theme (already present in Darwin) that the tree of life turns out to be a bush. Well-documented cases of anagenesis (in which a single lineage is gradually transformed) are quite rare. Paleontological reconstructions typically show modifications associated with (branching) speciation. Hence, in studying the mammal-like reptiles, there's no surprise in finding lots of closely related species, not just parents and daughters but sisters and cousins and aunts. Futuyma's point is that there are so many relatives in this family that it's hard to sort out the relationships, and, in particular, hard to tell which mammal-like reptiles are ancestral to the mammals.

So Johnson is quite wrong in thinking that there has to be a linear sequence linking all these fossils by ancestor-descendant relations: evolutionists would be quite surprised if that were so. But the rhetoric of his case depends on a skillful ambiguity. He insinuates doubts about whether jaw morphology is a reliable guide to relationship by talking of "species outside of any possible ancestral line." The suggestion, of course, is that evolutionists are committed to thinking of some of these species as *unrelated*, and that this undermines their claims that anatomical features (like the size and positioning of bones) are a good indicator of evolutionary relationships. But that's completely false. Those who study this transition don't believe that the fossils can be fitted into a single line of ancestors and descendants, but they do think of all of them as related. To repeat, there are daughters and sisters and cousins and aunts. The difficulty lies in assigning particular fossils particular degrees of relationship. But that difficulty doesn't interfere with the enterprise of revealing the reptile-mammal transition in the fossil record. Once again, a legal analogy may prove helpful. If the defense denies that any member of the Crebozo gang could have done the dirty deed, and the prosecution shows how Phil, Al, Pete, and Mike (Crebozos all) had motive, means, and access, the general claim that one of the Crebozos is guilty may be established without the prosecution's being able to tell which of the individual thugs delivered the decisive blow.

Johnson's entire book is filled with the sophistries I've been exposing, as he distorts the positions he opposes, shifts standards of evidence, quotes people out of context, and uses ambiguity to cover his argumentative gaps. Any well-trained philosopher with no particular axe to grind and a modest knowledge about evolutionary biology could hardly fail to see that rhetoric substitutes for sound argument on virtually every page—which is why the endorsements of Johnson by Plantinga and van Inwagen are so revealing. I'll close this part of the discussion by looking at one example that ought to strike philosophers as especially egregious, Johnson's invocation of Popper and the famous (infamous?) falsifiability criterion.

Creos old and new love Sir Karl. The country-bumpkin appeal to the falsifiability criterion proposed that the theory of evolution reduced to some principle of natural selection and that this principle turned out to be a tautology ("Those who survive, survive"—to be sung to the tune of *Che sera, sera*). That line of objection has been decisively refuted, both by pointing out that it misunderstands the principle of natural selection, and, far more importantly in my opinion, by showing how absurd it is to reduce the theory of evolution to the principle of natural selection. Johnson tries a different tack. He starts from the idea that the theory of evolution is akin to Marx's account of history or Freud's psychoanalysis in making no genuine predictions at all.

There are well-known philosophical objections to so simple-minded an invocation of the falsifiability criterion. Both Freud and Marx offered sufficiently rich bodies of doctrine that they could make predictions about individual events. When their predictions went wrong, they modified their *total* doctrines to preserve their central principles. Johnson points this out, and immediately cries "Foul!" (DOT 146). He fails to appreciate the fact that almost all scientists spend large portions of their time behaving in similar ways: they set up an experiment in light of their theoretical understanding, find that it fails to work, tinker a bit, revise their views about the situation, and, in some instances, make modifications of parts of the underlying theory to protect the central principles. To cite a familiar example, when the orbit of Uranus wasn't as predicted, astronomers didn't abandon Newton's gravitational theory (in favor of what? one wonders), but hunted for a new planet. Popper himself understood the nuances of testing far better than those whom he has inspired, and, in the wake of appeals to falsifiability, philosophers of science have shown how the refutation of a theory with broad scope proceeds, not by single decisive experiments but by the accumulation of cases that challenge defenders to find any way of supplementing the central principles.

The history of science is full of theories that came to grief because of the building up of difficulties. Darwin's theory of evolution isn't among the shipwrecks. Johnson charges that this is because "[t]he central Darwinist concept that later came to be called the 'fact of evolution'—descent with modification—was thus from the start protected from empirical testing" (DOT 149). His false allegation rests on a confusion: the fact that evolutionary theory has survived doesn't entail that it was *bound* to succeed. If we had discovered the world to be different in different respects then evolutionary theory would have been given up because it faced insuperable difficulties.

Suppose we had found that the similarities in structure on which claims of evolutionary relationship are based didn't correlate with observed patterns of descent: we claim that two organisms are descended from a common ancestor because they share some feature and then discover that the feature is present in organisms we know to be unrelated; this happens again and again. Imagine further that, as we move from level to level, the groupings by similarity vary: anatomical and physiological similarities are underlain by very different tissue and cellular types; cellular similarities don't correlate with chromosomal differences; all of the groupings at higher levels are quite different from those we find at the molecular level—humans, for example, turn out to be biochemically very similar to frogs and palm trees and radically different from chimpanzees. Claims about biogeographical distribution consistently founder on the inability of organisms to make the journeys that are hypothesized (recall Darwin's concern with this problem, which drove him to do experiments on the transport of seeds, thus confuting Johnson's claim that he proposed "no daring experimental tests" [DOT 149]). The fossil record turns out to be chaotic, with strata that are geologically ancient containing samples of all classes of organisms. Fundamental theories in physics imply that the earth is quite young. And so on and on.

My construction of this list is not particularly imaginative, for I've drawn on a large number of the ways in which creos, including Phillip Johnson, have tried to falsify evolutionary theory. (The example about similarities between frog proteins and human proteins was once a staple of old-time creationist entertainments; although it's been decisively rebutted, there may be a faint trace in Johnson's remarks about molecular variation in frogs [DOT 90].) The point is that Darwin's evolutionary theory could have gone the way of phlogiston chemistry, the corpuscular theory of light, blending inheritance, the universal ether, stabilist theories of the continents, and many other discarded theories. It didn't, not because evolutionary theorists are stubborn ideologues but because the kinds of observations that would have discredited it (occasionally, but wrongly, hailed as "facts" in the creationist literature) have not been made. Far from being "vacuous" or "unfalsifiable," evolutionary theory sticks its neck out again and again, denying the co-presence of human and dinosaur footprints at Paluxy, predicting the morphology of ancestral ants (subsequently discovered by E. O. Wilson, W. L. Brown, and F. M. Carpenter), ruling out the possibility that the chicken genome is more similar to the human genome than the latter is to the chimpanzee genome, and in a host of further commitments.

Johnson's fondness for claiming that evolutionary theory doesn't belong in the Popperian Temple of Proper Science actually rests on his repeated contention that the theory is insulated against one particular type of problem, the absence of transitional forms in the fossil record. Here's a typical version of the charge.

If Darwin had made risky predictions about what the fossil record would show after a century of exploration, he would not have predicted that a single "ancestral group" like the therapsids and a mosaic like *Archaeopteryx* would be practically the only evidence for macroevolution. Because Darwinists look only for confirmation, however, these exceptions look to them like proof. (DOT 153)

As I've already maintained, evolutionary theory does make a large number of predictions about the character of the fossil record, notably that fossils will occur in a particular order in strata worldwide. But what Johnson wants is a particular kind of prediction. If all these intermediates actually existed, he claims, we ought to find traces of them, and it's anti-Popperian dogmatic weaseling to be complacent when those traces aren't found. But this is simply to ignore an independently confirmed chunk of science, the geophysics and geochemistry of fossilization. If you couple evolutionary theory with an extremely bad theory, say, the Panglossian Theory of fossilization, whose claim is that every organism that has ever existed has a chance greater than 1/100 of being fossilized, then, of course, evolutionary theory predicts that the fossil record should contain myriad intermediates that we don't find. But the fact that a theory conjoined with a very bad auxiliary theory predicts lots of falsehoods cuts no ice whatsoever. What Johnson ought to do is to show that evolutionary theory plus a realistic theory of fossilization predicts the existence of far more intermediate forms than we've yet found.

To the best of my knowledge, nobody has done that. In principle, one could use our understanding of the vicissitudes of fossilization to compute the chances of finding remains of various types, calculate estimates of population sizes, and then use our best account of phylogenies to arrive at probabilities that the fossil record will reveal a particular sort of distribution. Whether or not that can be done in practice is unclear, for the information currently available might be too limited to narrow the range of probabilities. But if this is the sort of prediction that Johnson demands (ignoring the other kinds I've mentioned), then he could try to work out the details in a serious and responsible fashion. Of course, it's a lot quicker—and probably more to his taste—to invoke the Panglossian Theory and damn evolutionary theory unfairly.

I'll illustrate his tactic with another analogy. Consider the theory that the texts we print in volumes with "Holy Bible" on the spine have been produced by a historical process in which Ancient manuscripts in various languages have been copied and recopied, translated and retranslated. Imagine a fringe sect that challenges this theory, holding that alleged similarities among texts are based on misunderstandings of the ancient languages, that the works we have are fictions composed in the late Middle Ages, and that there were no intermediates linking them back to the Ancient world. Phillip Johnson, quite rightly, is moved to answer this criticism, but, in the ensuing debate, he finds himself facing the following objection. "Where are all those intermediate texts, Professor Johnson? Surely your theory predicts that the historical record should be full of copies showing the ways in which Ancient versions have been successively amended. We know that it isn't. Are you going to desert your Popperian principles and ignore this falsification?" We can easily guess how Johnson would reply. He'd point out that libraries are often looted and burned, that texts are lost and thrown away, that vandalism was omnipresent in the Ancient world, and so forth. He'd accuse his opponents of foisting on him an absurd auxiliary theory about historical preservation. And he'd be right. Yet the sophistries of his challengers are exactly his own.

## 6. Where's the Beef?

I come at last to the most basic difficulty with the neo-creo attack, its dim suggestions that the scientific world needs a shot of supernaturalism. The born-again creationists tread different paths to a common destination. Whether hedgehogs or foxes, they conclude that evolutionary theory is beset by problems—one very deep and systematic problem for Michael Behe, a whole scatter of troubles for Phillip Johnson—and they portray the establishment as dogmatic in its insistence on excluding creative design: given that the going story of life and its history is such a shambles, why are these evolutionists so obstinate in thinking that some "purely naturalistic process" produced people? When this conclusion is made explicit, there's a natural question to pose to the neo-creos. How exactly is the appeal to creative design supposed to help?

I've been contending throughout that the charges of "insoluble problems" are wildly overblown. But let's play along for a bit. Consider the difficulties that Behe and Johnson cite, and suppose that they really do need to be addressed. Why should we think that invoking creative design, with all its theological resonances, is just the ticket for solving them?

Behe and Johnson don't say. They've learned from the failures of pre-Reformation creationism, and they know much, much better than to put their literalist cards out on the table. Fine. But we ought to be a little curious about what sort of magic a creative design model might be able to work.

Let's start with Behe, and concede to him that we haven't a clue about how you can produce the bacterial flagellum or the clotting cascade in small steps. We might think we'd get some clues once developmental molecular genetics has developed a bit, but maybe Behe has a plausible proposal that will save us the wait and the trouble. What could it be? Well, it has to involve creative design, so we can assume that the unbridgeable gaps between the bacteria sans flagella and their fully equipped successors are transcended through the activities of some Creator or "creative force." Continuing to be generous, let's give Behe the personalized version.

So what does the Creator do? Option 1: He (we'll throw in patriarchy as well) arranges the selection regime for the hapless intermediates, directs the mutations, and so forth; so, in accord with a doctrine Behe has "no particular reason to doubt," organisms are linked by descent, and the Creator's work is devoted to making sure that just the right mutations arise in the right order and that the organisms on the way to the complex final state are protected against the consequences of having lots of useless spare parts that will be assembled at some final stage. Option 2: the Creator dispenses with a lot of the intermediate steps by cunningly arranging for lots of mutations to happen at once; if 183 new proteins are needed for the new structure, then zap! He strikes the appropriate loci with his magical mutating finger; or maybe he does it in two goes of 92 and 91 (with a protective environmental regime for the halfway stages); or in three interventions of 61 mutations a trick.... Here, again, organisms are related by descent with modification, although the "descent" and the "modifications" are a bit abnormal. Option 3: the Creator gives up on mutation and selection entirely, simply creating a bunch of organisms with the right molecular stuff *de novo*; of course, if Behe thinks that this is the way things worked, then he really does have doubts about descent with modification.

The first point to note is that there's absolutely no evidence in favor of any of these options—they are the kinds of things to which one would be driven only if one thought that Behe's Big Problem was so intractable that there was no alternative. But matters are actually much worse than that, as one can see by posing questions about the Creator's psychology. Why should anyone think that the kind of Creator for whom Behe and Johnson both want to make room would undertake any of these projects? In Option 1, we envisage a Creator with the power to direct mutations and contrive protective environments who prefers simulating natural selection with gerrymandered selection pressures to directing all the needed mutations at once. In Option 2, we envisage a Creator who has the power to create organisms, but who prefers to simulate descent by the magic of mass mutation rather than simply producing the kinds of organisms He wants (either successively or simultaneously). In Option 3, we envisage a Creator who creates all the kinds of organisms He wants, as He wants them, but equips them with the genomic junk found in organisms He's created earlier. I am no engineer, but these visions inspire me to echo Alfonso X on the complexities of the Ptolemaic account of the solar system—had the Creator consulted me at the Creation, I think I could have given him useful advice.

Perhaps I am being unfair. Maybe the project of design looks ludicrous because I have selected the wrong options for Creative intervention. Behe could easily answer my concerns by coming up with an alternative, one that would explain how creative design has figured in the history of life on our planet and how that creative design is part of a project worthy of his favorite Creator. I'm inclined to think that he won't do that, that the silence in neo-creo positive proposals will continue to be deafening. After all, positive doctrines and explanations have always been creationism's Achilles Heel.

Notice that the line of argument in which I'm now engaged isn't a defense of evolutionary theory. For the sake of argument, I've conceded that evolutionary theory faces deep and intractable problems, although I've spent most of my time arguing that that's totally false. To show that the problems alleged to face evolutionary theory can't be solved by appealing to creative design isn't to rehabilitate the theory, for one doesn't always have to adopt the better of two alternatives. But in demonstrating that evolutionary theory is clearly superior to the imaginable members of the creationist family I ought to sap the motivation of those who are drawn to creationism. Attacking evolutionary theory was supposed to make room for God, but, as we've seen, there's not much hope for an active role for the Deity in any successor to evolutionary theory.

Although it's hard to see just which of my three options Behe would choose, his position is less indefinite than Johnson's. What Johnson thinks actually happened in the history of life is deeply obscure. All he tells us is that the hallmark of creationism is the idea that "a supernatural Creator not only initiated the process but in some meaningful sense *controls* it in furtherance of a purpose" (DOT 4 n. 1). Controls it how and when? With what purpose? Johnson doesn't say. For one so enthusiastic about canonizing Sir Karl, Johnson's "creation model" is rather short on "risky predictions."

Suppose we were to concede Johnson's claims about the difficulties for evolutionary theory. It would be natural to expect, as his book puffs to its conclusion, that he would say something about how those difficulties vanish once one invokes the activities of the designing Creator. Consider those puzzles about the fossil record. What exactly do they indicate? Does Johnson believe that organisms whose fossilized remains are lower in the geological column were around long before those higher up? Which organisms does he think are related to which? And, if he denies the descent of major groups from the organisms evolutionary theorists identify as their ancestors, how does he think the later organisms were formed?

Here are the possibilities. Option 1: Johnson might claim that the fossil record is profoundly misleading, that there's been no succession of organisms; this would be to take over part of old-style creationism, claiming that the major kinds of organisms were all formed at once, and have inhabited the earth since the creation (except, of course, for those that have gone extinct); it would, however, remain uncommitted to whether or not the earth is old or young. Option 2: Johnson might claim that the fossil record really does show a sequence of organisms, with some appearing later in earth's history than others, claiming that the major kinds are created as they appear, and are not modified descendants of earlier organisms. Option 3: Johnson might propose that the history of life is one of descent with modification, but that the Creator has guided the processes (perhaps in the ways signaled in Behe's first two options, considered previously).

Now option 1 is highly problematic because it offers no account of the worldwide ordering of the fossils in geological strata and no account of the anatomical, physiological, developmental, and molecular affinities among organisms. The first of these is the familiar difficulty that led people early in the nineteenth century—including extremely devout naturalists—to abandon the idea of a single fixed creation in favor of a sequence of creations. Option 2 founders on the need to understand why later organisms take over features at all levels from earlier organisms, features that are often no longer functional. Are we to assume that the junk in our genomes and the vestigial bits and pieces of anatomy are just signs of the Creator's whimsy? Option 3 is of course an evolutionary account of life, one widely adopted in the later decades of the nineteenth century by theists who thought that there had to be a supernatural component to mechanisms of evolutionary change. It would require Johnson, like Behe, to explain just what it is that the Creator does, and why he does things that way. All three schematic creation models face large and familiar problems, which is why all the detailed versions of all of them have been abandoned by thinkers whose knowledge and intellectual integrity greatly exceed Johnson's.

But wait! Maybe Johnson has some gleaming new version that will put the general worries to rest? Alas, any reader who expected that would be disappointed. Toward the end of his book, he confesses

I am not interested in any claims that are based upon a literal reading of the Bible, nor do I understand the concept of creation as narrowly as Duane Gish does. If an omnipotent Creator exists He might have created things instantaneously in a single week or through gradual evolution over billions of years. He might have employed means wholly inaccessible to science, or mechanisms that are at least in part understandable through scientific investigation.

The essential point of creation has nothing to do with the timing or the mechanism the Creator chose to employ, but with the element of design or purpose. In the broadest sense, a "creationist" is simply a person who believes that the world (and especially mankind) was *designed*, and exists for a *purpose*. With the issue defined that way, the question becomes: Is mainstream science opposed to the possibility that the natural world was designed by a Creator for a purpose? If so, on what basis? (DOT 113)

Not only is no creation model offered here, but the definition of "creationism" is modified to make it compatible with orthodox Darwinism!

Johnson's original formulation of the position required that the Creator not merely set things up and let 'em roll, but that He actively intervene in the history of life. In this later passage, the commitment to continued intervention has been abandoned. Some Darwinians would be prepared to allow that the Creator fixed the initial conditions for the universe, although they'd contend that everything that has occurred since can be understood as the outcome of natural processes. For those theistically inclined, a view like this has often seemed superior to one on which the universe requires continual janitorial work—Leibniz chided Newton for hypothesizing that the Creator might have to tinker with His handiwork. Of course, there's a residual cluster of worries centering around the motivations of an omnipotent Creator for proceeding in so indirect a fashion.

Johnson's official line is that we ought not second-guess the Almighty. Confronting concerns about the apparent policy of letting later organisms inherit the junk of their ancestors, he chides evolutionary theorists for "speculating" about what a "proper Creator" would do. But, if the creation model is to be taken seriously as an account of life and its history, the character of processes and products must be full of clues to the attitudes of the Creator, and, on the basis of our observations, it's clear that the motivations of a Creator who let the evolutionary process unfold in the ways that it has in order to produce our own species are quite baffling. The more intimately the Creator becomes involved in the adjustments of the process, the greater the bafflement.

So Johnson leaves everything vague, hoping that nobody will notice that he's either committing himself to an extremely implausible hands-on Creator with purposes for which his means seem singularly ill designed or a slightly more credible hands-off Creator who produced the current world in just the ways cosmologists and Darwinian evolutionary theorists suggest. But his final question reveals his blindness to the historical fate of his options. Unless he does come forward with a new proposal for understanding the role of the Creator in the history of life, we're entitled to suppose that the only ways of articulating a creationist view are those that have been tried from the late eighteenth century to the present. Those are just the options I've canvassed, successively explored by ingenious, pious, but honest thinkers who rejected them because they were at odds with the record of historical and contemporary life. Coffin nails are driven deeper with the advance of fossil discoveries, the dissection of molecular relationships, our increased understanding of biogeography, and all the rest. In the end, the only answer one can give to Johnson's question—presumably intended as rhetorical—is that the best mainstream science can allow him is a Creator who set things up and let them unfold by natural processes. Whether this does more than pay lip service to the yearning for purpose and design is a matter I leave to theists.

The neo-creo model factory is strikingly out of new resources. For all the fancy rhetoric, all the academic respectability, all the accusations and gesticulations, born-again creationism is just what its country cousin was. A sham.

## Notes

I am extremely grateful to Dan Dennett and Ed Curley for sharing with me their unpublished discussions of the creationist writers I discuss here. I have also learned much from an illuminating essay by Niall Shanks and Karl Joplin, "Redundant Complexity: A Critical Analysis of Intelligent Design in Biochemistry," *Philosophy of Science*. 66, 1999, 268–282. Finally, I'd like to thank Robert Pennock for his editorial encouragement and for the insights of his own excellent treatment of the neo-creos in *Tower of Babel*.

1. I'll be quoting extensively from two creationist works, Michael Behe, *Darwin's Black Box: The Biochemical Challenge to Evolution*, New York: The Free Press, 1996 (cited as DBB) and Phillip Johnson, *Darwin on Trial*, Washington D.C.: Regnery Gateway, 1993 (cited as DOT).

2. I borrow this pithy formulation from Dan Dennett.

3. For further discussion of this issue, see my *Abusing Science: The Case against Creationism*, Cambridge MA.: MIT Press, p. 117.

# 12

# Biology Remystified: The Scientific Claims of the New Creationists

Matthew J. Brauer and Daniel R. Brumbaugh

## 1. Introduction

In an era of widespread concern about educational standards in public schools and international rankings in science and math (National Science Board 1998a,b), science and the science curriculum remain under continual attack nationwide by creationists (Greenwood and Kovacs North 1999; Holden 1999).[1] Biology, particularly the study of evolution, has taken the brunt of these assaults, but fundamental ideas in geology, physics, and astronomy are increasingly being challenged as well. In this paper, we discuss writings from three representatives of the neo-creationist "intelligent design" movement. We evaluate their critique of Darwinism and some of their ideas about science in general—setting both against the synthetic history and norms of science.

## 2. The Evolutionary Antisynthesis

During the period from the 1930s through the 1950s, biologists realized that the disparate fields of zoology, botany, embryology, systematics, genetics, and paleontology could be built on the theory of evolution. This "modern synthesis," accelerated by the cross-fertilization of thought among biological disciplines, served to (1) ground much of organismal biology in the formal rigor of evolutionary genetics and (2) shift it away from traditional typological concepts to an emphasis on variation within and between populations (Mayr and Provine 1980). As a result, discipline-specific theories of inheritance and evolution faded in importance, and persistent teleological notions of purpose- or progress-driven evolution were rejected (Simpson 1944; Mayr and Provine 1980; Mayr 1988).

Fifty years later there is a self-conscious and concerted attempt to reincorporate purpose into evolutionary thought (Dembski 1998). This neo-creationist movement (usually labeled by its members as "intelligent design" or "design theory") is not motivated by new scientific discoveries. Nor is it the result of a new synthesis of existing theories. Rather, the selling of design theory is motivated entirely by the religion and politics of a small group of academics who seek to defeat secular "modernist naturalism" by updating previously discredited creationist approaches. Although their conclusions derive significantly from their Judeo-Christian theology, their political goals require that they deny this. The proponents of intelligent design insist on the *scientific* validity of their arguments.[2]

Intelligent design, accepted at face value, is the opposite of the evolutionary synthesis of biological thought that started early in the 1930s. It is an "antisynthesis," whose goal is the devaluation and fracturing of the unified, methodologically naturalistic study of evolution. Because this effort is extremely unlikely to succeed among mainstream scientists familiar with the evidence, this movement employs a small number of dissenting scientists to target other, mostly nonscientific academic disciplines (e.g., philosophy and law) and certain portions of the public. These advocates assert that the neo-creationist design theory is a valid alternative to modern evolutionary biology because:

1. Evidence supporting the origin of new organic forms by means of natural selection and other naturalistic causes ("Darwinian evolution") is sparse, ambiguous, and contradictory;

2. Evidence supporting intelligent design as the primary source of organic diversity exists, and would be allowable in a more ecumenical "theistic science"; and

3. Portions of the "scientific establishment" are conflicted over the relative importance of these two classes of evidence. Only by methodologically disallowing nonnaturalistic evidence and explanation does Darwinism prevail over theistic science. Consequently, the scientific establishment represents a sort of "scientific priesthood," jealously guarding its power in a conspiracy to suppress evidence of intelligent design in favor of Darwinian evolution.

Neo-creationists use language resembling that used by mainstream scientists or philosophers of science. They liberally sprinkle jargon from such diverse fields as molecular genetics, chaos theory, and information theory throughout their texts, and quote from the philosophical and evolutionary biology literature in support of their arguments. They enthusiastically invoke Popper and Kuhn, as if to demonstrate the scientific inevitability of a creationist program. In response to these efforts, this essay examines the validity of the *scientific* claims of three popular design theory texts: Phillip Johnson's (1991) *Darwin on Trial*, Michael Denton's (1985) *Evolution: A Theory in Crisis*, and Michael Behe's (1996) *Darwin's Black Box: The Biochemical Challenge to Evolution*. We take this approach even though these texts would generally not be considered part of the scientific literature, even by their authors, although they do explicitly claim some measure of logical or scientific authority. Since there are no relevant scientific articles by them in the peer-reviewed literature, and since these writings are among the most popular and significant neo-creationist critiques of modern evolutionary biology, their scientific claims deserve scrutiny. As Johnson says, his "purpose is to examine the scientific evidence on its own terms, being careful to distinguish the evidence itself from any religious or philosophical bias that might distort our interpretation of that evidence" (Johnson 1991, p. 14). This essay will evaluate whether he, Denton, and Behe have achieved this goal. We will show that these influential creationist writers consistently misrepresent the scope, scale, and meaning of the scientific evidence they criticize. They also consistently overestimate the amount and the meaning of disagreement among evolutionary biologists.

## 3.   The Science Critic

As one of the most prominent proponents of anti-evolutionary thought, Johnson has had an enormous influence on the framing of recent neo-creationist rhetoric (Johnson 1991). He frequently assumes the populist mantle of William Jennings Bryan as he passionately warns of the social costs of unchecked "Darwinist Religion" and "Darwinist Education" (Johnson 1991, chaps. 10 and 11). Moreover, like the Great Commoner, he contends that scientific interpretation should not be left to the scientific elite: rather, anyone with critical faculties should be able to follow and reinterpret scientific evidence:

I am not a scientist but an academic lawyer by profession, with a specialty in analyzing the logic of arguments and identifying the assumptions that lie behind those arguments. This background is more appropriate than one might think, because what people believe about evolution and Darwinism depends very heavily on the kind of logic they employ and the kind of assumptions they make. Being a scientist is not necessarily an advantage when dealing with a very broad topic which cuts across many scientific disciplines and also involves issues of philosophy. Practicing scientists are of necessity highly specialized, and a scientist outside his field of expertise is just another layman. (Johnson 1991, p. 13)

Having thus justified his own credentials and devalued those of scientists, Johnson develops his cross-examinations of "evolutionist" witnesses in subsequent chapters. In doing so, he also presents his interpretation of the science of evolutionary biology.

And what is the main scientific point that Johnson advocates? It is simply that the genetic, populational, and species-level phenomena that we observe and experiment with have little connection to the patterns of diversity we see in the living world. These patterns, according to Johnson, could be better explained with a loosening of methodological constraints to include theistic purpose in science. It is thus an attack not only on the observations that biologists use as evidence, but also on the possibility of a logical, naturalistic connection between these data and the radiation of biological diversity.

### Categories of Evolutionary Evidence

We focus on two major categories of evidence that lead biologists to the conclusion that evolution by *descent with modification* is a good model of the real world. One of these derives from observations and experiments relating to the processes that drive changes within and between populations ("microevolution"). These experiments address such questions as:

1. How do genes get transmitted from generation to generation?
2. How do different types of changes (e.g., mutations and recombination) occur in genetic material, how frequent are these changes, and what effects do they have on specific traits?
3. How do the frequencies of alleles (i.e., different versions of a gene) shift over time or vary across geographical ranges?
4. What selective mechanisms or random events cause populations to change genetically in static and changing environments?

Since these observations occur over a human or historical time scale (years to thousands of years), we will refer to them as evidence of "contemporary" evolution. Such observations usually concern the mechanisms responsible for change along a lineage ("anagenesis") since these are the most common evolutionary dynamics. Occasionally though, these short-term observations shed light on the process of "cladogenesis," or the splitting of populations into species. Scientists sometimes refer to cladogenesis, and the pattern it helps create, as "macroevolution." Viewed as a range of evolutionary process and pattern, macroevolution extends from the origin of new species over hundreds to millions of years to the radiation and extinction dynamics of higher-level groups (such as genera, families, orders, etc.) over millions to tens of millions of years. Note here that the radiative scope of micro- and macroevolution overlap—making a strict dichotomy impossible. Because of the continual nature of evolutionary change, analyses of microevolution merge with those of macroevolution. Only the depths of time fundamentally prevent the *direct* observation of processes across the whole spectrum, forcing us to rely on inferences from various other sources of data.

The second line of evidence we discuss—the fossil record—comes from two sources. One is paleontological, consisting of the petrified remains of dead organisms. This has long been seen as a good record—imperfect as it might be—of the past history of life on earth (Desmond 1984; Rudwick 1985). Another, more recently exploited form of "fossil" evidence comes from the genetic sequences of living organisms. Just as fossil morphological data can be used for evolutionary reconstructions ("phylogenies"), these sequence data are the basis for molecular phylogenetic studies (Hillis et al. 1996). As with the paleontological evidence, molecular information results from the singular origin and development of life on earth. Moreover, like morphological evidence from fossils, slowly changing or "conserved" molecular data can be used to infer events across geological time. Notably, whether one is analyzing the differences among mammalian fossils, the genetic differences among living species, or the changes in viruses over years, the phylogenetic methods are exactly the same. Again, although it is impossible to observe the complete process that created these fossil records, one could rightly expect that if evolution were an important historical process, one would see evidence of it within them.

## Scientific Acceptance of the Evidence

In fact, both contemporary and fossil categories of evidence are available in such overwhelming abundance within the scientific literature that no single researcher can completely keep up with it. For example, using the keyword "evolution" and requesting currently published periodicals, a quick search of the online catalogs of the libraries of the American Museum of Natural History and the University and Texas at Austin revealed 43 different journals. These serials were just the ones with a major focus on evolution carried by these two institutions, and did not include an even larger number of general journals (in such areas as general science, biology, morphology, systematics, physiology, ecology, microbiology, biochemistry, and biophysics) that also contain many articles relating to evolution. Since the 43 journals themselves have multiple issues per year (from one to twelve), with each issue presenting multiple articles (from a few to dozens), the total number of evolution-relevant articles published each year (in this quick subsample alone) runs in the thousands. Johnson acknowledges that this literature is "voluminous," but nevertheless ludicrously underestimates it in writing that "most of the professional scientific literature is available in the premier scientific journals *Nature* and *Science*, the most prestigious scientific organs in Britain and America respectively, and at a somewhat more popular level in the British *New Scientist* and the *Scientific American*" (Johnson 1991, p. 13).

As noted by Johnson, one can imagine how this overabundance of scholarly work causes a degree of scientific specialization, just as in other academic areas like law and philosophy. What Johnson fails to realize, however, is that this abundance is an unmistakable sign of a robust and mature field of science. Only healthy, advancing disciplines radiate subspecialties as new avenues open themselves to researchers. Moreover, all new biological subdisciplines remain strongly rooted in evolutionary biology, with many new specialties representing new syntheses of preexisting areas (such as the emergence of "evolutionary development" or "evo-devo" from within mainstream evolutionary and developmental biology). Scientific specialization, consequently, should not imply fractionation: the vast majority of biologists (and many other scientists as well) remain well versed in very different types of evolutionary evidence (spanning various areas of physics, geology, and biology), and continue to engage in discussions with their colleagues across these disciplines. These professionals accept evolution as a reality (National Academy of Sciences 1999), just as most lawyers and philosophers recognize certain foundational principles in their disciplines, despite their specialties. This acceptance troubles Johnson. He believes that, rather than repeatedly testing the reality of evolution,

> ... the objective has been to find confirmation for a theory which was presumed to be true at the start of the investigation. (Johnson 1991, p. 99)

> If we assume that Darwinian evolution is basically true then it is perfectly reasonable to adjust the theory as necessary to make it conform to the observed facts. The problem is that the adjusting devices are so flexible that in combination they make it difficult to conceive of a way to test the claims of Darwinism empirically. (Johnson 1991, p. 30)

Of course this is wrong. In contrast to theistic design theories, there are presumably an infinite number of ways in which molecular or fossil evidence could disprove a hypothesis of evolution by common descent (National Academy of Sciences 1998). For example, if undisturbed and verifiable fossils were found to severely violate the predicted correlation between stratigraphic location and hypothesized evolutionary relationship, contemporary ideas of evolution would be refuted. Scientists regularly see, however, evidence that not only fails to reject an evolutionary hypothesis but overwhelmingly confirms and refines it (National Academy of Sciences 1999). One wonders how much larger an accumulation of evidence would be necessary before Johnson accepts the Darwinian interpretation. More important, what positive evidence can he or other neo-creationists offer in refutation of Darwinism or in support of their beliefs? In the following sections, we present a small number of scientific studies, using both contemporary and fossil evidence, in order to demonstrate how Johnson routinely misconstrues and evades scientific arguments. While we have chosen the examples to emphasize certain themes in Johnson's text, they are otherwise representative of other work within evolutionary biology.

## Ongoing Evolution

The following studies of rapid adaptation to different resources help demonstrate why we think speciation—one of the mantles of evolution—works the way it does. In each, recent geological or geographical events place a limit on the amount of time that has been available for the accumulation of evolutionary change. The change observed is sometimes subtle, but the hallmarks of differentiation are unmistakable. And when multiplied over longer "geological" periods of time, the differentiation is more than enough to account for the diversification seen in the morphological and molecular fossil records.

In recent years, Feder and colleagues (Feder et al. 1988, 1990; Feder et al. 1994) have studied the evolving ecological relationships of the apple maggot fly *Rhagoletis pomonella*, an agricultural pest native to eastern and central North America. Prior to the introduction of apples by humans, the fly exclusively parasitized the fruits of native hawthorns, just as its close relatives the flowering dogwood and sparkleberry flies specialize exclusively on single species of dogwood fruits and blueberries (Berlocher 1999). All these flies are highly dependent on their host fruits: they mate on them, lay eggs in them, and eat and develop within them as larvae. As a consequence, individuals have evolved intrinsic preferences for their fruit types and will eat and breed only in that fruit in the wild. When humans introduced apples to North America—less than 300 years ago—they provided a new resource for hawthorn-infesting flies. Some of these flies colonized the new apple resource, and eventually specialized on it. Because of their genetically based fruit preferences, apple- and hawthorn-infesting flies mate almost exclusively with flies having the same preferences, and have become recognized as a new host race (Feder et al. 1994).

What are the implications for speciation in this scenario? The different fruit trees, although they overlap in range and are frequently found in close proximity to each other, represent different ecological niches for the flies. After a subpopulation changed its preference to exploit the newly flourishing apples, it became specialized on that resource. This had a measurable genetic effect. Although the different host-preferring populations are morphologically indistinguishable to humans, they are significantly different in their frequencies of genetic markers. The behavioral and ecological barrier to mating between these niche-specialists, therefore, is high enough that they are already diverging after just 300 years—an evolutionary blink.[3]

In another case of fly evolution that also takes advantage of a recent introduction, we see rapid morphological evolution from natural selection in just a decade (Huey et al. 2000). The fruit fly *Drosophila sub-*

*obscura*, in its ancestral Old World range, has a geographical gradation ("cline") in wing size—with wings getting longer with increasing latitude. Since laboratory populations maintained at low temperature also evolve larger size, this cline is thought to reflect a selective response to temperature gradients across latitudes. In the late 1970s, the species was introduced into the New World, probably from a series of small founder populations on imported fruit. Throughout the 1980s, as the introduced flies spread over fifteen degrees of latitude in temperate North America, they maintained uniform wing length. Within the last ten years, though, a wing-length cline has become established, almost exactly convergent with that of the native European population. The basis for this cline is genetic rather than environmental since flies raised under common conditions show the wing type of their parents. However, the genetic mechanism responsible for wing elongation in the North American population turns out to be different from that of the European ancestral stock. These results demonstrate that organisms can respond rapidly to selection, potentially by more than one mechanism, even *without* speciation.

Of course, such studies may not show the evolution of a new "kind" of fly as demanded by some neo-creationists. To scientists, however, such a concern is simply irrelevant since evolution necessarily generates higher-level patterns from lower-level processes. In other words, because evolution incrementally produces large morphological divergences from the addition of smaller changes over time, all anyone could *possibly* observe would be a series of changes within populations (as in *Drosophila subobscura*), or in special circumstances, species-level divergences (as in *Rhagoletis pomonella*). No one should expect to see the divergence of higher level groups after so few years.

Such studies do indicate, however, the rate and power of evolution to modify forms in a short period of time, and the variety of means by which organisms in natural populations can respond to new conditions. Given the changes seen over mere decades in *Drosophila subobscura* and over centuries in *Rhagoletis pomonella*, what might we expect over longer periods of time? In Hawaii, considered a "hotspot" of evolution, a founding population of another *Drosophila* fruit fly species seems to have colonized a volcanic island more than 5 million years ago (Carson and Clague 1995). Since this colonization, islands have both appeared and disappeared and the founding lineage has evolved into at least 700 extant species. On the island of Hawaii alone, at least 100 species have evolved in the 450,000-year life span of the island (Kaneshiro et al. 1995; Kaneshiro 1998). Across the whole archipelago, *Drosophila* have very different behaviors, different morphologies, and rarely interbreed in the wild (Carson 1996; Freeman and Herron 1998). For example, size varies by more than twentyfold among Hawaiian species, and particular ones, such as *Drosophila heteroneura*, have evolved strikingly novel head, mouthpart, wing, and leg morphologies (Templeton 1977; Val 1977).

Finally, in another example of incipient speciation, Rundle et al. (2000) found that populations of small freshwater fish called threespine sticklebacks (*Gasterosteus* spp.) are in the process of differentiating based on their partitioning of food resources within lakes. At the end of the Pleistocene era (about ten thousand years ago) retreating glaciers exposed low-elevation freshwater lakes in coastal British Columbia, which marine sticklebacks independently colonized. These lakes have two general habitat types: open water and near-shore bottom. Independently, in each of at least three lakes, the colonizing sticklebacks have differentiated into a pair of species. The slimmer "limnetic" species have longer snouts, narrower mouths, and more and longer gill rakers—all apparently suited for feeding in schools on plankton in the open water. The larger "benthic" species have deeper bodies, shorter and broader snouts, wider mouths, and fewer and shorter gill rakers. They are solitary browsers, feeding on large bottom-dwelling invertebrates. Although the limnetics from all lakes resemble each other more than they resemble benthics (and vice versa), genetic data indicates that all the species independently evolved from the ancestral marine threespine stickleback, *Gasterosteus aculeatus* (Taylor 1999; Rundle et al. 2000). These species pairs interbreed only rarely in their home lakes (about 1–2% of adults appear to be hybrids; McPhail 1994) and there are many major genetically based characteristics distinguishing them. Again, by a number of criteria, it is reasonable to say that speciation is in progress, repeatedly and in parallel across these lakes.

After questioning the ability of small mutations and natural selection to produce new species (Johnson 1991, chaps. 2 and 3), Johnson responds to these types of studies with both a grudging acceptance and a dismissive shift in the critique. First he writes:

In some cases, convincing circumstantial evidence exists of evolution that has produced new species in nature. Familiar examples include the hundreds of fruitfly species in Hawaii and the famous variations among "Darwin's Finches" on the Galapagos Islands. (Johnson 1991, p. 19)

After this acknowledgement, however, he dismisses the importance of speciation.

Success in dividing a fruitfly population into one or more populations that cannot interbreed would not constitute evidence that a similar process could in time produce a fruitfly from a bacterium. (ibid.)

Some accounts credit the fruitfly experiments with producing new species, in the sense of populations that do not breed with each other; others dispute that the species border has in reality been crossed. Apparently the question turns on how narrowly or broadly one defines a species, especially with respect to populations that are inhibited from inbreeding but not totally incapable of it. I am not interested in pursuing the question, because what is at issue is the capacity to create new organs and organisms by this method, not the capacity to produce separated breeding populations. In any case, there is no reason to believe that the kind of selection used in the fruitfly experiments has anything to do with how fruitflies developed in the first place. (ibid., p. 159)

With these dismissals, Johnson conflates the phenomenon of speciation with that of the origin of higher-level taxa. By his criteria, the only acceptable evidence of speciation would be that which demonstrates a sudden, large difference in morphology (as between insect orders, for example) that even a nonspecialist would recognize. This kind of saltational speciation is more reminiscent of special creation and anachronistic ideas of "hopeful monsters," and plays no significant role in modern evolutionary theory (Futuyma 1998). Therefore, it is unreasonable and irrelevant to demand such evidence as confirmation of evolution. Johnson's criticism is further unwarranted because it demands that a demonstration of one phenomenon (incipient speciation) be equally adept at demonstrating many other phenomena (such as the origins of higher insect taxa). Because of this confusion, he virtually guarantees that there is no observational evidence that he can accept. In fact, having conceded microevolution and even macroevolution at the level of species radiations, he basically redefines "Darwinism" to include only the deeper phylogenetic patterns that are most difficult to unravel in detail.

### What Is a Species?

Throughout his book, Johnson dismisses the point that should be central to any scientific criticism of evolution. This is the question of what constitutes a species. If, according to Johnson, microevolution can act only to cause variation within species (or really "kinds" or "types"), what exactly are the boundaries of those entities? Moreover, when does the radiation of species, the actual foundation of macroevolution, become relevant in his caricature of descent with modification ("Darwinism")?

There are, as Johnson implies, various "species concepts" used among biologists. However, the differences among them are more in the kinds of epistemological and operational criteria used rather than the scope or "narrowness" of definition. For example, one concept (the biological species concept) stresses the importance of potential interbreeding, while another (the phylogenetic species concept) uses the criterion of completely fixed (behavioral, morphological, or genetic) differences. Under a strict interpretation of the first definition, coyotes and gray wolves would be considered a single species, since they may routinely interbreed under certain circumstances. The second definition recognizes that, despite their ability to hybridize, coyotes and gray wolves have unique identities that have arisen over evolutionary time. This variety of concepts also reflects the preoccupations of different biological disciplines that use different types of data in their analyses. For example, behavioralists, population geneticists, and paleobiologists all emphasize different sorts of data. More fundamentally, however, the range of species concepts reflects the continuous nature of the speciation process, and the difficulty in identifying a discrete point at which speciation can be considered complete (and therefore irreversible). Because the microevolutionary transition from separate populations to new species is generally thought to occur through the accumulation of small changes, difficulties in defining emerging species are consistent with evolutionary theory and should be expected. Again, Johnson's misrepresentation of these complex and subtle factors is summarized by his observation that "apparently the question turns on how narrowly or broadly one defines a species...." He then dismisses the importance of the issue saying "I am not interested in pursuing the question...."

Instead of understanding the details of evolutionary process, Johnson seems more interested in subjective and scientifically anachronistic "types" or "kinds." Speciation remains uninteresting to him because its products (species) do not meet his threshold of *significant morphological* (*versus* behavioral, biochemical, or genetic) innovation. Despite contradictory suggestions elsewhere in his book, Johnson claims that everyone agrees that microevolution occurs, including creationists. Even creation-scientists concur, not because they "have tightened their act," but because their doctrine has always been that God created basic kinds, or types, which sub-sequently diversified.... The point in dispute is not whether microevolution happens, but whether it tells us anything important about the processes respon-sible for creating birds, insects, and trees in the first place. (1991, p. 68)

In other words, change within a population or radiation within a genus of flies is meaningless because they're still just flies. Only at arbitrary higher levels of classification does Johnson seem to get interested in the differences—wondering how the fish type transforms into the amphibian type, or how the reptile type becomes the mammal type. Unfortunately, used in this way, without an appreciation for the details of speciation and classification, such identities become purely typological and immutable by *definition*. Moreover, the type transitions he tends to focus on, such as the vertebrate ones leading to humans, are often quantitatively less divergent than ignored transitions in other groups—such as the radiation within Diptera of horse flies, fruit flies, crane flies, and mosquitos.

Presumably *some* creationist might be interested in where one "type" of organism leaves off and another begins. Unfortunately, there are only arbitrary criteria with which to determine which organisms are signifi-cantly different in type from others. One person's subjective conclusion that lions and tigers are different types may be challenged by another's claim that, to the contrary, they are the same "big cat" type. The second creationist's assertion that cats, dogs, and bears are different, in turn, could be challenged by a third who says they all belong to the "mam-malian carnivore" kind. This lack of objective criteria is a direct con-sequence of two problems for creationists. First, they lack any sort of theory that would guide an objective definition of "kind" or "type." Second, evolutionary theory predicts a range of evolved differences across divergence times that is perfectly consistent with the nested varia-tion observed among organisms. Differences among populations merge into differences among species; differences among species ultimately merge into differences among genera and families. At higher levels (e.g., orders, classes, and phyla), there are larger differences among groups, but such increases in magnitude follow naturally from a hierarchical, evolu-tion-based classification system.[4] Ongoing extinctions will also tend to make some evolutionary branches appear more distinctive, much like when a storm tears certain limbs from a tree—leaving large gaps among the remaining branches. Consequently, much of the phenomenological interest in differences among types or kinds is simply a consequence of ongoing radiations and extinctions.

Although greater morphological divergence is expected at higher levels of classification, the contemporary evolutionary model (in contrast to Johnson's caricature of Darwinism) does not predict a tight correlation between the origin of species and higher groups and the appearance of major morphological changes. First, levels of taxonomic *rank* (vs. the hierarchical clustering itself) above the species-level are arbitrary, such that one specialist may describe the same group of species as an order (or as a suborder, superorder, etc.), while another may describe it as a class. This situation parallels, albeit with much more formal rigor, the arbitrary designations of type and kind among creationists.

More important, though, is that evolution proceeds at various rates in different groups and places. Although sometimes discussed by scientists when contrasting different patterns of evolution, the constant gradualism depicted by Johnson as the essence of Darwinism has long been viewed as an overly simplistic model (Jackson and Cheetham 1999). Relatively younger species radiations, such as those of Lake Victoria cichlid fish (Bouton et al. 1998; Stiassny and Meyer 1999) or the Hawaiian *Drosophila* (Kaneshiro et al. 1995; Freeman and Herron 1998), often demonstrate rapid morphological shifts, while older radiations may remain largely undetected ("cryptic") because of their relatively slow rates of morphological change (e.g., Cheetham 1986; Jackson and Cheetham 1994). These species dynamics are scientifically interesting and informative to our overall understanding of evolutionary pattern, despite Johnson's insistence on changes in kind and his avoidance of any connections between micro- and macroevolution. Indeed, the evolution of new species is at the heart of the issue regarding the supposed differences between microevolution, which creationists now accept, and macroevolution, which they persist in rejecting (Pennock 1999, p. 155). Even earlier advocates *within* mainstream creationism recognized the continuity of process between micro- and macroevolution. When the Religion and Science Association debated accepting limited speciation within types, the creationist Byron C. Nelson stated that acceptance of speciation would open "the door of evolution so wide that I, for one, don't see a place to shut it" (quoted in Numbers 1992, p. 114). In the light of these worries, neo-creationist obsessions with types can be seen as nothing more than an expression of their interest in artificially separating "accepted" processes of microevolutionary change from the realm of macroevolutionary pattern. By their arbitrary designation, types and kinds are just those fixed entities not *directly* traceable through microevolution and speciation events.

## The Fossil Record

When confronted with evidence of speciation, Johnson replies that these processes can't explain the bigger transformations among his "kinds" of organisms. With the fossil evidence, the situation is somewhat reversed. After being shown a record of the grand sweep of morphological evolution, Johnson disputes the evidence because it does not show the minutiae of the speciation process in great enough detail. For example, he claims that the evidence represented by *Basilosaurus* (a proto-whale showing vestigial hind-limbs) fossils should be excluded from consideration because "none of these [evolutionary] accounts mentions the existence of any unresolved problems in the whale evolution scenario.... Even the vestigial legs present problems. By what Darwinian process did useful hind limbs wither away to vestigial proportions ...?" (Johnson 1991, pp. 84–85). That is to say, we should discount the evidence because the model explaining it is not completely specified. So Johnson discards a fossil series, replete with transitional anatomical features, leading to modern whales.[5]

Such dismissals form a pattern. In his presentation of the fossil evidence (chap. 6), Johnson often comes close to accepting the validity of the evolutionary interpretation. He then always contrives to find a way, however, in which it supposedly comes up short. In discussing the well-documented intermediates in jaw morphologies between reptiles and mammals, he concedes "the narrow point" but goes on to require unfossilizable details of the "all-important reproductive systems." Despite abundant comparative developmental evidence that these jaw bone details are homologous (i.e., not convergent), Johnson concludes that "convergence in skeletal features between two groups does not necessarily signal an evolutionary transition" (1991, p. 76). He allows that "*Archaeopteryx* is on the whole a point for the Darwinists," but not without the unanswered qualification: "but how important is it?" (1991,

p. 79) (Note here that Johnson rejects this case regardless of its fitting his criteria!) He excludes the example of *Basilosaurus* because "expressions like 'found in direct association with' and 'undoubtedly' whet [Johnson's] curiosity." He wonders "is it certain that *Basilosaurus* had shrunken limbs, or is it only certain that fossil foot bones were found reasonably close to *Basilosaurus* skeletons?" (1991, p. 178). His responses to the fossil data seldom rise above such thinly veiled accusations of fraud and incompetence.

Another technique neo-creationists use to cast doubt on the standard paleontological interpretation is to emphasize the lack of certainty among paleontologists in evaluating the direct lineal relationship of fossil representatives to potential extant descendants. For Johnson, anything less than proven direct ancestry has no bearing whatsoever on the evolutionary interpretations of the fossil record. In contrast, given the unavoidable incompleteness of the fossil record, paleontologists more realistically hope to find approximate ancestors in their search for fossils. These fossils are used to estimate evolutionary relationships, which are then tested by the acquisition of new data and reanalysis.

In the discussion of therapsids (mammal-like reptiles that are precursors to mammals), Johnson accepts only fossils that can be unambiguously shown to lie on the line connecting modern mammals with the ancestor they shared with reptiles (Johnson 1991, p. 76). This stringency creates a false, straw-man requirement that ignores the unavoidably complex picture resulting from the incomplete sampling of diverse extinct taxa. To provide a numeric context, first contemplate the large number of species living at any given time. These are actually just a small fraction of the species that have ever lived, since the average estimated "life-span" of a species is just 5–10 million years and the history of macroscopic life spans more than 600 million years. In fact, using just animals and plants (and ignoring microbial bacteria, fungi, and protozoans), scientists calculate that documented fossil species ($\sim 250,000$) are likely to represent less than 1% of the total number of species that have ever lived (May et al. 1995; Futuyma 1998). Moreover, since fossils are best preserved and represented among shelled marine organisms, the fraction is much higher for these taxa, and much lower for many others. To illustrate this graphically, consider figure 12.1, representing a schematic phylogeny of Amniota that includes real fossil groups as well as

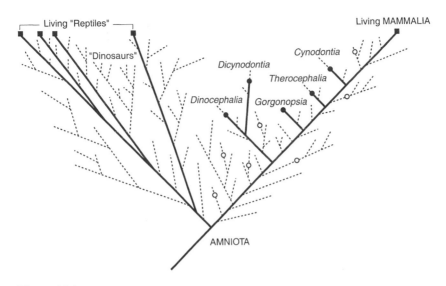

**Figure 12.1**
A schematic amniote phylogeny, illustrating both known and hypothetical "ghost" (undetected) extinct lineages between living reptiles and mammals (indicated with solid squares). Though populations of evolving amniotes existed continuously along every branch of this evolutionary "bush," only a very small fraction of these will have left representative fossils (symbolized by circles in the therapsid "transition" zone between reptiles and mammals). Of these fossil records, only a small minority in turn have been found and analyzed (shown here by solid circles representing some known therapsid groups). Consequently, the "transition" between reptiles and mammals will necessarily be relatively poorly sampled, and what fossils exist are highly unlikely to lie on the actual "line" connecting living reptiles and mammals. Despite these difficulties, researchers can make reasonable estimates of evolutionary relationships among known taxa (the solid lines). Moreover, transitional patterns can be inferred despite the missing data. Note that although time generally progresses as one moves upward along a lineage here, different branches have not been drawn according to a constant time scale. Also note that "reptiles" and "dinosaurs" are not monophyletic groups because they traditionally exclude birds that are descended from a lineage within dinosaurs (modified from Kemp 1988; Dingus and Rowe 1998; Pough et al. 1999).

hypothetical, undetected "ghost" lineages. Because of (1) the branching complexity of an evolving group, (2) the large number of extinct representatives, and (3) the fact that only a miniscule fraction of only certain species are ever preserved as fossils, one can never be certain that a haphazard sample of fossils will lie exactly on the line leading to modern mammals. In fact, with respect to actual genetic lineages within species, the chances of ever finding "direct ancestors" for any descendent are effectively zero.

Whereas a group such as therapsids can be confidently stated to be ancestral to mammals, pinpointing the actual ancestral therapsid population, species, or lineage is all but impossible. When it serves his purpose, however, Johnson is quick to adopt this unrealistic stringency:

The notion that mammals-in-general evolved from reptiles-in-general through a broad clump of diverse therapsid lines is not Darwinism. Darwinian transformation requires a single line of ancestral descent. (1991, p. 76)

Of course paleontologists do not think that "mammals-in-general" evolved from "reptiles-in-general." Scientists work with the data resolution at hand and use methodologies that allow interpolation when possible. Although such efforts can lead to errors, these will generally be caught with further data collection and analysis. This scientific approach is far more objective than the speculative, lawyerly doubt-casting approach demonstrated by Johnson:

If our hypothesis is that mammals evolved from therapsids only once ... then most of the therapsids with mammal-like characteristics were not part of a macroevolutionary transition. If most were not then perhaps all were not. (ibid., p. 77)

Despite Johnson's assertions, there *is* a tremendous amount to be learned about ancestral lineages by examining the "dead-end" offshoots. If all therapsids are not part of *the* transition to mammals, don't they nevertheless tell us about other macroevolutionary transitions? (see fig. 12.1) If *Archaeopteryx* is not precisely *the* ancestor of modern birds, does it lack informative value? Many lineages of dinosaurs living during the Jurassic were moving toward a birdlike morphology and way of life (Chiappe 1995). Most of these eventually went extinct, while one continued and evolved into contemporary birds. By studying the characteristics of both extinct and living groups, and using these characters to reconstruct the evolutionary relationships among these species, scientists can actually estimate certain characteristics of undiscovered ancestors. In other words, such phylogenetic methods not only "work around" the absence of direct ancestors but also predict what these ancestors shared in common with their descendents (Pagel 1999).

To provide a genetic illustration of this, imagine a room filled with thousands of people at a large family reunion. Within this family are many generations of participants ranging from the family matriarchs and patriarchs to their infant great, great, great, great grandchildren. Using genetic information from only the infants and modern phylogenetic methods for the analysis, scientists can estimate, with high confidence, the genetic make-up of all members of the older generations. This is exactly like constructing evolutionary relationships of ancestral lineages using just living species (i.e., "the infants"). Now imagine that scientists at the family reunion are also given permission to study the genes of various great aunts and uncles who had no descendents. Then the estimates of genetic make-ups from the unsampled members of the party would be even better. This parallels the inclusion of nonancestral extinct species in the analysis of evolutionary relationships with the living species.[6] The same techniques are used whether analyzing the evolution of viruses over days and years, the evolution of genetic conditions in humans over decades and centuries, or the evolution of species radiations over thousands and millions of years (Hillis et al. 1994; Freeman and Herron 1998).

From Johnson's misrepresentation of the fossil record, a casual reader would get the unavoidable impression that (1) examples of well-established fossil series showing actual transitional forms are few, and (2) these examples all suffer from unavoidable flaws that render them useless or misleading. While more careful reading reveals the repeated side-stepping and evasion of evidence, only by turning to other sources would one realize the extent to which Johnson's selectivity of examples constitutes a serious error of omission. For example, in Johnson's chapter on the vertebrate sequence no mention is made of the fossil record of horses, or of skull and skeletal character evolution in titanotheres. In his chapter on fossils, no mention is made of major fossil taxa such as diatoms, foraminiferans, ammonoids, or bryozoans. Yet he still contends that the fossil record is impoverished. These examples, available to Johnson through one of his major sources (Futuyma 1986), provide more detailed if less well-known examples of evolutionary "transitions." Ignoring everything but *Archaeopteryx*, he concludes that:

Persons who come to the fossil evidence as confirmed Darwinists will see a stunning confirmation, but skeptics will see only a lonely exception to a consistent pattern of fossil disconfirmation. If we are testing Darwinism rather than merely looking for a confirming example or two, then a single good candidate for ancestor status is not enough to save a theory that posits a worldwide history of continual evolutionary transformation. (1991, p. 79)[7]

Despite Johnson's assessment, there are literally dozens of detailed, well-studied examples, which, if given even the grudging and qualified acceptance Johnson gives the therapsid, *Archaeopteryx*, and *Basilosaur* fossils, provide overwhelming support for evolution. This paleontological evidence is, in turn, corroborated by the molecular fossils and studies from plate tectonics and biogeography. To most fair-minded people, such multiple, independent lines of evidence—all consistent with a naturalistic model of evolution—would be more than enough to convince them of the likelihood of evolutionary events. Similarly, the evidence suggesting naturalistic microevolution within populations and its importance in speciation dynamics leading to molecular, morphological, and taxonomic diversification is undeniable, in the end, even to Johnson. Confronted by all this evidence—evidence that allows and suggests a robust synthesis of processes at various scales—he nevertheless refuses to concede the connections. Instead, Johnson repeatedly attempts to deconstruct this synthesis through factual misrepresentation, tactical shifts in his criteria for "evidence," and spurious suggestions of fraud and conspiracy. Ultimately, these efforts result in nothing more than a retreat to that creationist, "God of the gaps" refuge where valid scientific questions are relegated to the realm of theistic mysteries. For someone professing to "examine the scientific evidence on its own terms, being careful to distinguish the evidence itself from any religious or philosophical bias," such a vantage point seems risky.

## 4.   Molecular Revelations

Molecular biology has become a favorite source of quotes for the design theorists. The fields are new enough, and the jargon is enough in the public ear, that they can be cited as sources of new evidence against

Darwinian evolution. In his book *Evolution: A Theory in Crisis?* Michael Denton writes about the development of gene sequencing:

Armed with this new technique, biology at last possessed a strictly quantitative means of measuring the distance between two species and of determining the patterns of biological relationships. If it is true, as typology implied, that all members of one type, however superficially divergent, always conform exactly to the basic cidos of thcir typc, all posscssing cqually and in full mcasurc all the defining character traits of their type and all standing therefore equidistant in all important aspects of their biological design from the members of other types, might this principle of equidistance be revealed by these new molecular studies? If the divisions in the nature were really as orderly as early nineteenth-century biologists insisted, might this overall orderliness be confirmed by the new field of comparative biochemistry? On the other hand, the new molecular approach to biological relationships could potentially have provided very strong, if not irrefutable, evidence supporting evolutionary claims. Armed with this new technique, all that was necessary to demonstrate an evolutionary relationship was to examine the proteins in the species concerned and show that the sequences could be arranged into an evolutionary series. (Denton 1985, p. 276)

In the context of modern biology, this paragraph is nearly uninterpretable. Denton's speculations distinguish between a "typological" (i.e., pre-Darwinian) worldview and a "sequential" worldview that he attributes to Darwinian evolution. The typological view, he explains, sees species (or "types") as fundamentally discrete, nonoverlapping entities. This model also seems to rely on the ad hoc "principle of equidistance." How this "principle" could possibly be used to distinguish between evolutionary and nonevolutionary models is never clearly stated. It could be as simple as saying that each child in a family, for example, is equally related to each of their cousins.

Leaving aside for a moment this confusion, Denton's main point seems to be that this new technology should provide the opportunity to find (if it exists) irrefutable evidence of ancestor-descendant relationships. Denton goes on to describe the result he claims was actually found:

However, as more protein sequences began to accumulate during the 1960s, it became increasingly apparent that the molecules were not going to provide any evidence of sequential arrangements in nature, but were rather going to reaffirm the traditional view that the system of nature conforms fundamentally to a highly ordered hierarchic scheme from which all direct evidence for evolution is emphatically absent. Moreover, the divisions turned out to be more mathematically perfect than even most die-hard typologists would have predicted. (ibid., p. 277)

In the remainder of the chapter Denton explains his interpretation of the molecular data as upholding the "typological" versus the "sequential" concept of life. He claims that Darwinian evolution requires intermediate forms between so-called primitive and advanced organisms. Aside from the fact that Denton's argument is being placed in the domain of molecular biology, it is fundamentally equivalent to the ancient claim that the one should expect (if Darwinian evolution is correct) to see animals intermediate in form between, for example, frogs and humans. As discussed above with respect to Johnson, this expectation is an absurd caricature of evolutionary biology, and Denton's insistence on restating it demonstrates a fundamental lack of understanding of ancestor-descendant relationships. In order to appreciate the depth of Denton's misunderstanding and the incoherence of his appeal, we will briefly reproduce his argument.

Proteins are the universal units of biological structure and function. They are composed of sequences of linked units—amino acids—that, along with other factors, determine the shape and functionality of the proteins. Across species, these sequences often vary, such that different taxa have different versions of individual protein types.[8] For example, *cytochrome c* is a protein involved in cellular metabolism. It is composed (in humans) of 105 amino acid subunits. There is fair bit of cytochrome c sequence variation among species, as demonstrated by the sequence excerpts in table 12.1 (Denton 1985).

In the early years of molecular biology, known sequences for a wide variety of organisms were compiled into huge tables, such as Dayhoff's *Atlas of Protein Structure and Function* (Dayhoff 1972). The differences between two proteins can be expressed as a percent of sequence divergence or percent of sequence difference. Thus, as Denton writes, "between horse and dog (two mammals) the divergence [in cytochrome c sequence] is six percent, between horse and turtle (two vertebrates) the divergence is eleven percent, and between horse and fruit fly (two animals) the divergence is twenty-two percent" (1985 p. 278). He makes the point that "every sequence can be unambiguously assigned to a particular subclass. No sequence or group of sequences can be designated as intermediate with respect to other groups. All the sequences of each subclass are equally isolated from members of another group. Transitional or intermediate classes are completely absent from the matrix [of sequences from 33 species]." In a series of figures, Denton illustrates the concept of equivalent sequence divergence between all members of a group and any

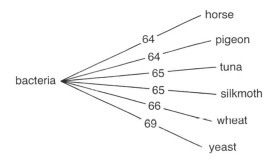

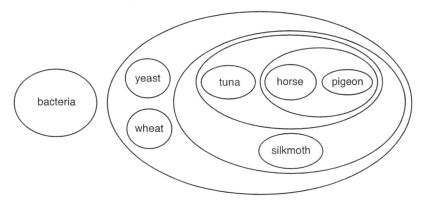

**Figure 12.2**
Similarity in protein sequence distances between outgroup and ingroup members (modified from Denton 1985).

**Figure 12.3**
The hierarchical clustering resulting from a consideration of protein distance data (modified from Denton 1985).

given organism from outside of the group (fig. 12.2). This, according to Denton, reiterates the "isolated and unique" nature of each class:

The classification system that is derived from these comparative molecular studies is a highly ordered non-overlapping system composed entirely of groups within groups, of classes which are inclusive or exclusive of other classes. There is a total absence of partially inclusive or intermediate classes, and therefore none of the groups traditionally cited by evolutionary biologists as intermediate gives even the slightest hint of a supposedly transitional character. (ibid., p. 286)

A series of Venn diagrams then illustrates the hierarchical clustering of organismal groups (fig. 12.3). Denton summarizes the end result of all of this evidence in this way:

This new area of comparative biology illustrates just how erroneous is the assumption that advances in biological knowledge are continually confirming the traditional evolutionary story. There is no avoiding the serious nature of the challenge to the whole evolutionary framework implicit in these findings. For if the *ancient representatives of groups* within amphibia, lungfish, cyclostomes and reptiles manufactured proteins *similar to those manufactured by their living relatives* today, and if, therefore, *the isolation of the main divisions of nature was just the same in the past as it is today*, if for example ancient lungfish and ancient amphibia were *as separate from each other as their present day descendants are*, then the whole concept of evolution collapses. (ibid., p. 291, italics added)

A number of errors leap from this passage. First, although "ancient representatives" of various taxa probably made similar proteins to those of their living descendents, evolutionary biologists think that these proteins would have *changed* over time. In other words, the "isolation of the main divisions" of organisms, as judged by distances among their proteins, is *not* static. Instead, molecular distances are roughly correlated with time of divergence. Second, because contemporaneous organisms cannot be descended from one another, modern extant fish species are certainly *not* the ancestors of modern amphibians, nor are contemporary reptiles the ancestors of contemporary birds or mammals. Rather, evolutionary biology holds that modern fish share a common ancestor with modern frogs, salamanders, reptiles, and mammals. Much like Johnson's, Denton's idea of evolutionary transitions absurdly harkens back to eighteenth-century concepts of *scala naturae*: fish → amphibian → reptile → mammal (Panchen 1992). (The *scala naturae* held that fish are below amphibians, which are below reptiles, and so on, up to humans and sometimes beyond to heavenly beings. Evolution by this scheme entails the transformation of organisms up this hierarchy.) These writers miss the fact that living fish are no less modern than living mammals—each has had exactly the same amount of time of evolution since their last common ancestor. That the ancestor more closely resembled the fish than the mammal is immaterial.

Using his Venn diagrams, Denton argues for a hierarchical, clustered grouping scheme for living taxa as an alternative to the "sequential" scheme he erroneously insists is a feature of evolutionary biology. In fact, such nested diagrams have long been viewed as equivalent to the hierarchically branching structures (i.e., phylogenetic trees) that modern biologists use (fig. 12.4), though the meanings vary among epochs and

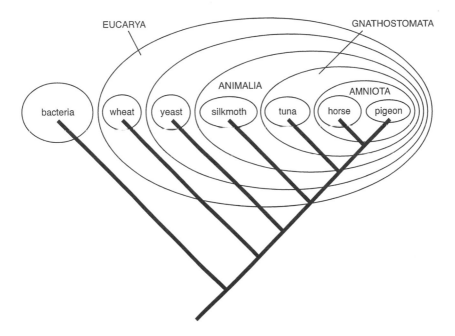

**Figure 12.4**
The hierarchical Venn classification of vertebrate groups overlaid on a phylogenetic tree (see Dingus and Rowe 1998; Pough et al. 1999).

authors (Panchen 1992). The set of all organisms sharing a particular common ancestor (called a "clade") is recognized relative to an "outgroup": a representative of a group not sharing that common ancestor. Mammals are a clade relative to other tetrapods, or relative to other amniotes. As such, every living member of the mammalian clade is equally distant, evolutionarily, from any outgroup member, regardless of whether that outgroup is a nonmammalian amniote (such as a pigeon), a nonmammalian gnathstostome (such as a tuna), or a nonmammalian animal (such as a silkmoth). Likewise, every member of the clade *amniota* (for example, horses and pigeons) is equally distant from nonamniotes. It should be clear why Denton's incredulity that "man is as close to lampreys as are fish!" (Denton 1985, p. 284) is misplaced. Jawed fish and humans share a common ancestor that is more recent than the ancestor of lampreys, fish, and humans together.

In summary then, Denton's assertions about the expectations of evolutionary biology are patently false. The actual expectations are in fact well supported by the evidence Denton claims as his own. Far from being a "theory in crisis," evolutionary theory continues to make and test empirical predictions. As new technologies and fields of study emerge (such as molecular genetics), the theory incorporates them into its growing body of understanding.

## 5. Is It Irreducible?

A third popular book in the anti-evolution literature, *Darwin's Black Box,* claims credibility from the fact that the author is a biochemistry professor at Lehigh University. In his book, Michael Behe attempts to update arguments against evolution that have emphasized the complexity of biological structures.

… neo-Darwinism must be reconsidered in light of advances in biochemistry. The scientific disciplines that were part of the evolutionary synthesis are all non-molecular. Yet for the Darwinian theory of evolution to be true, it has to account for the molecular structure of life. It is the purpose of this book to show that it does not.

Although appearing to accept the genetic, systematic, and paleontological evidence of the evolutionary synthesis, Behe claims that the molecular evidence is somehow different. Within the molecular realm, in particular, he claims so-called irreducible complexity challenges or overturns Darwinism:

What type of biological system could not be formed by "numerous, successive, slight modifications?" Well, for starters, a system that is irreducibly complex. By irreducibly complex I mean a single system composed of several well-matched, interacting parts that contribute to the basic function, wherein the removal of any one of the parts causes the system to cease functioning. An irreducibly complex system cannot be produced directly … by slight, successive modifications of a precursor system, because any precursor to an irreducibly complex system that is missing a part is by definition nonfunctional. (Behe 1996, p. 39)

Behe's definition and its purported refutation of a model of evolution (defined as "numerous, successive, slight modifications") are not only circular, but demonstrate a fundamental confusion between the concepts "simpler precursor to the current system" and "proper subset of the current system." It turns out that Behe imagines a precursor to an "irreducibly complex" system as being equivalent to the complex system minus a part or two.

As an example, consider a modern city in the United States—similar, in some ways, to a complex organism. If the sewage system of New York

City or the freeways of Los Angeles County were suddenly removed, those cities would, without a doubt, cease to function as they do now. But this does not imply that, at every step, freeways and sewage systems were integral to the developing proto-cities. The fact that the system now operates as a cohesive whole, dependent on all of its parts, does nothing to refute the evolution of the system. The statement "any precursor to an irreducibly complex system that is missing a part is by definition non-functional" is argument by fiat. Given the current state of an "irreducibly complex" system as defined by Behe, *of course* one cannot remove a component and be left with a working, if suboptimal, system. This is merely Behe's definition. However, complex systems may frequently have antecedents that operate less efficiently or robustly with fewer parts (e.g., Wimmer et al. 2000). That the more efficient, more complex system has developed a dependence on the complexity says nothing about the steps leading to its development.

## Complexity in Isolation

Complexity can be made to seem "irreducible" by examining a system in isolation. Behe does this in his discussion of the molecular cascade involved in vision. He presents a clear description of a complicated molecular pathway, in which a photon triggers a cascade resulting in the firing of a retinal nerve cell. This system, he suggests, is the molecular equivalent of one of the earliest challenges to evolution—the complexity of the metazoan eye. His discussion of the molecular pathway of vision is representative of his approach throughout the book. In essence, his strategy is to describe in detail a complex biochemical pathway. He then asserts that, since all elements of the pathway are necessary to its functioning, and since the pathway is too complex to have arisen *de novo*, that an evolutionary origin for the pathway is not plausible. Behe's description of the visual cascade emphasizes its marvelous complexity. But his discussion leaves out some additional complex aspects of the system that go a long way toward explaining how such a system might have developed. By focusing on a cellular subsystem in isolation, Behe makes the subsystem and its origins seem *more* mysterious. It is analogous to considering the functioning of a series of escalators in a twenty-story building, in isolation and without reference to any other structure. The picture of the escalators, alone and unsupported, is clearly one to marvel at. How could escalators just exist like that? How could the treads and motors be assembled so that the whole structure was self-supporting and functional in isolation? The answer, of course, is that no such structure actually could exist: the escalator functions only within the context of the building. Moreover, since their origin in 1892 as a tilted conveyor belt, the development of the escalator has taken place in a piecewise fashion. As technology allowed, stairways were replaced by various versions of moving stairs. In the proper context, isolated systems seem less miraculous.

How does increasing the number of subsystems simplify the evolution process for the overall structure? The answer derives from the fact that the subsystems are all made of variations of the same components, modified so that the specific function arises from a variant of a generic structure. Behe's example of the visual-pigment signaling cascade provides a good example. One would never know from his detailed description of the molecular steps in the pathway that they are in fact *ubiquitous* features of many different signaling processes in most cells. Specific signaling pathways are frequently derivatives of a generic signaling mechanism. The proximal step of visual perception is part of the GTP-coupled receptor signaling pathway. This pathway forms a generic "whole-cell" response signal, whereby a stimulus from outside the cell can quickly and efficiently change the chemical state of the cell. The photo-receptor apparatus merely modifies one of these steps to accommodate a photon as the source of the external signal. Other cells modify the same pathway as well. In ova, for example, the GTP-coupled pathway is triggered by the adhesion of a sperm cell. There exists, in essence, a standing pool of complexity within the cell that can be exploited for adaptation to new functions. Although this argument appears to sidestep the *origin* of this standing pool of complexity, it in fact suggests a mechanism by which the complex interactions may have arisen. In this scenario, simple systems are modified by the stepwise addition of simple elements. Over time, the entire pathway becomes dependent on all of the elements.

## Plasticity of Molecular Function

There are several well-known cases of molecules acquiring and then adapting to secondary functions. For example, eye lens proteins, generically known as *crystallins*, exemplify the conundrum presented by Behe:

by what possible sequence of evolutionary events could an organism evolve a new protein for lens construction on the way to evolving an eye? Actually, an evolving lineage does not necessarily evolve a new protein, but co-opts and gradually modifies the function of an existing one. In nearly all cases studied, taxon-specific crystallins show a remarkable sequence similarity to proteins of very different function. Furthermore, the identity of the similar protein differs among taxa. Among vertebrates there are crystallins showing similarity to the enzymes quinone reductase, lactate dehydrogenase, argininosuccinate lyase, and alpha-enolase (Hendriks et al. 1987; Gonzalez et al. 1995). In cephalopods, there is a crystallin similar to the enzyme glutathione-S-transferase (Tomarev et al. 1995). Some of these cases show evidence of a gene-duplication event, in which the gene copy was recruited to its secondary crystallin function, and thereafter lost its original activity.

A similar process is seen in the evolution of specialized antifreeze glycoproteins (AFGPs) in Antarctic notothenioid fish (DeVries and Wohlschlag 1969; DeVries 1971; Chen et al. 1997a,b). These proteins lower the freezing point of the blood serum to below that of the surrounding water, inhibiting ice crystal growth and allowing the fish to survive in subfreezing water. The structure of an AFGP tells the story of its origins: the sequence and structure of its ends closely match those of the digestive protein trypsinogen, which normally occurs in the intestine. The noncoding regions responsible for the pattern of gene expression of AFGP and trypsinogen are also nearly identical. Within the coding portions of the AFGP gene ("exons"), a small unit has been duplicated many times, relative to the trypsinogen gene—suggesting that these repeats may be important for the new function.

About 20 million years ago, notothenioid fishes colonized Antarctic waters, which subsequently dropped in temperature dramatically. Various lines of evidence suggest that the trypsinogen gene was incidentally duplicated, which, because it then expressed the protein at high levels in the gut, protected the gut contents from freezing. Over time, the expression pattern of the gene shifted, so that it became expressed in the liver as well, and thereby led to the expression of antifreeze proteins in the blood serum. Along the way, a series of duplications of a small unit (potentially from replication slippage and other errors) resulted in a protein better suited for antifreeze activity. Note that no part of this scenario invokes any unknown process: gene duplication and the amplification of repeats have been observed in the laboratory, and mechanisms by which expression patterns can change are well understood (Li 1997). If one looks at the amount of sequence divergence between trypsinogen and AFGP, one can calculate a rough approximation of divergence time for the two genes. This has been estimated as 5 to 14 million years ago. The waters of the Antarctic are estimated (from the isotopic analyses of sediment) to have frozen about 10 to 14 million years ago. Independent genetic evidence suggests that AFGP-expressing Antarctic notothenioid fishes diverged from other notothenioids about 5 to 21 million years ago (Bargelloni et al. 2000). Even accounting for the large variance, these three independent lines of evidence support an adaptive, "neo-Darwinian" hypothesis for the evolution of a new protein function in response to a novel environmental challenge.

One might argue that perhaps the structure of trypsinogen is also coincidentally optimal for lowering the freezing point of blood serum. If that were the case, one could expect that other cold-water marine fishes would have similar antifreeze proteins. To test this possibility, Chen et al. (1997b) examined the corresponding antifreeze gene from distantly related Arctic cod. The antifreeze genes of these fish show certain similarities to those of the Antarctic notothenioids: both are composed of polyprotein genes linked by short cleavage sites. The entire sequence of the genes reveals independent origins, however, as there is no sequence similarity between the cod AFGP's and trypsinogen. The intron-exon structures of the cod and notothenioid genes differ, as do the sequence of the repeats between the two groups of fishes. Together these comprise strong evidence for convergence of protein function, with independent origins from different ancestral genes. Two different fish taxa, at different times and places, evolved similar solutions to the similar challenges facing them. As with visual-pigment signaling cascades and lens crystallins, AFGP's evolved from whatever genetic precursors were available. All that was required was that, at each step, the resulting individuals be somewhat better (or at least no worse) off than the rest of their cohorts lacking the change. No doubt the initial AFGPs were not ideally suited to their new function, but they had millions of years over which to improve.

This ad hoc approach to biological function, in which existing features are co-opted into new functions, is ubiquitous at all scales of the organism. It is possible because molecular and anatomical features are often very plastic in function. An enzyme involved in digestion might also have a moderate antifreeze property that can, via selection, improve to very high efficiency. The phenomenon has been seen repeatedly in laboratory evolution experiments. A protein exquisitely adapted to act with high efficiency on a specific substrate can also act on suboptimal substrates, albeit with low efficiency (Li 1997). The efficiency can be improved by selection, however, and the path from poor catalyst to excellent catalyst can often be remarkably short.

## 6.  The Advancing Field

Modern biology is a vibrant and active discipline. Given how rapidly the field is developing, it is hard to guess at the form data will take in coming decades. Neo-creationists such as Behe and Denton believe that it is only a matter of time until technological advances prove the falsehood of evolutionary theory. In the century since Mendel, however, the opposite has happened: the evolutionary synthesis has proceeded unhindered, while the creationist program has failed to demonstrate any relevance to the science of biology. What follow are two very brief synopses of studies from the recent literature. These hint at some of the research paths molecular biologists are following, and highlight their implications for evolutionary biology.

### Interspersed Genetic Elements

The first study discusses a recently discovered class of "molecular fossils." Many organisms have large classes of "retrotransposons"—quasi-independent segments of DNA that can copy and reinsert themselves into many places in the genome. Vertebrates, for instance, contain groups of "SINEs" and "LINEs" (for short- and long interspersed elements) (Verneau et al. 1998; Hillis 1999; Nikaido et al. 1999; Shedlock and Okada 2000). These transposable elements exist both as "master elements" that still actively copy themselves, and as large numbers of copies, most of which are defective in copying activity. The master element functions by copying itself through a series of enzymatic steps, and then inserting the copy into another location in the genome.[9] Once a faulty copy of a LINE is transposed, it begins to accumulate mutations that would be deleterious to the active master element. These mutations serve to "tag" the copy, such that all descendants will contain the changes accumulated by their parent copy. Sometimes a master element copies itself *in toto*, retaining its activity. This functional copy can also accumulate neutral mutations that distinguish it and its descendants from the original master and its descendants.

What do the properties of these transposons tell us about the evolutionary history of their carriers? In essence they provide a traceable lineage of DNA sequences present in a given group (for example, mammals). The history of these DNA sequence lineages would be expected to mirror that of the organisms carrying them. As an example, assume that the ancestor of all living mice (genus *Mus*) had a particular LINE element, with several copies of that element. Its offspring and the independent populations and species derived from them would also have these LINE elements. But at the time of splitting off from the ancestral population, each descendant lineage would also begin accumulating its own profile of master and copy elements. These profiles give *unambiguous* testimony to the relatedness of the organisms carrying them. Evidence from LINEs and SINEs has agreed very well with our previous understanding of mammalian relationships. These elements are beginning to be used as an exquisitely sensitive tool for resolving phylogenetic relationships.

The only way that neo-creationists have addressed this kind of evidence in the past is to suggest that the neutral mutations seen in introns, pseudogenes, retrotransposons, and the like are in fact not neutral, but rather are strongly selected for.[10] Different mice species share these sequences because the mouse environment forces them to converge. However, SINE and LINE elements are present in tens of thousands of copies in the mammalian genome. It therefore requires an enormous leap of faith (without any supporting evidence) to imagine that all of these copies are being tandemly selected so strongly as to converge with such complete agreement.

### Visual Pigment Evolution

Shozo Yokoyama and colleagues have used an elegant evolutionary argument to predict the functional importance of the different amino acid units in the visual protein *opsin* (Yokoyama et al. 1993; Yokoyama 1994, 1995; Nei et al. 1997; Radlwimmer and Yokoyama 1998;

Yokoyama and Radlwimmer 1998; Yokoyama et al. 1999; Chang and Donoghue 2000). They reasoned that since similarities in sequence within a related group (such as a New World monkeys) are likely due to relatedness, divergences should be caused by selection for different functions (in this case, for peak sensitivities to different wavelengths of light). After making the predictions, the researchers engineered changes to the protein, and measured the wavelength of peak light absorbance. In this way, they identified the specific parts of the protein responsible for variations in the protein's function. Their laboratory results agreed closely to the predictions derived from their evolutionary hypotheses.

Yokoyama's work also has implications for the evolutionary relationships of primates, including humans. For example, if we momentarily ignore substantial amounts of other data and adopt a creationist perspective, we *could* claim that the observation that humans share a large degree of sequence similarity with nonhuman primates provides evidence not of shared ancestry, but of convergence. According to this argument, chimps and humans have similar ways of making a living, and the shared genetic material is due either to a microevolutionary adaptive convergence that has occurred in the time since chimps and humans arose independently, or to some fundamental constraint on the function of the gene product. Since molecular convergence is thought to have occurred in many cases, the neo-creationist claim can not be excluded prima facie. In fact, it represents a coherent scientific model accounting for a set of real observations (albeit in isolation)—in marked contrast to most of other creationist claims. The convergence-vs.-shared-descent distinction is therefore one that, in principle, could generate some testable predictions. In terms of visual pigments, convergence implies that humans and chimps see similarly, and that their shared sequences are in some sense optimized for that mode of seeing by external, selective constraints on the protein shape of the visual pigment. Does this hypothesis make sense in light of what we know about these proteins?

If there were some strong constraint leading to sequence similarity among primates, one could expect that changing the protein in the most similar regions would diminish the effectiveness of the protein. Yokoyama and his colleagues have shown this to not be the case, however. In fact, the protein engineering evidence demonstrates that only a few amino acids are necessary to "tune" the protein to a specific wavelength, with other sites being relatively unimportant to the protein function. With respect to this measurable function, therefore, sequence similarity appears to *not* be caused by convergence. There are other, more general constraints on protein structure, of course. For example, the protein is made up of seven membrane-spanning "alpha-helices." This is true of many different proteins in this class of membrane proteins, however, and the sequence limitations necessary for this structure are very well understood. In conclusion then, shared sequences in this protein (and similarly, in many other proteins not discussed here) occur in the regions that are *not* critical for the specific light absorption functioning of the protein. Rather, the pattern of sequence similarity corresponds closely to our understanding of the evolutionary relationships among living organisms, as derived from completely independent scientific sources.

Perhaps, though, there are other functions for which the protein is responsible. Perhaps the developmental program of the eye relies on a certain structure in the "nonfunctional" regions of the protein. Perhaps the developmental constraints of primates are similar enough that this hypothetical function has led to convergence in sequence in regions currently thought to be neutral. Or perhaps there is truly evidence of design in these molecules. Admittedly, there is an enormous amount that we do *not* know about the details of protein structure, function, and evolution. If neo-creationists were seriously committed to solving these issues rather than simply advocating a theistic philosophy, proponents of "creation-science" or "design-based biology" would adopt experimental systems like this one and use them to test others' and their own theories.

## 7.   Manufacturing Dissent

The "scientific" claims of such neo-creationists as Johnson, Denton, and Behe rely, in part, on the notion that these issues are the subject of suppressed debate among biologists. After all, if their claims had any validity whatsoever, one would expect that some reasonable number of biologists would agree with them, thereby causing the "debate." Consequently, according to neo-creationists, the apparent absence of this discussion and the nearly universal rejection of neo-creationist claims[11] must be due to the conspiracy of suppression among professional biologists instead of a lack of scientific merit. Johnson frequently invokes and emphasizes this supposed conspiracy,[12] while Behe claims that there is a submerged debate and suggests that there is in fact a groundswell of doubt just under the surface. He writes: "a raft of evolutionary biologists examining whole organisms wonder just how Darwinism can account for their observations" (Behe 1996, p. 28).

Let us examine in greater detail some of the evidence Behe offers in support of his claim that the "[scientific] natives are restless" (Behe 1996, p. 28). One citation comes from the evolutionary geneticists Allen Orr and Jerry Coyne, in a paper on the genetics of adaptation (Orr and Coyne 1992). Behe's quote from the paper is shown below in roman type: in his version, the sentence ends with a period after the word "weak." The remainder of the sentence, as actually written, is appended to the quote, and shown here in italics:

We conclude—unexpectedly—that there is little evidence for the neo-Darwinian view: its theoretical foundations and the experimental evidence supporting it are weak, *and there is no doubt that mutations of large effect are sometimes important in evolution.* (Orr and Coyne 1992, p. 725; Behe 1996, p. 29)

We see in Behe's selective quotation some common themes. The first is an overstatement of the scope of the disagreements among biologists, reinforced by the provocative conflation of the concepts of "neo-Darwinian evolution" with "Darwinism." Behe uses the words as synonyms. What Orr and Coyne quite clearly critique is *not* the evolution of new forms by means of molecular and population genetic phenomena or the caricature of descent with modification discussed above (i.e., "Darwinian" evolution, or, as labeled by Behe and Johnson, "Darwinism"). Rather, they address the naïve and once widely held view that all changes in a population's phenotype necessarily come through only gradual accumulation of many mutations of small and nearly equal effect ("neo-Darwinism"). That Orr and Coyne intended this to be the substance of their argument would have been clear to any reader given the opportunity to see the complete sentence as originally written. Behe's selective editing (done without even adding ellipses to indicate that the pruning had taken place) serves only to confuse the issue and emphasize his agenda. In a similar manner, Behe misrepresents the meaning of the following quote, also from Orr and Coyne (1992):

Although much is known about mutation, it is still largely a "black box" relative to evolution. Novel biochemical functions seem to be rare in evolution, and the basis for their origin is virtually unknown. (Behe 1996, p. 29)

The remainder of this paragraph reads:

*Is there a difference in the kinds of mutations producing minor modifications of function and those producing completely novel functions at a biochemical level? What is the relative importance of point mutations, regulatory mutations, exon shuffling, and reading through formerly untranscribed sequences in the production of novel biochemical functions?* (Orr and Coyne 1992, p. 725)

How does the intent of this sentence, as actually written, depart from that attributed to it by Behe? The sentence is clearly the opening of a discussion on the importance of different processes to evolutionary phenomena as a whole, and is certainly not the concession of defeat Behe implies. Finally, it is educational to read the opening paragraph of this paper—cited by Behe as evidence for his "raft" of disgruntled evolutionary biologists:

In the past 20 years a number of acrimonious debates have occurred in evolutionary biology. Unfortunately, these arguments tended to fragment the field and to obscure our understanding of the evolutionary processes in the fashion of the parable of the blind men and the elephant. It is time to try to glimpse evolution as a whole. Evolution consists not of one or two all-important processes (which one depending on the writer!) but rather of an aggregate of processes of various sorts affecting different taxa differently. It is our purpose to identify some of the major processes in organismal evolution and to point out some major gaps in our knowledge. (Orr and Coyne 1992, p. 725)

## 8.   Cracking the Box

As the linguist Noam Chomsky has said, we can distinguish the variety of our ignorance by whether it is a *mystery* or merely a *problem*. We can see how the design theorists deviate from the scientific approach by the ways in which they treat the gaps still present in our understanding. For many centuries, life had been the ultimate mystery, the grandest and blackest of black boxes. Now, with the help of an evolutionary framework, much of our ignorance about life has been reduced to the status of problems—we still do not know all of the answers, but we have an idea about what the answers will look like, and how we might go about finding them. Darwinian evolution may be a flawed model to the extent that, like any model, it is incomplete. However, the opportunities for improvement are vast and continue to be taken advantage of. Creationism and design theory, on the other hand, have had centuries of opportunity to demonstrate their workability as models and have consistently gone nowhere.

It is as if, prior to evolutionary biology, we had been walking at the bottom of a canyon, hoping to understand the topology of the land through which the river flows. Until Darwin, we had a view in one dimension, twisted and obscure. Darwin's insights allowed us to climb out of the canyon and to perceive the vastness of the vista. Parts of the landscape are still obscured, or too distant to make out, but we can see the lay of the land. Furthermore, we can see the hills surrounding us that could provide better vantage points.

The creationists do not see science operating this way. Behe's salient metaphor is rather that of the black box—the complexity of life that is *in principle* not to be understood except in light of some mysterious force. This is an apt metaphor for the neo-creationist aim as well. Given a mechanism of great complexity, they choose to label it a black box, declaring it off limits to further investigation. The greatest cost of invoking design, therefore, is shutting off problems from the light of reason and accumulated understanding. Further travel along these paths is simply halted by authoritarian assertion. Design theory shares much in this regard with other pseudoscientific theories (e.g., astrology, paranormal forces, and alien visitation and abduction). In all such cases, while "evidence" of mysterious forces is asserted, it has not been scientifically substantiated and further scientific avenues are thwarted given the slipperiness of the claimed phenomena.

Like other people, scientists hold various religious beliefs about the origins of the universe, life, and the human mind (Grinnell 1992; Larson and Witham 1997). These beliefs are not "scientific," however, just because they are held by scientists. Rather, scientific judgments are withheld from areas beyond the realm of naturalistic, scientific investigation and inference (National Academy of Sciences 1998). Conversely, when it comes to unraveling scientific problems, most practicing scientists, regardless of their religious beliefs, refuse to invoke the existence of unknown supernatural forces—even in the absence of known naturalistic mechanisms. This suspended judgment is accompanied by the hope that human minds will eventually find a crack in the apparently impenetrable surface of the mystery. Such a crack will lead to new approaches to the problem and, eventually, to its solution. In contrast to this model of scientific behavior, Behe periodically speaks to his desire to *remystify*, rather than demystify biological processes. For example, he approvingly underscores the contention that since "... the information needed to begin life could not have developed by chance ... life [should] be considered a given, like matter or energy" (Behe 1996, p. 29). If life is to be considered "a given," then what possible role could a science of biology have? This position illustrates the absurd paradox of the call for a theistic or creationist science: in the eyes of Behe, Johnson, and colleagues, the most important, intriguing mysteries *must* remain mysterious, untouched by scientific inference. Even in the domains where we believe we've begun to make problems of mysteries, Behe and Johnson would have us abandon both what we've already gained and whatever promise exists for the future. In return, they offer us the fragmented insights of their parochial vitalism.

## 9.    What's at Stake?

Seventy-six years after the Scopes trial, the U.S. public remains uncertain about the relationship between evolutionary biology and alternative ideas of creationism (Larson 1997). Controversies regarding the teaching of evolution (as well as other science) and creationism continue to appear regularly in school districts across the nation and recent polls indicate substantial ambivalence about the teaching of evolution and creationism (Moore 1999; DYG 2000). Although a majority of Americans supports the teaching of evolution in science classes, many are unclear about what "evolution" means, as well as its status within science. The public generally believes that creationism should *not* be taught as a scientific alternative to evolution, though approximately half still consider their opinions subject to change. Many people also find no inconsistency between evolution and a belief that "God created humans and guided their development," and, relative to other fundamental theories (e.g., relativity), they question evolution's status as a proven "theory" to a much greater extent (DYG 2000). With such results, one can conclude that the public simply lacks detailed understanding of both the relevant scientific and religious issues. What is the significance of this conflicting embrace of science and religious parochialism, modernity and tradition? Does it matter that the public support for science seems wide but shallow?

Since contemporary human welfare depends so much on science and technology, and both science and science education in turn depend on public financial support, the stakes are large (Futuyma et al. 1998). To the extent that the science is misrepresented in the classroom and misunderstood by the public, support for it and its products are jeopardized. Various advocates for "scientific creationism" attempt to foster and take advantage of this uncertainty in different ways, though all seem to believe that their religious agendas are strengthened as the public understanding of science wanes. In this essay, we have focused on several neo-creationist writers who pursue this objective somewhat more subtly than many of their ideological allies (Numbers 1992; Kitcher, this volume). Seeming to embrace the scientific literature, they claim to judge naturalistic science by its own standards of evidence and logic. In reality, however, they seek to return biology to its disconnected roots in past centuries through misrepresentation, caricature, innuendo, and mysticism. Thinly disguising their ideological commitments through design theory, they declare the evidence supporting naturalistic evolution to be terminally weak. The only alternative, they then suggest, is for evolutionary science to adopt intelligent design and purpose: instead of a naturalistic synthesis, we need a theistic antisynthesis. Despite the new garb, such logic parallels that of their discredited creationist predecessors, and if extended, might return astronomy to the Ptolemaic system, geology to Noachian catastrophism, and medicine to bloodletting. Given the enormous challenges facing humanity in a rapidly changing environment, there is abundant need for spiritual problem-solving in the world. Similarly, the need for greater scientific understanding remains, and we must not let tired, erroneous, and dogmatic creationist assertions distract science from the successful methodologies and logic used to solve scientific problems.

## Notes

1. See the "Updates" section in issues of *Reports of the National Center for Science Education* for recent educational challenges to evolutionary science. The NCSE web site also provides headlines regarding current policy actions relating to evolution: http://www.natcenscied.org/headlines.htm.

2. For more discussion of the neo-creationists, see Pennock (1999) and Kitcher, this volume.

3. Admittedly, such examples of "sympatric" speciation (i.e., speciation where the diverging populations live in the same place) are relatively uncommon.

"Allopatric" or geographically separated populations (as in Hawaiian *Drosophila* in different valleys or sticklebacks invading isolated lakes) are thought to speciate much more easily owing to the absence of gene flow from interbreeding ("hybridization") that may keep speciating groups together. Although uncommon, however, sympatric cases are the best demonstrations of the potential power of natural selection to drive speciation. Even in the face of hybridization, given multiple resources that allow specialization, selection can quickly drive subpopulations to diverge into separate races and then species.

4. Note that many "types" mentioned by Johnson, such as trees, fishes, and reptiles, are not even recognized as natural, monophyletic groups by contemporary scientists (Futuyma 1998). Monophyletic groups are those that include all descendants from a common ancestor. Trees have evolved multiple times in many different plant families and would therefore be considered a functional, polyphyletic category by botanists. Polyphyletic groups are taxonomic entities that include species derived from multiple, distinct ancestral lineages. Fishes and reptiles are considered to be paraphyletic functional grades since these groups do not typically include all descendants of their most recent common ancestors. Reptilia, for example, only becomes a monophyletic group when it includes all descendant birds and mammals.

5. See more discussion of this passage in Kitcher, this volume.

6. Although this example is hypothetical, it reflects actual genetic and genealogical studies underway. For example, Ashkenazic Jews, because of their relative lack of mixing with other populations, have long been used by medical researchers to trace the evolution and inheritance of certain genetic conditions (see, e.g., Risch et al. 1995; Laken et al. 1997). Additionally, Icelanders, constituting another relatively homogeneous group, are currently being studied for similar genetic information (Marshall 1997; Arngrímsson et al. 1999).

7. See further comments on this passage in Kitcher, this volume.

8. An organism may also have several related versions of a protein gene that may vary by sequence and precise function.

9. Other types of molecular fossils are well known. In any stretch of DNA in which the precise sequence is not under selection (for example, in the "intron" spacer regions within a gene, or in "pseudogenes," nonfunctional relicts of gene copies), mutations may accumulate without cost to the organism. These mutations contain a wealth of information about a lineage's evolutionary history.

10. Behe addresses the pseudogene question by saying, essentially, that maybe they have a cryptic function that we haven't yet discovered. He also suggests that the mechanism for duplicating genes, and thereby allowing for the origin of pseudogenes, is another "irreducibly complex" system that could not have arisen by itself. Finally, he gives the answer that even if pseudogenes are nonfunctional, and can be duplicated by existing cellular or viral machinery, that they still could represent the visible decay of the system that has occurred over the past several billion years since the original act of creation. Here, Behe's first two arguments are incompatible with his third. Except for his "hypothetical cryptic function" hypothesis (which still fails to account for the precision matching between

humans and chimps of the intron's insertion points), Behe's objections do not even address the substance of the evidence offered.

11. See the National Center for Science Education web site for a list of statements from scientific organizations regarding creationism: http://www.natcenscied.org/voicont.htm # sciorg.

12. Speaking of Darwinists, Johnson refers to this suppression: "to reject that possibility [of 'some orienting force'] because it is 'disturbing' is to imply that [they think] it is better to stick with a theory which is against the weight of the evidence than to admit that the problem is unsolved" (Johnson 1991, p. 159).

# References

Arngrímsson, R., S. Sigurardóttir, M. L. Frigge, R. I. Bjarnadóttir, T. Jónsson, H. Stefánsson, Á. Baldursdóttir, A. S. Einarsdóttir, B. Palsson, S. Snorradóttir, A. M. A. Lachmeijer, D. Nicolae, A. Kong, B. T. Bragason, J. R. Gulcher, R. T. Geirsson, and K. Stefánsson. 1999. A genome-wide scan reveals a maternal susceptibility locus for pre-eclampsia on chromosome 2p13. *Human Molecular Genetics* 8: 1799–1805.

Bargelloni, L., S. Marcato, L. Zane, and T. Patarnello. 2000. Mitochondrial phylogeny of notothenioids: A molecular approach to Antarctic fish evolution and biogeography. *Syst. Biol.* 49: 114–129.

Behe, M. J. 1996. *Darwin's Black Box: The Biochemical Challenge to Evolution.* Free Press, New York, N.Y.

Berlocher, S. H. 1999. Host race or species? Allozyme characterization of the "flowering dogwood fly," a member of the *Rhagoletis pomonella* complex. *Heredity* 83: 652–662.

Bouton, N., F. Witte, J. J. M. van Alphen, A. Schenk, and O. Seehausen. 1998. Local adaptations in populations of rock-dwelling haplochromines (*Pisces: Cichlidae*) from southern Lake Victoria. *Proc. R. Soc. Lond.* B 266: 355–360.

Carson, H. L. 1996. Sexual selection: A driver of genetic change in Hawaiian *Drosophila*. *J. Heredity* 88: 343–352.

Carson, H. L. and D. A. Clague. 1995. Geology and biogeography of the Hawaiian Islands. In W. L. Wagner and V. A. Funk (eds.), *Hawaiian Biogeography: Evolution on a Hot Spot Archipelago*. Smithsonian Institution Press, Washington, D.C.: 14–29.

Chang, B. S. W. and M. J. Donoghue. 2000. Recreating ancestral proteins. *Trends. Ecol. Evol.* 15: 109–114.

Cheetham, A. H. 1986. Tempo of evolution in a Neogene bryozoan: Rates of morphologic change within and across species boundaries. *Paleobiology* 12: 190 202.

Chen, L., A. L. DeVries, and C.-H. C. Cheng. 1997a. Convergent evolution of antifreeze glycoproteins in Antarctic notothenioid fish and Arctic cod. *Proc. Natl. Acad. Sci.* USA 94: 3817–3822.

Chen, L., A. L. DeVries, and C.-H. C. Cheng. 1997b. Evolution of antifreeze glycoprotein gene from a trypsinogen gene in Antarctic notothenioid fish. *Proc. Natl. Acad. Sci.* USA 94: 3811–3816.

Chiappe, L. M. 1995. The first 85 million years of avian evolution. *Nature* 378: 349–355.

Dayhoff, M. O. 1972. *Atlas of Protein Structure and Function*. National Biomedical Research Foundation, Silver Spring, Md.

Dembski, W. 1998. The intelligent design movement. Discovery Institute, Center for the Renewal of Science and Culture, Seattle, Wash. (http://articles.discovery. org/CRSCdbEngine.php3?id=121)

Denton, M. 1985. *Evolution: A Theory in Crisis*. Adler & Adler, Bethesda, Md.

Desmond, A. 1984. *Archetypes and Ancestors: Palaeontology in Victorian London 1850–1875*. University of Chicago Press, Chicago, Ill.

DeVries, A. L. 1971. Glycoproteins as biological antifreeze agents in Antarctic fishes. *Science* 172: 1152–1155.

DeVries, A. L. and D. E. Wohlschlag. 1969. Freezing resistance in some Antarctic fishes. *Science* 163: 1073–1075.

Dingus, L. and T. Rowe. 1998. *The Mistaken Extinction: Dinosaur Evolution and the Origin of Birds*. W. H. Freeman, New York, N.Y.

DYG. 2000. Evolution and creationism in public education: An in-depth reading of public opinion (results of a comprehensive, national survey). People for the American Way Foundation, Washington, D.C. (http://www.pfaw.org/issues/ education/creationism-poll.pdf)

Feder, J. L., C. A. Chilcote, and G. L. Bush. 1988. Inheritance and linkage relationships in the apple maggot fly. *J. Heredity* 80: 277–283.

Feder, J. L., C. A. Chilcote, and G. L. Bush. 1990. The geographic pattern of genetic differentiation between host associated populations of *Rhagoletis pomonella* (*Diptera: Tephritidae*) in the eastern United States and Canada. *Evolution* 44: 570–594.

Feder, J. L., S. B. Opp, B. Wlazlo, and K. W. Reynolds. 1994. Host fidelity is an effective pre-mating barrier between sympatric races of the Apple Maggot fly *Rhagoletis pomonella*. *Proc. Natl. Acad. Sci.* USA 91: 7990–7994.

Freeman, S. and J. C. Herron. 1998. *Evolutionary Analysis*. Prentice Hall, Upper Saddle River, N.J.

Futuyma, D. J. 1986. *Evolutionary Biology* (2nd ed.). Sinauer Associates, Sunderland, Mass.

Futuyma, D. J. 1998. *Evolutionary Biology* (3rd ed.). Sinauer Associates, Sunderland, Mass.

Futuyma, D. J., T. R. Meagher, M. J. Donoghue, C. H. Langley, L. Maxson, A. F. Bennett, H. J. Brockmann, M. W. Feldman, W. M. Fitch, L. R. Godfrey, J. Hanken, D. Jablonski, C. B. Lynch, L. Real, M. A. Riley, J. J. Sepkoski, Jr., and V. B. Smocovitis. 1998. Evolution, Science and Society: Evolutionary Biology and the National Research Agenda. A. P. Sloan Foundation, National Science Foundation, Washington, D.C. (http://www.rci.rutgers.edu/~ecolevol/fulldoc.html)

Gonzalez, P., P. V. Rao, S. B. Nuñez, and J. S. Zigler, Jr. 1995. Evidence for independent recruitment of *zeta*-crystallin/quinone reductase (CRYZ) as a crystallin in camelids and hystricomorph rodents. *Mol. Biol. Evol.* 12: 773–781.

Greenwood, M. R. C. and K. Kovacs North. 1999. Science through the looking glass: Winning the battles but losing the war? *Science* 286: 2072–2078.

Grinnell, F. 1992. *The Scientific Attitude* (2nd ed.). Guilford Press, New York, N.Y.

Hendriks, W., J. Leunissen, E. Nevo, H. Bloemendal, and W. W. DeJong. 1987. The lens protein *alpha*-crystallin of the blind mole rat, *Spalax ehrenbergi*: Evolutionary change and functional constraints. *Proc. Natl. Acad. Sci. USA* 84: 5320–5324.

Hillis, D. M. 1999. SINEs of the perfect character. *Proc. Natl. Acad. Sci. USA* 96: 9979–9981.

Hillis, D. M., J. P. Huelsenbeck, and C. W. Cunningham. 1994. Application and accuracy of molecular phylogenies. *Science* 364: 671–677.

Hillis, D. M., C. Moritz, and B. K. Mable. 1996. *Molecular Systematics* (2nd ed.). Sinauer Associates, Sunderland, Mass.

Holden, C. 1999. Breakdown of the year: Creationists win in Kansas. *Science* 286:2242.

Huey, R. B., G. W. Gilchrist, M. L. Carlson, D. Berrigan, and S. Luís. 2000. Rapid evolution of a geographic cline in size in an introduced fly. *Science* 287: 308–309.

Jackson, J. B. C. and A. H. Cheetham. 1994. Phylogeny reconstruction and the tempo of speciation in cheilostome Bryozoa. *Paleobiology* 20: 407–423.

Jackson, J. B. C. and A. H. Cheetham. 1999. Tempo and mode of speciation in the sea. *Trends. Ecol. Evol.* 14: 72–77.

Johnson, P. E. 1991. *Darwin on Trial*. Regnery Gateway, Washington, D.C.

Kaneshiro, K. T. 1998. Hawaiian Drosophilidae: Distribution, species groupings, and taxonomic treatment. Unpublished.

Kaneshiro, K. Y., R. G. Gillespie, and H. L. Carson. 1995. Chromosomes and male genitalia of Hawaiian *Drosophila*. In W. L. Wagner and V. A. Funk (eds.), *Hawaiian Biogeography: Evolution on a Hot Spot Archipelago*. Smithsonian Institution Press, Washington, D.C.: 57–71.

Kemp, T. S. 1988. Interrelationships of the Synapsida. In M. J. Benton (ed.), *The Phylogeny and Classification of the Tetrapods*, vol. 2. Clarendon Press, Oxford, U.K.

Laken, J. J., G. M. Petersen, S. B. Gruber, C. Oddoux, H. Ostrer, F. M. Giardiello, S. R. Hamilton, H. Hampel, A. Markowitz, D. Klimstra, S. Jhanwar, S. Winawer, K. Offit, M. C. Luce, K. W. Kinzler, and B. Vogelstein 1997. Familial colorectal cancer in Ashkenazim due to a hypermutable tract in APC. *Nature Genetics* 17: 79–83.

Larson, E. J. 1997. *Summer for the Gods: The Scopes trial and America's continuing debate over science and religion.* BasicBooks, New York, N.Y.

Larson, E. J. and L. Witham. 1997. Scientists are still keeping the faith. *Nature* 386: 435–436.

Li, W.-H. 1997. *Molecular Evolution.* Sinauer Associates, Sunderland, Mass.

Marshall, E. 1997. Human genetics: Tapping Iceland's DNA. *Science* 278:566.

May, R. M., J. H. Lawton, and N. E. Stork. 1995. Assessing extinction rates. In J. H. Lawton and R. M. May (eds.), *Extinction Rates.* Oxford University Press, Oxford, U.K.: 1–24.

Mayr, E. 1988. *Toward a New Philosophy of Biology: Observations of an Evolutionist.* Harvard University Press, Cambridge, Mass.

Mayr, E. and W. B. Provine. 1980. *The Evolutionary Synthesis: Perspectives on the Unification of Biology.* Harvard University Press, Cambridge, Mass.

McPhail, J. D. 1994. Speciation and the evolution of reproductive isolation in the sticklebacks (Gasterosteus) of south-western British Columbia. In M. A. Bell and S. A. Foster (eds.), *The Evolutionary Biology of the Threespine Stickleback.* Oxford University Press, Oxford, U.K.: 399–437.

Moore, D. W. 1999. Americans support teaching Creationism as well as evolution in public schools, Gallup News Service (30 August 1999). (http://www.gallup.com/poll/releases/pr990830.asp)

National Academy of Sciences. 1998. Teaching About Evolution and the Nature of Science. National Academy Press, Washington, D.C. (http://bob.nap.edu/html/evolution98/)

National Academy of Sciences. 1999. *Science and Creationism: A View from the National Academy of Sciences* (2nd ed.). National Academy Press, Washington, D.C. (http://stills.nap.edu/html/creationism/)

National Science Board. 1998a. Failing our children: Implications of the Third International Mathematics and Science Study. NSB-98-154. National Science Foundation, Arlington, Virginia (http://www.nsf.gov/nsb/documents/1998/nsb98154/nsb98154.htm)

National Science Board. 1998b. Science & Engineering Indicators. NSB-98-1. National Science Foundation, Arlington, Virginia (http://www.nsf.gov/sbe/srs/seind98/start.htm)

Nei, M., J. Zhang, and S. Yokoyama. 1997. Color vision of ancestral organisms of higher primates. *Mol. Biol. Evol.* 14: 611–618.

Nikaido, M., A. P. Rooney, and N. Okado. 1999. Phylogenetic relationships among cetartiodactyls based on insertions of short and long interspersed elements: Hippopotamuses are the closest extant relatives of whales. *Proc. Natl. Acad. Sci.* USA 96: 10261–10266.

Numbers, R. L. 1992. *The Creationists.* A. A. Knopf, New York, N.Y.

Orr, H. A. and J. A. Coyne. 1992. The genetics of adaptation: A reassessment. *Amer. Nat.* 140: 725–742.

Pagel, M. 1999. Inferring the historical pattern of biological evolution. *Nature* 401: 877–884.

Panchen, A. L. 1992. *Classification, Evolution, and the Nature of Biology.* Cambridge University Press, Cambridge, U.K.

Pennock, R. T. 1999. *Tower of Babel: The Evidence against the New Creationism.* MIT Press, Cambridge, Mass.

Pough, F. H., C. M. Janis, and J. B. Heiser (eds.). 1999. *Vertebrate Life* (5th ed.). Prentice Hall, Upper Saddle River, N.J.

Radlwimmer, F. B. and S. Yokoyama. 1998. Genetic analyses of the green visual pigments of rabbit (*Oryctolagus cuniculus*) and rat (*Rattus norvegicus*). *Gene* 218: 103–109.

Risch, N., D. de Leon, L. Ozelius, P. Kramer, L. Almasy, B. Singer, S. Fahn, X. Breakefield, and S. Bressman. 1995. Genetic analysis of idiopathic torsion dystonia in Ashkenazi Jews and their recent descent from a small founder population. *Nature Genetics* 9: 152–159.

Rudwick, M. J. S. 1985. *The Great Devonian Controversy: The Shaping of Scientific Knowledge among Gentlemanly Specialists.* University of Chicago Press, Chicago, Ill.

Rundle, H. D., L. Nagel, J. W. Boughman, and D. Schluter. 2000. Natural selection and parallel speciation in sympatric sticklebacks. *Science* 287: 306–308.

Shedlock, A. M. and N. Okada. 2000. SINE insertions: Powerful tools for molecular systematics. *BioEssays* 22: 148–160.

Simpson, G. G. 1944. *Tempo and Mode in Evolution.* Columbia University Press, New York, N.Y.

Stiassny, M. L. J. and A. Meyer. 1999. Cichlids of the Rift Lakes. *Sci. Amer.* 280: 64–69.

Taylor, E. B. 1999. Species pairs of north temperate freshwater fishes: Evolution, taxonomy, and conservation. *Reviews in Fish Biology and Fisheries* 9: 299–324.

Templeton, A. R. 1977. Analysis of head shape differences between two interfertile species of Hawaiian *Drosophila*. *Evolution* 31: 630–641.

Tomarev, S. I., S. Chung, and J. Piatagorsky. 1995. Glutathione-S-transferase and S-crystallins of cephalopods: Evolution from active enzyme to lens-refractive proteins. *J. Mol. Evol.* 41: 1048–1056.

Val, F. C. 1977. Genetic analysis of the morphological differences between two interfertile species of Hawaiian *Drosophila*. *Evolution* 31: 611–629.

Verneau, O., F. Catzeflis, and A. V. Furano. 1998. Determining and dating recent rodent speciation events by using L1 (LINE-1) retrotransposons. *Proc. Natl. Acad. Sci.* USA 95: 11284–11289.

Wimmer, E. A., A. Carleton, P. Harjes, T. Turner, and C. Desplan. 2000. *bicoid*-Independent Formation of Thoracic Segments in *Drosophila*. *Science* 287: 2476–2479.

Yokoyama, S. 1994. Gene duplications and evolution of the short wavelength-sensitive visual pigments in vertebrates. *Mol. Biol. Evol.* 11: 32–39.

Yokoyama, S. 1995. Amino acid replacements and wavelength absorption of visual pigments in vertebrates. *Mol. Biol. Evol.* 12: 53–61.

Yokoyama, S. and F. B. Radlwimmer. 1998. The "five-sites" rule and the evolution of red and green color vision in mammals. *Mol. Biol. Evol.* 15: 560–567.

Yokoyama, S., W. T. Starmer, and R. Yokoyama. 1993. Paralogous origin of the red- and green-sensitive visual pigment genes in vertebrates. *Mol. Biol. Evol.* 10: 527–538.

Yokoyama, S., H. Zhang, F. B. Radlwimmer, and N. S. Blow. 1999. Adaptive evolution of color vision of the Comoran coelacanth (*Latimeria chalumnae*). *Proc. Natl. Acad. Sci. USA* 96: 6279–6284.

# V

## Plantinga's Critique of Naturalism and Evolution

This section extends the debate about scientific naturalism begun in section II, and to a lesser extent continues discussion of some of the questions raised in section III. In the first article. Alvin Plantinga argues against *methodological* naturalism. He looks at some of the reasons that have been given in support of it and finds them wanting. He does find a couple of reasons that he agrees support it in part, but argues that these do not also support the suggestion that science is religiously neutral. He concludes that "there is little to be said for methodological naturalism" and recommends that Christians pursue their own "Augustinian science" that takes seriously Augustine's image of the struggle between the City of God and the City of Man, and is "devoted to the worship and service of the Lord." This article is Plantinga's condensed version of a paper that he originally presented at the 1997 Naturalism, Theism, and the Scientific Enterprise Conference organized by Robert Koons (see Barbara Forrest's paper in section I), and which Koons picked for the special issue of *Perspectives on Science & Christian Faith* that he guest-edited that was devoted to the conference.

In the second essay, written for this volume, philosopher of biology Michael Ruse gives his own take on the difference between metaphysical and methodological naturalism and responds to Plantinga's arguments. He defends methodological naturalism and argues that it is not a philosophy opposed to theism. Ruse goes on to criticize Plantinga's proposal for an Augustinian science and argues that, despite Plantinga's claims to the contrary, it is a kind of "God-of-the-gaps" theology and is not a good assumption for either scientist or theologian. He concludes that Plantinga "has given us no reasons to give up on methodological naturalism, or inasmuch as he has it has been only because of his prior commitment to his own version of Christian theism."

The third and fourth articles critique other arguments Plantinga made against naturalism and evolution in his book *Warrant and Proper Function* (Oxford, 1993) and in other papers. Plantinga argues that a certain kind of Christian theism can provide what he calls "proper warrant" for human knowledge, while an epistemology based on metaphysical naturalism combined with evolution cannot.

In "Plantinga's Case against Naturalistic Epistemology," philosopher Evan Fales looks at Plantinga's argument that naturalists are committed to a neo-Darwinian account of human origins, and that the reliability of our cognitive faculties is improbable or unknown relative to that theory, whereas theism underwrites cognitive reliability. Fales argues, to the contrary, that neo-Darwinism provides strong reasons for expecting general cognitive reliability, whereas the likelihood of that relative to theism is unknowable.

In "Plantinga's Probability Arguments against Evolutionary Naturalism," philosophers Branden Fitelson and Elliott Sober focus on the probability arguments Plantinga gives for his conclusion. In a preliminary argument, Plantinga tries to show that the conjunction of evolution and naturalism is probably false, given that our psychological mechanisms for forming beliefs are generally reliable. In a second argument, he tries to show that the conjunction of evolution and naturalism is self-defeating. Fitelson and Sober show that both arguments contain serious errors.

# 13

## Methodological Naturalism?

Alvin Plantinga

Unmatched for sweep and eloquence, St. Augustine's *De Civitas Dei* is a magnificently powerful expression of a view of human history that has been taken up by a host of later Christians.[1] According to that view, human history involves a struggle, a contest, a battle between what he calls the *Civitas Dei*, the City of God, on the one hand, and, on the other, the City of the World or the City of Man. The former is devoted to the worship and service of the Lord; the latter serves quite a different master. Augustine believes that all of human history is to be understood in terms of this struggle, and nearly any cultural endeavor of any size or significance is involved in it. Now modern natural science is an enormously important aspect of contemporary intellectual life. There are of course those naysayers who see in it no more than technology, no more than a means of serving such practical ends as fighting disease and building bridges or space vehicles. But surely they are wrong. Science has indeed done these important things, but it has done more: it has also given us powerful insights into ourselves and into the world God has created. Science has transformed our intellectual landscape; it is difficult even to imagine what our intellectual life would be without it. If we follow Augustine, we should therefore expect that science, too, plays an important role in the contest he describes.

According to an idea widely popular ever since the Enlightenment, however, science (at least when properly pursued) is a cool, reasoned,

The original version of the paper was published in *Facets of Faith Science, Vol. 1: Historiography and Modes of Interaction*, ed. Jitse M. van der Meer (Lanham, MD: University Press of America, 1996, 177–221). This condensed version appeared in *Perspectives on Science and Christian Faith* (1997, vol. 49, no. 3), and in *Origins and Design* (1997, vol. 18, nos. 1 and 2).

wholly dispassionate[2] attempt to figure out the truth about ourselves and our world, entirely independent of ideology, or moral convictions, or religious or theological commitments. Of course this picture has lately developed some cracks. It is worth noting, however, that sixteen centuries ago Augustine provided the materials for seeing that this common conception cannot really be correct. It would be excessively naive to think that contemporary science is religiously and theologically neutral, standing serenely above that Augustinian struggle and wholly irrelevant to it. Perhaps *parts* of science are like that: the size and shape of the earth and its distance from the sun, the periodic table of elements, the proof of the Pythagorean Theorem—these are all in a sensible sense religiously neutral. But many other areas of science are very different; they are obviously and deeply involved in this clash between opposed worldviews. There is no neat recipe for telling which parts of science are neutral with respect to this contest and which are not, and of course what we have here is a continuum rather than a simple distinction. But here is a rough rule of thumb: the relevance of a bit of science to this contest depends upon how closely that bit is involved in the attempt to come to understand ourselves as human beings. Perhaps there is also another variable: how "theoretical" the bit in question is, in the sense of being directed at *understanding* as opposed to control.

It would be of great interest to explore this area further, to try to say precisely what I mean in saying that science is not religiously neutral, to see in exactly what ways Christianity bears on the understanding and practice of the many relevantly different sciences and parts of science. The first is not the focus of this paper, however; and the second question (of course) requires vastly more knowledge of science than I can muster. That is a question not just for philosophers, but for the Christian community of scientists and philosophers working together. What I shall do instead is vastly more programmatic. I shall argue that a Christian academic and scientific community ought to pursue science in its own way, *starting from* and taking for granted what we know as Christians. (This suggestion suffers from the considerable disadvantage of being at present both unpopular and heretical; I shall argue, however, that it also has the considerable advantage of being correct.) Now one objection to this suggestion is enshrined in the dictum that science done properly necessarily involves "methodological naturalism" or (as Basil Willey calls it) "provisional atheism."[3] This is the idea that science, properly so-called, cannot involve religious belief or commitment. My main aim in this paper is to explore, understand, discuss, and evaluate this claim and the arguments for it. I am painfully aware that what I have to say is tentative and incomplete, no more than a series of suggestions for research programs in Christian philosophy.

## Weak Arguments for Methodological Naturalism

The natural thing to think is that (in principle, at any rate) the Christian scholarly community should do science, or parts of science, in its own way and from its own perspective. What the Christian community really needs is a science that takes into account what we know as Christians. Indeed, this seems the rational thing in any event; surely the rational thing is to use *all* that you know in trying to understand a given phenomenon. But then in coming to a scientific understanding of hostility, or aggression, for example, should Christian psychologists not make use of the notion of sin? In trying to achieve scientific understanding of love in its many and protean manifestations, for example, or play, or music, or humor, or our sense of adventure, should we also not use what we know about human beings being created in the image of God, who is himself the very source of love, beauty, and the like? And the same for morality? Consider that enormous, and impressive, and disastrous Bolshevik experiment of the twentieth century, perhaps the outstanding feature of the twentieth century political landscape: in coming to a scientific understanding of it, should Christians not use all that they know about human beings, including what they know by faith?

True: there could be *practical* obstacles standing in the way of doing this; but in principle, and abstracting from these practical difficulties (which in any event may be more bark than bite), the right way for the Christian community to attain scientific understanding of, say, the way human beings are and behave, would be to start from what we know about human beings, including what we know by way of faith. Hence the sorts of hypotheses we investigate might very well involve such facts (as the Christian thinks) as that we human beings have been created by God in his image, and have fallen into sin. These "religious" ideas might take a place in our science by way of explicitly entering various hypotheses.

They might also play other roles: for example, they might be part of the background information with respect to which we evaluate the various scientific hypotheses and myths that come our way.

I say this is the natural thing to think: oddly enough, however, the *denial* of this claim is widely taken for granted; as a matter of fact, it has achieved the status of philosophical orthodoxy. Among those who object to this claim are Christian thinkers with impressive credentials. Thus Ernan McMullin:

> But, of course, methodological naturalism does not restrict our study of nature; it just lays down which sort of study qualifies as *scientific*. If someone wants to pursue another approach to nature—and there are many others—the methodological naturalist has no reason to object. Scientists *have* to proceed in this way; the methodological of natural science gives no purchase on the claim that a particular event or type of event is to be explained by invoking God's creative action directly.[4]

Part of the problem, of course, is to see more clearly what this methodological naturalism *is*. Precisely what does it come to? Does it involve an embargo only on such claims as that a particular event is to be explained by invoking God's creative action *directly*, without the employment of "secondary causes"? Does it also proscribe invoking God's *indirect* creative action in explaining something scientifically? Does it pertain only to scientific *explanations*, but not to other scientific assertions and claims? Does it also preclude using claims about God's creative action, or other religious claims as part of the background information with respect to which one tries to assess the probability of a proposed scientific explanation or account? We shall have to look into these matters later. At the moment however, I want to look into a different question: what reason is there for accepting the claim that science does indeed involve such a methodological naturalism, however exactly we construe the latter? I shall examine some proposed reasons for this claim and find them wanting. I shall then argue that nevertheless a couple of very sensible reasons lie behind at least part of this claim. These reasons, however, do not support the suggestion that science is religiously neutral.

Well then, what underlies the idea that science in some way necessarily involves this principle of methodological naturalism? First, and perhaps most important: this conception of science is an integral and venerable part of the whole conception of faith and reason we have inherited from the Enlightenment. I do not have the space to treat this topic with anything like the fullness it deserves; but the central idea, here, is that science is objective, public, sharable, publicly verifiable, and equally available to anyone, whatever their religious or metaphysical proclivities. We may be Buddhist, Hindu, Protestant, Catholic, Muslim, Jew, Bahai, none of the above: the findings of science hold equally for all of us. This is because proper science, as seen by the Enlightenment, is restricted to the deliverances of *reason* and *sense* (perception) which are the same for all people. Religion, on the other hand, is private, subjective, and obviously subject to considerable individual differences. But then if science *is* indeed public and sharable by all, then of course one cannot properly pursue it by starting from some bit of religious belief or dogma.

One root of this way of thinking about science is a consequence of the modern foundationalism stemming from Descartes and perhaps even more importantly, Locke. Modern classical foundationalism has come in for a lot of criticism lately, and I do not propose to add my voice to the howling mob.[5] And since the classical foundationalism upon which methodological naturalism is based has run aground, I shall instead consider some more local, less grand and cosmic reasons for accepting methodological naturalism.

### Methodological Naturalism Is True by Definition

So *why* must a scientist proceed in accordance with methodological naturalism? Michael Ruse suggests that methodological naturalism or at any rate part of it is *true by definition*:

Furthermore, even if Scientific Creationism were totally successful in making its case as science, it would not yield a *scientific* explanation of origins. Rather, at most, it could prove that science shows that there can be *no* scientific explanation of origins. The Creationists believe that the world started miraculously. But miracles lie outside of science, which by definition deals only with the natural, the repeatable, that which is governed by law.[6]

Ruse suggests that methodological naturalism is true by definition of the term "science" one supposes; Ruse apparently holds there is a correct definition of "science," such that from the definition it follows that science deals only with what is natural, repeatable, and governed by law. (Note that this claim does not bear on the suggestions that a Christian scientist can propose hypotheses involving such "religious" doctrines as, say, original sin, and can evaluate the epistemic probability of a scientific hypothesis relative to background belief that includes Christian belief.) Ruse's claim apparently rules out hypotheses that include references to God: God is a supernatural being, hypotheses referring to him therefore deal with something besides the natural; hence such hypotheses cannot be part of science.

Three things are particularly puzzling about Ruse's claim. First, enormous energy has been expended, for at least several centuries, on the "demarcation problem": the problem of giving necessary and sufficient conditions for distinguishing science from other human activities.[7] This effort has apparently failed; but if in fact there *were* a definition of the sort Ruse is appealing to, then presumably there would be available a set of necessary and sufficient conditions for something as being science. Ruse does not address the many and (I think) successful arguments for the conclusion that there is no such set of necessary and sufficient conditions, let alone such a definition of the term "science"; he simply declares that—by definition—science has the properties he mentions.

Second, Ruse here proposes three properties that he says are by definition characteristic of any bit of science: that bit deals with things that (a) are repeatable, (b) are merely natural, and (c) are governed by natural law. But take repeatability, and consider this passage by Andrei Linde: speaking of the Big Bang, he says, "One might think it very difficult to extract useful and reliable information from the unique experiment carried out about $10^{10}$ years ago."[8] According to Linde, the Big Bang is unique and therefore, presumably, unrepeatable—at any rate it *might* turn out to be unrepeatable. If so, would we be obliged to conclude that contemporary cosmological inquiries into the nature of the Big Bang and into the early development of the universe are not really part of science?

Consider next the property of being governed by law. The first point, here, would be that the very existence of natural law is controversial; Bas van Fraassen, for example, has given an extended and formidable argument for the conclusion that there are no natural laws.[9] There are *regularities*, of course, but a regularity is not yet a law; a law is what is supposed to *explain* and *ground* a regularity. Furthermore, a law is supposed to hold with some kind of *necessity*, typically thought to be less stringent than broadly logical necessity, but necessity nonetheless.[10] This idea of lawfulness, I think, is an inheritance of Enlightenment deism (see below); and perhaps here as elsewhere Enlightenment deism misses the mark. Perhaps the demand for law cannot be met. Perhaps there are regularities, but no laws; perhaps there is nothing like the necessity allegedly attaching to laws. Perhaps the best way to think of these alleged laws is as universally or nearly universally quantified counterfactuals of divine freedom.[11] So suppose van Fraassen is right and there are no natural laws: would it follow by definition that there is not any science? That seems a bit strong. Further, it could be, for all we know, that there are some laws, but not everything is governed by them (or wholly governed by them). Perhaps this is how it is with earthquakes, the weather, and radioactive decay. Would it follow that one could not study these things scientifically?

The third puzzling thing about Ruse's claim: it is hard to see how anything like a reasonably serious dispute about what is and is not science could be settled just by appealing to a *definition*. One thinks this would work only if the original query were really a *verbal* question—a question like *Is the English word "science" properly applicable to a hypothesis that makes reference to God?* But that was not the question: the question is instead *Could a hypothesis that makes reference to God be part of science? That* question cannot be answered just by citing a definition.

### "Functional Integrity" Requires Methodological Naturalism?

Diogenes Allen, John Stek, and Howard Van Till give answers of that sort. According to Van Till, God has created a world characterized by "functional integrity":

By this term I mean to denote a created world that has no functional deficiencies, no gaps in its economy of the sort that would require God to act immediately, temporarily assuming the role of creature to perform functions within the economy of the created world that other creatures have not been equipped to perform.[12]

Note first that Van Till seems to be directing his fire at only one of the several ways in which Christians might employ what they know by faith in pursuing natural science; he is arguing that a scientific hypothesis cannot properly claim that God does something or other *immediately* or *directly*. (Note also that the claim here is not that such a hypothesis would not be *scientific*, but that it would be *false*.) What he says seems to be consistent, so far as I can tell, with the claim (say) that in doing their psychology Christian psychologists can properly appeal to the fact that human beings have been created in the image of God, or are subject to original sin.

So suppose we turn to Van Till's proscription of hypotheses to the effect that God has done something or other immediately or directly. This idea of direct action conceals pitfalls and deserves more by way of concentrated attention than I can give it here.[13] The basic idea, however, is fairly clear. An example of *indirect* divine creation would be my building a house; we may say that *God* creates the house, but does so indirectly, by employing *my* activity as a means. So God acts indirectly if he brings about some effect by employing as a means the activity of something else he has created. God acts directly, then, if and only if he brings about some effect, and does not do so by way of employing as a means the activity of some created being.

Now Van Till suggests that God does nothing at all in the world *directly*; only *creatures* do anything directly. But no doubt Van Till, like any other theist, would agree that God directly conserves the world and all its creatures in being; he is directly active in the Big Bang, but also in the sparrow's fall. Were he to suspend this constant conserving activity, the world would disappear like a dream upon awakening. And no doubt Van Till would also agree (on pain of infinite regress) that if God does anything in the world indirectly, he also does something directly: presumably he cannot cause an effect indirectly without also, at some point, acting directly, creating something directly. Van Till must therefore be understood in some other way. Perhaps his idea is that God created the universe at some time in the *past* (acting directly at that time) but since then he never acts directly in the world, except for conserving his creation in being, and miracles connected with salvation history. But why think a thing like that? Consider the fact that Christians as diverse as Pope Pius XII and John Calvin have thought that God created human souls directly; can we simply assume without argument that they are mistaken? What is the warrant for supposing that God no longer acts directly in the world?

Van Till appeals for support, for this theological position, to Allen and Stek; Allen asserts that

God can never properly be used in scientific accounts, which are formulated in terms of the relations between the members of the universe, because that would reduce God to the status of a creature. According to a Christian conception of God as creator of a universe that is rational through and through, there are no missing relations between the members of nature. If in our study of nature, we run into what seems to be an instance of a connection missing between members of nature, the Christian doctrine of creation implies that we should keep looking for one.[14]

Allen's suggestion seems to imply, not just that Christians cannot properly propose, as part of science, that God has done something directly, but also that it would be out of order to appeal, in science, to such ideas as that human beings have been created in God's image. For this idea is not a matter of saying how things in the world are related to each other; it is instead a matter of saying how some things in the world—we human beings—are related to God. Allen believes that scientific accounts must always be formulated in terms of the relationships between members of the created universe (and if that is true, then perhaps, as he says, referring to God in science would be to reduce him to a creature). Taken at face value, however, this seems hasty. A textbook on astronomy may tell you what the diameter of Jupiter is (or how old the earth, or the sun, or the Milky Way is). This does not tell you how things in the world stand related to each other, but instead just tells you something about one of those things; it is science nonetheless.

Allen's main point, of course, is that a scientific account cannot properly be formulated in terms of the relationship of anything to God. But why not? What is the authority for this claim? Does not it seem arbitrary? Consider the truth that human beings have been created in the image of God, but have also fallen into sin. This dual truth might turn out to be very useful in giving psychological explanations of various phenomena. If it is, why should a Christian psychologist not employ it? Why would the result not be science? It could be that investigation would suggest that God created life directly; that it did not arise through the agency of other created things. If that is how things turn out, or how things appear at a given time, why not say so? And why not say so as part of science? As a Christian you believe, of course, that God made the world and could have done so in many different ways; why not employ this knowledge in evaluating the probability of various hypotheses (for example, the Grand Evolutionary Myth)? Christians also have beliefs about what is rational in Simon's sense—i.e., about what sorts of goals a properly functioning human being will have. Christians also have beliefs about what sorts of actions are in their own or someone else's best interests. Why not employ these beliefs in making a scientific evaluation of the probability of, say, Simon's account of altruism, or in giving her own account of these phenomena?

Finally, consider John Stek:

> Since the created realm is replete with its own economy that is neither incomplete (God is not a component within it) nor defective, *in our understanding of the economy of that realm so as to exercise our stewardship over it*—understanding based on both practical experience and scientific endeavors—*we must methodologically exclude all notions of immediate divine causality.* As stewards of the creation, we must methodologically honor the principle that creation interprets creation; indeed, we must honor that principle as "religiously" as the theologian must honor the principle that "Scripture interprets Scripture"—or, since Scripture presupposes general revelation, that revelation interprets revelation. In pursuit of a stewardly understanding of the creation, we may not introduce a "God of the gaps," not even in the as-yet mysterious realm of subatomic particles. We may not do so (1) because God is not an internal component within the economy of the created realm, and (2) because to do so would be to presume to exercise power over God—the presumptuous folly of those in many cultures who have claimed to be specialists in the manipulation of divine powers (e.g., shamans in Russian folk religion and medicine men in primitive cultures).[15]

Stek insists that "we must methodologically exclude all notions of immediate divine causality" in our understanding of the economy of the created realm. One of his reasons seems to be that to appeal to a notion of immediate divine causality would be to introduce a "God of the gaps," and to do *that* would be to presume to exercise power over God. But am I really presuming to exercise power over God by, for example, concurring with John Calvin and Pope Pius XII, and many others, that God directly creates human beings? Or in claiming that he created life specially? At best, this requires more argument.

As Stek says, God is not an internal component within the created realm. It hardly follows, however, that he does not act immediately or directly in the created realm; like any theist, Stek too would agree that God directly and immediately conserves his creation in existence. And would not he also agree that if God creates anything indirectly, then he creates some things directly? So I am not sure why Stek thinks that we must observe this methodological naturalism. Why think that God does not do anything directly or create anything directly? What is the reason for thinking this? Scripture does not suggest it; there do not seem to be arguments from any other source; why then accept it?

These reasons, then, for the necessity or advisability of methodological naturalism do not seem strong; and since they *are* so weak, it is perhaps reasonable to surmise that they do not really represent what is going on in the minds of those who offer them. I suggest that there is a different and unspoken reason for this obeisance to methodological naturalism: *fear and loathing of God-of-the-gaps theology.* As we saw above, Stek declares that "In pursuit of a stewardly understanding of the creation, we may not introduce a 'God of the gaps'"; he, together with the other three authors I have cited in this connection (McMullin, Van Till and Allen), explicitly mention God-of-the-gaps theology and explicitly connect it with methodological naturalism via the suggestion that God has done this or that immediately. The idea seems to be that to hold that God acts directly in creation is to fall into, or anyway lean dangerously close to this sort of theology. But is this true? Precisely what *is* God-of-the-gaps theology?

There is not anything that it is *precisely*; it is not that sort of thing. Somewhat vaguely, however, it can be characterized as follows. The God-of-the-gaps theologian is an Enlightenment semideist who thinks of the universe as a vast machine working according to a set of necessary and inviolable natural laws. (Perhaps a God has created the universe: but if he did, it is now for the most part self-sufficient and self-contained.) These natural laws, furthermore, have a kind of august majesty; they are necessary in some strong sense; perhaps not even God, if there is such a person, could violate them; but even if he could, he almost certainly would not. (Hence the otherwise inexplicable worry about miracles characteristic of this sort of thought.) Natural science investigates and lays out the structure of this cosmic machine, in particular by trying to discover and lay bare those laws, and to explain the phenomena in terms of them. There seem to be *some* phenomena, however, that resist a naturalistic explanation—so far, at any rate. We should therefore postulate a deity in terms of whose actions we can explain these things that current science cannot. Newton's suggestion that God periodically adjusts the orbits of the planets is often cited as just such an example of God-of-the-gaps theology.

The following, therefore, are the essential points of God-of-the-gaps theology. First, the world is a vast machine that is almost entirely self-sufficient; divine activity in nature is limited to those phenomena for which there is no scientific, i.e., mechanical and naturalistic explanation. Second, the existence of God is a kind of large-scale hypothesis postulated to explain what cannot be explained otherwise, i.e., naturalistically.[16] Third, there is the apologetic emphasis: the best or one of the best reasons for believing that there is such a person as God is the fact that there are phenomena that natural science cannot (so far) explain naturalistically.

Now McMullin, Stek, Van Till, and Allen all object strenuously to God-of-the-gaps theology—and rightly so. This line of thought is at best a kind of anemic and watered-down semideism that inserts God's activity into the gaps in scientific knowledge; it is associated, furthermore, with a weak and pallid apologetics according to which perhaps the main source or motivation for belief in God is that there are some things science cannot presently explain. A far cry indeed from what the Scriptures teach! God-of-the-gaps theology is worlds apart from serious Christian theism. This is evident at (at least) the following points. First and most important, according to serious theism, God is constantly, immediately, intimately, and directly active in his creation: he constantly upholds it in existence and providentially governs it. He is immediately and directly active in everything from the Big Bang to the sparrow's fall. Literally nothing happens without his upholding hand.[17] Second, natural laws are not in any way independent of God, and are perhaps best thought of as regularities in the ways in which he treats the stuff he has made, or perhaps as counterfactuals of divine freedom. (Hence there is nothing in the least untoward in the thought that on some occasions God might do something in a way different from his usual way—e.g., raise someone from the dead or change water into wine.) Indeed, the whole *interventionist* terminology—speaking of God as *intervening* in nature, or *intruding* into it, or *interfering* with it, or *violating* natural law—all this goes with God-of-the gaps theology, not with serious theism. According to the latter, God is already and always intimately acting in nature, which depends from moment to moment for its existence upon immediate divine activity; there is not and could not be any such thing as his "intervening" in nature.

These are broadly speaking metaphysical differences between Christian theism and God-of-the-gaps thought; but there are equally significant epistemological differences. First, the thought that there is such a person as God is not, according to Christian theism, a hypothesis postulated to *explain* something or other,[18] nor is the main reason for believing that there is such a person as God the fact that there are phenomena that elude the best efforts of current science.[19] Rather, our knowledge of God comes by way of *general* revelation, which involves something like Aquinas's general knowledge of God or Calvin's *sensus divinitatis*, and also, and more importantly, by way of God's *special* revelation, in the Scriptures and through the church, of his plan for dealing with our fall into sin.

God-of-the-gaps theology, therefore, is every bit as bad as McMullin, Van Till, Stek, and Allen think. (Indeed, it may be worse than Van Till and Stek think, since some of the things they think—in particular their ban on God's acting directly in nature—seem to me to display a decided list in the direction of such theology.) Serious Christians should indeed resolutely reject this way of thinking. The Christian community knows that God is constantly active in his creation, that natural laws, if there are any, are not independent of God, and that the existence of God is certainly not a hypothesis designed to explain what science cannot. Furthermore, the Christian community begins the scientific enterprise already believing in God; it does not (or at any rate need not) engage in it for apologetic reasons, either with respect to itself or with respect to non-Christians. But of course from these things it does not follow for an instant that the Christian scientific community should endorse methodological naturalism. The Christian community faces these questions: How shall we best understand this creation God has made, and in which he has placed us? What is the best way to proceed? What information can we or shall we use? Well, is it not clear initially, at any rate, that we should employ whatever is useful and enlightening, including what we know about God and his relationship to the world, and including what we know by way of special revelation? Could we not sensibly conclude, for example, that God created life, or human life, or something else specially? (I do not say we *should* conclude that: I say only that we *could*, and should if that is what the evidence most strongly suggests.) Should we not use our knowledge of sin and creation in psychology, sociology, and the human sciences in general? Should we not evaluate various scientific theories by way of a background body of belief that includes what we know about God and what we know specifically as Christians? Should we not decide what needs explanation against that same background body of beliefs?

Well, why not? That certainly seems initially to be the rational thing to do (one should make use of all that one knows in trying to come to an understanding of some phenomenon); and it is hard to see anything like strong reasons against it. We certainly do not fall into any of the unhappy ways of thinking characteristic of God-of-the-gaps theology just by doing one of these things. In doing these things, we do not thereby commit ourselves, for example, to the idea that God does almost nothing directly in nature, or that the universe is something like a vast machine in whose workings God could intervene only with some difficulty; nor are we thereby committed to the idea that one of our main reasons for belief in God is just that there are things science cannot explain, or that the idea of God is really something like a large-scale hypothesis postulated to explain those things. Not at all. Indeed, the whole God-of-the-gaps issue is nothing but a red herring in the present context.[20]

### Two Stronger Arguments for Methodological Naturalism

These arguments, therefore, are not very convincing; but there are two quite different, and I think, stronger arguments or lines of reasoning for embracing methodological naturalism in the practice of science. The first of these really deserves a paper all to itself; here, unfortunately, I shall have to give it relatively short shrift.

### Duhemian Science

We can approach this argument by thinking about some striking passages in Pierre Duhem's *The Aim and Structure of Physical Theory*.[21] Duhem was both a serious Catholic and a serious scientist; he was accused (as he thought) by Abel Rey of allowing his religious and metaphysical views as a Christian to enter his physics in an improper way.[22] Duhem repudiated this suggestion, claiming that his Christianity did not enter his physics in an improper way, because it did not enter his physics in any way at all.[23] Furthermore, he thought the *correct* or *proper* way to pursue physical theory was the way in which he had in fact done it; physical theory should be completely independent of religious or metaphysical views or commitments.

He thought this for two reasons. First, he thought religion bore little relevance to physical theory: "Was it not a glaring fact to us, as to any man of good sense, that the object and nature of physical theory are things foreign to religious doctrines and without any contact with them?"[24]

But there is something else, and something perhaps deeper. Although Duhem may have thought that *religious* doctrines had little to do with physical theory, he did not at all think the same thing about metaphysical doctrines. In fact he believed that *metaphysical* doctrines had often entered deeply into physical theory. Many theoretical physicists, as he saw it, took it that the principal aim of physics is to *explain* observable phenomena. Explanation is a slippery notion and a complex phenomenon; but here at any rate the relevant variety of explanation involves giving an account of the phenomena, the appearances, in terms of the nature or constitution of the underlying material reality. He goes on to give a striking illustration, recounting how atomists, Aristotelians, Newtonians, and Cartesians differ in the explanations or accounts they give of the phenomena of magnetism: atomists give the requisite explanation, naturally enough, in terms of atoms; Cartesians in terms of pure extensions; and Aristotelians in terms of matter and form.[25] The differences among these explanations, he says, are metaphysical; they pertain to the ultimate nature or constitution of matter. But of course if the aim is to explain the phenomena in terms of the ultimate nature or constitution of matter, then it is crucially important to get the latter right, to get the right answer to the metaphysical question "What *is* the nature or constitution of matter?" In this way, he says, physical theory is subordinated to metaphysics: "*Therefore, if the aim of physical theories is to explain experimental laws, theoretical physics is not an autonomous science; it is subordinate to metaphysics.*"[26]

Well, what is the matter with that? The problem, says Duhem, is that if you think of physics in this way, then your estimate of the worth of a physical theory will depend upon the metaphysics you adopt. Physical theory will be dependent upon metaphysics in such a way that someone who does not accept the metaphysics involved in a given physical theory cannot accept the physical theory either. And the problem with *that* is that the disagreements that run riot in metaphysics will ingress into physics, so that the latter cannot be an activity we can all work at together, regardless of our metaphysical views:

> Now to make physical theories depend on metaphysics is surely not the way to let them enjoy the privilege of universal consent.... If theoretical physics is subordinated to metaphysics, the divisions separating the diverse metaphysical systems will extend into the domain of physics. A physical theory reputed to be satisfactory by the sectarians of one metaphysical school will be rejected by the partisans of another school.[27]

So here we have another argument for methodological naturalism, and a simple, commonsense one at that: it is important that we all— Christian, naturalist, creative antirealist, whatever—be able to work at physics and the other sciences together and cooperatively; therefore we should not employ in science views, commitments, and assumptions only some of us accept—that is, we should not employ them in a way that would make the bit of science in question unacceptable or less acceptable to someone who did not share the commitment or assumption in question.[28] But then we cannot employ, in that way, such ideas as that the world and things therein have been designed and created by God. Proper science, insofar as it is to be common to all of us, will have to eschew any dependence upon metaphysical and religious views held by only some of us; therefore we should endorse methodological naturalism. We do not, of course, have to be metaphysical naturalists in order to pursue Duhemian science; but if science is to be properly universal, it cannot employ assumptions or commitments that are not universally shared.

Duhemian science, therefore, is maximally inclusive; we can all do it together and agree on its results. But what about those who, like Simon, for example, think it is important also to do a sort of human science which starts, not from methodological naturalism, but from metaphysical naturalism? And what about those who, like the atomists, Cartesians, and Aristotelians think it is important to pursue a sort of science in which the aim is successful explanation in terms of underlying unobservable realities? And what about Christians or theists, who propose to investigate human reality employing all that they know, including what they know as Christians or theists? So far as Duhem's claims go, there is nothing improper about any of this. Should we call this kind of activity "science"; does it deserve that honorific term? There is no reason in Duhem for a negative answer. It is important, to be sure, to see that science of this sort is not *Duhemian* science and does not have the claim to universal assent enjoyed by the latter; but of course that is nothing against it.

According to the fuller Duhemian picture, then, we would all work together on Duhemian science; but each of the groups involved—naturalists and theists, for example, but perhaps others as well—could then go on to incorporate Duhemian science into a fuller context that includes the metaphysical or religious principles specific to that group. Let us call this broader science "Augustinian science." Of course the motivation for doing this will vary enormously from area to area. Physics and chemistry are overwhelmingly Duhemian[29] (of course the same might not be true for *philosophy* of physics); here perhaps Augustinian science would be for the most part otiose. The same goes for biological sciences: surely much that goes on there could be thought of as Duhemian science. On the other hand, there are also non-Duhemian elements in the neighborhood, such as those declarations of certainty and the claims that evolutionary biology shows that human and other forms of life must be seen as a result of chance (and hence cannot be thought of as designed). In the human sciences, however, vast stretches are clearly non-Duhemian; it is in these areas that Augustinian science would be most relevant and important.

So return to our central question: should the Christian scientific community observe the constraints of methodological naturalism? So far as this argument is concerned, the answer seems to be: yes, of course, in those areas where Duhemian science is possible and valuable. But nothing here suggests that the Christian scientific community should not also engage in non-Duhemian, Augustinian science where that is relevant. There is nothing here to suggest that "if it ain't Duhemian, it ain't science."

### Science Stoppers?

There is still another reason for methodological naturalism; this one, too, is common sense simplicity itself. God has created this whole wonderful and awful (both taken in their etymological senses) world of ours. One of the things we want to do as his creatures is to understand the world he has made, see (to the extent that we can) how it is made, what its structure is, and how it works. This is not, of course, the only thing God's children must do with the world; we must also appreciate it, care for it, love it, thank the Lord for it, and see his hand in it. But understanding it is valuable, and so is understanding it in a theoretical way. One way of understanding something is to see how it is made, how it is put together, and how it works. That is what goes on in natural science. The object of this science is nature; for Christians, its aim (one of its aims) is to see what the structure of this world is and how it works; this is a way of appreciating God's creation, and part of what it is to exercise the image of God in which we have been created.

But there will be little advance along this front if, in answer to the question, Why does so and so work the way it does? or What is the explanation of so and so? we regularly and often reply "Because God did it that way" or "Because it pleased God that it should be like that." This will often be true,[30] but it is not the sort of answer we want at that juncture. It goes without saying that God has in one way or another brought it about that the universe displays the character it does; but what we want to know in science are the answers to questions like "What is this made out of? What is its structure? How does it work? How is it connected with other parts of God's creation?" Claims to the effect that God has done this or that (created life, or created human life) *directly* are in a sense science stoppers. If this claim is true, then presumably we cannot go on to learn something further about how it was done or how the phenomenon in question works; if God did it directly, there will be nothing further to find out. How does it happen that there is such a thing as light? Well, God said, "Let there be light" and there was light. This is of course true, and of enormous importance, but if taken as science it is not helpful; it does not help us find out more about light, what its physical character is, how it is related to other things, and the like. Ascribing something to the direct action of God tends to cut off further inquiry.

Of course this is a reason for only *part* of methodological naturalism. There are several *different* ways in which Christianity might enter into the texture of science: (1) stating and employing hypotheses according to which God does things directly; (2) stating and employing hypotheses according to which he does something indirectly; (3) evaluating theories with respect to background information that includes Christian theism;

(4) employing such propositions as *human beings have been created in God's image*, either directly or as background; (5) doing the same for such doctrines as that of original sin, which do not involve any direct mention of God at all; and (6) deciding what needs explanation by way of referring to that same background. The considerations cited in the last paragraph are at best a reason for a proscription of (1).

But they are not even much of a reason for that. The claim that God has directly created life, for example, may be a science stopper; it does not follow that God *did not* directly create life. Obviously we have no guarantee that God has done everything by way of employing secondary causes, or in such a way as to encourage further scientific inquiry, or for our convenience as scientists, or for the benefit of the National Science Foundation. Clearly we cannot sensibly insist in advance that whatever we are confronted with is to be explained in terms of something *else* God did; he must have done *some* things directly. It would be worth knowing, if possible, which things he *did* do directly; to know this would be an important part of a serious and profound knowledge of the universe. The fact that such claims are science stoppers means that as a general rule they will not be helpful; it does not mean that they are never true, and it does not mean that they can never be part of a proper scientific theory. (And of course it does not even bear on the other ways in which Christianity or Christian theism can be relevant to science.) It is a giant and unwarranted step from the recognition that claims of direct divine activity are science stoppers to the insistence that science must pretend that the created universe is just there, refusing to recognize that it is indeed *created*.

So there is little to be said for methodological naturalism. Taken at its best, it tells us only that Duhemian science must be metaphysically neutral and that claims of direct divine action will not ordinarily make for good science. And even in these two cases, what we have reason for is not a principled proscription but a general counsel that in some circumstances is quite clearly inapplicable. There is no reason to proscribe a question like: "Did God create life specially?"; there is no reason why such a question cannot be investigated empirically;[31] and there is no reason to proscribe in advance an affirmative answer.

Christian thought (particularly since the High Middle Ages) as opposed to Greek (and in particular Aristotelian thought)[32] contains a strong tendency to see the world as through and through *contingent*. The world need not have existed; that is, God need not have created it. The world need not have had just the structure it does have; that is, God could have created it differently. This sense of the contingency of nature has been one important source of the emphasis upon the *empirical* character of modern science. As a sort of rough rule of thumb, we can say that it is by *reason*, by a priori thought, that we learn of what cannot be otherwise; it is by the senses, by way of a posteriori inquiry that we learn about what is contingent.[33] But the world as God created it is full of contingencies. Therefore we do not merely think about it in our armchairs, trying to infer from first principles how many teeth there are in a horse's mouth; instead we take a look. The same should go for the question how God acts in the world: here we should rely less upon a priori theology and more upon empirical inquiry. We have no good grounds for insisting that God *must* do things one specific way; so far as we can see, he is free to do things in many different ways. So perhaps he did create human life specially; or perhaps he has done other things specially. We cannot properly rule this out in advance by way of appeal to speculative theology; we should look and see.

My main point, therefore, can be summarized as follows. According to Augustine, Kuyper, and many others, human history is dominated by a battle, a contest between the *Civitas Dei* and the City of Man. Part of the task of the Christian academic community is to discern the limits and lineaments of this contest, to see how it plays out in intellectual life generally, and to pursue the various areas of intellectual life as citizens of the *Civitas Dei*. This naturally suggests pursuing science using all that we know: what we know about God as well as what we know about his creation, and what we know by faith as well as what we know in other ways. That natural suggestion is proscribed by the Principle of Methodological Naturalism. Methodological naturalism, however, though widely accepted and indeed exalted, has little to be said for it; when examined cooly in the light of day, the arguments for it seem weak indeed. We should therefore reject it, taken in its full generality. Perhaps we should join others in Duhemian science; but we should also pursue our own Augustinian science.

By way of conclusion, I call attention to something else John Stek has said:

Theology must take account of all that humanity comes to know about the world, and science must equally take account of all that we come to know about God. In fact, we cannot, without denying our being and vocation as stewards, pursue theology without bringing to that study all that we know about the world, nor can we, without denying our being and vocation as stewards, pursue science without bringing to that study all that we know about God.[34]

Just so.

## Notes

1. For example, many Reformed Christians follow Abraham Kuyper in holding that intellectual endeavor in general and natural science in particular are not independent of religious commitment. Perhaps the credit for this idea should go not to Augustine, but to Tertullian. Tertullian has suffered from a bad press; one of his major emphases, however, is that scholarship—intellectual endeavor—is not religiously neutral.

2. The idea is not, of course, that a scientist will not be passionate either about science generally, or his favorite theories, or his reputation; it is rather that none of these properly enters into the evaluation of a scientific theory or explanation.

3. "Science must be provisionally atheistic or cease to be itself." B. Willey, "Darwin's Place in the History of Thought," in *Darwinism and the Study of Society*, edited by M. Banton (Chicago: Quadrangle Books, 1961), 1–16. Willey does not mean, of course, that one who proceeds in this way is properly accused of atheism. In the same way, to call this procedure or proscription "methodological naturalism" is not to imply that one who proceeds in this way is really a naturalist. See E. McMullin, "Plantinga's Defense of Special Creation," *Christian Scholar's Review* 21 (September, 1991): 57.

4. McMullin, "Plantinga's Defence," 57.

5. I have argued elsewhere that one condition of rationality laid down by modern classical foundationalism is in fact self-referentially incoherent. See, for example, A. Plantinga, "Reason and Belief in God," in *Faith and Rationality*, edited by A. Plantinga and N. Wolterstorff (Notre Dame: University of Notre Dame Press, 1983), 60ff.

6. Ruse, *Darwinism Defended* (Reading: Addison-Wesley, 1982), 322 (my italics).

7. See, for example, L. Laudan, "The Demise of the Demarcation Problem," in *But is it Science?* edited by M. Ruse (New York: Prometheus Books, 1988), 337–50.

8. A. D. Linde, "Particle Physics and Inflationary Cosmology," *Physics Today* (September, 1987): 61.

9. See B. van Fraassen, *Laws and Symmetry* (Oxford: Oxford University Press, 1989), chaps. 2–5.

10. See, for example, D. Armstrong, *What is a Law of Nature?* (Cambridge: Cambridge University Press, 1983), 39ff.

11. That is, propositions that state how God (freely) treats the things he has made, and how he would have treated them had things been relevantly different. "Nearly universally quantified": if we think of them this way, we can think of miracles as going contrary to law without thinking of them (inconsistently) as exceptions to some universal and necessary proposition.

12. H. J. Van Till, "When Faith and Reason Cooperate," *Christian Scholar's Review* 21 (September, 1991): 42.

13. See, for example, W. P. Alston, "Divine and Human Action," in *Divine and Human Action: Essays in the Metaphysics of Theism*, edited by T. Morris (Ithaca: Cornell University Press, 1988), 257–80.

14. D. Allen, *Christian Belief in a Postmodern World* (Louisville: Westminster/ John Knox Press, 1989), 45.

15. J. H. Stek, "What Says the Scriptures?" in *Portraits of Creation: Biblical and Scientific Perspectives on the World's Formation*, edited by H. J. Van Till, R. E. Snow, J. H. Stek, and D. A. Young (Grand Rapids: William B. Eerdmans Publishing Company, 1990), 261.

16. I do not mean to suggest that one who espouses or advocates God-of-the-gaps theology herself believes in God only as such a hypothesis: that is quite another question.

17. In addition, most medieval Christian thinkers have also insisted on a separate divine activity of God's; any causal transaction in the world requires his *concurrence*. Problems arise here; to some ears it sounds as if this doctrine is motivated less by the relevant evidence than by a desire to pay metaphysical compliments to God.

18. See my "Is Theism Really a Miracle?" *Faith and Philosophy* 3, no. 2 (1986): 132ff.

19. A further problem with this way of thinking: as science explains more and more, the scope for God's activity is less and less; it is in danger of being squeezed out of the world altogether, thus making more and more tenuous one's reasons (on this way of thinking) for believing that there is such a person as God at all. (Of course it must also be acknowledged on the other side that things sometimes go in the opposite direction; for example, it is much harder now than it was in Darwin's day to see how it could be that life should arise just by way of the regularities recognized in physics and chemistry.)

20. Further, Newton seems to me to have suffered a bum rap. He suggested that God made periodic adjustments in the orbits of the planets: true enough. But he did not propose this as a reason for believing in God; it is rather that (of course) he already believed in God, and could not think of any other explanation for the movements of the planets. He turned out to be wrong; he could have been right, however, and in any event he was not endorsing any of the characteristic ideas of God-of-the-gaps thought.

21. P. Duhem, (1906) *The Aim and Structure of Physical Theory*, translated by P. P. Wiener, with the foreword by Prince Louis de Broglie (Princeton: Princeton University Press, 1954).

22. A. Rey, "La Philosophie Scientifique de M. Duhem," *Revue de Métaphysique et de Morale* 12 (July, 1904): 699ff.

23. See the appendix to *The Aim and Structure of Physical Theory*, which is entitled "Physics of a Believer" and is a reprint of Duhem's reply to Rey; it was originally published in the *Annales de Philosophie Chrétienne* 1 (October and November, 1905): 44f. and 133f.

24. Duhem, 278.

25. Duhem, 10–18.

26. Duhem, 10.

27. Duhem, 10.

28. This would not preclude, of course, employing such ideas in theories proposed, not as true, but only as empirically adequate.

29. The Principle of Indifference is non-Duhemian, but it is not easy to find other examples. (I am assuming that *interpretations* of quantum mechanics [as opposed to quantum mechanics itself] belong to philosophy rather than physics.)

30. Though not always: if the question is "Why was there such a thing as WW II?" the answer is not "Because it pleased God to do things that way." God of course *permitted* World War II to take place; but it was not pleasing to him.

31. Why could a scientist not think as follows? God has created the world, and of course has created everything in it directly or indirectly. After a great deal of study, we cannot see how he created some phenomenon P (life, for example) indirectly; thus probably he has created it directly.

32. See Aristotle, *Posterior Analytics*, bk. I, 1–2, 4, where Aristotle declares that *scientia* is a matter of seeing what necessarily follows from what one sees to be necessarily true. (Of course Aristotle's own practice is not always easy to square with this suggestion.)

33. Of course, this is at best a rough and general characterization: we can obviously learn of necessities a posteriori (for example, by using computers to prove complicated theorems) and perhaps also of contingencies a priori. This question of the connection between the a priori and the necessary, on the one hand, and the contingent and the a posteriori on the other (the question of the relationship between the a priori/a posteriori distinction and the necessary/contingent distinction) is as deep as it is fascinating.

34. Stek, 260–1.

# 14

# Methodological Naturalism under Attack

Michael Ruse

In the last decade, Biblical literalism—especially that version known as "creationism," concerned to deny evolution and affirm the truth of the early chapters of Genesis—has shown considerable vigor. Moreover, it has gained a remarkable respectability, for whereas previously the supporters of creationism—although often qualified in various areas of science or the humanities—had little standing in the academic community, we now find enthusiasts among people of deservedly renowned prestige from the very best institutions. With this rise in status has come a new way of approaching the problem. A new way that I suspect is part cause, and part effect, of the rise. Twenty years ago, the creationists' chief appeal was to their understanding of the facts of science—the fossil record, homologies, and so forth—but now philosophy has come to the fore.

It is true that this is not an entirely new phenomenon. Indeed, in the celebrated Arkansas Creation Trial of 1981, when a federal judge ruled that creationism could not be taught in the biology classes in publically funded schools in the United States, I myself—a professional historian and philosopher of science—was an expert witness for the plaintiffs. After the event, I felt thus encouraged to put together a collection on the philosophical issues in the dispute—*But Is It Science? The Philosophical Issue in the Creation/Evolution Controversy* (1988)—showing among other things that philosophical questions were being raised even before Charles Darwin published his *On the Origin of Species* in 1859. Yet, truly, philosophy was but a side issue. Now, however, the new infusion of creationists have taken up the philosophical issues in a major way and in many respects it is these that are at the front of the dispute.

In particular, perhaps realizing that a straight frontal scientific approach will not succeed—although, as you will learn, ultimately I am not sure that the science is now quite as absent as everyone pretends—the new creationists are making much of the claim that *the* essential difference between evolutionists on the one hand and creationists or "theists" (as they prefer to call themselves) on the other hand is one of conflicting philosophies (Johnson 1991, 1995). Evolutionists, supposedly, are committed to the secular atheistic materialistic philosophy of *naturalism* and from this evolution follows as a consequence. Creationists, on the other hand, are committed to some form of spiritual or religious philosophy of *theism* and from this follows their commitment to a Genesis-based world perspective. (To be fair, most of the new creationists seem willing to forsake a very young earth. But they stand rock firm on such things as the instantaneous miraculous creation of life from nonlife, a denial of evolution, and particularly an affirmation of a special and privileged place for the arrival and status of human life. Unlike earlier creationists, the new ones talk little about the Flood. This may be because the new creationists tend not to be dispensationalists and hence feel no need of an earlier catastrophe balancing the predicted Armageddon. See Numbers 1992 for more on the significance of the Flood in traditional literalist thinking.)

Evolutionists do not speak with a unified voice, but my impression is that generally in important respects they are inclined to agree with their opponents: they do think that naturalism, somehow defined, is indeed an important underpinning to their position. However, where they disagree with the creationists is in the implication that this means that evolutionism is simply a manifestation of an atheistic world philosophy. At least, those concerned with the fight between evolution and creationism are unwilling to make this concession. There are evolutionists—notably Richard Dawkins (1995, 1996), Daniel Dennett (1995), and William Provine (1989)—who are materialists, atheists, naturalists, and evolutionists, and who see everything as a united package deal. But these men do not speak for all evolutionists or all naturalists. Those of us—for I am one—who are unwilling to be pinned into the corner of atheistic evolutionism point historically to the fact that there have been distinguished evolutionists who were practicing Christians. In this century, notably the two leading evolutionists, Sir Ronald Fisher (1950) in England

and Theodosius Dobzhansky (1967) in America, were both absolutely and completely committed to the idea of Jesus as their Savior. Philosophically, those of us who would separate atheism and evolutionism suggest that simply using a catch-all term "naturalism" conceals subtleties in peoples' approaches. Once these subtleties are uncovered, the clash between evolution and creationism is no longer seen to be the simple black and white philosophical matter that the creationists claim.

Picking up on this last point, evolutionists who want to divorce their science from supposedly atheistic implications invite one to draw a distinction between two forms of naturalism. On the one hand, one has what one might call "metaphysical naturalism": this indeed is a materialistic, atheistic view, for it argues that the world is as we see it and that there is nothing more. On the other hand, one has a notion or a practice that can properly be called "methodological naturalism": although this is the working philosophy of the scientist, it is in no sense atheistic as such. The methodological naturalist is the person who assumes that the world runs according to unbroken law; that humans can understand the world in terms of this law; and that science involves just such understanding without any reference to extra or supernatural forces like God. Whether there are such forces or beings is another matter entirely and simply not addressed by methodological naturalism. Hence, although indeed evolution as we understand it is a natural consequence of methodological naturalism, given the facts of the world as they can be discovered, in no sense is the methodological naturalist thereby committed to the denial of God's existence. It is simply that the methodological naturalist insists that, inasmuch as one is doing science, one avoid all theological or other religious references. In particular, one denies God a role in the creation.

This is not to say that God did not have a role in the creation, but simply that, *qua* science, that is *qua* an enterprise formed through the practice of methodological naturalism, science has no place for talk of God. Just as, for instance, if one were to go to the doctor one would not expect any advice on political matters, so if one goes to a scientist one does not expect any advice on or reference to theological matters. The physician may indeed have very strong political views, which one may or may not share. But the politics are irrelevant to the medicine. Similarly, the scientist may or may not have very strong theological views, which one may or may not share. But inasmuch as one is going to the scientist for science, theology can and must be ruled out as irrelevant.

Naturally enough the new creationists have responded to this line of argumentation. The way taken by the well-known critic of Darwinism, Phillip Johnson (1995)—an academic lawyer on the faculty at Berkeley— denies that one can thus separate methodological and metaphysical naturalism; at least, Johnson thinks that any such separation is bound to be unstable. In his opinion, methodological naturalism—however well-intentionally formulated—inevitably collapses within a very short time into metaphysical naturalism. Hence, even though one may claim that evolutionism has no materialistic, atheistic, philosophical underpinnings, in fact it is impossible to keep such underpinnings out of the picture.

Frankly, intellectually, I am not sure about the depth of this response. Certainly, those of us with philosophical training and inclinations ask for more. And this we do now have, thanks particularly to America's most distinguished philosopher of religion, Alvin Plantinga (1991a,b,c, 1993, 1994, 1995). Although serving on the Notre Dame faculty, Plantinga is a reformed Christian, a Calvinist who believes strongly that evolutionism is resting on shaky philosophical foundations. The appeal to methodological naturalism fails, and Plantinga has therefore taken it upon himself to expose its inadequacy. It is this attack that is the focus of discussion here: a focus given a personal savor by the fact that I am one of the chief objects of Plantinga's attack.

## Methodological Naturalism as Incoherent

Plantinga has a number of critiques of methodological naturalism, one of the first of which—directed against me—is of the very coherence of the notion of methodological naturalism. I have characterized the notion— as I did above, as indeed I did in the Arkansas Creation Trial (and as was picked up by the judge in that trial, and used as a support of his judgment against creationism)—as an approach to the empirical world that demands understanding in terms of unbroken law (Ruse 1982, 1984, 1988, 1995, 1996). That is to say, it requires understanding in terms of regularities, which in some way or another we feel are more than mere contingencies, but rather part of the necessary succession of the empirical world. Neither I, nor anyone else, has ever insisted in our characterization of methodological naturalism that the necessity of law be interpreted in and only in some particular way. I myself have endorsed a neo-Humean position, seeing the necessity of laws as a natural regularity on which one imposes an evolutionarily derived psychological construction. But were someone to see the necessity in some other (non-god-invoking) way, I would not deny them the status of methodological naturalist.

Plantinga objects to this characterization on a number of grounds, one of which involves repeatability. He writes as follows:

But take repeatability, and consider this passage by Andrei Linde: speaking of the Big Bang, he says, "One might think it very difficult to extract useful and reliable information from the unique experiment carried out about 10 years ago. According to Linde, the Big Bang is unique and therefore, presumably, unrepeatable—at any rate it *might* turn out to be unrepeatable. If so, would we be obliged to conclude that contemporary cosmological inquiries into the nature of the Big Bang and into the early development of the universe are not really part of science? (Plantinga 1997, 146 [344])

As a matter of fact, Plantinga here is raising an objection that has often been raised by critics of the claim that scientific understanding involves reference to law. His point, as was theirs, is that there are many unique events that science must surely try to cover and understand, but that given the uniqueness of these events, in some sense this precludes lawful understanding. But, as many critics of the critics have countered, there surely has to be something wrong with this argument. Take for instance the demise of the dinosaurs at the end of the Cretaceous. This was in itself a unique phenomenon and unrepeatable; but, uniqueness notwithstanding, the demise was made up of many factors that can individually be brought under lawful understanding. Today, it seems most probable that an asteroid or a comet or some such thing hit the earth. This was no unique phenomenon, nor was the hitting of the earth by the asteroid or comet such that the normal laws of nature—that is to say Galileo's laws of motion—could not be applied. Then, it is believed that there was a huge dust cloud raised and the earth became dark. Again, even if this was a unique phenomenon—and the dust cloud in the last century after the explosion of Krakatau makes one doubt this—one can still apply laws. One has all sorts of experience of dusk causing darkness; then of darkness cutting off photosynthesis of plants and of the dying of plants; and then of the consequent starvation of animals that are part of the ecological food chain depending on plants. In other words, although the dinosaurs existed only once and will never reappear—so their demise was certainly something unique—the various components involved in the extinction of the dinosaurs are such that they can be brought beneath regularity. In principle, we have nothing different from any frequently repeatable phenomenon, like the death of annual plants at the end of every growing season. Hence, here, Plantinga does not offer us reason to give up on methodological naturalism.

Nor does Plantinga's second objection carry a great deal of weight. He argues that the whole notion of scientific "law" is itself in some sense untenable. He writes: "Bas van Fraassen, for example, has given an extended and formidable argument for the conclusion that there are no natural laws. There are *regularities*, of course, but a regularity is not yet a law; a law is what is supposed to *explain* and *ground* a regularity. Furthermore, a law is supposed to hold with some kind of *necessity*, typically thought to be less stringent than broadly logical necessity, but necessity nonetheless." (Plantinga 1997, 146; [344] in this volume) Plantinga goes on to suggest that maybe the whole notion of law is just an unfortunate legacy of the Enlightenment and that perhaps it can be given up.

Now, although I am the last person to belittle the formidable philosophical powers of Bas van Fraassen—as eminent in the field of the philosophy of quantum mechanics as Plantinga is in the field of the philosophy of religion—what is being extrapolated here is far stronger than van Fraassen or anyone else would want (should want) to claim. Certainly there are questions about how one might interpret the necessity of laws: I myself have allowed that already. But neither van Fraassen nor anyone else is going to deny that there are certain sorts of regularities of some kind and that these are presupposed in the activity of science. At least, if this denial is at the heart of van Fraassen's thinking, then I can only say that the response of the average scientist will be: "News to me!"

Nor is it particularly helpful to try to belittle the appeal to law by connecting it with the Enlightenment, or—as Plantinga does specifically—to Enlightenment deism. Perhaps the notion of scientific law did indeed originate in such deism (a conclusion that would have shocked deeply Christian philosophers and scientists of the past two hundred years), but here surely the fallacy of "psychologism"—the confusion of the context of discovery with the context of justification—has some bite. The most venerable of scientific concepts—for instance, work, force, and cause—have theological backgrounds. Indeed, if the eminent historian of physics Richard S. Westfall (1982) is correct, Newtonian gravitation has its roots in alchemic speculations. But today, one can use these notions without in any sense having to confess that they are still theological or alchemical. In the same way, even if indeed it is the case that law does have its roots in Enlightenment deism, there is absolutely no reason that we cannot ourselves today use it in an entirely secular way: the way of the methodological naturalist. Here again, Plantinga's objections fall to the ground.

A third objection that Plantinga brings to my characterization of methodological naturalism is that, at some level, it is unacceptably fuzzy or loose: to such a degree that it really is unworkable.

First, energy has been expended, for at least several centuries, on the "demarcation problem": the problem of giving necessary and sufficient conditions for distinguishing science from other human activities. This effort has apparently failed; but if in fact there *were* a definition of the sort Ruse is appealing to, then presumable there would be available a set of necessary and sufficient conditions for something as being science. Ruse does not address the many and (I think) successful arguments for the conclusion that there is no such set of necessary and sufficient conditions, let alone such a definition of the term "science"; he simply declares that—by definition—science has the properties he mentions. (Plantinga 1997, 145 [344])

Again, I fail to see that this is an effective counter. It is certainly the case that there are subjects that are on the borderline between science and nonscience, judged from the perspective of methodological naturalism. Indeed, I myself have spent the last ten years on a massive study of the history of evolutionary theory, through the two and a half centuries of its life. A major theme of my now-published labors, *Monad to Man: The Concept of Progress in Evolutionary Biology* (1996), is that evolutionism grew up from being a pseudoscience, through being a popular science, to being what I term a mature or "professional" science. At various stages along this process, one sees a transformation as evolution becomes more subject to the strict dictates of methodological naturalism. But, of course, part of my thesis is that evolutionism itself has evolved, and there are times when really it sits or rather sat on the fence between something that really would satisfy the criteria of good science, and something that would not.

In other words, what I argue is that there were times when one could not bring forward absolutely tight conditions showing that evolution was in or out of the category of good science or even genuine science at all. But this does not mean that the notion of good science—and evolutionism being in or out of this notion—is thereby rendered otiose or impossible to apply. The point is there were borderline cases here, as elsewhere in life: the fact that there is no clean demarcation between science and nonscience is no argument against the very idea of methodological naturalism.

(There is a lot more that could be said on this particular issue. Let me just simply say that Plantinga is simply mistaken when he says: "Ruse does not address the many and [I think] successful arguments for the conclusion that there is no such set of necessary and sufficient conditions" [1997, 145] for methodological naturalism. In fact, in the collection I have already mentioned, *But Is It Science? The Philosophical Question in the Creation/Evolution Controversy*, from which Plantinga is himself drawing my discussion, I do offer arguments. I may not be successful in my counter-reply, but I certainly take them up and would refer the interested reader to these.)

## Augustinian Science

Let us move on to the second round of objections that Plantinga has to methodological naturalism. Plantinga objects to my very attempt to characterize science as something that is marked by the methodological naturalistic approach. He writes:

Ruse suggests that methodological naturalism is true by definition of the term "science" one supposes; Ruse apparently holds there is a correct definition of "science," such that from the definition it follows that science deals only with what is natural, repeatable, and governed by law. (Note that this claim does not bear on the suggestions that a Christian scientist can propose hypotheses involving such "religious" doctrines as, say, original sin, and can evaluate the epistemic probability of a scientific hypothesis relative to background belief that includes Christian belief.) Ruse's claim apparently rules out hypotheses that include references to God: God is a supernatural being, hypotheses referring to him therefore deal with something besides the natural; hence hypotheses cannot be part of science. (1997, p. 145 [343–344])

Then he faults me here, saying that I simply have no right to invoke a mere definition to achieve my end. He writes:

The ... puzzling thing about Ruse's claim: it is hard to see how anything like a reasonably serious dispute about what is and is not science could be settled just by appealing to a *definition*. One thinks this would work only if the original query were really a *verbal* question—a question like *Is the English word "science" properly applicable to a hypothesis that makes reference to God?* But that was not the question: the question is instead *Could a hypothesis that makes reference to God be part of science? That* question cannot be answered just by citing a definition. (ibid., p. 146 [345])

It is true that there is something puzzling here, but not necessarily my argument. It would indeed be very odd were I simply trying to characterize "science" as something that, by definition, employs methodological naturalism: and then simply leaving things like that. My victory would be altogether too easy to achieve. Apart from anything else, I would simply be ruling religion out of science by fiat. But, this is not what I am doing now or have done in the past. I am certainly not trying to offer an analytic definition of what one means by "science," just as for instance one might offer an analytic definition of (to use the philosopher's old favorite) "grue" as meaning "green before time *t* and blue after time *t*." This is a definition that is analytic or stipulative. What I am trying to do is to offer a lexical definition: that is to say, I am trying to characterize the use of the term "science." And my suggestion is simply that what we mean by the word "science" in general usage is something that does not make reference to God and so forth, but which is marked by methodological naturalism.

I am not saying anything at all about whether or not God exists, or has any role in the world or anything like that. I am simply saying that science does not allow for this possibility, judged *qua* science. I think I am on pretty strong grounds and I am comforted to find that my opinion is shared by Ernan McMullin, who is not only an eminent philosopher of science, but also a Roman Catholic Priest. One cannot accuse him of being unsympathetic to religion! In the course of a discussion (incidentally directed against Plantinga), McMullin concurs completely with my understanding of the meaning of "science." Of Plantinga's claim that one should not restrict "science" simply to that which is governed by methodological naturalism, and that one should allow a more extended "science" that perhaps tries to understand not only in terms of law, but also in terms of God's intervention—what Plantinga at that point called "theistic science"—McMullin writes as follows:

I do not think ... that theistic science should be described *as* science. It lacks the universality of science, as that term has been understood in the later Western tradition. It also lacks the sort of warrant that has gradually come to characterize a properly "scientific" knowledge of nature, one that favours systematic observation, generalization, and the testing of explanatory hypotheses. Theistic science appeals to a specifically Christian belief, one that lays no claim to assent from a Hindu or an agnostic. It requires faith, and faith (we are told) is a gift, a grace, from God. To use the term "science" in this context seems dangerously misleading; it encourages expectations that cannot be fulfilled. (McMullin 1991 [chap. 8])

McMullin's point is precisely mine, namely that we should not use the word "science" for activities that go beyond the bounds of methodological naturalism, however worthy such activities and their products may be.

But here we come to what I think is the important part of Plantinga's position. He is certainly too good a philosopher to think that everything is just a matter of definition. I suspect that he does not really think that even I think it is all just a question of definition. Rather, Plantinga believes that whether or not conventional science (judged by adherence to methodological naturalism) is satisfactory, you should open up the inquiry to something broader. We ought to open it up to an understanding of the world that allows, not only working through law, but also the intervention of God in various ways: ways that, in conventional language, we would characterize as "miraculous." This is Plantinga's "theistic science"; although now he prefers the term "Augustinian science," because he thinks that this is something that would be acceptable to Saint Augustine. (I am not sure that he is right here, but one quibble about the use of terms per discussion is all I can handle.)

One's initial response is that, if Plantinga wants to extend his understanding of the notion of science beyond what we can call "science" without qualification to some kind of "science" that includes reference to miracles, that which he calls "Augustinian science," then he is at perfect liberty to do so. Indeed who can or should quarrel with what he is doing? One may not feel the extension is a particularly useful one, but that is another matter entirely. The point is that science without qualification is left untouched. But, of course, there is more at issue here: the significant fact is that Plantinga wants to go on using the term "science," whether he puts in the qualification or not. He does not want to speak simply of Augustinian knowledge or some such thing, but of Augustinian *science*.

As McMullin points out, this is a significant move, because clearly at some level Plantinga wants to give his extended science the status or authenticity of science without qualification. He wants to suggest that, in some way, his science is as good (in fact he would want to say probably a lot better) as science without qualification. His science therefore ought to be eligible for such things as grants and university support, not to mention be permitted to be put in classrooms during science courses. (Plantinga does not draw out all of these consequences explicitly, although he does make mention of grants and so forth. The line of argument is there, nevertheless, and even if he were not to pick it up personally, one can see how it could be used as a tool for fighting on the evolution/creation front.)

The question now is whether this extended enterprise—what Plantinga calls "Augustinian science"—has any right to the name of "science," in some sense or another. Here, Plantinga goes beyond definition and offers an argument. He suggests that the reason people (including Christians) are uncomfortable with extending the notion of "science" beyond that which is discovered using methodological naturalism, to that which may involve God's direct action in His Creation, is because it seems like a move of desperation. In particular, it seems like an invocation of so-called God-of-the-gaps theology. One supposes that the laws of nature go along as best they can, but that every now and then the laws break down, and so (*faute de mieux*) one has to plug in God. Supposedly, this strategy goes back to Newton, who thought that his laws of motion would not do everything that was needed to keep the planets running as they undoubtedly do, and so invoked God's help every now and then to give things a bit of a shove or an adjustment. Clearly, there is something rather desperate about this tactic: both scientifically and theologically. God apparently could not do the job properly through laws, so has to keep tinkering with His Creation. This is no happy assumption for either scientist or theologian.

As it happens, Plantinga agrees with those who would deny the extension of the name of "science" on the grounds of a God-of-the-gaps type of argument. However, he argues that as a Christian one has no need for such an argument at all. I quote him again:

God-of-the-gaps theology is worlds apart from serious Christian theism. This is evident at (at least) the following points. First and most important, according to serious theism, God is constantly, immediately, intimately, and directly active in his creation: he constantly upholds it in existence and providentially governs it. He is immediately and directly active in everything from the Big Bang to the sparrow's fall. Literally nothing happens without his upholding hand. Second, natural laws are not in any way independent of God, and are perhaps best thought of as regularities in the ways in which he treats the stuff he has made, or perhaps as counterfactuals of divine freedom. (Hence there is nothing in the least untoward in the thought that on some occasions God might do something in a way different from his usual way—e.g., raise someone from the dead or change water into wine.) Indeed, the whole *interventionist* terminology—speaking of God as *intervening* in nature, or *intruding* into it, or *interfering* with it, or *violating* natural law—all this goes with God-of-the-gaps theology, not with serious theism. According to the latter, God is already and always intimately acting in nature, which depends from moment to moment for its existence upon immediate divine activity; there is not and could not be any such thing as his "intervening" in nature. (1997, p. 149 [350])

Plantinga's position is that, properly understood theologically, God's interventions and the running of law are a seamless whole of the same logical type. Therefore, from a Christian theistic point of view, there is absolutely no reason to deny the possibility of miraculous interventions. Indeed, Plantinga's position rather is that, as a Christian, one ought to expect God to be intervening: not out of a failure to do the job properly in the first place, but because God is always sustaining His Creation. Elsewhere, Plantinga has added to this argument by pointing out that Christians believe that miracles are ongoing all of the time. For instance, Catholics believe that in the mass there is the miraculous transubstantiation of the bread and wine into the body and blood of Christ. Similar miracles occur when human souls are created individually, whether you believe this occurs when a person is conceived or when a person is born. So, since miracles are common phenomena, there is no real reason to deny that they may be occurring continuously in other cases, for instance with the origination of new species. (See Plantinga 1991a,b; b reprinted in this volume, chap. 9.)

Let me make three points in connection with Plantinga's argument. First, not all Christians believe that God's constant sustaining of His creation means that one should expect God to be intervening with miracles on a regular basis. McMullin points out that there is a whole tradition, going back at least to Augustine, that looks on the world as developing from potentials set by God. Not that Augustine was a biological evolutionist, or anything like that. But certainly his position was that God prefers to work through some sort of developmental, probably law-bound process, rather than through breaking off every now and then from His sustaining laws and doing things by hand as it were. Hence, argues McMullin, there is little reason to think that there is some sort of presumption, from a Christian perspective, that God would combine law with miracle. (Note how important this claim is for Plantinga. If he can establish the case that law and miracle are of the same logical type, then his hoped-for extension of science goes through much more readily than otherwise. Already in "science" without qualification, one has God's action of one kind. Simply extending that to "Augustinian science" is not demanding actions of a logically different kind. One has less a qualitative shift, as it were, and more a quantitative shift.)

Casting the discussion specifically in terms of our thinking about the origin of new organisms, with respect to the Genesis story, McMullin writes as follows:

The issue, be it noted, is not whether God *could* have intervened in the natural order; it is presumably within the power of the Being who holds the universe at every moment in existence to shape that existence freely. The issue is rather, whether it is antecedently *likely* that God would do so, and more specifically whether such intervention would have taken the form of special creation of ancestral living kinds. Attaching a degree of *likelihood* to this requires a reason; despite the avowed intention not to call on Genesis, there might appear to be some sort of residual linkage here. In the absence of the Genesis narrative, would it appear likely that the God of the salvation story would also act in a special way to bring the ancestral living kinds into existence? It hardly seems to be the case. (McMullin 1991 [chap. 8])

McMullin's point is that Plantinga is only arguing that miracles are, at some level, as likely as laws because he has in the first place made a fairly literalistic reading of the Bible. But this now raises a second objection: it is by no means obvious to one working from a Christian position that one must agree that God works almost indifferently through law and through miracle. First, it is only if one has already made a priori a fairly literalistic reading of the Bible that one would think that God's miracles are going to be as frequent as Plantinga rather implies. If one interprets, say, the Abraham and Isaac story, not so much as a literal case of rather difficult relations between father and son, but of symbolic in some sense of Israel's faith toward the law, then the whole question of frequent miracles by God becomes rather more problematic. Second, one can (as

McMullin and others point out) make a distinction between the order of nature and the order of grace, that is, between what is known as "cosmic history" and what is known as "salvation history." To quote McMullin again:

> The train of events linking Abraham to Christ is not to be considered an analogue for God's relationship to creation generally. The Incarnation and what led up to it were unique in their manifestation of God's creative power and a loving concern for the created universe. To overcome the consequences of human freedom, a different sort of action on God's part was required, a transformative action culminating in the promise of resurrection of the children of God, something that (despite the immortality claims of the Greek philosophers) lies altogether outside the bounds of nature.
>
> The story of salvation is a story about men and women, about the burden and the promise of being human. It is about free beings who sinned and who therefore *needed* God's intervention. Dealing with the human predicament "naturally," so to speak, would not have been sufficient on God's part. But no such argument can be used with regard to the origins of the first living cells or of plants and animals. The biblical account of God's dealings with humankind provides no warrant whatever for supposing that God would have brought the ancestors of the various kinds of plants and animals to be outside the ordinary order of nature. (ibid.)

One has no expectations from God's use of miracle in a certain special set of events that God will be using miracles as frequently or indifferently as Plantinga rather implies.

Third, surely there is an intellectual sleight of hand in Plantinga's highlighting of the miracles that supposedly are occurring frequently today. Plantinga knows as well as anybody else that transubstantiation is not supposed at all to be a miracle that in any way violates, or goes as an alternative, to the ordinary course of nature. If one cuts up the bread and wine after the mass, one is not going to see bleeding flesh oozing blood. The change is in the essence of the substance, rather than in the accidental properties. The same is true of the soul, whether or not one believes this to be an entirely coherent notion or indeed just a throwback to pre-Christian Platonic ideas. One is not going to find souls as material phenomena. Certainly, the ongoing creation or insertion of souls, if indeed this be true, is not something that goes as an alternative to law—in the sense of either one or the other but not both. Hence, here again, one has no reason for accepting Plantinga's extension of the notion of "science."

All in all, therefore, one can say that Plantinga has not made his case about the likelihood of miracles occurring as often as laws for the Christian theist.

## Science Stopping

A third and final argument offered by Plantinga strikes at what I suspect many defenders of methodological naturalism would take to be its strongest point: the pragmatic argument that nothing succeeds like success. Let us grant that methodological naturalism (so says its supporter) is an approach or an attitude, rather than something that is necessarily true a priori. Why then should one endorse it? Why in particular should one insist always in going with methodological naturalism, even when (as is surely the case) there are unsolved problems? Why, to take a particularly difficult problem, should one assume that there must be a naturalistic account of life's origins? Everything we know about life is incredibly complex. However it started, there must have been a number of intricate moves of a kind that one would not normally expect to find happening naturally. Why then persist in believing in the natural origins of life, simply because this is demanded by the methodologically naturalistic position, when prima facie such a natural origin seems so very nonobvious? Why in particular should one refuse to rule out miracles and interventions by God? That is to say, why should we assume that methodological naturalism is so very successful, so very important that we must go with it, even in the face of challenge?

The answer that the methodological naturalist gives here is that, in the past, the methodologically naturalistic approach yielded fantastic dividends. As Thomas Kuhn (1962) says about paradigms, because scientists have persisted in taking a methodologically naturalistic approach, problems that hitherto seemed insoluble have eventually given away to solutions. Take an example from biology. For many years, indeed ever since Darwin, there was much debate about how insect sociality could have evolved. How is it that the worker ants, for instance, devote their whole lives to the nest, despite the fact that they do not reproduce themselves? People had no answer but did not give up. They persisted, and finally in the early 1960s the then graduate student, William Hamilton (1961a,b), provided an explanation (invoking what came to be known as "kin selection") that accounted for insect sociality in terms of individual genetic selfishness. (Briefly, the answer is that in the hymenoptera—the bees, the ants, and the wasps—females are more closely related to sisters than they are to daughters. Thus they improve their genetic success by raising fertile sisters, rather than by raising fertile daughters.) The methodological naturalist says that this is a moral for us all: although there are indeed many unsolved problems, notably the origin of life, past experience suggests that these problems will be solved eventually by a methodologically naturalistic approach. Therefore, one should persist, no matter how improbable the finding of a solution seems today.

Plantinga challenges this. While he agrees that giving up on methodological naturalism is in some sense what he calls a "science stopper"—something that brings methodologically naturalistic science to an end—he claims that as Christians, we have no reason to think that such science stopping events do not happen.

The claim that God has directly created life, for example, may be a science stopper; it does not follow that God *did not* directly create life. Obviously we have no guarantee that God has done everything by way of employing secondary causes, or in such a way as to encourage further scientific inquiry, or for our convenience as scientists, or for the benefit of the National Science Foundation. Clearly we cannot sensibly insist in advance that whatever we are confronted with is to be explained in terms of something *else* God did; he must have done *some* things directly. It would be worth knowing, if possible, which things he *did* do directly; to know this would be an important part of a serious and profound knowledge of the universe. The fact that such claims are science stoppers means that as a general rule they will not be helpful; it does not mean that they are never true, and it does not mean that they can never be part of a proper scientific theory. (And of course it does not even bear on the other ways in which Christianity or Christian theism can be relevant to science.) It is a giant and unwarranted step from the recognition that claims of direct divine activity are science stoppers to the insistence that science must pretend that the created universe is just there, refusing to recognize that it is indeed *created*. (Plantinga 1997, pp. 152–153; [357] in this volume)

Let me make two points in response to this argument. First, Plantinga is making his case from an already-established theistic position. Already, he accepts that there are going to be exceptions to laws, or at least that there have been such exceptions. Hence, it is at least possible that there will be such exceptions in the future. If he did not make this assumption, at best one would have ignorance, in which case the methodological naturalist will say that one simply does not know whether methodological naturalism will work all the way. It is just that as a matter of general policy one has no choice, but to go with it—in the light of the fact that, in the past, employing methodological naturalism has been a very successful strategy. To argue otherwise, to argue as does Plantinga, is to assume the very theism whose necessity is at issue in the first place.

Second, Plantinga altogether underestimates the power and success of methodological naturalism. He can be so slighting of its potential only because he does not take modern science seriously. This is a strong claim to make and I am sure that Plantinga would challenge it vigorously. It is true, nevertheless. In his writings over the past decade, Plantinga has frequently made reference to evolutionary theory and to its supposed inadequacies. The ways in which he has done this show unambiguously that Plantinga's mind has been made up before he starts. Certainly the kinds of arguments he brings against science, particularly against the science of animal and plant origins, are the kinds that, were similar ones brought in a philosophical context, he himself would agree that the proponent should not be taken seriously—or at least that the person putting forward these arguments has already made a prior commitment to their falsity.

To make this point, let me refer you to an extended discussion of evolutionary theory offered by Plantinga at the beginning of this decade. About some parts of the evolutionist's thesis, broadly construed, Plantinga is so far from offering argument that he is just contemptuously dismissive. For instance, about the claim that life may have originated from nonlife, Plantinga has only the following to say:

Finally, there is the ... the Naturalistic Origins Thesis, the claim that life arose by naturalistic means. This seems to me to be for the most part mere arrogant bluster; given our present state of knowledge, I believe it is vastly less probable, on our present evidence, than is its denial. Darwin thought this claim very chancy; discoveries since Darwin, and in particular recent discoveries in molecular biology, make it much less likely than it was in Darwin's day. I can't summarize the evidence and the difficulties here. (Plantinga 1991d [128])

I simply just do not see that this is an argument at all. Suppose I recast it in philosophical terms. Suppose, following Plantinga, I said:

Finally there is the ontological argument, the claim that the definition of God yields his existence. This seems to me to be for the most part mere arrogant bluster. Given our present state of knowledge I believe it is vastly less probable, on our present evidence, than is its denial. Aquinas thought this claim very chancy; discoveries since Aquinas, and in particular recent discoveries in modal logic, make it much less likely than it was in Aquinas's day. I cannot summarize the evidence and the difficulties here.

I expect Plantinga would at least smile, even scornfully, at this argument. A blush would be more appropriate. He is offering no more himself.

Plantinga is not much better when he looks at the bulk of evolutionary theorizing. As he himself seems to be aware, the real claim for the fact of evolution (as opposed to the mechanisms of evolution) is the consilience that Darwin offers: the argument that many different areas of biology— embryology, animal instinct, biogeography, paleontology, systematics— all point to the fact of evolution, and conversely are given meaning through this fact. But, aware or not of this consilience, Plantinga makes absolutely nothing of it whatsoever.

The arguments from vestigial organs, geographical distribution, and embryology are suggestive, but of course nowhere near conclusive. As for the similarity in biochemistry of all life, this is reasonably probable on the hypothesis of special creation, hence not much by way of evidence against it, hence not much by way of evidence for evolution. (Plantinga 1991d [132])

Again, I convert this into its philosophical equivalent.

The arguments from miracles, causes, and ontology are suggestive, but of course nowhere near conclusive. As for the similarity in biochemistry of all life, this is reasonably probable on the hypothesis of evolution, hence not much by way of evidence against it, hence not much by way of evidence for God.

Once again, one assumes that Plantinga would shudder at this kind of argumentation. Why therefore should one not feel the same way about his treatment of the case for evolution?

I could continue. For instance, there is Plantinga's naive and somewhat arrogant attitude toward the fossil record. Deliberately, he turns his back on some of the strongest pieces of evidence, as given through the fossil record, in favor of evolution. One thinks of the detailed connections now discovered linking the marine mammals to land precursors, or the evolution of humankind. A hundred years ago, no one had much knowledge of human evolution: in particular, as to whether brains got big before humans got up on their hind legs, or conversely. Now, however, thanks to fabulous finds in East Africa—most particularly, Lucy (*Australopithecus afarensis*)—we know that human beings started walking before their brains grew in size. And from *Australopithecus afarensis*, some 3 or 4 million years ago, we have a very detailed record: from *Homo habilis* to *Homo erectus*, and from *Homo erectus* to our own species, *Homo sapiens* (Johanson and Edey 1981; Ruse 1982). To say that the fossil record is not adequate to support evolution is to show ignorance—or more.

There are times when I do think more than ignorance is involved. Consider Plantinga on the question of the Cambrian explosion, that huge increase in life just over half a billion years ago.

There is the Cambrian explosion. The fossil record displays unicellular life going all the way back, so they tell us, to 3 or 3.5 billion years ago—only a billion years or so after the formation of the Earth itself, and much less than a billion years after the Earth cooled sufficiently to permit life. There is no fossil record of skeletal animals until about 530 million years ago, 2.5 or 3 billion years after the appearance of uni-cellular life. Then there is a veritable explosion of invertebrate life, a riot of shapes and anatomical designs, with ancestors of the major contemporary forms and all the marine invertebrate phyla represented, together with a lot of forms wholly alien in the contemporary context. None of this was known in Darwin's day, and would surely have given him pause. And now in a recent issue of *Science* we learn that the time during which this explosion took place was much shorter than previously thought; it all happened during a period of no more than 5 or 10 million years ..., a period that seems much too short to accommodate such furious evolutionary creativity, at least with respect to any known mechanisms. On balance, it is likely that if Darwin knew what we now know about the complexity of such organs as the mammalian eye and the human brain, the enormous intricacy revealed by biochemistry and molecular biology (including the astonishing complexity of the simplest forms of life), the Cambrian explosion, the lack of closure in the fossil record, and so on, he would have been neither a Darwinian nor a devotee of [the Common Ancestry Thesis]. (Plantinga 1991c, p. 753)

In fact Darwin did know about the Cambrian explosion and he did as a matter of fact worry about it. (It was known in his day as the bottom of the Silurian system.) Today, however, we have a much more detailed knowledge of the explosion and possible suggestions have been put forward to account for it. Most notably the American paleontologist, J. John Sepkoski, Jr., has shown through computer simulation that the Cambrian explosion is just the kind of exponential growth that we would expect given the rates of speciation that are known to have been occurring at that time. Sepkoski has been able to map exactly the explosion itself against his computer models and has reduced the whole problem to one of readily understandable and acceptable mathematics. It may of course be the case that Sepkoski is wrong, or that his position will need modifying, but this is as it may be. The point is that he and other paleontologists have been putting forward answers to explain the explosion, answers that are highly plausible, given modern paleontology and modern mathematics. (See Sepkoski 1978, 1979, 1984; Ruse 1999.)

Plantinga has no knowledge whatsoever of this and ploughs on regardless. Once again, using the philosophical analogy, how would he feel if one were simply to say that the ontological argument is as worrisome as it was back at the time of Aquinas, because everybody knows that existence is not a predicate? And if this claim were made in total ignorance of the kinds of arguments that, for the past three decades, Plantinga and other sophisticated philosophers of religion have been putting forward in favor of the ontological and related arguments? It is not a question of whether or not Plantinga and others are right in their thinking about the ontological argument and its fellows. It is rather that competent scholarship requires that one take note of this, and this holds even if one is working outside one's own field. One must not simply dismiss carefully thought-out positions, with a sneer, from a position of ignorance.

I am sorry if I sound indignant about all of this, but if one is going to argue about important issues—and Plantinga and I are certainly united in thinking that these are important issues—then one ought to take the opposition seriously. This Plantinga does not do, and it is why I am not convinced by his third argument against methodological naturalism. It may be indeed that methodological naturalism does not succeed in doing everything that it sets out to do. It may be that it never will. But to assume that there are going to be "science stoppers," and that this should lead one to pull back from a commitment of methodological naturalism, is to reveal that one has another agenda. We know that Plantinga's agenda is Christianity. That is fair enough. But it is an agenda backed by a deliberate ignorance of work that is going on today in science. Plantinga is able to talk so confidently about science stoppers only because he has not and apparently will not look at what scientists are saying and achieving. These people may not be right, but they do deserve more of a hearing than Plantinga gives them.

## Conclusion

These then are the arguments that Plantinga brings against methodological naturalism and my responses to him. I would argue that he has given us no reasons to give up on methodological naturalism, or inasmuch as he has it has been only because of his prior commitment to his own version of Christian theism. So I see no reason why one should not continue to draw the distinction between methodological and metaphysical naturalism; to argue that the two can be separated; and to argue that, whatever may be the philosophical and theological basis underlying metaphysical naturalism, it is not the case that the methodological naturalist has to adopt the same position. This all being so, then, although I am happy to accept that methodological naturalism leads today to a belief in evolution, I am not prepared to accept that methodological naturalism is a philosophy opposed to theism. I see no reason at all to deny the Christian access to methodological naturalism, saying that it is untenable for the Christian to insist that in our understanding of the natural world one employ only the methodologically naturalistic approach. Evolution and Christianity should not be separated in this way.

## References

Box, J. F. 1978. *R. A. Fisher: The Life of a Scientist*. New York: Wiley.

Dawkins, R. 1995. *A River Out of Eden*. New York: Basic Books.

———. 1996. *Climbing Mount Improbable*. New York: Norton.

Dennett, D. C. 1995. *Darwin's Dangerous Idea*. New York: Simon and Schuster.

Dobzhansky, T. 1967. *The Biology of Ultimate Concern*. New York: The New American Library.

Fisher, R. A. 1950. *Creative Aspects of Natural Law: The Eddington Memorial Lecture*. Cambridge: Cambridge University Press.

Greene, J. C. and M. Ruse. 1996. On the nature of the evolutionary process: The correspondence between Theodosius Dobzhansky and John C. Greene. *Biology and Philosophy* 11: 445–491.

Hamilton, W. D. 1964a. The genetical evolution of social behaviour I. *Journal of Theoretical Biology* 7: 1–16.

Hamilton, W. D. 1964b. The genetical evolution of social behaviour II. *Journal of Theoretical Biology* 7: 17–32.

Hull, D. L. and M. Ruse, editors. 1998. *Readings in the Philosophy of Biology: Oxford Readings in Philosophy*. Oxford: Oxford University Press.

Johanson, D. and M. Edey. 1981. *Lucy: The Beginnings of Humankind*. New York: Simon and Schuster.

Johnson, P. E. 1991. *Darwin on Trial*. Washington, D.C.: Regnery Gateway.

————. 1995. *Reason in the Balance: The Case Against Naturalism in Science, Law, and Education.* Downers Grove, Ill.: InterVarsity Press.

Kuhn, T. 1962. *The Structure of Scientific Revolutions.* Chicago: University of Chicago Press.

McMullin, E. 1991. Plantinga's defense of special creation. *Christian Scholar's Review* 21, no. 1: 55–79. Reprinted as chap. 8 in this volume.

Numbers, R. 1992. *The Creationists.* New York: A. A. Knopf.

Plantinga, A. 1991a. An evolutionary argument against naturalism. *Logos* 12: 27–49.

————. 1991b. Evolution, neutrality, and antecedent probability: A reply to Van Till and McMullin. *Christian Scholar's Review* 21, no. 1: 80–109. Reprinted as chap. 9, this volume.

————. 1991c. Reply to McMullin. Reprinted in Hull, D. L., and Ruse, editors. 1998. *Readings in the Philosophy of Biology.* Oxford: Oxford University Press.

————. 1991d. When faith and reason clash: Evolution and the Bible. *Christian Scholar's Review* 21, no. 1: 8–32. Reprinted as chap. 6, this volume.

————. 1993. *Warrant and Proper Function.* New York: Oxford University Press.

————. 1994. Naturalism defeated. *Unpublished Manuscript.*

————. 1995. Methodological naturalism. In Van der Meer (ed.), *Facets of Faith and Science.* Lanham, Md.: University Press of America.

————. 1997. Methodological naturalism. *Perspectives on Science and Christian Faith* 49, no. 3: 143–154. Reprinted as chap. 13 in this volume.

Provine, W. 1989. Evolution and the foundation of ethics. *Science, Technology, and Social Progress.* Bethlehem, Pa.: Lehigh University Press.

Ruse, M. 1975. The relationship between science and religion in Britain, 1830–1870. *Church History* 44: 505–522.

————. 1979. *The Darwinian Revolution: Science Red in Tooth and Claw.* Chicago: University of Chicago Press.

————. 1982. *Darwinism Defended: A Guide to Evolutionary Controversies.* Reading, Mass.: Benjamin/Cummings Pub. Co.

————. 1984. A philosopher's day in court. In A. Montagu, (ed.) *Science and Creationism.* 311–342. New York: Oxford University Press.

————. 1986a. *Taking Darwin Seriously: A Naturalistic Approach to Philosophy.* Oxford: Blackwell.

————. 1988. *But is it Science? The Philosophical Question in the Creation/Evolution Controversy.* Buffalo: Prometheus.

————. 1995. *Evolutionary Naturalism: Selected Essays.* London: Routledge.

————. 1996. *Monad to Man: The Concept of Progress in Evolutionary Biology.* Cambridge, Mass.: Harvard University Press.

————. 1999. *Mystery of Mysteries: Is Evolution a Social Construction?* Cambridge, Mass.: Harvard University Press.

————. 2000. *Can a Darwinian be a Christian? One Person's Answer*. Cambridge: Cambridge University Press.

Sepkoski, J. J. 1978. A kinetic model of Phanerozoic taxonomic diversity I. Analysis of marine orders. *Paleobiology* 4: 223–251.

————. 1979. A kinetic model of Phanerozoic taxonomic diversity II. Early Paleozoic families and multiple equilibria. *Paleobiology* 5: 222–252.

————. 1984. A kinetic model of Phanerozoic taxonomic diversity III. Post-Paleozoic families and mass extinctions. *Paleobiology* 10: 246–267.

van Fraassen, B. 1989. *Laws and Symmetry*. Oxford : Oxford University Press.

Westfall, R. S. 1982. *Never at Rest*. Cambridge: Cambridge University Press.

# 15

# Plantinga's Case against Naturalistic Epistemology

Evan Fales

In the closing chapter of his recent book *Warrant and Proper Function*, Alvin Plantinga mounts a sustained attack upon the cluster of views generally associated with the term "naturalized epistemology." I shall argue here that the strategy of Plantinga's last chapter fails, and indeed backfires against his own position. To say that Plantinga attacks naturalistic epistemology is a bit misleading, however; for Plantinga considers his own epistemology to be a naturalized one. What it shares most conspicuously with its counterparts—besides an emphasis upon the idea that knowledge consists in beliefs reliably generated by cognitive mechanisms operating in a suitable environment—is its externalism. That is, Plantinga denies that it is a condition on knowledge that one *knows* one's beliefs to have been thus reliably generated.

Nevertheless, there is a profound difference between Plantinga's naturalized epistemology and most other positions which fall under that rubric. According to Plantinga, naturalized epistemology can only succeed when hitched to a supernaturalistic—specifically, a theist—ontology. Thus, Plantinga's target is not naturalized epistemology *simpliciter*, but rather naturalized epistemology coupled with ontological naturalism. I shall take naturalism[1] to be the thesis that there exists nothing other than spatiotemporal beings embedded within a space-time framework. Thus, naturalism denies the existence of abstract entities such as propositions and numbers, it rejects Platonic universals, and it does not countenance disembodied minds, gods, and the like.

Plantinga's strategy is to argue that naturalism requires the neo-Darwinian theory of evolution (or something closely similar) to account

Originally published in *Philosophy of Science*, 63 (Sept. 1996, pp. 432–451).

for our existence, but that this theory undermines a naturalistic epistemology in this way: if the neoDarwinian account is true, then it is unlikely or at least not demonstrably likely that our cognitive faculties are reliable generators of true beliefs. Hence, if naturalism-plus-naturalized epistemology (hereafter NNE) is correct, then we could not be in a position to know that it is, and so could not know that our sensory faculties and inferential processes yield knowledge. In that case, NNE is an unstable position: it gives way to skepticism.

On the other hand, says Plantinga, theism plus naturalized epistemology is a stable position that can resist skepticism. For if God created us, either directly or indirectly through directed evolutionary means, then we have a guarantee that our cognitive faculties, when normal and properly used, are reliable.

Plantinga's argument, then, hinges on three crucial claims: that naturalism is committed to a neoDarwinian theory of our origins; that neoDarwinian evolutionary processes would, for all we know, most likely yield cognitive processes which are unreliable in the relevant epistemic sense; and that theistic creation, in contrast, would yield reliable cognitive mechanisms.

Before these three theses are subjected to critical scrutiny, something should be said about the nature of the epistemological enterprise, as Plantinga (in concert with most other externalists) sees it. Traditionally, the task of epistemology was understood to be that of securing the epistemic credentials of broad classes of beliefs, both of common sense and of refined theorizing, while sharpening the critical tools needed to sift out falsity, superstition, and the like. Along the way, epistemologists have had a good deal to say about what it is for a proposition to be known or justifiably believed; but this analytical enterprise was clearly subservient to the larger goal of getting clear about what we *do* know—and, by and large, underwriting certain pre-philosophical beliefs about that.

In an externalist epistemology, this conception of the philosophical enterprise must be inverted. Externalism forces the focus to shift to the analytic task, the task of saying just what knowledge and justification *are*. The underwriting of particular beliefs or classes of beliefs must be secondary: after all, the core thesis of externalism is that we might not *know* that a belief or set of beliefs is known or justified for us, even though it in fact *is*. Of course, it would be nice to have this second-order knowledge (though it, too, would presumably be a matter of having reliably produced second-order beliefs), but externalists typically insist that knowing one knows (or is justified) is not a necessary condition of knowing (being justified). Thus it is ironic that externalists have nevertheless focused so much attention upon showing that various cognitive processes *are* reliable.

It is just here that many supporters of NNE have seen in the neo-Darwinian theory an important ally. They have assumed that natural selection would tend to select for reliable cognitive mechanisms. Natural as this assumption appears, rejecting it lies at the heart of Plantinga's attack on NNE. If Plantinga can make good his claim that NNE is (probabilistically) *in*coherent, whereas theistic naturalized epistemology (hereafter TNE) is coherent, he can claim a significant advantage for TNE. I shall show that this challenge fails. But since my own ontological and epistemological views are not naturalistic, it is not my purpose here to praise NNE. I come to disarm TNE, not to praise NNE. I shall take up each of Plantinga's central theses in turn, but my discussion of the first and third will be rather brief.

## 1.   Are Naturalists Committed to Darwinism?

Plantinga's attack upon neoDarwinism is not confined to *Warrant and Proper Function*. In a related paper, "When Faith and Reason Clash: Evolution and the Bible,"[2] Plantinga outlines what he takes to be both empirical and specifically Christian reasons to reject neoDarwinism.[3] Both there and in *Warrant and Proper Function*, Plantinga repeatedly emphasizes the claim that, for naturalists, neoDarwinism is "the only game in town." He emphasizes this because he wants to show that naturalism falls if neoDarwinism does. But although he is in a sense quite right about this, his way of putting the point is tendentious, in two respects.

First, it is certainly not true that the falsity of neoDarwinism *entails* the falsity of naturalism. There is no *a priori* connection between the two. So far as the *a priori* possibilities go—as Hume was right to insist—anything might be the cause of life as we observe it. However, Plantinga's point can reasonably be taken to be that, *relative to what we* (think we) *know about non-biological processes in the universe*, neoDarwinism holds out the only reasonable hope for a non-purposive account of the existence of life. And this is plausible. If we allow also that a purposive account would have to have recourse to some supernatural agent, then Plantinga has his conclusion.

At the same time, it is striking that Plantinga says almost nothing about the *converse* thesis *viz.*, that Christian theism (at least Plantinga's favored version of it) becomes implausible if neoDarwinianism is *true*. He emphasizes the stake that naturalists have in the truth-value of neo-Darwinism, while remaining virtually silent about the risks for TNE. This is the second (and more serious) way in which Plantinga's point is tendentiously put.

By way of restoring the balance, let us list—without discussion—some of the ways in which the truth of neoDarwinism would present challenges for Christian theism. Obviously, the force of some of these challenges varies with the version of Christian theism one adopts, especially with the extent to which one takes certain Biblical texts to be literally true.[4] Here, then, is a quick reminder of why neoDarwinism so arouses the passions of many theists.

1. It undermines traditional versions of the Argument from Design.
2. It suggests that the great panorama of life on this earth is not directed toward some determinate end, and that the existence of *Homo sapiens* is an accident.
3. It fits well with materialism, but uneasily with dualism. If human beings have souls, just when, how, and why did they enter the picture?
4. It suggests that much more of our behavior than we had thought has biological determinants, and so places in question a libertarian conception of human freedom. For many theists, therefore, it threatens the notion of moral autonomy.
5. It strongly supports the idea that the Garden of Eden Story is a fable, thus undermining the doctrine of original sin, fundamental to Christian soteriology, as it is understood by many Christians.
6. It is difficult to square the existence of a benign Creator, whose ultimate goal is human salvation, with a means to that end which involves a process, red in tooth and claw and rife with suffering, spanning nearly four billion years prior even to the appearance of our species upon the earth.

There are Christian responses—even intellectually respectable ones—to each of the items on this truncated list. Even so, it is not hard to see that plausible responses place severe strains on the credibility of conservative Christian ideology.[5] Since there is no need to press this familiar point, we may proceed to Plantinga's second thesis.

## 2. Is the Reliability of Human Cognitive Faculties Improbable, Given NeoDarwinism?

My main aim in this paper is to argue that Plantinga has not demonstrated that this is so. If Plantinga's argument fails here, then he will not have shown that NNE is probabilistically incoherent. Therefore he will not have shown that a consistent naturalistic epistemology must be framed within a supernaturalistic ontology.

Plantinga proceeds as follows. First he softens the enemy position by citing the work of philosophers—most notably Stephen Stich (1990)—who have argued, on various grounds, that evolutionary processes should not be expected regularly to lead (and in fact often do not) to reliable cognitive mechanisms. (Stich, in turn, relies heavily upon the findings of Kahneman, Slovic, and Tversky (1982), and Nisbett and Ross (1980), which show—quite embarrassingly—how robustly certain fallacious forms of inference are entrenched within the human information-processing system. I shall have more to say about these findings presently.)

A central point for Plantinga, as for Stich, is that natural selection does not directly favor true belief. Rather, it favors appropriate action. Since true beliefs need not engender successful action, nor false beliefs be fatal, naturalists need to show that, on average, reliable belief-forming mechanisms confer an advantage over various alternative possibilities. Indeed, Plantinga is able to haul out a passage in a letter of Darwin's to W. Graham that supposedly reflects this worry:[6]

With me the horrid doubt arises whether the convictions of man's mind, which have been developed from the mind of the lower animals, are of any value or at all trustworthy. Would anyone trust in the convictions of a monkey's mind, if indeed there are any convictions in such a mind?

The citation allows Plantinga to dub this source of the problem for NNE, "Darwin's Doubt."[7]

In the light of this general reflection, Plantinga proceeds as follows. First he outlines five different evolutionary scenarios, consistent with neoDarwinism, which would result in our cognitive faculties being un-reliable, and which are compatible with most of our beliefs being false. Then, in a preliminary argument, Plantinga suggests that, relative to neoDarwinism, the objective probability of some one of these sce-narios having been realized in our own case is larger than that of a scenario under which evolution has conferred on us reliable belief-generating mechanisms. Therefore, we ought to believe, if we are neo-Darwinians, that our cognitive mechanisms are probably unreliable.

A second argument—Plantinga calls it his main argument—concedes that estimating the objective probabilities here is a risky business. Per-haps the probabilities are not as Plantinga supposes: perhaps, objectively speaking, the mechanisms of neoDarwinian evolution do tend to favor the development of reliable cognitive mechanisms, when they favor the evolution of cognition at all. Admitting that, Plantinga reasons to a weaker conclusion: that, in that case, we at least are not in a position to *know* that the probability of reliable cognitive faculties is high, given neoDarwinism. Weaker; but strong enough. If the reliability of our cog-nitive powers remains seriously in doubt, given neoDarwinism, then we cannot appeal to that theory to support a naturalistic epistemology that certifies our most cherished beliefs. In particular, our ground for confi-dence in neoDarwinism is itself undermined, and the position becomes probabilistically incoherent or unstable.

Now, in response to this, a naturalist has available two strategies. He or she can, first, simply refuse to address the second-order question, the question of whether our cognitive faculties indeed *are* reliable. Natural-ists can insist that they do not know whether those faculties are reliable or not; nor must this be known, in order for it to be the case that many (or most) first-order beliefs are known to be true. That, after all, is just what externalism permits.

However, if a naturalist accepts neoDarwinism (as is likely), then this response will no doubt not be very satisfying. For one thing, externalists typically *do* want to go on to certify the reliability of our cognitive faculties. For another, even though this response addresses Plantinga's "main" argument, it will not do as a reply to his preliminary argument. If Plantinga has succeeded in *showing* that, given neoDarwinism, it is unlikely that our cognitive faculties are reliable, then a neoDarwinian externalist epistemology is indeed in trouble, not with respect to the analytic task of saying what it *is* to know or be justified, but with respect to providing a framework in which our beliefs—from common sense to science—can be coherently affirmed to *be* justified.

The naturalist's best strategy, therefore, is to show that Plantinga is *wrong* in his estimate of the implications of neoDarwinism for our evaluation of our cognitive powers. And that is what I shall now argue.

There are, Plantinga suggests, five different ways in which neoDarwinian evolution can produce creatures with belief-generating mechanisms that would offer no comfort to reliabilist conceptions of knowledge.

(1) The first possibility is that there might be no causal connection between beliefs and action at all. Since the only thing survival and procreation demand are getting one's body parts into the right places at the right times, we can imagine creatures whose adaptive responses to their environment are handled in an entirely cognition-free way, while their beliefs are on permanent holiday. In such creatures, the unreliability of belief-forming mechanisms would be no liability at all (nor would reliability confer a selective advantage).

(2) A second, related possibility is that beliefs are causally connected with behavior, but only by way of being effects of that behavior, or "side-effects" of the causes of behavior. Here again, the truth or falsity of a belief will play no role in the appropriateness of the behavior it is linked to; and so truth will confer no selective advantage.

(3) Beliefs might indeed causally affect behavior, but do so in a way that is sensitive only to their syntax, not to their content or semantics. Then once again, the *truth-value* of a belief would be irrelevant to its role in producing adaptive behavior.

(4) Perhaps beliefs could be causally efficacious, and their content materially relevant to the behavior they help generate, while the behavior thus generated is *maladaptive*. No organism, perhaps, is *perfectly* efficient, perfectly attuned to its environment. Maladaptive characteristics are a burden which can be borne provided they are not *too* maladaptive. Natural selection can even *favor* maladaptive traits, when these are closely linked on the genome with genes which confer strongly adaptive traits—so that the benefits of inheriting that region of a genome outweigh the costs. This can happen when the two genes lie closely adjacent on a chromosome, so that they tend to travel together during genetic reshuffling; or it can happen by way of pleiotropy—a single gene coding for two or more traits. Indeed, adjacency and pleiotropy can be invoked to provide mechanisms that would explain possibilities (1) and (2) as well.

The present point is that an organism may be able to hobble along with lots of false beliefs and misdirected actions.

(5) Finally, evolution might produce organisms in which false belief leads to *adaptive* action. As Plantinga points out, this can happen in several ways. Freddy the caveman may believe that saber-toothed tigers make great pets, may want to tame the one that has just appeared, and believe that running away from it as fast as he can is the best way to corral it. Or Freddy may want to be eaten, believe correctly that this cat will eat him if given the chance, and believe falsely that shoving a fire-brand in its face maximizes the chances for his desired fate. The general point is that when beliefs cause action, the action that results is the combined result of various desires and beliefs. Thus, a false belief can "cancel out" another false belief, or "cancel out" a destructive desire, to produce a beneficial action. Why should we suppose, then, that if nature selects for creatures whose actions are guided by their beliefs (and desires), it will be *true* beliefs that confer the greatest selective advantage?

Here we have five ways in which the random processes of neoDarwinian evolution might produce creatures with belief-generating mechanisms whose reliability one would not want to bet on. Perhaps we are such creatures. The other possibility is the evolution of creatures whose actions are guided by constructive desires informed by largely true beliefs and sound inferences. But so far, the chances of this happening are perhaps only one to five—unless the odds favoring possibilities (1)–(5) are small.

Let us turn, then, to an evaluation of those odds. Since rather similar considerations apply to (1), (2), and (4), I shall take up these possibilities first, then turn to (3) and (5).

The neural systems by means of which organisms generate and manage their beliefs are biologically expensive.[8] Both in terms of the genetic coding required, and in terms of energy expenditure devoted to growth and maintenance, neural systems—brains in particular—are costly devices.[9] It is hard to see how it could be otherwise, given the nature and complexity of their functions. This appears to be especially true of the mechanisms for belief formation and processing. First, beliefs must be generated—in our case, by distinct mechanisms linked to the dozen or more sensory, proprioceptive, and introspective modalities with which we are equipped. Next, to be of use, they must be catalogued and stored for efficient retrieval, they must constantly be squared with one another to insure as much consistency and inductive coherence as possible, and there must be inferential mechanisms devoted to the production of further beliefs.

Of course, many organisms get along without all of this—or any of it; but *having* these capacities, as we clearly do, is an expensive proposition, biologically speaking. That gives the neoDarwinian a *prima facie* reason for assigning a low probability to the development of such mechanisms unless they confer a decided selective advantage. The selective advantage of intelligence, when linked in an appropriate way to action, can hardly be denied. While many ecological niches can be successfully filled without the benefit of intelligence (witness the cockroach), Plantinga will agree that *Homo sapiens* has, more than any other species, specialized in intelligence as a survival strategy. We have few other biological advantages; most of our eggs are in that basket.[10] Our heavy investment in big brains and otherwise mediocre bodies makes it all the more unlikely that resources would be wasted on elaborate belief-forming and processing mechanisms that have no practical utility.

Let us remind ourselves that the question before us is this: given neoDarwinism, and given the cognitive mechanisms found in *Homo sapiens* and their ecological context, what is the probability that those mechanisms are either maladaptive—(4)—or adaptively irrelevant—(1) or (2)? Very small, say I. But perhaps, as Plantinga suggests, there is a significant probability that Darwinian evolution would produce one of these possibilities via the mechanisms of association or pleiotropy. Yet even granting that association and pleiotropy are common, the occurrence of costly traits that ride piggyback upon beneficial ones is quite rare, thanks to selective pressures.

Even worse, human intelligence is not a single trait—that is, it is not encoded by a single gene, but by many, many genes. What is the likelihood that these genes, each finely tuned to work in concert with the others to produce an integrated cognitive apparatus, would *all* be getting free rides from other genes and thus be relatively insulated from adaptive pressures?

If fact, there is *no* plausible neoDarwinian scenario to back up (1), (2), or (4): they are wildly improbable. For the development of something as complex as the human brain, only the slow, tortuous route of selection in virtue of positive step-by-step adaptive change is at all plausible.

What, then, of possibility (3)? (3) allows that beliefs play a role in action; but that this role is quite independent of their content and truth-value. The issues raised by this possibility are more complex. Indeed, there is at present a lively debate over the functional utility of consciousness, and over the question whether it confers a selective advantage not otherwise achievable. Velmans, for example, has argued that from a third-person perspective, it appears that pre-conscious processes do all the work; consciousness is a mere epiphenomenon. Van Gulick, on the other side, argues that consciousness plays a essential integrative role in the mind's handling of information. Flanagan (1992, Chapter 7) allows that evolution could quite possibly have produced organisms which perform all the mental functions we perform, only nonconsciously. But he agrees with Van Gulick concerning the importance of consciousness in the human mental economy, and concerning the selective advantages it confers.[11] To be sure, the object of this debate is consciousness, but it is the cognitive usefulness of consciousness which occupies center stage. It seems clear that the jury is still out here. Perhaps we shall have to wait some decades before the matter is clarified. But given that we *have* beliefs, whose content is at least causally linked to neuronal coding (presumably in terms of some compositional algorithm), there is more to be said about the reliability of those beliefs.

The further issue is whether it is conceptually possible that semantic content would fail to map onto syntax so as to preserve truth. That is, we must ask in virtue of what it is the case that a given mental representation denotes or picks out high temperature (say) as opposed to moderate temperature. Here much will depend upon one's views about intentionality. But an entirely reasonable view, from a naturalistic perspective, is that mental representations get their content in virtue of being caused in the right way by items in the environment; and that this is a *conceptual* truth. Thus if a mental representation is caused in the right way by heat, then it is a representation of heat; and if it is not so caused, then it is not a representation of heat. So long as representations are causally linked to the world via the syntactic structures in the brain to which they correspond, this will guarantee that syntax maps onto semantics in a generally truth-preserving way. The details here are unavoidably complex; they have to do with how reference to particulars and to properties initially gets fixed, and with what it is for a representation to be caused "in the right way."[12] But from a naturalistic perspective, the probability of (3), while harder to assess, may well be no higher than that of (1), (2), or (4).

The most interesting case, in a way, is (5). According to (5), beliefs are (partly) responsible for action, but in a way that permits actions to be on target while the producing beliefs are wildly wrong. How plausible is this?

It appears that Plantinga has in mind here a conception of the connection between beliefs, desires, and action that is roughly that deeply embedded in folk psychology. The beliefs and desires function as premises in a bit of practical reasoning; the outcome is a decision to act (or the action itself). Keeping this inference-performing part of our rational processes intact, we are to imagine the production of felicitous action from corrupt beliefs. Of course, there is no reason to assume the inferential mechanisms are sound: maybe they are corrupt us well. Either way, do we have here a workable action-guiding system?

I think not. In the first place, Plantinga's examples, like my Freddy cases, work only because the beliefs which supply the immediate doxastic input to Freddy's practical syllogism are perceptual—that is, non-inferential—beliefs, or simple inductive generalizations. That makes it easy to imagine a cognitive mechanism that takes input information, systematically reverses truth-value, and thereby produces *systematically* false beliefs. But what happens when deductive inference comes into play? *True* premises guarantee true conclusions: so a system that relies consistently upon true inputs to guide inference and action can employ general rules and hope to get things (i.e., action) right. But when a deductive argument employs false premises, the truth-value of the conclusion is *random*. Thus there *cannot* be any set of *general* algorithms which get a creature to use the conclusions of such arguments in a way that reliably promotes successful action. A cognitive system which is not *extremely* limited in the inferential procedures it employs must either give up all hope of successfully directing action or become unintelligibly complex and *ad hoc* in its procedures for connecting belief to action.

But this conclusion applies *even when the beliefs in play* are non-inferential or based only on enumerative induction. Freddy, who is carrying a heavy rock he falsely believes to be light and soft, nearly steps on a Puff Adder. Believing that being hit by something light and soft will be fatal for the adder (also false), he quickly drops the rock on it, and lives to see another day. So far so good for Freddy. Continuing on with his rock, Freddy encounters an angry warthog on the trail. Still believing the rock to be light as a feather, and believing (falsely) that dancing upon something light deters warthogs, Freddy proceeds to do a two-step on top of the rock directly in the path of the charging pig. The moral of this fable is plain: there are no effective algorithms connecting false belief to appropriate action, as there are when the input is true beliefs and the rules of inference employed are valid or inductively sound. Intelligent action is hard enough for a brain to manage; burdening it with ever-changing, completely arbitrary principles would make the task impossible. Freddy may survive the adder, but he will not live long. Nor will his genetic heritage.

Here let me quote the work of Katherine Milton, who has investigated the influence of diet upon the evolution of intelligence in monkeys. Milton found that getting adequate nutrition from the forest canopy is not nearly as easy a task for monkeys as had been generally assumed. Monkeys have two principle options: either forage primarily for the abundant but low-calorie and hard-to-digest leaves,[13] or specialize in high-energy, easily digested fruit and very young leaves, which are far less abundant, highly seasonal, and scattered in complex ways through the canopy.[14] The former strategy requires morphological adaptations—specialization of the digestive tract—whereas the latter requires behavioral adaptations—an ability to locate nuts and fruit. Milton was able to study howler and spider monkeys in the same forest setting. These species are similar in size and fairly closely related, but whereas howlers are mainly leaf-eaters, spider monkeys specialize in fruit. Milton says:

These digestive findings fascinated me, but a comparison of brain size in the two species yielded one of those "eurekas" of which every scientist dreams. I examined information on the brain sizes of the howler and spider monkeys because the spider monkeys in Panama seemed "smarter" than the howlers—almost human. Actually, some of them reminded me of my friends. I began to wonder whether spider monkeys behaved differently because their brains were more like our own ... The spider monkey brain weighs almost twice that of howlers.

Now, the brain is an expensive organ to maintain; it usurps a disproportionate amount of energy (glucose) extracted from food. So I knew natural selection would not have favored development of a large brain in spider monkeys unless the animals gained a rather pronounced benefit from the enlargement. Considering that the most pronounced difference between howler and spider monkeys is their diets, I proposed that the bigger brain of spider monkeys may have been favored because it facilitated the development of mental skills that enhanced success in maintaining a diet centered on ripe fruit.

A large brain would certainly have helped spider monkeys to learn and, most important, to remember, where certain patchily distributed fruit-bearing trees were located and when the fruit would be ready to eat. Also, spider monkeys comb the forest for fruit by dividing into small, changeable groups. Expanded mental capacity would have helped them to recognize members of their particular social unit and to learn the meaning of the different food-related calls through which troop members convey over large distances news of palatable items. (Milton 1993, 90)

To Plantinga or anyone else who may think that spider monkeys could have achieved these adaptations via belief-generating mechanisms that yield a large number of false beliefs and memories, I offer the following challenge. Construct *in detail* an account according to which a monkey, presented, say, with a fruit bearing tree, forms (obeying some information-processing algorithm, we must assume) a false belief (*which* false belief?) such that it meets the following conditions: (1) combined with other present false beliefs and/or destructive desires, it leads (with good probability) to felicitous action; (2) when combined on other occasions with yet *other* false beliefs/bad desires, it *still* is likely to produce correct action; and (3) if destructive desires are invoked, a plausible Darwinian story can be told about how they evolved from the action-guiding desires of the pre-rational ancestors of the monkey, or in some other way. Now construct a system of algorithms that will achieve this for the monkey's beliefs generally. I say it cannot be done.

To this it should be added that one of the distinctive capacities of human beings is the ability to engage in linguistic communication. The adaptive advantages of this development can scarcely be overestimated. Yet linguistic communication is impossible unless the agents participating in the practice have beliefs which systematically correspond, on the whole, with what is true of the domain about which they are communicating (or at least with what is true of the contexts which initially fix the meaning or reference of the terms employed), and only provided they generally say what they believe (as has been argued by Davidson and others).

But perhaps I have overstated the case. Surely—as Nisbett, Kahneman, and others have amply demonstrated—it is not true that human sensory faculties and inferential processes are in all circumstances reliable. The literature on the fallibility of sense perception is vast and fairly familiar; here, let me focus on inference. What the experiments of Nisbett et al. have shown is that certain sorts of inferences, though fallacious, are regularly made. The regularities are quite robust: even highly intelligent people fall prey to them (which gives Nisbett and Ross occasion to wink at their fellow academics). Of course, it is not the case that *everyone* succumbs to the fallacies *all* the time; if *that* were so, we would never know the reasoning to be fallacious. But nevertheless, Stich's question remains: if neoDarwinian processes of evolution are so effective at winnowing epistemic wheat from chaff, why are these fallacious inferences apparently "hard-wired" into our cognitive makeup? I shall suggest that there are plausible neoDarwinian strategies for answering that question. But before coming to that, let's have a look at Plantinga's own supernaturalistic version of naturalistic epistemology. It is time to turn the tables on TNE.

### 3.   Is Human Cognitive Reliability Likely on TNE?

Having claimed that the naturalistic epistemologist cannot accept metaphysical naturalism without either inductive incoherence (if the probabilities of (1)–(5) above are high on neoDarwinism) or skepticism (if the probabilities are unknown), Plantinga argues briefly that metaphysical supernaturalism—specifically, theism—escapes the difficulty. This is because, if we have been made by a theistic God, as creatures special to him, then divine providence can be counted on to supply us with reliable faculties, cognitive and otherwise. So Plantinga:

[The theist] has no corresponding reason for doubting that it is a purpose of our cognitive systems to produce true beliefs,... He may indeed endorse some form of evolution guided and orchestrated by God. And *qua* traditional theist ... he believes that God is the premier knower and has created us human beings in his image, an important part of which involves his endowing them with a reflection of his powers as a knower. (1993, 236)

Of course, having rejected Cartesian foundationalism, Plantinga thinks it would be circular to employ our reasoning faculties to argue *to* the existence of such a divine guarantor of our reasoning abilities. But he is prepared to claim that the belief that there is such a God and Maker of us, unlike metaphysical naturalism, supports a confidence in our cognitive powers. So the theist is not forced into incoherence or skepticism.

Unfortunately for TNE, Plantinga has no obvious right to this quick conclusion, unless he simply *builds it in* to his theistic hypothesis that God has created us reliable knowers. The appeal to our having been created "in the image of God" will not secure that conclusion; for that could mean any number of things, and clearly we do not "reflect" God's nature as a knower very closely, or even remotely as closely as would be possible in creatures God could make.

We could, of course—and this is surely what Plantinga intends—understand our being created in God's image as implying that our cognitive faculties are reliable, at least in the domains in which they are properly employed. Would this help? Consider a parallel strategy on the part of the neoDarwinian. Worried about Plantinga's argument, the neoDarwinian might respond by tacking onto his theory some proposition such that, relative to the emended theory, the evolution of reliable belief-forming mechanisms becomes highly probable in organisms that make their living by reasoning. Whatever the details, there is every reason to suppose that the additional hypothesis, call it $H$, will not be inconsistent with standard neoDarwinism ($D$).

Suppose that there is no significant independent evidence *for H* (nor against it), but that it is (roughly) probabilistically independent of $D$. (If it were probable relative to $D$, it would not be required as an independent hypothesis.) Now our evidence $R$—that human belief-forming mechanisms are quite reliable—is very probable relative to $D$ & $H$; and so it appears that Plantinga's dilemma has been deflected. But in fact, the neoDarwinian has conveniently ignored the fact that this maneuver comes at a price. In the absence of independent evidence for $H$, the prior probability of $D$ & $H$ is lower than that of $D$ alone; and whether or not one give this a Bayesian analysis, a clear constraint is that this prior must be *enough* lower that the probability of $R$ relative to naturalism cannot be raised by such a maneuver.[15] Failure to observe such a constraint would be to license cheating.

Let us turn now to theism. Traditionally, theism has been understood as the doctrine that there is just one God, creator of the universe, omnipotent, omniscient, perfectly good, and possessing essentially a variety of other perfections. But Plantinga does not understand what he calls theism in this way. Theism, as Plantinga takes it, includes all this but more as well. It includes, in particular, the view that God created humans in His own image. Perhaps it includes a good deal else; perhaps Plantinga thinks you are not a theist unless you hold that God's name is YHWH and that Moses was His prophet. However Plantinga constructs his version of theism, let us distinguish it from the traditional definition by calling the latter unvarnished theism (*UT*) and the former varnished theism (*VT*).

Now when it comes to showing that *R* is likely relative to theism, it is the doctrine that humans are created in the image of God (call this *I*) that does most of the work. So we can think of *VT* as just *UT* & *I*. As I shall show in a moment, *UT* cannot by itself be known to render *I* likely, so *I* is not a redundant part of *VT*. How much independent evidence is there, then, for *I*? So far as I can see, very little. The only generally recognized independent source for this doctrine is Gen. 1:26–27, which is echoed in various passages in the Old Testament and the New.[16] On the face of it, *I* makes a factual claim about an event (or series of events?) which occurred prehistorically. In terms of the traditional distinction between natural theology and revelation, it is clear that whatever evidence there is for *I* must come via the latter route; indeed, it is hard to see how natural theological arguments *could* provide grounds for such a claim as *I*.

How seriously, then, should one take the testimony of Gen. 1:26–27? That topic truly requires more space than I can give it, so a few general observations will have to suffice. First there is the familiar difficulty that the creation accounts in Gen. 1 and Gen. 2 are contradictory, and specifically, conflict with respect to the creation of human beings. Then, there is the generally mythical character of Genesis; many of the themes in the first eleven chapters are borrowed from, or influenced by, the myths of other ancient Near Eastern cultures. Further, there is the archaeological evidence that the story of the Egyptian captivity and Exodus are mythical, not historical accounts.[17] All these considerations undermine the reliability of Genesis as a source of historical information. Finally, there are manifold problems of interpretation: just what did the author of Gen. 1 mean in saying that we were created in God's image? Did that author actually have our cognitive faculties in mind?

I said above that *I* does most of the work in securing the likelihood of *R* on *VT*; that is, just as God's cognitive faculties are perfect, so we, as imperfect images of God, can count on our cognitive faculties being good ones in their limited way. But unless *R* is actually built into *I*, we have no right to conclude even this.

The difficulty is this. Even if God has the traditional theistic attributes and made us in His image, we have no right to draw conclusions about what God would do or would be likely to do, in making us, unless we had a much firmer grasp on God's purposes and judgments concerning what is good and what is not, than we evidently do. Even if theists have no independent reason to suspect their cognitive faculties of serious fallibility in more ordinary contexts, *this* context gives them eminently good reasons for doubt.

The problem of evil is itself sufficient to raise this doubt. Theists can argue that the logical incompatibility of observed evil with theism has not been demonstrated. But unless they can show that observed evil is *likely* given theism, their only defense will be to claim that we are radically ignorant of God's purposes—too ignorant even to conceive what sort of justifying reasons God has for permitting some horrific evils. But then, it appears that we are not really very good knowers, when it comes to grasping the divine scheme of things.[18] Indeed, *one* of the apparent evils which we cannot explain (as Descartes saw) is why God made us such poor knowers, so fallible and so unable to understand Him and His moral order better. Perhaps—for some reason we doubtless cannot fathom—God has even made us very poor at receiving and interpreting His revelations.[19] That would be unsurprising, on the supposition that God saw fit to make us defective in these ways. Given all this, the theist has no right to conclude that, relative to *VT*, her cognitive faculties are as good as she would like to think they are. If God can see fit to allow small children to die of terrible diseases for some greater good we cannot imagine, might He not have given us radically defective cognitive systems, and allowed us to be lulled into thinking them largely reliable, also for some unimaginable reason? Plantinga is quite correct in observing that *VT* provides no reason to think our cognitive reliability is low. But it also gives no reason for thinking it is high—and that is the claim Plantinga needs. It is the theist, not the metaphysical naturalist, who is in trouble here.

But suppose now that theists do interpret *I* in such a way that *R* is rendered likely by *VT*. Would that help? Clearly, the theist now has an escape from the loop of skeptical doubts—if she has good reason to think that theism (in this sense) is true. Recall however the naturalist. He, too, can escape the skeptical loop by adding *H* to *D*—*if* he has good independent grounds for thinking that $D$ & $H$ is true. But his grounds for thinking that were (we supposed) no better than his grounds for $D$ alone. So if $D$ alone gives no good reason for thinking that $R$ is true, $D$ & $H$ can hardly do so—$H$ will be as unlikely on the available evidence as $R$ itself is. But the same applies to the theist's $I$, as now understood. If $R$ has a low or unknown likelihood on $UT$, then, because $I$, as now understood, makes $R$ likely, and because the theist has precious little independent evidence for $I$, it will follow that $I$—and hence $UT$ & $I$—has a low or unknown likelihood, relative to $UT$ plus all the evidence. But then, even if the theist has good reasons for believing $UT$, she has no good reason for believing $VT$ under the present interpretation of $I$. And that means that she, too, has no way of escaping the skeptical loop.

## 4.  Human Fallibility

I have not yet explained how the findings in cognitive psychology alluded to above are to be explained. But it should be clear, first, that they are equally an embarrassment for the theist. Descartes was concerned to show that our cognitive faculties were so designed by a beneficent God that, properly employed, they would not deceive. But the fallacy-generating mechanisms uncovered by psychologists suggest a different story. For most of us, in the relevant circumstances, the correct inferences are cognitively unavailable, even upon reflection. Where they are available, the impulse to reason incorrectly still appears to be strong and innate. Why?

NeoDarwinians have several natural ways to answer that question. In part, they can use the very strategies Plantinga has charged against them. They can point out (1) that evolution is opportunistic—it can only work with the materials at hand, with lucky mutations, and with selective pressures that are common and persistent; and (2) that evolution can permit some degree of maladaptiveness, especially when associated benefits outweigh costs. To use Herbert Simon's term, evolution is a satisficer; and so are we. How might Darwinians marshall these two points?

When it comes to details, it is evident that no single explanatory strategy can apply to every cognitive limitation; Darwinians will have to proceed on a case-by-case basis. But some illustrative remarks will serve to indicate the procedure. I shall discuss briefly two types of cognitive failure: visual illusions and inductive inferences to beliefs about cause-effect relationships.

Visual illusions typically achieve their effects by taking advantage of such visual signal processing mechanisms as edge detection, etc., which under normal circumstances are essential to our being able to make visual sense of our surroundings. Illusions trick these mechanisms by co-opting the very information-processing algorithms whose automatic operation serves us faithfully in the environment for which they are adapted. Take size constancy. When two one-inch human figures are drawn along-side railroad tracks receding into the distance, the "nearer" figure looks smaller; the perspectivally drawn tracks trigger the size-constancy mechanism. But that is just as it should be in real life, where it is more important to gauge the size and distance of an approaching saber-toothed cat, than to interpret a psychologist's drawing.

Causal judgments provide an example from the sphere of inferential processes. As Hume supposed, we commonly make such judgments by noticing correlations between instances of one type of event, $C$, and instances of another, $E$. But humans often use the wrong procedure in estimating the statistical relevance of $C$ to $E$, by tending to ignore the frequency with which $E$ occurs when $C$ has not occurred. And this can lead to mistaken causal judgments.[20]

Why are we so addicted to an incorrect procedure for estimating causal relevance, since incorrect causal judgments can sometimes be costly? A Darwinian answer, I suggest, will rely upon two points. First, on an evolutionary time scale, the invention of arithmetic is extremely recent; computation of the relevancy of $C$ to $E$ would, for most of our biological existence, have been carried out by less explicit processes. And that may not be such an easy task. Computing in one's head such a simple function as the difference between two fractions is hard even *with* arithmetic. Second, it may ordinarily not have mattered much: for most purposes, a rough estimate of $P(E/C)$ is good enough. More than that: estimating $P(E/C) - P(E/\sim C))$ is far more time-consuming, and in many cases, survival may have depended upon the *rapid* making of causal judgments. Thus, and in general, the costs—temporal and otherwise—of accurate, loophole-free reasoning must be weighed against the benefits of improved reliability. Evolution tends to home in on organisms that can get things right and get things fast, sacrificing only as much of each for the other as necessary for effective action. For creatures such as humans who have found their ecological niche by specializing heavily in reasoning and the flexibility that a powerful capacity to learn affords, this conclusion applies in spades—certainly for the handling of more mundane beliefs.

Finally, how should we estimate the reliability of *theoretical* inferences, especially those in science, given neoDarwinism? That is too large a subject to address here, but in a nutshell, the answer is surely, first, that such inferences are indeed somewhat more risky; and secondly, that they rely very largely on reflective refinements of the same computational strategies that have been tested and improved by eons of evolutionary development for the ordinary purposes of survival and procreation. Many of *those* inferences are also quite theoretical: witness the young swain's earnest attempts to guess what his intended is thinking.

## Notes

1. Plantinga calls this view metaphysical naturalism. In this paper, I only consider Plantinga's two epistemological arguments, presented in Chapter 12 of *WPF*, for the conclusion that naturalism is false or at least irrational to accept. My remarks are directed primarily toward the latter argument, which Plantinga calls his main argument against naturalism, but bear equally against the first. In Chapter 11, Plantinga has already mounted an argument against naturalism. It runs from the claims that knowledge ascriptions to a being presuppose that being satisfies a design plan, and that design involves normative characteristics of which no naturalistic account can be given, to the conclusion that the existence of a knower requires the existence of a designer, i.e. God. This argument founders on the problem that if it is sound, then either God is no knower, or else He was designed. It is also incorrect, I believe, in claiming that no suitable naturalist reduction of the requisite normative notions can be given. For details, see Fales 1994.

2. This is the lead paper of a symposium published in *Christian Scholars Review* 21 (Sept. 1991). The issue was devoted to Plantinga's paper, replies by Howard J. Van Till, Pattle Pun, and Ernan McMullin; and Plantinga's response, "Evolution, Neutrality, and Antecedent Probability: A Reply to McMullin and Van Till".

3. To simplify, I shall use the term 'neoDarwinism' to refer to any theory of biological evolution that has strong affinities to Darwin's and incorporates molecular genetics: specifically, that does not invoke any supernatural agent or purposive cause directing the process, and that takes random variation, the genetic transmission of traits, and natural selection to be central mechanisms.

4. It is perhaps fair to say, on the rather sketchy evidence in Plantinga 1991 and elsewhere, that Plantinga is quite committed to saving, if possible, the historicity of something "like" the (perhaps parabolic) story of the Fall in Genesis, the

general accuracy of the biblical history of Israel from the Patriarchs through Exodus, the Judges, Kings and Prophets, many of the miracle stories, and of course the ministry, death, and resurrection of Jesus.

5. These strains are significantly reinforced by the results of the last 150 years of textual analysis of the Bible, and the last 50 years—especially the last two decades—of Biblical archaeology. But that is another story.

6. Darwin 1881, V. 1, p. 281. Plantinga quite unfairly omits mention of the context of this passing remark of Darwin's, which occurs under the heading "Religion." Darwin is commenting on a book by Graham which presents a version of the Argument from Design that infers the existence of a creator from the existence of laws of nature. Darwin responds that he "cannot see this"; nevertheless, after offering an objection, he says, "But I have no practice in abstract reasoning, and I may be all astray. Nevertheless you have expressed my inward conviction, though far more vividly and clearly than I could have done, that the Universe is not the result of chance. But then with me the horrid doubt, etc." (A footnote here records Darwin's reaction to a similar suggestion made by the Duke of Argyle that biological intricacies must be the product of a mind. Darwin replied, "Well, that often comes over me with overwhelming force; but at other times it seems to go away."

The context makes it quite clear that the kind of "convictions" Darwin has in mind are general theoretical hunches supported by intuitions of some sort, rather than conclusions clearly reasoned from evidence. His references to the "convictions" of monkeys is therefore best seen as irony. There is no support here for the view that Darwin suspected our cognitive faculties (or those of our simian forbearers), of gross unreliability when engaged in their customary activities. Even a cursory examination of Darwin's *The Descent of Man* shows the reverse to be true.

7. As we shall see, there are in fact good empirical reasons to trust, in general, the beliefs of monkeys.

8. It is controversial how many animals *have* beliefs, and how sophisticated reasoning based on those beliefs is. My own view is that many mammals and birds have action-guiding beliefs (or something rather like beliefs) and inferential capacities of varying sophistication. Although it is, notoriously, difficult to formulate the criteria we use to ascribe beliefs to other minds, it is certain that our belief-ascribing practices extend naturally, albeit with some attenuation, to many non-human species. There is ample evidence to support the view that mental faculties generally, and cognitive abilities in particular, are expensive in terms of brain capacity and complexity. To avoid undue controversy, I shall confine my discussion to the primates.

9. See, e.g., Milton 1993, 90.

10. It is generally agreed that the advantages conferred by our erect posture and manual dexterity are closely linked to our intellectual capacities. Our investment in intelligence as a species is extensive. It includes two related and biologically very costly adaptations, namely, giving birth at an earlier stage of fetal development than is the case for any other mammal (so the large skull can pass through

the birth canal), and a greater expenditure of energy in childrearing than characterizes any other mammal.

11. See Velmans 1991 and the responses. Velmans adopts the peculiar view that, although epiphenomenalism is correct from a third-person perspective, the causal efficacy of consciousness is correct from a first-person point of view. I cannot make sense of this. Van Gulick's views can be found in his (1994) and elsewhere.

12. If the semantic content of beliefs were not hooked in some way to the syntactic structure of neural states, then (causally speaking) beliefs would float free, and possibility (3) would degenerate into (1) or (2). For further details on how content is conceptually linked to the causal story, see Fales 1990, Chapter 12.

13. Though many tropical trees produce compounds which render their leaves unpalatable.

14. Monkeys are primarily vegetarian, but supplement their protein intake by eating insects.

15. Plantinga (unpublished), in a reply to a number of philosophers who have proposed this sort of defense of naturalism (he mentions Fred Dretske, Carl Ginet, Timothy O'Connor, Richard Otte, John Perry, Ernest Sosa, and Stephen Wykstra), makes just this point, though he poses the question in terms of the existence of defeaters for the warrant of a belief.

16. In the O.T., references to this idea outside of Genesis are very rare. The Hebrew term *tselem* is used mainly to refer to idols, that is, representations of foreign gods. Moreover, in Gen. 1:26–27 it is the "gods" (Elohim) who fashion man in *their* image—whatever that might mean. The N.T. passages further complicate matters. Rom. 8:29, II Cor. 3:12–4:4, and Col. 3:10 all imply that we have lost the divine image and need to recover it through Christ; I Cor. 15:49 goes so far as to suggest that we never had it and will not acquire it until we enter the Kingdom of Heaven. To make matters worse, I Cor. 11:7 appears to imply that men, but not women, were made in the divine image.

In any event, the Adamic fall creates a special difficulty for those theists, like Calvin, who hold that human beings have inherited from this event nor merely moral or volitional depravity, but cognitive depravity as well. It is not clear how far this liability is supposed to go in depriving us of powers once possessed by Adam and Eve, but Calvin presumably did not believe—because it would have been foolhardy to do so—that conversion to Christianity (or to Calvinism) somehow guarantees the restoration of the original human cognitive powers and faculties. Of course, it might seem entirely derelict on God's part for Him to have so arranged things that a sin on the part of Adam and Eve could be transmitted to the rest of humanity. But then, what do we know?

Plantinga (unpublished) is more expansive in finding sources which warrant theism; he mentions Calvin's *sensus divinitatis*, Aquinas' natural but confused knowledge of God, the authority of the Church, and the internal testimony of the Holy Spirit in addition to Scripture as warrants for theism. Plantinga does not say which, if any of these, he takes to provide independent warrant for *I*. Needless to say, the theist will have to show what independent reasons there are for the reliability—or even the existence—of these alleged sources. Of course, the theist who believes that God has *told* her that her cognitive faculties are in good shape cannot be faulted on the score of having (in the way at issue) an incoherent or self-undermining set of beliefs; but neither can the schizophrenic who believes that God has told him that he is the *parousia*. I discuss the epistemic status of revelations in greater detail in Fales (forthcoming a) and (forthcoming b). Plantinga himself promises more on this score in his forthcoming *Warranted Christian Belief*; further discussion of the issue will have to await that volume.

17. See Shanks et al. 1992 for details.

18. Plantinga admits as much in his (1985), pp. 34–36.

19. We have abundant evidence of that (if evidence counts for anything under these circumstances) in the very small number of revelations, out of all the purported ones, that could possibly be true, given the conflicts between them.

20. See Kahneman, Slovic, and Tversky 1982, especially Chapters 8 and 10. For a further discussion of these findings and a survey of the literature, see Fales and Wasserman 1992.

## References

Darwin, C. (1919), *The Life and Letters of Charles Darwin*, Vol. I. New York: D. Appleton & Co.

Fales, E. (1990), *Causation and Universals*. New York: Routledge.

———. (1994), "Review of Alvin Plantinga, *Warrant and Proper Function*," *Mind* 103: 391–395.

———. (forthcoming a), "Mystical Experience as Evidence," *International Journal for Philosphy of Religion*.

———. (forthcoming b), "Scientific Explanations of Mystical Experience, Parts I and II," *Religious Studies*.

Fales, E. and E. A. Wasserman (1992), "Causal Knowledge: What Can Psychology Teach Philosophers?" *The Journal of Mind and Behavior* 13: 1–27.

Flanagan, O. (1992), *Consciousness Reconsidered*. Cambridge, MA: MIT Press.

Kahneman, D., P. Slovic, and A. Tversky (eds.), (1982), *Judgment Under Uncertainty: Heuristics and Biases*. Cambridge: Cambridge University Press.

Milton, K. (1993), "Diet and Primate Evolution," *Scientific American* 269: 86–93.

Nisbett, R. E. and L. Ross (1980), *Human Inference: Strategies, and Shortcomings of Social Judgment*. Englewood Cliffs, NJ: Prentice-Hall.

Plantinga, A. (1985), "Self-Profile," in J. E. Tomberlin and P. Van Inwagen (eds.), *Alvin Plantinga*. Boston: D. Reidel.

———. (1991), "When Faith and Reason Clash: Evolution and the Bible," *Christian Scholars Review* 21: 8–32 and 81–109.

———. (1993), *Warrant and Proper Function*. Oxford: Oxford University Press.

————. (unpublished), "Nature Defeated."

Shanks, H., W. Dever, B. Halpern and P. K. McCarter, Jr. (1992), *The Rise of Ancient Israel*. Washington, D.C.: Biblical Archaeology Society.

Stich, S. (1990), *The Fragmentation of Reason*. Cambridge, MA.: MIT Press.

Van Gulick, R. (1994), "Deficit Studies and the Function of Phenomenal Consciousness," in G. Graham and G. L. Stephens (eds.), *Philosophical Psychopathology*. Cambridge, MA: MIT Press.

Velmans, M. (1991), "Is Human Information Processing Conscious?" *Behavioral and Brain Sciences* 14: 651–669.

# 16

# Plantinga's Probability Arguments against Evolutionary Naturalism

Branden Fitelson and Elliott Sober

In Chapter 12 of *Warrant and Proper Function*, Alvin Plantinga constructs two arguments against evolutionary naturalism, which he construes as a conjunction *E&N*. The hypothesis *E* says that "human cognitive faculties arose by way of the mechanisms to which contemporary evolutionary thought directs our attention" (p. 220).[1] With respect to proposition *N*, Plantinga (p. 270) says "it isn't easy to say precisely what naturalism is," but then adds that "crucial to metaphysical naturalism, of course, is the view that there is no such person as the God of traditional theism." Plantinga tries to cast doubt on the conjunction *E&N* in two ways. His "preliminary argument" aims to show that the conjunction is probably false, given the fact (*R*) that our psychological mechanisms for forming beliefs about the world are generally reliable. His "main argument" aims to show that the conjunction *E&N* is self-defeating—if you believe *E&N*, then you should stop believing that conjunction. Plantinga further develops the main argument in his unpublished paper "Naturalism Defeated" (Plantinga 1994). We will try to show that both arguments contain serious errors.

## 1. The Preliminary Argument

Plantinga constructs his preliminary argument within a Bayesian framework. His goal is to establish that $\Pr(E\&N\,|\,R)$—the probability of *E* and *N*, given *R*—is low. To do this, Plantinga uses Bayes' Theorem, which says that this conditional probability is a function of three other quantities:

Originally published in *Pacific Philosophical Quarterly* (1998, vol. 79, pp. 115–129).

$$\Pr(E\&N \mid R) = \frac{\Pr(R \mid E\&N) \cdot \Pr(E\&N)}{\Pr(R)}.$$

Plantinga says you should assign to $\Pr(R)$ a value very close to 1 on the grounds that you believe $R$ (p. 228). He argues that $\Pr(R \mid E\&N)$ is low. Although Plantinga doesn't provide an estimate of the prior probability $\Pr(E\&N)$, he says that it is "comparable" to the prior probability of traditional theism $(TT)$ (p. 229), meaning, we take it, that their values aren't far apart.

This last claim should raise eyebrows, not just among evolutionary naturalists who reject the idea that their theory and traditional theism are on an equal footing before proposition $R$ is taken into account, but also among critics of Bayesianism, who doubt that there is an objective basis for such probability assignments. Plantinga says (p. 220, footnote 7) that his probabilities can be interpreted either "epistemically" or "objectively," but that he prefers the objective interpretation. However, Bayesians have never been able to make sense of the idea that prior probabilities have an objective basis. The siren song of the Principle of Indifference has tempted many to think that hypotheses can be assigned probabilities without the need of empirical evidence, but no consistent version of this principle has ever been articulated. The alternative to which Bayesians typically retreat is to construe probabilities as indicating an agent's subjective degree of belief. The problem with this approach is that it deprives prior probabilities (and the posterior probabilities that depend on them) of probative force. If one agent assigns similar prior probabilities to evolutionary naturalism and to traditional theism, this is entirely consistent with another agent's assigning very unequal probabilities to them, if probabilities merely reflect intensities of belief.

Although Plantinga's Bayesian framework commits him to making sense of the idea that the conjunction $E\&N$ has a prior probability, his argument does not depend on assigning any particular value to this quantity. As Plantinga notes (p. 228), if $\Pr(R) \approx 1$ and $\Pr(R \mid E\&N)$ is low, then $\Pr(E\&N \mid R)$ also is low, no matter what value $\Pr(E\&N)$ happens to have.

## 1.1   Proposition $R$

For the sake of clarity, it is worth spelling out proposition $R$ more precisely. What does it mean for our psychological mechanisms for forming beliefs to be "generally" reliable? In his unpublished manuscript, Plantinga says that $R$ means that "the great bulk" of our beliefs are true (Plantinga 1994, p. 2). Aside from questions about how beliefs are to be counted, we don't want to challenge the truth of this summary statement. However, it drastically underspecifies the data that need to be explained. For the fact of the matter is that our cognitive mechanisms are reliable on some subjects, unreliable on others, and of unknown reliability on still others. We should divide our beliefs into categories and associate a characteristic degree of reliability with each of them.[2] Perhaps certain simple perceptual beliefs are very reliable, while beliefs about other subjects are less so. Rather than trying to obtain a summary statement about all these mechanisms and the beliefs they generate, it would be better to consider a conjunction $R_1 \& R_1 \& \ldots \& R_n$, which specifies the degree of reliability that human belief formation devices have with respect to different subject matters, or in different problem situations. Plantinga (pp. 216–217, 227, 231–232, and in a personal communication) does not object to this partitioning and uses it himself to discuss the probability that $E\&N$ confers on $R$.

If $R$ is true, why should one bother to spell it out in more detail? This wouldn't matter if Plantinga's argument were deductive. A sound argument stays sound when the premises are supplemented with more (true) details. However, probability arguments don't have this property. Even if $\Pr(R \mid E\&N)$ is less than $\Pr(R \mid TT)$, it remains to be seen whether $\Pr(R_1 \& \ldots \& R_n \mid E\&N)$ is less than $\Pr(R_1 \& \ldots \& R_n \mid TT)$.

Before we get to that comparative question, let's consider whether the conditional probability $\Pr(R_1 \& R_2 \& \ldots \& R_n \mid E\&N)$ is high or low. Suppose that evolutionary naturalism does a good job of predicting each of the conjunct $R_i$'s, conferring on each a probability, say, of 0.99. It still could turn out that $E\&N$ confers a low probability on the conjunction. If

$$\Pr(R_1 \& R_2 \& \ldots \& R_n \mid E\&N)$$

$$= \Pr(R_1 \mid E\&N) \cdot \Pr(R_2 \mid E\&N) \cdots \Pr(R_n \mid E\&N)$$

(*i.e.*, if the $R_i$'s are probabilistically independent of each other, conditional on $E\&N$), then the left-hand term may have a low value, even though each product term on the right has a high value. Multiply 0.99 times itself sufficiently often and you get a number close to zero. This can happen to any good theory; it may confer a low probability on a massive conjunction of observations even though it confers a very high probability on each conjunct.

Once we decompose proposition $R$ into a conjunction of claims, it is far from obvious that evolutionary theory does a worse job of predicting this conjunction than traditional theism does. Plantinga says the traditional theist "believes that God is the premier knower and has created us human beings in his image, an important part of knower and has created us human beings in his image, an important part of which involves his endowing them with a reflection of his powers as a knower" (p. 237). However, an influential point of view in cognitive science asserts that human reasoning is subject to a variety of biases. It isn't just that people occasionally make mistakes, but that the human reasoning faculty seems to follow heuristics that lead to systematic error (Kahnemann, Tversky, and Slovic 1982). It would be no surprise, from an evolutionary point of view, if human beings had highly reliable devices for forming beliefs about practical issues that affect survival and reproduction, but are rather less gifted when it comes to matters of philosophy, theology, and theoretical science. Does traditional theology also predict this result? No doubt, a theology can be specified that makes any prediction one wants. However, it is not at all clear that Plantinga's traditional theology does a good job predicting the varying levels of reliability that the human mind exhibits. Plantinga must address the same problem that Paley's design argument faces: Why would an omniscient, omnipotent, and benevolent deity produce organisms who seem to be so manifestly imperfect in the adaptations they exhibit (Sober 1993)?

## 1.2   Setting $\Pr(R) \approx 1$

We mentioned earlier that Plantinga sets $\Pr(R) \approx 1$ because he believes proposition $R$. Within the context of Bayesian confirmation theory, assigning the evidence a probability close to unity has a peculiar consequence, as we now will show.

Bayesians define confirmation in terms of probability raising; an observation $O$ confirms a hypothesis $H$ if and only if the posterior probability $\Pr(H \mid O)$ is greater than the prior probability $\Pr(H)$. If we rewrite Bayes' theorem as follows

$$\frac{\Pr(H \mid O)}{\Pr(H)} = \frac{\Pr(O \mid H)}{\Pr(O)},$$

it is clear that $O$ cannot confirm $H$, if $\Pr(O) = 1$. With this assignment, the right-hand ratio can't be greater than unity, so the left-hand ratio can't either. Plantinga's stipulation that $\Pr(R)$ is *close* to unity doesn't quite insure that $R$ can't confirm a hypothesis $H$. After all, it is possible that $\Pr(R \mid H)$ should be even closer to unity than $\Pr(R)$ is. Let us say that a hypothesis $H$ is *quasi-deterministic* with respect to $R$ if $\Pr(R \mid H) > \Pr(R) \approx 1$. If evolutionary naturalism isn't quasi-deterministic in this sense, then $R$ can't confirm it, given Plantinga's assignment. Proposition $R$ may leave the probability of $E\&N$ unchanged, or it can lower its probability; there is nowhere to go but down. Unless traditional theism is quasi-deterministic with respect to $R$, it too cannot be confirmed by proposition $R$, if $\Pr(R) \approx 1$.

Bayesians like to point out that it is a consequence of Bayes's theorem that an observation is incapable of confirming a hypothesis when the observation is completely unsurprising. However, for most predictions of any interest, a Bayesian agent isn't certain in advance that they will come true; when surprising predictions *do* come true, they provide confirmation. The wet sidewalk ($W$) confirms the hypothesis that it has been raining. The fact that you believe that the sidewalk is wet shouldn't lead you to assign this evidence a probability of unity. A reasonable assignment of value to $\Pr(W)$ is given by the fact that the sidewalk is rarely wet. The observation is therefore somewhat surprising, and so is capable of confirming the hypothesis. Thus, Plantinga's claim that $\Pr(R)$ is close to unity is very odd; it is crucial to his argument that $\Pr(E\&N \mid R)$ is low. Plantinga needs a better reason for this assignment than the fact that he believes $R$.

Plantinga's preliminary argument might be replaced by a different argument, one that seeks to establish that $\Pr(TT \mid R) > \Pr(E\&N \mid R)$. The goal now is to compare two posterior probabilities, not to estimate their absolute values. This inequality is true precisely when

$$\Pr(R \mid TT) \cdot \Pr(TT) > \Pr(R \mid E\&N) \cdot \Pr(E\&N).$$

Notice that the value of $\Pr(R)$ is now irrelevant. The argument might begin with the assertion that it is more probable that our psychological mechanisms for forming beliefs are reliable if $TT$ is true than would be the case if $E\&N$ were true. If God exists and intervenes in natural processes to guarantee that human beings end up with reliable cognitive faculties, this makes $R$ more of a sure thing than would be the case if chancy natural processes are the only causes of the mental equipment we possess.[3] (Here we ignore the problems adumbrated in Section 1.1 concerning how $R$ should be spelled out.) If $TT$ and $E\&N$ are assigned the same prior probabilities, it then follows that $TT$ has the higher posterior probability.

The present argument provides a recipe for replacing any nondeterministic theory in the natural sciences. If quantum mechanics predicts that a certain experimental outcome was merely very probable, why not accept instead the theistic hypothesis that this outcome was the inevitable outcome of God's will? Theism can be formulated in such a way that it renders what we observe as probable as you please. Those who feel the need to appeal to God's intervention in the case of human mentality should explain why they do not do so across the board.

### 1.3   Is $\Pr(R\,|\,E\&N)$ Low? Rethinking "Darwin's Doubt"

Plantinga (pp. 223–228) argues that $\Pr(R\,|\,E\&N)$ is low by enumerating several logically conceivable scenarios that describe how beliefs and actions might be related. For each of them, he contends that it is very unlikely that the cognitive mechanisms that evolve should be highly reliable.[4] Here are the possibilities that Plantinga considers: (*i*) beliefs are not causally connected with behavior; (*ii*) beliefs don't cause behavior, but are effects of behavior or are effects of events that also cause behavior; (*iii*) beliefs cause behavior, but do so in virtue of their syntax, not by virtue of their semantics; (*iv*) the semantic properties of belief cause behavior, but the behaviors are maladaptive; (*v*) beliefs cause adaptive behavior. In this last category, the adaptive behaviors may be caused by true beliefs, but they also can be caused by false ones. To illustrate this point, Plantinga describes a prehistorical hominid named Paul who manages to avoid being eaten by tigers even though he desires that they should consume him. Paul gets what's good for him by desiring what is bad; he stays out of trouble because his beliefs are false in just the right way. In each of these scenarios, Plantinga says that it is improbable that our cognitive faculties should have evolved to be highly reliable. So $\Pr(R\,|\,E\&N)$ is low.

In the body of his more recent unpublished manuscript, Plantinga also says that $\Pr(R\,|\,E\&N)$ is low under scenario (*ii*) (epiphenomenalism); however, he draws a different conclusion in footnote 15 (p. 8). If beliefs and actions have neural events as common causes, then Plantinga concludes that the probability is "inscrutable," meaning that he can't figure out what value it should be assigned. We take this to be Plantinga's present considered view on the matter. Although Plantinga is developing the main argument against evolutionary naturalism in this manuscript and is not talking about the preliminary argument, it is worth noting that this conclusion undercuts the preliminary argument, which depends on assigning a low value to Pr($R \mid E\&N$).

Whether or not Plantinga's considered view now *is* that Pr($R \mid E\&N$) is inscrutable under scenario (*ii*), this is what his view *should* be, given the information he considers. Assuming that beliefs don't cause actions is not the same as assuming that they are wholly unrelated. Resistance to malaria doesn't cause anemia, nor does anemia cause malaria resistance; yet, the traits are correlated in a number of human populations because they are phenotypes caused by the same gene. There is no way to tell *a priori* how probable R is under scenario (*ii*).

When Plantinga turns his attention to category (*v*)—the case in which beliefs cause adaptive action—he argues that false adaptive beliefs are just as likely to evolve as true adaptive beliefs. The reason is that the behaviors produced by a set of true beliefs also could be produced by a set of false ones. The example of Paul shows one way this could be true. Plantinga describes another in the unpublished manuscript. If "that is a tree" is a true belief, then "that is a witch-tree" is a false belief that would lead to the same behavioral consequences, and so be equally fit. Plantinga's mistake here is that he ignores the fact that the probability of a trait's evolving depends not just on its fitness, but on its *availability*. The reason zebras don't have machine guns with which to repel lion attacks is not that firing machine guns would have been less adaptive than simply running away; the trait didn't evolve because it was not available as a variation on which selection could act ancestrally (see, also, Fodor, 1997).

This means that Plantinga's argument that Pr($R \mid E\&N$) will be low in category (*v*) situations is inadequate. Plantinga might reply that witch-beliefs and other systems of adaptive false beliefs were available ancestrally. However, we don't see any reason to think that this substantive claim about the past can be successfully defended. By ignoring the question of availability, Plantinga, in effect, assumes that natural selection acts on the set of *conceivable* variants. This it does not do; it acts on the set of *actual* variants.

In general, the way to have two (logically independent) properties be well-correlated is to have one cause the other, or to have each trace back to a common cause. If belief and action failed to be causally connected in either of these two ways, then it *would* be surprising for selection on action to lead cognitive mechanisms to evolve that are highly reliable. However, if belief and action are causally connected, then it takes a more detailed argument than Plantinga provides for concluding that reliable belief formation devices are unlikely to evolve via selection on actions. Proposition *R is* improbable under scenario (*i*), but that's about all one can say.

### 1.4   The Principle of Total Evidence

Suppose Plantinga is right in saying that $\Pr(E\&N \mid R)$ is low. It does not follow that *E&N* is improbable relative to *all* relevant evidence. Evolutionary naturalists can happily accept the idea that the conditional probability just mentioned is low; it does not follow that they should have less confidence than they presently do in the truth of evolutionary theory and naturalism.

If you draw a card at random from a standard deck of cards, the probability is only 1/52 that you will draw the 7 of diamonds. If you do draw this card, that doesn't mean that you should conclude that the deck isn't standard or that the card wasn't drawn at random. If you have independent evidence that the deck is standard and that the draw was random, you simply accept the fact that some of the things that happen don't have high probabilities. Even if it turns out that there are features of human cognitive makeup that are improbable on the hypothesis that human beings evolved, there is lots of evidence that the human mind is a product of evolution. In this light, the sensible thing to do is to accept evolutionary theory and come to terms with the fact that evolutionary processes sometimes have improbable outcomes.[5]

As for the separate hypothesis of naturalism, by which Plantinga means atheism together with some other claims that he does not spell out, these too must be evaluated in the light of *all* the evidence, not just with respect to proposition *R*.

### 1.5   A Contradiction and Two Ways Out

We have mentioned that Plantinga thinks $\Pr(R)$ is close to 1, $\Pr(R \mid E\&N)$ is low, $\Pr(R \mid TT)$ is high, and that $\Pr(E\&N)$ and $\Pr(TT)$ are "comparable." Plantinga's preliminary argument also includes the assumption that $N$ and $TT$ are the only two "significant alternatives" (p. 228).

If the claim that $\Pr(E\&N)$ and $\Pr(TT)$ are "comparable" means that their values are not far apart, and if the claim that $E\&N$ and $TT$ are the only two "significant alternatives" means that they are the only possibilities that have non-negligible prior probabilities, then this set of probability claims is contradictory. To see why, let's expand $\Pr(R)$:

$$\Pr(R) = \Pr(R \mid E\&N) \cdot \Pr(E\&N) + \Pr(R \mid TT) \cdot \Pr(TT).$$

If $E\&N$ and $TT$ are exhaustive, then Plantinga's claim that they have "comparable" priors means that they are each close to 0.5. Substituting this and the other values that Plantinga assigns to the component expressions, we obtain

$$1 \approx (\text{low}) \cdot (0.5) + (\text{high}) \cdot (0.5).$$

This is impossible—a contradiction of the axioms of probability.

There are a couple of ways out of this difficulty. One is to retain all the probability assignments, but deny that $E\&N$ and $TT$ exhaust the significant alternatives. If a third possibility, theory $X$, is countenanced, then $\Pr(R)$ expands to

$$\Pr(R) = \Pr(R \mid E\&N) \cdot \Pr(E\&N) + \Pr(R \mid TT) \cdot \Pr(TT)$$
$$+ \Pr(R \mid X) \cdot \Pr(X).$$

If $\Pr(R) \approx 1$, $\Pr(E\&N)$ and $\Pr(TT)$ are about the same, and $\Pr(R \mid E\&N)$ is low, then Plantinga must assign $\Pr(E\&N)$ and $\Pr(TT)$ negligible probabilities, so that $\Pr(X)$ is close to 1. He also must assign $\Pr(R \mid X)$ a value close to unity. This revision in Plantinga's argument thus requires the existence of an alternative to traditional theism that is vastly more probable *a priori*, and which entails that proposition $R$ is very probable indeed. The effect of these assignments is to make $\Pr(E\&N \mid R)$ and $\Pr(TT \mid R)$ low and $\Pr(X \mid R)$ high. If a low value for $\Pr(E\&N \mid R)$ suffices to reject evolutionary naturalism, then one should reject traditional theism as well. This revision of his argument indicates that the most acceptable alternative is theory $X$.

Another way to avoid contradiction is to reinterpret what it means for $\Pr(E\&N)$ and $\Pr(TT)$ to be "comparable." Plantinga has suggested to us (personal communication) that this should be taken to mean that the agent doesn't believe that the two theories have very different probabilities. For example, suppose that the agent places E&N and TT in the same wide interval of probabilities (say, between 0.05 and 0.95) and isn't able to be more specific than this.

Plantinga's various claims can be rendered consistent by this revision. To see why, let's return to the expansion of $\Pr(R)$ and see what happens when we let $\Pr(E\&N)$ and $\Pr(TT)$ be unknown, save for the fact that each must fall between 0.05 and 0.95:

$$\Pr(R) = \Pr(R \mid E\&N) \cdot \Pr(E\&N) + \Pr(R \mid TT) \cdot \Pr(TT)$$

$$1 \approx (\text{low}) \cdot (?) + (\text{high}) \cdot (?)$$

The insertion of question marks insures that no contradiction arises. However, values for the question mark quantities are not left open; $\Pr(E\&N)$ must be very close to zero and $\Pr(TT)$ must be very close to unity. The argument is now consistent, but is entirely deprived of its probative force. Consistency (and the revised interpretation of what "comparable" means) require the assumption that traditional theism is virtually certain *a priori*, and that evolutionary naturalism is almost certainly false, again *a priori*. Those not already convinced before proposition R is considered that traditional theism is vastly more probable than evolutionary naturalism will reject the argument at the outset. In addition, this revision of Plantinga's preliminary argument undermines its original motivation. Plantinga's thought was to develop what he calls "Darwin's doubt"—that $\Pr(R \mid E\&N)$ is low. However, once $\Pr(R)$ is set close to 1, and $\Pr(E\&N)$ is assumed to be small, it *automatically* follows that $\Pr(E\&N \mid R)$ is still smaller, no matter what value $\Pr(R \mid E\&N)$ happens to have.

## 2.   Plantinga's Main Argument against *E&N*

The argument just described is preliminary to the main event, in which Plantinga (pp. 234–235) argues that E&N is self-defeating. The main argument doesn't aim to show that the conjunction E&N is probably false (or that E&N is less probable than TT), but that people shouldn't believe E&N:

1. $\Pr(R \mid E\&N)$ is low or its value is inscrutable.
2. Therefore, E&N is a defeater of R—if you believe E&N, then you should withhold assent from R.

3. If you should withhold assent from *R*, then you should withhold assent from anything else you believe.

4. If you believe *E&N*, then you should withhold assent from *E&N* (*E&N* is self-defeating).

∴ You should not believe *E&N*.

Plantinga goes further, in both the book and the manuscript. He argues that naturalism by itself is self-defeating. In the book, he says that "if naturalism is true, then, surely so is evolution" (p. 236). In the manuscript, he says that Pr(*E* | *N*) is high (p. 11). Neither of these claims is right. Recall that proposition *E* adverts to the mechanisms described in *contemporary* evolutionary theory. If that theory were found wanting, it would not entail the falsehood of naturalism; naturalists could quite consistently cast about for a better scientific theory. In the manuscript Plantinga makes the point that *E* is "the only game in town" for a naturalist; this may or may not be true (now), but that hardly shows that naturalism on its own makes 1990's evolutionary theory probable.

## 2.1 Problems That the Main Argument Inherits from the Preliminary Argument

We have already discussed why we are unconvinced by Plantinga's argument that Pr(*R* | *E&N*) is low. We also argued that even if Pr(*E&N* | *R*) were low, that would not entail that *R* suffices to reject *E&N*. The symmetrical point that pertains to the main argument is that even if Pr(*R* | *E&N*) were low, that would not oblige people who believe *E&N* to withhold belief from *R*. After all, people who believe *E&N* might have other reasons for believing *R*. For example, they might argue that *R* is a basic proposition that does not need theoretical support, or that *R* derives its epistemic credentials from something other than the thesis of evolutionary naturalism.

The same point holds if you don't know what value to assign to Pr(*R* | *E&N*). People who believe *E&N* should not regard the fact that this probability is "inscrutable" to them as a reason to reject *R*. We suspect that many people who are well acquainted with the theory of special relativity and who think that birds fly still don't know what value to assign to Pr(Special relativity | birds fly), especially if probability has to be an objective quantity; however, that doesn't show that they should withhold belief in special relativity. The Principle of Indifference is flawed because it claims to obtain probabilities from ignorance; the start of Plantinga's main argument makes the complementary mistake of holding that ignorance of probabilities is a guide to belief.

In the light of these points, consider the following passage from *Warrant and Proper Function* (p. 229, our italics and brackets) in which Plantinga justifies the first step of the main argument:

> Someone who accepts E&N and also believes that the proper attitude toward $\Pr(R \mid E\&N)$ is one of agnosticism [or, one of low degree of belief] clearly enough has good reason for being agnostic about [or, having a low degree of belief with respect to] R as well. *She has no other information about R ... but the source of information she does have gives her no reason to believe R and no reason to disbelieve it.*

Notice that Plantinga assumes that evolutionary naturalists have no basis for deciding what to think about R, other than the proposition E&N itself. This crucial assumption is never defended in either *Warrant and Proper Function* or "Naturalism Defeated."

### 2.2    What Defeating R Means

In the second step of the main argument, Plantinga says that E&N's defeat of R means that evolutionary naturalists should withhold assent from anything else they believe—for example, from E&N itself. This goes beyond what the defeat of proposition R really entails. Proposition R says that "the great bulk" of the beliefs we have are true (Plantinga 1994, p. 2). If evolutionary naturalists should withhold assent from R, this does *not* mean that they should withhold assent from most of what they believe, much less from everything they believe. Even if E&N defeats the claim that "at least 90% of our beliefs are true," it does not follow that E&N also defeats the more modest claim that "at least 50% of our beliefs are true." Plantinga must show that E&N not only defeats R, but also defeats the claim that "at least a non-negligible minority of our beliefs are true."

### 2.3    Conditional Probability and Defeat

Although we, like a number of other commentators, have interpreted the main argument of *Warrant and Proper Function* as asserting that a low value for $\Pr(R \mid E\&N)$ suffices for E&N to defeat R, Plantinga (1994) denies that this is what he meant and tries to develop an account of defeat that clarifies how the argument is supposed to go. However, Plantinga still spends time arguing in this later manuscript that $\Pr(R \mid E\&N)$ is low or inscrutable, so he presumably still holds that the value of this probability is relevant to establishing that $E\&N$ defeats $R$.

Plantinga develops three principles that he thinks govern the defeat relation. Although he explains why he thinks these principles are correct, he never explains how they are relevant to establishing that $E\&N$ defeats $R$. In fact, their logical form renders them incapable of closing the gap between premises 1 and 2. The first and third principles assert *sufficient* conditions for $X$'s *not* defeating $Y$. The second states a *necessary* condition for $X$'s defeating $Y$.

In addition, Plantinga's "First Principle of Defeat" apparently helps establish that $E\&N$ is *not* a defeater of $R$. Substituting $R$ and $E\&N$ for $A$ and $B$ in this principle yields

If $S$ rationally believes that the warrant that $E\&N$ has for him is derivative from the warrant that $R$ has for him, then $E\&N$ is not a defeater, for him, of $R$.

We suspect that many evolutionary naturalists rationally believe that their warrant for believing $E\&N$ depends on their being warranted in believing that their cognitive faculties are highly reliable.[6]

Not only is a low value for $\Pr(X \mid Y)$ not *sufficient* for $Y$'s defeating $X$; it also is not *necessary*, if defeaterhood is to ground the idea of *self-defeat*. The reason is that $\Pr(Y \mid Y) = 1$, for all $Y$. And as difficult as it is to connect low probability to defeaterhood, it seems even harder to see why the inscrutability of $\Pr(X \mid Y)$ should help establish that $Y$ defeats $X$.

In the preliminary argument, Plantinga assigns to $\Pr(R)$ a value close to unity because he believes $R$ to be true. In the main argument as formulated in *Warrant and and Proper Function*, he gives the impression that he thinks evolutionary naturalists should withhold belief in $R$ because $E\&N$ fails to confer a sufficiently high probability on $R$. These are two ways of expressing the same sentiment: *high probability is necessary for rational belief*. However, the more recent manuscript "Naturalism Defeated" repudiates the idea that there is any such simple relation between probability and acceptance.

What Plantinga is coming up against here is a close relative of the phenomenon that Kyburg's (1961) lottery paradox made vivid. Suppose there are 10,000 tickets in a fair lottery; one ticket will win and each has the same chance of winning. Suppose you adopt the following criterion for belief—you accept a proposition if you think it has a high probability. If so, you will accept each proposition of the form "ticket $i$ won't win." However, the conjunction of these contradicts the starting assumption that the lottery is fair. Therefore, high probability is not sufficient for rational belief. A similar counterexample can be constructed to show that high probability is also not *necessary* for rational belief. Consider any $n$ propositions $P_1, \ldots P_n$ such that (*i*) you accept each of the $P_i$, and (*ii*) each of the $P_i$ is very highly probable. The conjunction $P_1 \& \ldots \& P_n$ may turn-out to be quite *im*probable (see section 1.1 for a concrete example of this probabilistic phenomenon). Nonetheless, it apparently would be rational for you to accept the conjunction $P_1 \& \ldots \& P_n$. Hence, high probability is not necessary for rational belief (see, also, Maher 1993, section 6.2.4). Philosophers of probability have extracted from these paradoxes one of two lessons—either the concepts of acceptance and rejection are suspect, or they are more subtly related to the concept of probability than the threshold criterion just described.

This connection with the lottery paradox suggests that the task of repairing the main argument is formidable. That argument begins with claims about probability, moves to claims about defeat, and then concludes with a claim about self-defeat. Each step along the way requires principles quite different from the ones that Plantinga has so far described. Whether plausible principles exist that forge the requisite connections we leave to the reader to conjecture.

## 3.    A Question for Evolutionism

Although Plantinga's arguments don't work, he has raised a question that needs to be answered by people who believe evolutionary theory and who also believe that this theory says that our cognitive abilities are in various ways imperfect. Evolutionary theory does say that a device that is reliable in the environment in which it evolved may be highly unreliable when used in a novel environment. It is perfectly possible that our mental machinery should work well on simple perceptual tasks, but be much less reliable when applied to theoretical matters. We hasten to add that this is *possible*, not *inevitable*. It may be that the cognitive procedures that work well in one domain *also* work well in another; *Modus Ponens* may be useful for avoiding tigers *and* for doing quantum physics.

Anyhow, if evolutionary theory does say that our ability to theorize about the world is apt to be rather unreliable, how are evolutionists to apply this point to their own theoretical beliefs, including their belief in evolution? One lesson that should be extracted is a certain humility—an admission of fallibility. This will not be news to evolutionists who have absorbed the fact that science in general is a fallible enterprise. Evolutionary theory just provides an important part of the explanation of why our reasoning about theoretical matters is fallible.

Far from showing that evolutionary theory is self-defeating, this consideration should lead those who believe the theory to admit that the best they can do in theorizing is to do the best they can. We are stuck with the cognitive equipment that we have. We should try to be as scrupulous and circumspect about how we use this equipment as we can. When we claim that evolutionary theory is a very well confirmed theory, we are judging this theory by using the fallible cognitive resources we have at our disposal. We can do no other.

Plantinga suggests that evolutionary naturalism is self-defeating, but that traditional theism is not. However, what is true is that neither position has an answer to hyperbolic doubt. Evolutionists have no way to justify the theory they believe other than by critically assessing the evidence that has been amassed and employing rules of inference that seem on reflection to be sound. If someone challenges all the observations and rules of inference that are used in science and in everyday life, demanding that they be justified from the ground up, the challenge cannot be met. A similar problem arises for theists who think that their confidence in the reliability of their own reasoning powers is shored up by the fact that the human mind was designed by a God who is no deceiver. The theist, like the evolutionary naturalist, is unable to construct a non-question-begging argument that refutes global skepticism.

## Notes

Gordon Barnes, Matt Davidson, Ellery Eells, Malcolm Forster, Patrick Maher, Ernan McMullin, Alvin Plantinga, and Dennis Stampe provided valuable criticisms and suggestions. We are grateful to them for their help.

1. All page references are to *Warrant and Proper Function* (*viz.*, Plantinga 1993), unless otherwise noted.

2. An even better strategy would be to associate a characteristic degree of sensitivity with a mental faculty in a given environmental setting. Roughly speaking, sensitivity is a world-to-head relation, measured by the probability that the agent will believe *p*, conditional on *p*'s being true. In contrast, reliability is a head-to-world relation, measured by the probability that *p* will be true, conditional on the agent's believing *p*. Sensitivity tends to be a more stable property of measurement devices than reliability. See Sober (1994, essays 3 and 12) for discussion.

3. Here we go along with Plantinga's usage of the term "traditional theism," according to which this doctrine makes different observational predictions than evolutionary naturalism. However, there is room to argue whether this way of seeing matters is the only one that is available from various religious traditions (McMullin 1993). Plantinga understands traditional theism to be the idea, not just that God sets evolutionary processes in motion (where these are understood in terms of the best theories that science now provides), but that he occasionally intervenes in them to insure certain outcomes. The idea that God does the former, but not the latter, confers on proposition *R* precisely the same probability that evolutionary theory by itself confers on *R*. This can be seen more clearly by considering figure 16.1. If evolutionary processes (*E*) "screen off" God's activity (*G*) from what we can observe (*O*), then $Pr(O \mid E\&G) = Pr(O \mid E\&\text{not-}G)$. Plantinga thinks this equality is false; he holds that atheistic evolutionism confers on the observations (specifically, on proposition *R*) a probability different from the one provided by theistic evolutionism. This means that Plantinga is thinking of God as not simply acting *through* natural evolutionary processes, but as affecting the world by a separate, "miraculous" pathway.

4. In discussing "Darwin's doubt," Plantinga (1994, p. 4) quotes with approval a point made by Churchland (1987, p. 548) to the effect that natural selection "cares" only about how adaptive the behaviors are that a set of beliefs causes; it does not care, in addition, whether those beliefs are true. Plantinga interprets this to mean that true beliefs are no more likely to evolve than false ones, but a probabilistic representation of Churchland's point (which is about *conditional independence*) shows that this does not follow. Churchland's point is that

Pr(Belief set *B* evolves | *B* produces adaptive behaviors & *B* is true)

   = Pr(Belief set *B* evolves | *B* produces adaptive behaviors & *B* is false).

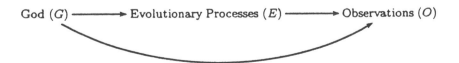

**Figure 16.1**
The possible relationships between *G*, *E*, and *O*.

However, from this it does not follow that

Pr(Belief set *B* evolves | *B* is true) = Pr(Belief set *B* evolves | *B* is false).

In just the same way, although it is true that

Pr(it will rain tomorrow | a storm is approaching & the barometer

reading is low) = Pr(it will rain tomorrow | a storm is approaching & the

barometer reading is high),

it does not follow that

Pr(it will rain tomorrow | the barometer reading is low)

= Pr(it will rain tomorrow | the barometer reading is high).

5. The point we are making here accords with what Plantinga (1994) calls "the perspiration objection," which he attributes to "Wykstra, DePaul, and others." Although Plantinga discusses how the objection should be formulated, he does not, as far as we can see, provide an answer to it.

6. Of course, one can't deduce *E&N* from *R* alone. But, evolutionary naturalists might reasonably maintain that *R* is one of several premises which underwrite their non-deductive inferences concerning the plausibility of *E&N*. It is worth pointing out that Plantinga himself makes use of this kind of non-deductive "warrant derivation." On page 39 of "Naturalism Defeated," he says, in the context of discussing an objection to his argument, that the warrant that *P* has for you (where *P* is such that Pr[*R* | *N&E&P*] is high, but *P* is *logically indepen-dent* of *R*) is "... derivative from the warrant *R* has for you ... it is hard to see what other source [of warrant] there could be [for *P*]." We see no reason why it would be irrational for evolutionary naturalists to say the same thing about the warrant that *E&N* has for them.

# References

Churchland, P. (1987). Epistemology in the Age of Neuroscience. *Journal of Philosophy* 84, 544–553.

Fodor, J. (1997). Is Science Biologically Possible? Unpublished manuscript of the 1998 Benjamin Lecture at the University of Missouri.

Kahnemann, D., P. Slovic, und A. Tversky (eds.) (1982). *Judgment Under Uncertainty: Heuristics and Biases*. Cambridge University Press.

Kyburg, H. (1961). *Probability and the Logic of Rational Belief* Wesleyan University Press.

Maher, P. (1993). *Betting on Theories*. Cambridge University Press.

McMullin, E. (1993). Evolution and Special Creation. *Zygon* 28, 299–335.

Plantinga, A. (1993). *Warrant and Proper Function*. Oxford University Press.

Plantinga, A. (1994). Naturalism Defeated. Unpublished manuscript.

Sober, E. (1993). *Philosophy of Biology*. Westview Press.

Sober, E. (1994). *From a Biological Point of View*. Cambridge University Press.

# VI

## Intelligent Design Creationism vs. Theistic Evolutionism

Must one chose between evolution and belief in God? The answer to this question depends, of course, on the details of evolution and on one's conception of God. An ironic feature of the creation/evolution controversy is that creationists and strong atheists agree in answering this question in the affirmative, while most theologians answer it in the negative. Pope John Paul II recently reiterated the established position of the Catholic Church that there is no conflict between evolution and Christian faith. (Stephen Jay Gould discusses the Pope's statement in section VIII of this volume, in his article "Nonoverlapping Magisteria.") Most mainstream Protestant theologians agree. Creationists, however, are typically fundamentalist or evangelical Christians and reject this as an unacceptable "accommodation" to evolution; they believe that features of evolutionary theory directly contradict basic religious commitments.

In the first essay, originally published in *First Things*, the journal of The Institute on Religion and Public Life, Phillip Johnson does not attack evolution directly, but rather those "influential Christian intellectuals" who defend it. He sets out what he takes to be the core commitment of creationism: that we do not owe our existence to a blind materialistic process but rather to a purposeful Creator. Responding to the essays in section III by Van Till and McMullin, Johnson suggests that their line of reasoning exhibits a "remarkable convergence" with the scientific naturalism his movement opposes, and he labels them "theistic naturalists." He argues that the best theistic naturalists can do is to wall off the supernatural miracles reported in Scripture as a "salvation history" separate from natural history, but that any such "compromise with naturalism" raises direct conflicts with the details of Genesis as well as the New Testament. In particular he cites a passage from the first chapter of Romans that states that God's invisible nature is clearly perceived in Creation, so that fools who still reject God "are without excuse." Darwinism and theism, he concludes, "are fundamentally incompatible."

Fuller Theological Seminary professor Nancey Murphy, like other theistic evolutionists, accepts the idea of God's "Design" of the world, but disagrees with IDC's narrow understanding of this. In her essay she gives an assessment of Johnson's *Darwin on Trial*. She praises some aspects of Johnson's book, but overall finds his arguments to be "dogmatic," "unconvincing," and "fallacious." She concludes that he has added to the confusion between evolution and atheistic philosophical naturalism by failing to recognize the distinction between the latter and what she calls "methodological atheism." She points out that the move to exclude natural theology from science was originally advocated for theological reasons to *protect* theology.

The third essay is by Arthur Peacocke, whom Murphy cites as one example of someone who has articulated a way in which Christian thought can make biology its own. Peacocke is a biochemist turned theologian, who is now a priest in the Church of England. William Dembski has written that "Design theorists are no friends of theistic evolution," but in his article Peacocke suggests that Darwinism should be welcomed as a "disguised friend" of theism. Peacocke first delivered this paper as the Idreos Lecture at Harris Manchester College in 1997 and slightly updated it for inclusion in this volume.

In the fourth essay in this section, "The Creation: Intelligently Designed or Optimally Equipped?" physicist Howard Van Till extends the critique that he began in his exchange with Alvin Plantinga (in section III, this volume). Van Till has also debated Phillip Johnson in print ("God and Evolution: An Exchange," *First Things* vol. 34, June/July 1993), writing that Johnson presents "just enough truth to mislead persuasively." In his current article, he criticizes not only Johnson, but also Michael Behe, William Dembski, Paul Nelson, Stephen Meyer, and others of the ID movement. He agrees with them in rejecting "evolutionary naturalism" (which he uses in the metaphysical sense to mean the explicit "denial of a Creator-God"), and in this article criticizes philosopher Daniel Dennett as a purported representative of that position. However, he rejects the "simple either/or debate" that he says ID theorists promote in the same way that evolutionary naturalists do. From the point of view of his own Calvinist religious heritage, he develops the idea of a "robust formational economy of nature" to show how a Christian can embrace evolution without adopting a form of deism.

The final essay in this section, "Is Theism Compatible with Evolution?" was written for this volume by philosopher of religion Roy Clouser as a general rebuttal to Johnson's and other creationists' forms of incompatibilism and natural theology. Clouser articulates a quite different compatibilist position from the other sorts of theistic evolution we have seen, drawing on the work of Dutch philosopher Herman Dooyeweerd. Clouser spends the first part of his article defending a "'literal' (but not literal*ist*)" interpretation of Genesis to show that creationists are wrong in reading it as being inconsistent with evolution, which he fully accepts. His goes on to criticize the intelligent design movement from a unique perspective, in that he agrees with them that science is a materialist religion that makes a priori metaphysical assumptions (under his broad notion of what it means to be a religious view and to presuppose something as "divine"), but he rejects their contention that science and the "orderliness" of nature can in any way either entail God's existence or even make it probable.

# 17

## Creator or Blind Watchmaker?

Phillip E. Johnson

As a notorious critic of Darwinism, I enjoy reading a newsletter called *Basis*, which is published by an organization calling itself the San Francisco Bay Area Skeptics. There self-styled skeptics take a very dim view of anyone who suggests that the Darwinian theory of evolution might be an appropriate subject for skeptical inquiry, and on that account their editorial ire is sometimes aimed in my direction. A recent issue of *Basis* reported on a local meeting at which the featured speaker was a woman identified as "a religious person and science teacher at a Catholic school." This science teacher was assuring her audience that despite the religious affiliation of her school, she taught evolution and not creationism in her science classes. A questioner from the audience put her on the spot by asking: "Do you think that evolution is directed?" The newsletter reports that this question was followed by a "dramatic pause," after which the teacher replied with what it called a "battled 'No.'" The reporter for *Basis* commented: "I would have expected a more rapid answer, but the battle between her curriculum and her beliefs had a few more moments of unrest left to settle."

The appearance of that story coincided with the release of a new Gallup Poll, reporting on the state of American opinion regarding evolution and creation. According to this survey, approximately 47 percent of Americans can be described as creationists, in that they say they believe that God created mankind in pretty much our present form sometime within the last 10,000 years. (The wording of the question did not rule out a long period of animal evolution before the appearance of man, however.) Another 40 percent agreed with the following statement:

Originally published in *First Things* (1993, no. 29, pp. 8–14).

"Man has developed over millions of years from less advanced forms of life, but God guided this process, including man's creation." Only 9 percent of the sample said that they accepted the naturalistic view of evolution, which in Gallup's wording was that man has developed over millions of years from less advanced forms of life, with God having no part in this process.

Against that background of public opinion, we can see why the voice from the audience was asking exactly the right question, and also why we might expect a science teacher at a Christian institution to take a deep breath before answering in a quavering voice. When Darwinists speak of "evolution," they mean the creed of the 9 percent. Science educators frequently obscure this point in order to avoid further arousing political opposition to the teaching of evolution as fact in the public schools, but they are perfectly explicit about it when candor suits their purpose. For example, one of the founders of the neo-Darwinian synthesis, Harvard paleontologist George Gaylord Simpson, explained the "meaning of evolution" in the following widely quoted language:

Although many details remain to be worked out, it is already evident that all the objective phenomena of the history of life can be explained by purely naturalistic or, in a proper sense of the sometimes abused word, materialistic factors. They are readily explicable on the basis of differential reproduction in populations (the main factor in the modern conception of natural selection) and of the mainly random interplay of the known processes of heredity.... Man is the result of a purposeless and natural process that did not have him in mind.

The literature of evolutionary biology contains countless statements to the same effect. "Evolution," honestly understood, is not just a gradual process of development that a purposeful Creator might have chosen to employ. It is, by Darwinist definition, a purposeless and undirected process that produced mankind accidentally. By saying that she taught "evolution," the Catholic school teacher had said only that she did not teach the sudden, special creation of each species. She was not with the 47 percent who (perhaps) reject evolution altogether, but she might still be with the 40 percent who think there is a compromise position that combines creation and evolution. From a Darwinist viewpoint, however, this soft form of creationism is merely a relatively advanced kind of misunderstanding. Did the teacher really explain to her students and their parents that biological evolution is not God-directed but rather a purposeless process that produced mankind by accident? If so, how could she or her superiors possibly reconcile this teaching with their official commitment to Christian theism?

Now, through an educational system insistent upon uncritical acceptance by students at all levels of the claim that purposeless material mechanisms were responsible for the creation of all forms of life, scientific naturalism is becoming the officially established religion of America. There is no mystery about why atheists and agnostics welcome this kind of education. What is mysterious, however, is the relative lack of opposition to the establishment of naturalism from Christian intellectuals. There are, of course, theologians who have embraced naturalism with enthusiasm and proceeded to try to "save" Christianity by purging it of supernaturalism and mythology; one can be a Christian, or at least a professor at certain liberal Christian seminaries or divinity schools, and be as opposed to the existence of a supernatural Creator as any atheist. But Christians and other theists, who really believe in a personal God standing outside nature and ruling it—how can they accommodate the dictates of a scientific establishment that absolutely insists that all creation resulted from undirected evolution?

Some people don't accommodate, but simply reject the authority of that establishment in toto. Those who regard Scripture as more authoritative than scientific theories, and who are confident that they know the correct way to interpret it, may choose to defend the Genesis account as literally true and employ scientific argument to discredit the alternatives. Fundamentalist creationists of this kind make up perhaps half of the 47 percent that the Gallup poll defined as creationist. Unfortunately, the commitment of this large group to a literal interpretation of Genesis has confused and divided the Christian world, and even played into the hands of the evolutionary naturalists. Darwinists assiduously promote the notion that the only possible alternatives are six-day Genesis literalism on the one hand, and fully naturalistic, neo-Darwinistic evolution on the other. Given such an understanding of the alternatives, anyone who suspects that the cosmos may be billions of years old, or that life may have been created through some long-term process of development, becomes an "evolutionist"—who by definition rejects "creationism." Under Darwinist auspices, science education, in the media as well as the schools, consequently aims to enlighten such persons about what evolution really means, and to wash the lingering effects of creationism from their minds. "Properly" educated people gradually learn that forces like mutation and selection were adequate to perform all the work of biological creation, and that the notion of a purposeful Creator is therefore superfluous, discardable without loss.

By controlling the terminology, then, Darwinists have given the world the impression that the significant divide in public opinion about evolution is that between the Genesis literalists and everybody else. This is a sorry misunderstanding. For the fundamental disagreement is not over the age of the earth or the method of creation; it is over whether we owe our existence to a purposeful Creator or a blind materialistic process. The 47 percent in the 1991 Gallup Poll who say that God created suddenly and the 40 percent who say that God created gradually are basically in agreement—in comparison to the 9 percent who say that God did not create at all. When the majority finally understands this, it will become possible to challenge the monopoly of evolutionary naturalism both in the media and the educational system.

What the situation requires is a critique of evolutionary naturalism that puts aside the biblical issues for the time being and concentrates on the scientific and philosophical weaknesses in the established Darwinist orthodoxy. Unfortunately, many of the most influential Christian intellectuals have themselves been so strongly influenced by naturalistic philosophy that they have tried to baptize it. Their position, which I call theistic naturalism, starts from the premise that God refrains from interference with those parts of reality that natural science has staked out as its own territory. Theistic naturalists concede to Darwinism the role of telling the true history of the development of life, and limit the Creator to activity in a metaphysical realm outside the reach of science. Theistic naturalism is more often implicit than explicit in religious discourse—as befits a philosophy so dominant in intellectual circles that people hardly ever have to think about it in any detail. Princeton Theological Seminary Professor Diogenes Allen's 1989 book *Christian Belief in a Postmodern World: The Full Weight of Conviction* provides a particularly thorough and thoughtful explication of theistic naturalism. Allen explains the division between the realms of science and theology by saying that there are questions a naturalistic science cannot purport to answer.

Our natural sciences seek to describe and explain the relations *between* the members of the universe, not their origin. The existence of the universe and its basic constituents are taken for granted by our sciences.... When we consider the whole of nature, the relations we find within nature cannot tell us why the universe exists nor why it is the kind of universe it is. The continuing increase of scientific knowledge, which discovers the relations that exist within our universe, does not get us closer to an answer to either question.

The role of religion, then, is to explain the ultimate features of the universe that remain when science has done everything of which it is capable. Up to that point, naturalism in scientific explanations is not only permitted but actually *required* by a correct view of God's role and nature. According to Allen.

God can never properly be used in scientific accounts, which are formulated in terms of the relations between the members of the universe, because that would reduce God to the status of a creature. According to a Christian conception of God as creator of a universe that is rational through and through, there are no missing relations between the members of nature. If, in our study of nature, we run into what seems to be an instance of a connection missing between members of nature, the Christian doctrine of creation implies that we should keep looking for one.

When the philosopher Alvin Plantinga expressed skepticism towards Darwinism in a lecture published in a special issue of the *Christian Scholar's Review*, and proposed that Christians engage in a "theistic science" to counter the domination of naturalism, commentators employed reasoning similar to Allen's to throw cold water on the idea. Calvin College physics professor Howard Van Till argued that "the world created by the God who reveals himself in Scripture" must be characterized by what Van Till calls "functional integrity." This term denotes "a created world that has no functional deficiencies, no gaps in its economy of the sort that would require God to act immediately, temporarily assuming the role of creature to perform functions within the economy of the creation that other creatures have not been equipped to perform." Notre Dame philosophy professor Ernan McMullin argued in the same vein that it is improbable that an omniscient creator would find it necessary to intervene in the cosmic order to "plug the gaps."

This line of reasoning establishes a remarkable convergence of Christian theism and scientific naturalism. Consider, for example, the attitude that scientific naturalists take towards theories of chemical evolution, which aspire to explain the origin of life on earth. Until very recently, the popular and professional literature of this field tended to be wildly optimistic, suggesting that science was on the verge of providing a satisfactory theory of how life evolved from nonliving chemicals. The current literature is much more sober, and some commentators candidly describe the field as dominated by contradictory speculations and lack of experimental success. One of these commentators is New York University chemistry professor Robert Shapiro, author of the excellent popular book *Origins: A Skeptic's View of the Creation of Life on Earth*. Shapiro realizes that a satisfactory theory of chemical evolution may be a long time coming, but his faith in naturalistic explanation is equal to the challenge:

Some future day may yet arrive when all reasonable chemical experiments run to discover a probable origin for life have failed unequivocally. Further, new geological evidence may indicate a sudden appearance of life on the earth. Finally, we may have explored the universe and found no trace of life, or processes leading to life, elsewhere. In such a case, some scientists might choose to turn to religion for an answer. Others, myself included, would attempt to sort out the surviving less probable scientific explanations in the hope of selecting one that was still more likely than the remainder.

The theistic naturalists seem to share this fervent faith that a naturalistic explanation for the origin of life simply *must* be there to be found. To suppose that God may have played some direct, active role in creating the first life on earth would reduce God to the status of a creature, would posit an impossible missing relation between the members of nature, and would deny the functional integrity of the universe. One might almost say that it would constitute blasphemy.

A basis for a convergence between theism and naturalism may be found in the writings of George Gaylord Simpson. Simpson, one of the founders of the neo-Darwinian theory, was a thoroughgoing scientific naturalist who derided Christianity as superstition. Nonetheless, he was willing to acknowledge that science could not explain everything, and that some kind of place for religion might be found in that philosophical territory which is beyond the reach of scientific investigation. As he put it:

There is neither need nor excuse for postulation of nonmaterial intervention in the origin of life, the rise of man, or any other part of the long history of the material cosmos. Yet the origin of that cosmos and the causal principles of its history remain unexplained and inaccessible to science. Here is hidden the First Cause sought by theology and philosophy. The First Cause is not known and I suspect it will never be known to living man. We may, if we are so inclined, worship it in our own ways, but we certainly do not comprehend it.

Simpson's First Cause, like the God of theistic naturalism, is responsible for the ultimate origin and original characteristics of the cosmos, but thereafter leaves creation entirely to material processes that go their own way. Simpson thought that the First Cause could be mentioned and then forgotten, because he assumed that science would eventually explain nearly everything of consequence to human beings. The role left to the First Cause would then be so small, and so remote from human concerns, that only the incorrigibly religious would have any interest in it. Recently, the physicist Stephen Hawking has moved to take away even that vestigial refuge, predicting that physicists will soon discover a "theory of everything" that will permit them to "know [i.e., become] the mind of God."

It is more likely, however, that there are important questions that can never be answered by scientific inquiry. For example, many persons who are adamant about the exclusive authority of science in explaining the physical world are willing to concede that questions about morality or the purpose of life have to be addressed in some other way. As the realm of science diminishes in our imagination, the potential importance of the First Cause correspondingly increases. All we have to do is take up Simpson's invitation to worship the First Cause and it becomes the God of theistic naturalism, which is responsible for the ultimate origin of the universe and its initial boundary conditions, but which never intervenes to take a part in natural history.

But can this First Cause fill the role of "the God who reveals Himself in Scripture"? The biblical God seems to engage in a lot of creaturely behavior. The New Testament tells us that the Second Person of the Trinity was born as a human baby in a stable in Bethlehem, and after growing to adulthood performed some very immediate miraculous acts before dying a very creaturely death on the cross. He changed water into wine, healed leprosy and blindness, fed multitudes on scraps of food, raised Lazarus from the dead, and eventually rose from the dead Himself and ascended to heaven. Is this a God who would never, in Van Till's words, "temporarily [assume] the role of creature to perform functions within the economy of the creation that other creatures have not been equipped to perform"? And how can so much supernatural activity be reconciled with a naturalistic philosophy of creation?

There are two schools of thought among theistic naturalists about how to deal with this problem. One approach, characteristic of the liberal theology that culminated in Tillich and Bultmann, is to interpret the miracles as mythology. This is at least consistently naturalistic, but it relegates Christianity to the role of a human ethical system or existential choice. The other approach is to treat the miracles as real events, but to restrict them to a "salvation history" that is walled off from the natural history over which science claims exclusive authority. Diogenes Allen takes this latter approach:

In general we may say that God creates a consistent set of lawlike behaviors. As part of that set there are the known physical laws. These laws apply to a wide variety of situations. But in certain unusual situations such as creating a chosen people, revealing divine intentions in Jesus, and revealing the nature of the kingdom of God, higher laws come into play that give a different outcome than normal physical laws which concern different situations. The normal physical laws do not apply because we are in a domain that extends beyond their competence.

A division of this kind is about the best a Christian theologian can do—if he wants to respect the authority of naturalistic science within its own sphere. From a biblical standpoint, however, it is not only the events of salvation history that create difficulties for any compromise with naturalism. One is faced not simply with the details of the Genesis account, but with New Testament passages that reflect the fundamental logic of Christianity. For example, the first chapter of Romans tells us that

the wrath of God is revealed from heaven against all ungodliness and wickedness of men, who by their wickedness suppress the truth. For what can be known about God is plain to them, because God has shown it to them. Ever since the creation of the world, his invisible nature, namely, his eternal power and deity, has been clearly perceived in the things that have been made. So they are without excuse; for although they knew God they did not honor him as God or give thanks to him, but they became futile in their thinking and their senseless minds were darkened. Claiming to be wise, they became fools, and exchanged the glory of the immortal God for images resembling mortal man or birds or animals or reptiles.

That passage does not speak of a nature that merely raises questions that a naturalistic science cannot answer, but of a nature that points directly and unmistakably toward the necessity of a creator. And if nature does no more than raise questions, how can men be blamed for coming to the wrong conclusions about what to worship? If God stayed in that realm beyond the reach of scientific investigation, and allowed an apparently blind materialistic evolutionary process to do all the work of creation, then it would have to be said that God furnished us with a world of excuses for unbelief and idolatry.

There is an even more important passage to consider. The most important statement about creation in the Bible is not in Genesis; it is in the opening verses of the first chapter of John.

In the beginning was the Word, and the Word was with God, and the Word was God. He was in the beginning with God, all things were made through him, and without him was not anything made that was made.

The direct opposite of that statement is creation by purely materialistic, undirected evolution. Accommodationists can throw out a lot of nonessentials and still keep the substance of Christianity, but creation through the Word is one of the essentials.

Biblical considerations aside, theistic naturalism has a fatal logical weakness that stems from the fact that it attempts to reconcile two fundamentally inconsistent ways of thinking. Theism asserts that God rules everything; naturalism asserts that nature proceeds on its own, without supernatural influence. The two could be reconciled if we had some reason to believe that God chose to leave nature alone after establishing the initial conditions—except, of course, for the exceptional events of salvation history. It is hard to imagine any purpose for such a self-limitation other than the need to make a treaty of peace between science and theology, which is a consideration hardly likely to influence God. In any case, Darwinistic evolution would be a most peculiar creative method for God to choose, given the Darwinistic insistence that biological evolution was *undirected*. That requirement means that God neither programmed evolution in advance nor stepped in from time to time to pull it in the right direction. How then did God ensure that humans would come into existence so that salvation history would have a chance to occur? Once this logical difficulty is recognized, the attempt to reconcile Darwinism and theism collapses. Either God rules creation—which means that He somehow directed evolution to produce humans—or He doesn't. The former isn't Darwinism, and the latter isn't theism.

Of course, God *could* make some use of random mutation and natural selection in a fundamentally directed creative process. God can act freely as He chooses: that is just the problem for those who would constrain God by philosophy. God could employ mutation and selection or act supernaturally, whether or not His choice causes inconvenience for scientists who want to be able to explain and control everything. Once we allow God to enter the picture at all, there is no reason to be certain a priori that natural science has the power to discover the entire mechanism of creation. Maybe science can discover how living things were made, and maybe it can't. Consistent theists will therefore accept Darwinist claims for the creative power of mutation and selection only insofar as those claims can be supported by evidence. That isn't very far at all.

It isn't easy to assess the adequacy of the evidence, however, because Darwinists employ a vocabulary that systematically distracts attention from the central issue. "Evolution" is a thoroughly confusing term that may be used in a broad or narrow sense. As we have seen, it sometimes simply means no more than that creation took a lot longer than six days. In debates with those opposed to their views, Darwinists typically define the "fact of evolution" as meaning merely that "the earth has a history," or that "change has occurred." So defined, evolution is neither theologically important nor scientifically interesting. Having established this empty concept of evolution, however, Darwinists immediately go on to use the same term as if it meant neo-Darwinistic evolution, with an all-encompassing creative role for natural selection. Unpack the terminology and you will find the following illogic: change has occurred, and therefore mutation and selection caused the change to occur.

Darwinists use "evolution" to mean *both* minor variations and major creative innovations, and also *both* change that is directed by intelligence and change that is presumed to be undirected. That is why they think that domestic animal breeding, which employs purposeful human intelligence to achieve variation within the biological species, somehow illustrates how a purposeless material process created animals in the first place. "Evolution" in Darwinist usage stands for many different things, and an illustration of evolution in any sense proves the entire system. To get away from this tradition of confusion, we have to employ a terminology that keeps attention on the really important claim of Darwinism—*which is that biological creation could and did occur by known material mechanisms without the need for supernatural assistance.*

This central doctrine might be called the "blind watchmaker hypothesis," after the title of the famous book by Oxford University zoologist Richard Dawkins. Dawkins begins his book with the observation that "biology is the study of complicated things that give the appearance of having been designed for a purpose." That sounds like a creationist argument, and indeed Dawkins is happy to concede that each and every plant or animal cell nucleus contains "a digitally coded database larger, in information content, than all 30 volumes of the *Encyclopedia Britannica* put together." Undirected material processes do not write encyclopedias, no matter how much time is available. Of course, Dawkins is no creationist. He wrote *The Blind Watchmaker* to convince the public of something that Darwinists take for granted: namely, that the appearance of purposeful design in biology is misleading, because all living organisms, including ourselves, are the products of a natural evolutionary process employing random variation and natural selection. As Dawkins explains:

Natural selection is the blind watchmaker, blind because it does not see ahead, does not plan consequences, has no purpose in view. Yet the living results of natural selection overwhelmingly impress us with the appearance of design as if by a master watchmaker, impress us with the illusion of design and planning.

In short, the watchmaker is not only blind, but unconscious.

Is the blind watchmaker hypothesis *true*? From the naturalistic standpoint of Darwinists like Dawkins, the question really doesn't arise. Instead of truth, the important concept is *science*, which is understood to be our only (or at least by far our most reliable) means of attaining knowledge. Science is then defined as an activity in which only naturalistic explanations are considered, and in which the goal is always to improve the best existing naturalistic explanation. Supernatural creation— or God-guided evolution—is not a naturalistic explanation. The blind watchmaker hypothesis is therefore merely a way of stating the commitment of "science" to naturalism, and as such the existence of a blind watchmaker is a logical necessity. If a critic doesn't like Darwinism, his only permissible move is to suggest a better blind watchmaker. That a competent blind watchmaker doesn't exist at all is not a logical possibility.

That is why Dawkins and other Darwinists never prove the blind watchmaker hypothesis: instead, they illustrate it. A Darwinist only has to imagine how some complex organ or organism might have originated by mutation and selection, and the theory has another confirming example. For example, Dawkins imagines that small tree-dwelling mammals might have survived falls more frequently if they happened to grow flaps of skin between their digits that slowed their descent. Over many generations, these flaps might have grown into wings as further mutations accumulated, and so four-footed creatures might become flying bats. That sort of storytelling counts as a scientific explanation in Darwinist circles, and no one expects the storyteller to demonstrate that such a process either can or did occur.

There is no doubt that natural processes produce a degree of variation in existing genotypes, as illustrated by the differences between island species and their close mainland relatives. The question is whether similar processes have the power to furnish the genetic information required for (say) bacteria to become complex plants and animals, or for four-footed mammals to change into bats and whales. Darwinian evolution of the genuinely creative sort can't be observed in the field or the laboratory, or traced in the fossil record. Functional intermediates between distinct types often can't even be imagined. No matter: by Darwinist logic natural selection had to do the job anyway, or we wouldn't be here to carp at the lack of evidence. Critics may cite any number of objections, but at the end of the day the Darwinists still hold title to that term "science." As long as their rules are in force, their theory cannot be unseated.

When people ask whether Darwinism and theism are compatible, they normally take the Darwinism for granted and ask whether the theism has to be discarded. It is far more illuminating, however, to approach the question from the other side. Is there any reason that a person who believes in a real, personal God should believe that biological creation occurred by Darwinian evolution? The answer is clearly no. The sufficiency of any process of chemical evolution to produce life has certainly not been demonstrated, nor has the ability of natural selection to produce new body plans, complex organs, or anything else except variation within types that already exist. The fossil record notoriously does not evidence any continuous process of gradual change. Rather, it consistently shows that new forms appear suddenly and fully formed in the rocks, and thereafter remain fundamentally unchanged. That is why fossil experts from T. H. Huxley to Stephen Jay Gould have flirted with the heresy that biological transformations occurred in great (and therefore scientifically inexplicable) jumps. If Darwinian evolution is the only allowable source for life's diversity and complexity, then the shortage of evidence doesn't matter. The only question, to borrow Darwin's own words, is why "Nature may almost be said to have guarded against the frequent discovery of her transitional or linking forms." Atheists can leave the matter there, but theists have to go farther. If God exists, then Darwinian evolution is not the only alternative, and there is no reason for a theist to believe that God employed it beyond the relatively trivial level where the effects of variation and selection can actually be observed.

In short, the reason that Darwinism and theism are fundamentally incompatible is not that God could not have used evolution by natural selection to do his creating. Darwinian evolution might seem unbiblical to some, or too cruel and wasteful a method for a benevolent Creator to choose, but it is always possible that God might do something that confounds our expectations. No, the contradiction between Darwinism and theism goes much deeper. To know that Darwinism is true (as a general explanation for the history of life), one has to know that no alternative to naturalistic evolution is possible. To know *that* is to assume that God does not exist, or at least that God does not or cannot create. To infer that mutation and selection did the creating because nothing else was available, and then to bring God back into the picture as the omnipotent being who chose to create by mutation and selection, is to indulge in self-contradiction. That is why Darwin and his successors have always felt that theistic evolutionists were missing the point, although they have often tolerated them as useful allies.

Failure to understand that Darwinism is primarily a philosophical rather than an empirical doctrine has made theistic naturalists unduly fearful of incurring what is called the "god of the gaps" problem. This problem arises when we point to some gap in current scientific knowledge, and attribute unexplained events to a divine cause. The better theological position, of course, is that God is responsible for all events, and not just those for which scientific explanations are currently lacking.

Moreover, if science is in general on the right track, many gaps will eventually be filled with satisfactory explanations. In which case God—or rather, His supporters who underestimated the power of science—will have to make an embarrassing retreat. It is largely because of fear of the "god of the gaps" problem that theistic naturalists are reluctant to encourage any challenge to the validity of Darwinism.

The question that needs to be investigated, however, is not whether there are gaps in a fundamentally sound theory that has successfully explained a great deal. It is whether Darwinism is wrong *in principle* in assuming that marvelously complex structures like the human body, or even the bacterial cell, can be built up by an unguided material process. This is a question that deserves unbiased investigation, and not only for theological reasons. If mutation and selection cannot accomplish wonders of creativity, then science is on the wrong track and needs to be brought back to reality. As a start in the right direction, critics need to encourage Darwinists to stop hiding controversial assumptions by declaring them to be facts or incorporating them into the definition of "science," and to start replacing vague words like "evolution" with a precise set of terms that can be used consistently to illuminate the points of difficulty. Nobody on any side of the issue should object to clarifying the issues that way—nobody, that is, who really wants to find out the truth.

Attempts to accommodate theism and Darwinism are inherently futile, but the accommodation of theism and empirical science is quite another matter. In the long run theistic religion has nothing to fear from true science, because both are human understandings of an underlying reality rooted in the same divine source. Moreover, empirical science is limited by its methods, and can only tell us *how* things work rather than whether they were brought into existence in furtherance of a higher purpose. The fact that Darwinists continually claim to have a scientific answer to the question of purpose is itself an indication that they are engaged in something other than a true empirical science.

What kind of science is Darwinism? The best answer to that question is provided in the first chapter of Douglas Futuyma's widely used college evolutionary biology textbook. Professor Futuyma, an able and dedicated defender of Darwinism, explains Darwin's importance in these words:

By coupling undirected, purposeless variation to the blind, uncaring process of natural selection, Darwin made theological or spiritual explanations of the life processes superfluous. Together with Marx's materialistic theory of history and society and Freud's attribution of human behavior to influences over which we have little control, Darwin's theory of evolution was a crucial plank in the platform of mechanism and materialism—of much of science, in short—that has since been the stage of most Western thought.

Darwin, Marx, and Freud. These three giants of materialism head everyone's list of the most influential makers of the twentieth-century mind-set. Today there are still a few Freudians and Marxists left, but even they would be embarrassed to cite Freudianism or Marxism as examples of empirical science. We now know these ideologies for what they always were: imaginative stories told to advance a materialist worldview, and buttressed by a beguiling pretense of scientific methodology. We shouldn't let fear hold us back from investigating whether the same thing is true of Darwinism.

# 18

## Phillip Johnson on Trial: A Critique of His Critique of Darwin

Nancey Murphy

### Introduction

Phillip E. Johnson's recent book, *Darwin on Trial*[1] has attracted a fair amount of attention among conservative Christians. Yet it may create an inaccurate impression of the status of evolutionary biology—an impression that I hope to correct in this article. On the book's dust jacket it is said that Johnson, a professor of law at the University of California at Berkeley, took up the study of Darwinism because he judged the books defending it to be dogmatic and unconvincing. I, at least, find Johnson's own arguments dogmatic and unconvincing. The main reason is that he does not adequately understand scientific reasoning.

Many readers will be impressed, even overawed, by Professor Johnson's credentials. He is not a scientist but a lawyer, who claims that his law career, with "a specialty in analyzing the logic of arguments and identifying the assumptions that lie behind those arguments" well qualifies him for the task (p. 13). The fact that he is a professor at U.C. Berkeley certainly adds to his credibility in the eyes of many. But I wish you would bear with me in a little foolishness (cf. 2 Cor. 11:1). Is he from Berkeley? So am I. One of my doctorates was earned in the philosophy department at U.C. Berkeley, where I specialized in philosophy of science. Is he an expert in critical reasoning? So am I. I teach critical reasoning to seminary students (now at Fuller Theological Seminary) and have just completed a textbook on the subject. Most of my other research

Originally published in *Perspectives on Science & Christian Faith* (1993, vol. 45, no. 1, pp. 26–36).

and writing deals with methodological issues in theology, science, and the relations between the two.[2]

My plan is to describe some of the basic moves in scientific reasoning, and then examine in detail an important (and typical) passage in Johnson's book, explaining why it appears fallacious to one trained in scientific reasoning. Next, I shall describe some recent refinements in philosophers' understanding of scientific reasoning, and use them to describe the sort of study that would be required to make a fair assessment of the scientific standing of evolutionary biology.

Another issue that needs to be addressed is the very nature of science, and how it relates to religion. A bit of history will help us understand some of the positions taken by evolutionary biologists and excuse them from some of Johnson's criticisms.

I shall end with a few remarks on what I take to be the proper attitude for Christians toward evolutionary biology.

Before I proceed to the attack, however, I must say that Johnson's book has many good features. Johnson describes some of the failures and problems faced by evolutionary biology, and provides a valuable critique of popular writings that turn the science of evolutionary biology into an atheistic metaphysical system with many of the trappings of religion.

## Basic Scientific Reasoning

Francis Bacon's description of scientific reasoning has been influential for many years. In brief, he claimed that scientists must first rid their minds of all prejudice and preconceptions, then collect all the facts relevant to the issue at hand, and finally draw inductive inferences from the facts.[3] This view of scientific reasoning is inadequate, however, since it only accounts for our knowledge of observed regularities. An important advance in the philosophy of science of this century was the recognition of what has been called "hypothetico-deductive" reasoning.[4] This kind of reasoning frees science from dependence on direct observation, and accounts for all of our theoretical knowledge. It is called "hypothetical" because it relies on the formation of hypotheses to explain a given set of data or observations. It is called "deductive" because hypotheses must be tested by drawing conclusions from them and seeing if they are corroborated by further observation or experiment. So the test of a hypothesis is not by direct observation (most scientific hypotheses postulate un-observable entities or processes), but by asking what observable con-sequences follow from the hypothesis, and by testing these instead. Another way of putting the matter: a hypothesis is accepted on the basis of its ability to explain observations and results of experiments.

Consider the following analogy drawn from everyday experience. You come home from work and find the front door ajar and muddy tracks leading into the kitchen. You form a hypothesis: the kids were here. But of course, there could be other explanations, such as a prowler. To test the hypothesis, you make predictions based on your knowledge of the children's behavior. For example, you check to see if anyone has been into the cookies, or if their school clothes are on the floor upstairs. If your predictions are confirmed you do not need to see the children to know that your original hypothesis was correct.

So the form of hypothetico-deductive reasoning is as follows:

· We observe $O_1$.
· We formulate a hypothesis (H), which, if true, would explain $O_1$.
· Then we ask, if H is true, what additional observations ($O_2 \ldots O_n$) ought we be able to make?
· Finally, if $O_2$ through $O_n$ are observed, H is confirmed.

It is important to note that $O_2$ through $O_n$ are not equivalent to H; they are observable consequences that we deduce from H with the aid of additional assumptions—nibbled cookies and strewn clothing are not children.

Because the hypothesized entities or processes are unobservable, sci-entists often make use of *models*—observable entities or processes that are similar in important respects to the theoretical entities. A famous example is the billiard-ball model used to understand and account for the behavior of gasses in a closed container. Models are often helpful in deriving testable predictions from hypotheses (theories).

It is also important to note that hypothetical reasoning (like all rea-soning about matters of fact) can never amount to proof. The best that can be hoped for is a high degree of confirmation. Much of what phi-losophy of science is about is examination of the conditions under which a scientific theory can be said to be well-confirmed. So objecting that *any* scientific theory is "not proved" is empty—none can be.

The foregoing provides enough terminology to analyze some of Johnson's arguments, so we turn now to these.

## Johnson on Natural Selection

Chapter Two of *Darwin on Trial* is an examination of the thesis that natural selection, or survival of the fittest, (when combined with natural variation) provides an adequate account of macroevolution—that is, the evolution of all known species of living things from one or a few primitive ancestors. A crucial step in Johnson's overall criticism of evolutionary biology is his assessment of evidence for the efficacy of natural selection, so we must examine this short passage (pp. 17–20) with care. Johnson begins by noting that Darwin could not point to examples of natural selection in action, and so he had to rely heavily on an argument by analogy with artificial selection by breeders of domestic plants and animals.

However, Johnson replies to Darwin's argument as follows:

Artificial selection is not basically the same sort of thing as natural selection but rather is something fundamentally different. Human breeders produce variations among sheep or pigeons for purposes absent in nature, including sheer delight in seeing how much variation can be achieved. If the breeders were interested only in having animals capable of surviving in the wild, the extremes of variation would not exist....

What artificial selection actually shows is that there are definite limits to the amount of variation that even the most highly skilled breeders can achieve. Breeding of domestic animals has produced no new species, in the commonly accepted sense of new breeding communities that are infertile when crossed with the parent group....

In other words, the reason dogs don't become as big as elephants, much less change into elephants is not that we just haven't been breeding them long enough. Dogs do not have the genetic capacity for that degree of change, and they stop getting bigger when the genetic limit is reached. (p. 18)

Next, Johnson turns to evidence cited by contemporary evolutionists:

Darwinists disagree with that judgment, and they have some points to make. They point with pride to experiments with laboratory fruitflies. These have not produced anything but fruitflies, but they have produced changes in a multitude of characteristics. Plant hybrids have been developed which can breed with each other, but not with the parent species, and which therefore meet the accepted standard for new species. With respect to animals, Darwinists attribute the inability to produce new species to a lack of sufficient time.... In some cases, convincing circumstantial evidence exists of evolution that has produced new species in nature. Familiar examples include the hundreds of fruitfly species in Hawaii and the famous variations among "Darwin's Finches" on the Galapagos Islands....

Lack of time would be a reasonable excuse if there were no other known factor limiting the change that can be produced by selection, but in fact selective change is limited by the inherent variability in the gene pool. After a number of generations the capacity for variation runs out. It might conceivably be renewed by mutation, but whether (and how often) this happens is not known. (p. 19)

And now Johnson's conclusion, drawn from the above considerations:

Whether selection has ever accomplished speciation (i.e., the production of a new species) is not the point. A biological species is simply a group capable of interbreeding. Success in dividing a fruitfly population into two or more separate populations that cannot interbreed would not constitute evidence that a similar process could in time produce a fruitfly from a bacterium. If breeders one day did succeed in producing a group of dogs that can reproduce with each other but not with other dogs, they would still have made only the tiniest step towards proving Darwinism's important claims.

That the analogy to artificial selection is defective does not necessarily mean that Darwin's theory is wrong, but it does mean that we will have to look for more direct evidence to see if natural selection really does have a creative effect. (pp. 19–20)

## Analysis

What are we to make of this set of arguments? Before I begin a serious analysis, permit me another bit of foolishness: The series of steps in Johnson's argument recalls an old lawyer's joke about a defendant in a murder trial: "Your honor, I didn't kill him, and besides, it was an accident, and on top of that he really had it coming!" Similarly: artificial selection is not analogous to natural selection, and besides, selective breeders have not produced any new species, and on top of that they have only produced new plant species, but no new animal species.

More seriously, we must ask what observations or results are required to confirm (not prove) the scientific hypothesis that natural selection is capable of producing *radically different new species*.[5] Since we cannot directly observe natural selection at work, we need an observable model. Selective breeding has been proposed. (We will come back to the issue of the suitability of this model below.) What is at stake in testing the power of natural selection, then, is that our analogue to natural selection be shown to accomplish two things: First, we need to see that selection can produce *radical differences* within a population. Second, we need to see that selection can result in *speciation*—the development of one species out of another. The criterion here is incapacity to breed with the parent species.

Johnson seems to believe that both of these effects need to be observed in the same instance. He would have a point if there were something about one effect that precluded the other or made it less likely; for example, from the fact that you can pat your head and can also rub your stomach, I cannot infer with much confidence that you can do both at once. However, this does not appear to be such a case. The splitting of a population into two species isolates the gene pools, allowing them to diverge, and ultimately to manifest different physical characteristics. We can also imagine that a wide enough physical variation within a species would tend to isolate two or more gene pools and provide a necessary though not sufficient condition for speciation. Johnson notes, for example, that while dogs are all theoretically capable of interbreeding, size differences make it practically impossible.

Now, Johnson admits that we have examples of both of these changes as a result of intentional selective breeding. Regarding the first, he would like to see dogs as big as elephants, but the difference between a toy poodle and a great dane seems adequate to me. Regarding the second, there are instances from plant breeding and, he says, circumstantial evidence that many species of fruitflies have developed from one or a few species originally introduced to Hawaii. Yet his conclusion is that none of this is adequate evidence for the Darwinian thesis. In effect, he is claiming that because plant speciation and intra-species variation *are not equivalent to* macroevolution they provide no evidence for the power of natural selection. But recall that we never hoped to *observe* a case of macroevolution by means of natural selection. We were about the more modest task (and the only realistic task) of providing confirmatory evidence by means of a model—an analogous process—that macroevolution by means of natural selection is possible (given sufficient time and enough environmental pressure).

The form of the Darwinian reasoning is as follows:

- $O_1$ is observed (here, the patterns of speciation in existence today).

• A hypothesis (H) is formulated which, if true, would explain $O_1$ (here, the correlative hypotheses of variation, natural selection, and geographical isolation).

• If H is true, what additional observations $(O_2 \ldots O_n)$ ought we be able to make? (here, $O_1$: radical change within a population, and $O_2$: speciation).

• Finally, $O_2$ and $O_3$ have been observed, so H is confirmed.

Again, $O_2$ and $O_3$ are not equivalent to H; they are observable consequences that can be deduced from H with the aid of additional assumptions.

One of the crucial assumptions here is that selective breeding is like natural selection in relevant respects. It *is* like natural selection in that it operates by means of differential reproduction rates and within the variation that nature supplies. These seem to me to be the relevant factors. Johnson's claim that the characteristics breeders look for are different from the ones for which "Nature" selects seems to me beside the point. The issue is whether selective breeding can produce radical changes, including speciation; not the particular nature of those changes.

I believe it could be shown by examining other arguments that Johnson *consistently* fails to distinguish between evidence confirmatory of a hypothesis and a set of observations that together are equivalent to the hypothesis. For example, on pp. 25–7 he first lists six pieces of evidence that have been offered in support of the power of natural selection, then concludes *without explanation* that "none of these 'proofs' provides *any* persuasive reason for believing that natural selection can produce new species ..." (p. 27, emphasis mine). In Chapter 8, discussing theories about the origin of life, he concludes that because the synthesis of some of the components of living organisms does not actually amount to the production of life in the laboratory there is "*no reason* to believe that life has a tendency to emerge when the right chemicals are sloshing about in a soup" (p. 103, emphasis mine).

## Recent Philosophy of Science

I shall introduce in this section some of the refinements contributed by recent philosophers of science by commenting on further aspects of Johnson's arguments.

In the section quoted above, Johnson has said that there are definite limits to the amount of variation that even the most highly skilled breeders can achieve; that dogs do not get as big as elephants because they do not have the genetic capacity for that degree of change (p. 18); and that after a number of generations the capacity for variation runs out (p. 19). He then admits that *mutation* might renew the capacity for change, but claims that whether and how often this happens is not known (p. 19).

When Darwin proposed his theory of evolution, he speculated that there must be a mechanism that works predominantly to maintain the characteristics of a population from one generation to another, but that also allowed for some degree of fluctuation and for genuine novelty. At that time, of course, he did not know what that mechanism was. A great triumph for evolutionary theory, but one Johnson does not mention, came from the discovery of the role of genes in reproduction. The gene pool provides for variation within overall stability in most instances, but mutations allow for genuine novelty.

Johnson mentions mutation as though it is scarcely important at all, but in fact it is an essential "auxiliary hypothesis" for the evolutionary program, and it is simply not possible to draw Johnson's strong conclusions about the limits of variation without considering the frequency and kinds of mutations, and their potential contribution to viable changes in a population.[6]

This fact illustrates an important point stressed by philosophers of science. Theories (hypotheses) rarely or never face the test of experience standing alone. We are (almost) always faced with the testing of whole networks of theories and auxiliary hypotheses. This makes the falsification of a major theory very difficult—when negative evidence comes along, it can often be reconciled with the central theory by adding or changing lower-level (auxiliary) theories. If positive evidence is lacking, its absence can often be explained by the same strategy.

Johnson's book is full of examples of changes of this sort to make evolutionary theory consistent with the evidence (or the absence of evidence). For example, Darwin expected that the fossil record would soon provide evidence of species intermediate between known species and their ancestors (the "missing links"). When few such intermediates were found, later theorists proposed auxiliary hypotheses to explain their absence: for example, the fossil record is still only a small sample of all of the creatures that have existed, and it is to be expected that the intermediate species, being in between well-adapted forms, would not last long and would therefore leave little evidence behind in the form of fossils.

Theorizing of this sort is extremely common in science. Since major theories come along only rarely, most of scientific advance consists in the careful elaboration and qualification of major theories, fine-tuning them to fit the evidence. Several philosophers of science have noted, though, that there is a kind of fine-tuning that represents genuine improvement and growth in scientific knowledge, and another kind that is a mere face-saving device—linguistic tricks to protect a theory from falsification. So the important question is how to tell the difference.

Imre Lakatos (d. 1974) made a major contribution to philosophy of science by providing a criterion for distinguishing "progressive" from "degenerative" or "*ad hoc*" refinements of a network of theories.[7] The essence of his criterion is this: if a hypothesis that is added to the network not only explains the problems for which it was designed, but also leads to the prediction and corroboration of new facts of a different sort, then the modification is progressive. On the other hand, if it only takes care of the problem and is not independently confirmed by the successful prediction of novel facts, then it is a degenerative move. Lakatos made a double claim about this criterion. First, he claimed that it could account for the history of science better than other views,[8] in that history would show that scientists generally abandon research programs that are making mostly degenerative moves in favor of a more progressive rival. His second claim is that scientists *should* accept progressive programs and abandon degenerative ones—that this is what scientific rationality consists in.

### Application to Darwinism

Now, what consequences does this criterion of "progressiveness" have for evaluating evolutionary theory? It shows, first of all, that the only fair way to assess the program is by examining the auxiliary hypotheses that have been added to it to see whether each is a progressive or degenerative modification.

It is clear that Johnson is aware of the problem of *ad hoc* developments of a theory, as the following passage indicates:

Darwinists have evolved an array of subsidiary concepts capable of furnishing a plausible explanation for just about any conceivable eventuality. For example, the living fossils, which have remained basically unchanged for millions of years while their cousins were supposedly evolving into more advanced creatures like human beings, are no embarrassment to Darwinists. They failed to evolve because the necessary mutations didn't arrive, or because of "developmental constraints," or because they were already adequately adapted to their environment. In short, they didn't evolve because they didn't evolve.

Some animals give warning signals at the approach of predators, apparently reducing their own safety for the benefit of others in the herd. How does natural selection encourage the evolution of a trait for self-sacrifice? Some Darwinists attribute the apparent anomaly to "group selection." Human nations benefit if they contain individuals willing to die in battle for their country, and likewise animal groups containing self-sacrificing individuals may have an advantage over groups composed exclusively of selfish individuals.

Other Darwinists are scornful of group selection and prefer to explain altruism on the basis of "kinship selection." By sacrificing itself to preserve its offspring or near relations an individual promotes the survival of its genes. Selection may thus operate at the genetic level to encourage the perpetuation of genetic combinations that produce individuals capable of altruistic behavior. By moving the focus of selection either up (to the group level) or down (to the genetic level), Darwinists can easily account for traits that seem to contradict the selection hypothesis at the level of individual organisms.

Potentially the most powerful explanatory tool in the entire Darwinist armory is *pleiotropy*, the fact that a single gene has multiple effects. This means that any mutation which affects one functional characteristic is likely to change other features as well, and whether or not it is advantageous depends upon the net effect. Characteristics which on their face appear to be maladaptive may therefore be presumed to be linked genetically to more favorable characteristics, and natural selection can be credited with preserving the package.

I am not implying that there is anything inherently unreasonable in invoking pleiotropy, or kinship selection, or developmental constraints to explain why apparent anomalies are not necessarily inconsistent with Darwinism. If we assume that Darwinism is basically true than it is perfectly reasonable to adjust the theory as necessary to make it conform to the observed facts. The problem is that the adjusting devices are so flexible that in combination they make it difficult to conceive of a way to test the claims of Darwinism empirically. (pp. 29–30)

However, this passage also indicates that Johnson sees no difference between auxiliary hypotheses that are testable and those that are not. It is difficult to conceive a test for the hypothesis that the living fossils failed to evolve because they were already adapted to their environment—or to be more precise, it is hard to conceive of a way of showing this claim *false*. This seems to be an instance of a "linguistic trick" to protect the theory from falsification. But not so with all of the examples Johnson has cited here. For example, kinship selection is testable: if it is true, then there should be a direct relationship between the percentage of genes shared with another individual and the degree of "altruism" exhibited toward that individual—a prediction that has in fact been confirmed. In addition, genetic mapping makes the concept of pleiotropy empirically testable.

So it is clear that Johnson has failed to see the import of such cases. He does not understand their role in demonstrating that there are in fact ways "to test the claims of Darwinism empirically" (p. 30).

In general Johnson has given too little attention to the role genetic theory has played in the history of evolutionary biology. Genetics arose as a major new theory in complete independence of evolutionary biology. Initially there was strong antagonism between workers in the two fields. However, with the advent of population genetics under Fisher, Wright and Haldane, the two fields were reconciled. In Lakatos's terms, the entire genetic program came to function as an "auxiliary hypothesis" within the evolutionary program, providing a tremendous amount of fresh empirical evidence—evidence of exactly the sort that Lakatos has led us to expect from a progressive program. Another instance is "neutral allele" theory, with its associated phenomenon of molecular clocks.

Much remains to be done to provide an adequate assessment of the evolutionary program. There are a number of problems with the theory, but whether there are more than with other major theories, such as Big-Bang cosmology, remains to be seen. It must be emphasized, though, that the mere existence of problems does not disqualify a theory—good theories are always in process, and the question is whether they are progressing, overall, or degenerating. So the important question is *how* the evolutionary program deals with its problems; whether the auxiliary hypotheses needed to account for anomalies—for the absence of certain kinds of expected confirmatory evidence—can be independently tested and confirmed. Johnson does not pursue this question; nor can I do so here. Adequate treatment would require another book. But this is where the battle must be joined if we are to have a fair assessment of the evolutionary program.

It has been noted[9] that the kind of "novel facts" needed to provide independent confirmation of auxiliary hypotheses are usually rare, and get harder rather than easier to find as a program progresses. This suggests that the crucial evidence for evolutionary theory, if it can be produced, will not be massive. It will consist in a few confirmed predictions here and there. In this way, evolutionary biology will be entirely in line with many well-respected programs such as Big-Bang cosmology.

A major problem for anyone undertaking an assessment of the evolutionary program is that philosophy of science provides criteria for relative rather than absolute assessment. That is, the criteria we have been discussing are only capable of telling us which of two or more competing programs is the most acceptable. While there is competition within the evolutionary program, between punctuated equilibrium and gradualist theories of change, for instance, there is no major scientific competitor for the program as a whole. This being the case, there are limits to what critics of Darwinism can hope to accomplish. When a theory is the only one available, the burden of proof falls on those who wish to do away with it. It is simply a fact of the history of science that a theory is seldom—perhaps never—abandoned when there is no competitor to take its place. If criteria for rational choice are necessarily comparative, then this is a rational way to proceed. Beyond that, there is the practical question: what would evolutionary biologists *do* if there is no other conception of the field to guide their research?

## The Nature of Science

In this section I shall take up three issues raised by Johnson:

1. Evolutionary biology is not scientific because (according to Karl Popper) science is characterized by falsifiability, and the central ideas of Darwinism are held dogmatically.
2. Evolutionary theory is held dogmatically because it is the only account of life that fits with a naturalistic philosophy.
3. Evolutionary biologists ought to consider the possibility that life is the product of creative intelligence.

### Science and Falsifiability

In his final chapter Johnson adopts Karl Popper's criterion for distinguishing science from pseudoscience. Popper argued that what made science scientific was not its subject matter but the willingness of its proponents to allow their theories to be falsified.[10] In Johnson's words: "Progress is made not by searching the world for confirming examples, which can always be found, but by searching out the falsifying evidence that reveals the need for a new and better explanation" (p. 147).

Imre Lakatos was a colleague of Popper's at the London School of Economics. Lakatos treated Popper's claims about the natural of science as an empirical theory and argued that, as such, the history of science *falsified* Popper's account. His own theory, introduced above, was proposed as a "new and better explanation" of the course of the history of science. We have already seen his proposed criterion for distinguishing between acceptable and unacceptable (progressive and degenerating) research programs. Here it is relevant to introduce another feature of his account of science.

All scientific research programs, he concluded, include a central idea, called the hard core, which is usually too vague to be tested directly. In addition, there are the auxiliary hypotheses that mediate between the core theory and empirical data. Lakatos's study of the history of science convinced him that a certain amount of dogmatism with respect to the core of a program was both a regular feature of good science and a necessary strategy to allow for the development of scientific thought. His new version of falsificationism allows researchers to protect their core theory "dogmatically" so long as the program is progressive overall.[11]

From what has just been reported,[12] it follows that Johnson's criticism in the following quotation shows *not* the unscientific character of evolutionary biology, but rather that Johnson approaches it with an inadequate understanding of the philosophy of science:

The central Darwinist concept that later came to be called the "fact of evolution"—descent with modification—was thus from the start protected from empirical testing. Darwin did leave some important questions open, including the relative importance of natural selection as a mechanism of change. The resulting arguments about the process, which continue to this day, distracted attention from the fact that the all important central concept had become a dogma. (p. 149)

That is, the usual strategy in science is to hold on to a central idea—hold it "dogmatically," if you will—so long as the theoretical elaborations and additions that are necessary to reconcile it with the evidence lead to new discoveries rather than to blind alleys.

### Evolution and Naturalism
Johnson explains the evolutionists' dogmatism by attributing it, not to the usual processes of scientific development, but to an atheistic philosophical

naturalism. Johnson is quite right about this in some cases, and perhaps in most of the cases of *popular* books written in defense of evolution.

However, a subtle distinction needs to be made here. On the one hand there are the proponents of "a religion of scientific naturalism, with its own ethical agenda and plan for salvation through social and genetic engineering" (Johnson, p. 150). This religion is fair game for criticism by proponents of other religions, and ought not be allowed *establishment* in the curriculum of the public schools. On the other hand, there is what we might call *methodological atheism*, which is by definition common to all natural science. This is simply the principle that scientific explanations are to be in terms of natural (not supernatural) entities and processes.

Johnson is critical of biologists and philosophers who define science in this way. However, it is a fact of history (perhaps an accident of history) that this is how the institution of *natural* science is understood in our era. It is ironic, perhaps, that Isaac Newton and Robert Boyle, two of the scientists who led the move to exclude all natural theology from science (then called "natural philosophy") did so for *theological* reasons. Their Calvinist doctrine of God's transcendence led them to make a radical distinction between God the Creator and the operation of the created universe, and hence to seek to protect *theology* from contamination *by science*. The metaphysical mixing of science and religion, Boyle and Newton believed, corrupted true religion.[13]

So, for better or for worse, we have inherited a view of science as *methodologically* atheistic—meaning that science *qua* science, seeks naturalistic explanations for all natural processes. Christians and atheists alike must pursue scientific questions in our era without invoking a creator. The conflict between Christianity and evolutionary thought only arises when scientists conclude that if the only *scientific* explanation that can be given is a chance happening, then there is no other explanation at all. Such a conclusion constitutes an invalid inference from a statement expressing the limits of scientific knowledge to a metaphysical (or a-religious or anti-religious) claim about the ultimate nature of reality.

This is a subtle difference—one beyond the grasp of a fourth-grade science class (and perhaps beyond the grasp of some outspoken scientific naturalists as well?). For this reason I am sympathetic with Christians who object to the teaching of evolution in the public schools. But the answer is to help educators make the distinction, not to cooperate in blurring it as Johnson has done.[14]

### Creative Intelligence as a Scientific Hypothesis
Johnson writes:

> Why not consider the possibility that life is what it so evidently seems to be, the product of creative intelligence? Science would not come to an end, because the task would remain of deciphering the languages in which genetic information is communicated, and in general finding out how the whole system *works*. What scientists would lose is not an inspiring research program, but the illusion of total mastery of nature. They would have to face the possibility that beyond the natural world there is a further reality which transcends science. (p. 110)

The answer to Johnson's question is that anyone who attributes the characteristics of living things to creative intelligence has by definition stepped into the arena of either metaphysics or theology. Some might reply that the definition of science, then, needs to be changed. And perhaps it would be better if science had not taken this particular turn in its history. Could the nature of science change again in the near future to admit theistic explanations of natural events? There are a number of reasons for thinking this unlikely. A practical reason is the fact that much of the funding for scientific research in this country comes from the federal government. The mixing of science and religion would raise issues of the separation of church and state.

A second reason for thinking such a change unlikely is that many Christians in science, philosophy, and theology are still haunted by the idea of a "God of the gaps." Newton postulated divine intervention to adjust the orbits of the planets. When Laplace provided better calculations, God was no longer needed. Many Christians are wary of invoking divine action in any way in science, especially in biology, fearing that science will advance, providing the naturalistic explanations that will make God appear once again to have been an unnecessary hypothesis.

### Concluding Remarks

What, then, of the relation between Christianity and Darwinism? I hope I have made it clear that this question is ambiguous. One question is: How ought Christianity be related to evolutionary biology—the pure science? The other is: How ought Christianity be related to evolutionary metaphysics? The latter system of thought involves the use of scientific theory to legitimate a metaphysical-religious point of view, and it has been so successful that many cannot imagine Christian thought making its own, different use of biology. Nonetheless it can be done, and it has been done by the likes of biochemisttheologian Arthur Peacocke.[15]

Peacocke notes that the sciences can be organized in a hierarchy, with higher sciences studying more complex levels of organization in reality. For example, chemistry studies more complex organizations of matter than does physics; biochemistry more so than inorganic chemistry; within biology alone there is a hierarchy as we move from biochemistry to the study of cells, to tissues, organs, and finally to the functioning of entire organisms within their environments.

Each science has its recognized domain, and concepts and theories appropriate to its own level of interest. Yet each science is conditioned by the levels above and below. Lower levels set limits on the behavior of entities at higher levels—for example, chemical processes in nerves and muscles set limits on how high or fast an animal can jump. However, lower levels do not uniquely determine the behavior of entitles at higher levels—here one also has to take account of the environment. Thus, the animal's particular movements within the range permitted by chemistry and physics will be to some extent conditioned by ecological factors as well.

So any science alone provides an incomplete account of reality; it finds limits above and below, beyond which its explanatory concepts cannot reach. But what about the limits of the highest (or lowest) science in the hierarchy? Peacocke proposes that at the top of the hierarchy of the sciences we reach theology, the science that studies the most complex system of all—the interaction of God and the whole of creation.[16]

Peacocke's suggestion provides the groundwork for an exciting account of the relations between the sciences and theology. We can examine the kinds of relations that hold between two hierarchically ordered sciences, and then look for analogous relations between theology and one or more sciences. One relation we may expect to find is that when a science reaches an inherent limit, there may be a role for theology to play at that point. For example, it *may* be inherently impossible for science to describe what happened "before" the Big Bang.

Peacocke's understanding of the relation between science and theology means that we need not turn biology into theology, but we may and must bear in mind that there is a discipline "above" biology that answers questions that biology alone cannot answer. Is this discipline to be an atheistic metaphysic that elevates "Chance" to the role of ultimate explanation, or is it to be a theology of benevolent Design? The question calls for a careful comparison of the explanatory force of these two competing accounts of reality. The former has to explain (or explain away) all appearances of order and purpose; the latter has to explain a number of features of the world that (as biologists correctly point out) appear inconsistent with intelligent design.[17]

It looks to many as though these two explanatory systems are at a stand-off. For every feature that appears to be the product of design, there is another that appears to be the product of chance. However, I suspect that the design hypothesis, as the core of the theological research program, could be shown to be *more progressive* (in Lakatos's sense) than a research program based on chance. My guess is that while the atheistic program could explain (or explain away) all the evidence for design, it will have to do so by means of an assortment of *ad hoc* hypotheses. Besides this, the Christian program has at its disposal additional supporting evidence from a variety of domains: religious experience, history, the human sciences.

So there are two issues before us, both of which cry out for much more extensive and careful treatment than I have given them here: First, what is the true standing of evolutionary biology *as a science* and measured against the best criteria that have so far been proposed for evaluating scientific acceptability (truth). I make two claims with regard to Johnson's book: first, that he has allowed the Evolutionary Naturalists to confuse evolutionary science with something else and, second, that he has used too primitive a view of scientific methodology for his evaluation. I do not claim to have definitively refuted his claims against evolutionary science, but I hope to have undermined them, and to have shown the direction a definitive evaluation of evolutionary biology would have to take.

The second big issue is the clash of world views: evolutionary naturalism versus Christianity; Chance with a capital "C" versus Design. Settling this controversy is well beyond the capability of any single scholar on either side, but we do educators, school children, and perhaps even evolutionary biologists a great favor by carefully distinguishing this issue from the first.

An important effect of separating the theological-metaphysical issue from the scientific one may be to lessen the anxiety and heat of controversy that surrounds the latter. We want scientists to stop their attacks on Christianity, but all Bible-readers should know that the cessation of hostilities is not to be left to our opponents. Better to turn away wrath with a gentle word.[18]

## Notes

1. InterVarsity, 1991.

2. This sort of credential swapping is quite out of place in academic writing, but nonetheless it deserves a name. In practical reasoning, some arguments are called *ad hominem* (to or against the man); this argument I shall dub an *ab femina* argument (from the woman).

3. This view of science has been particularly influential in conservative American Christian circles. John Witherspoon promoted Bacon's views among the Princeton theologians, such as Charles Hodge, who have influenced American Fundamentalism. It is described and criticized at somewhat greater length by Johnson, pp. 146–47.

4. This term was coined by Carl Hempel. See his *Philosophy of Natural Science* (Prentice-Hall, 1966).

5. Actually, we are asking more of natural selection here than is required by the theory. Darwinian theory does not require that natural selection be directly responsible for reproductive isolation. The classical theory is that geographical isolation, followed by differential adaptation to different conditions, is the principal agent of speciation.

6. Johnson does take up this issue in the following chapter. My point is that the conclusions he draws in this chapter regarding the limits of variation are quite unwarranted because they cannot be made *independently* of the assessment of the possibilities for mutations.

7. See "Falsification and the Methodology of Scientific Research Programmes," in J. Worrall and G. Currie, eds., *The Methodology of Scientific Research Programmes: Philosophical Papers, Volume 1* (Cambridge University Press, 1978), pp. 8–101.

8. Such as Karl Popper's falsificationism.

9. By Alan Musgrave, in "Logical vs. Historical Theories of Confirmation," *British Journal for the Philosophy of Science* 25 (1974): 1–23.

10. Popper first elaborated this thesis in *Logik der Forschung* (Vienna, 1935); English translation, *The Logic of Scientific Discovery* (Harper, 1965).

11. There is insufficient space here to show that Lakatos's understanding of science is superior to Popper's. See my *Theology in the Age of Scientific Reasoning* (Cornell University Press, 1990), chapter 3; as well as Lakatos's "Falsification

and the Methodology of Scientific Research Programmes," op. cit.; and especially his "History of Science and Its Rational Reconstructions," also in *The Methodology os Scientific Research Programmes*, op. cit., pp. 102–138.

12. The same point is made by Thomas Kuhn in *The Structure of Scientific Revolutions* (University of Chicago, 1970); Paul Feyerabend in *Against Method* (New Left Books, 1975); and Larry Laudan in *Progress and its Problems* (University of California Press, 1977).

13. See Eugene Klaaren, *Religious Origins of Modern Science* (Eerdmans, 1977); and Frank Manuel, *A Portrait of Isaac Newton* (Harvard University Press, 1968).

14. For an excellent discussion of this and other issues, see Howard Van Till, Robert Snow, John Stek, and Davis Young, eds., *Portraits of Creation* (Eerdmans, 1990).

15. See *Creation and the World of Science* (Clarendon, 1979); *Intimations of Reality* (University of Notre Dame Press, 1985); or *Theology for a Scientific Age* (Basil Blackwell, 1990).

16. I elaborate and apply this view of the hierarchy of the sciences and their relation to theology in "Evidence of Design in the Fine-Tuning of the Universe," in Robert Russell, Nancey Murphy, and Chris Isham, eds., *Quantum Cosmology and the Laws of Nature: Scientific Perspectives on Divine Action* (The Vatican Observatory, forthcoming).

17. Peacocke's view is that God creates through exploration of the possibilities provided by chance as well as through lawgoverned design.

18. I wish to thank Philip Spieth at the University of California, Berkeley for helpful comments on an earlier draft of this paper.

# 19

# Welcoming the "Disguised Friend"— Darwinism and Divinity

Arthur Peacocke

Darwinism appeared, and, under the disguise of a foe, did the work of a friend. It has conferred upon philosophy and religion an inestimable benefit, by showing us that we must choose between two alternatives. Either God is everywhere present in nature, or He is nowhere.

—Aubrey Moore, in the 12th edition of Lux Mundi, 1891, p. 73

It would, no doubt, come as a surprise to many of the biologically cultured "despisers of the Christian religion," to learn that, as increasingly thorough historical investigations are showing,[1] the nineteenth-century reaction to Darwin in theological and ecclesiastical circles was much more positive and welcoming than the legends propagated by both popular and academic biological publications are prepared to admit. Furthermore, the scientific reaction was also much more negative than usually depicted, those sceptical of Darwin's ideas including initially *inter alia* the leading comparative anatomist of his day, Richard Owen (a Cuverian), and the leading geologist, Charles Lyell. Many theologians deferred judgment, but the proponents of at least one strand in theology in nineteenth-century England chose to interwine their insights closely with the Darwinian—I refer to that "catholic" revival in the Church of England of a stress on the doctrine of the Incarnation and its extension into the sacraments and so of a renewed sense of the sacramentality of nature and God's immanence in the world. More of the nineteenth-century theological reaction to Darwin was constructive and reconciling in temper than practically any biological authors today will allow.

Originally published in *The Idreos Lectures* (1997, pp. 5–23).

That is perhaps not surprising in view of the background to at least T. H. Huxley's aggressive propagation of Darwin's ideas and his attacks on Christianity, namely that of clerical restriction on, and opportunities for, biological scientists in England in the nineteenth century. His principal agenda was the establishment of science as a profession independent of ecclesiastical control—and in this we can sympathise. So it is entirely understandable that the present, twentieth-century *zeitgeist* of the world of biological science is that of viewing "religion" as the opposition, if no longer in any way a threat. This tone saturates the writings of the biologists Richard Dawkins and Stephen Gould and many others—and even philosophers such as Daniel Dennett.

This has polarised the scene, but what I do find even more surprising, and less understandable, is the way in which the "disguised friend" of Darwinism, more generally of evolutionary ideas, has been admitted (if at all) only grudgingly, with many askance and sidelong looks, into the parlours of Christian theology. (Only last year did the Pope finally admit that evolution is "more than a hypothesis"). I believe it is vital for this churlishness to be rectified in this last decade of the twentieth century if the Christian religion (indeed any religion) is to be believable and have intellectual integrity enough to command even the attention, let alone the assent, of thoughtful people in the beginning of the next millenium.

## Biological Evolution and God's Relation to the Living World

Let us consider various features and characteristics of biological evolution and any reflections about God's relation to the living world to which they may give rise.

### Continuity and Emergence

A notable aspect of the scientific account of the natural world in general is the seamless character of the web that has been spun on the loom of time: the process appears as continuous from its cosmic "beginning," in the "hot big bang," to the present and at no point do modern natural scientists have to invoke any non-natural causes to explain their observations and inferences about the past.

The processes that have occurred can be characterized as one of *emergence*, for new forms of matter, and a hierarchy of organization of these forms themselves, appear in the course of time. To these new organizations of matter it is, very often, possible to ascribe new levels of what can only be called "reality": In other words new kinds of reality may be said to "emerge" in time. Notably, on the surface of the Earth, new forms of *living* matter (that is, living organisms) have come into existence by this continuous process—that is what we mean by evolution.

What the scientific perspective of the world, especially the living world, inexorably impresses upon us is a *dynamic* picture of the world of entities and structures involved in continuous and incessant change and in process without ceasing. The scientific perspective of a cosmos, and in particular that of the biological world, as in development all the time must re-introduce into our understanding of God's creative relation to the world a dynamic element which was always implicit in the Hebrew conception of a "living God," dynamic in action—even if obscured by the tendency to think of "creation" as an event in the past. Any notion of God as Creator must now take into account, that God is continuously creating, continuously giving existence to, what is new; that God is *semper Creator*; that the world is a *creatio continua*. The traditional notion of God *sustaining* the world in its general order and structure now has to be enriched by a dynamic and creative dimension—the model of God sustaining and giving continuous existence to a process which has an inbuilt creativity, built into it by God. God is creating at every moment of the world's existence in and through the perpetually-endowed creativity of the very stuff of the world. God indeed makes "things make themselves," as Charles Kingsley put it in *The Water Babies*.

Thus it is that the scientific perspective, and especially that of biological evolution, impels us to take more seriously and more concretely than hitherto the notion of the immanence of God-as-Creator—that God is the Immanent Creator *creating in and through the processes of the natural order*. I would urge that all this has to be taken in a very strong sense. If one asks where do we see God-as-Creator during, say, the processes of biological evolution, one has to reply: "The processes themselves, as unveiled by the biological sciences, *are* God-acting-as-Creator, God *qua* Creator." (This is not pantheism for it is the *action* of God that is identified with the creative processes of nature, not God's own self.) The processes are not themselves God, but the *action* of God-as-Creator. God gives existence in divinely created time to a process that itself brings forth the new: thereby God is creat*ing*. This means we do not have to look for any extra supposed gaps in which, or mechanisms whereby, God might be supposed to be acting as Creator in the living world.

The model of musical composition for God's activity in creation is here, I would suggest, particularly helpful. There is no doubt of the "transcendence" of the composer in relation to the music he or she creates—the composer gives it existence and without the composer it would not be at all. So the model properly reflects, as do all those of artistic creativity, that transcendence of God as Creator of all-that-is which, as the "listeners" to the music of creation, we wish to aver. Yet, when we are actually listening to a musical work, say, a Beethoven piano sonata, then there are times when we are so deeply absorbed in it that, for a moment we are thinking Beethoven's musical thoughts with him. In such moments the

*music is heard so deeply*
*That it is not heard at all, but you are the music*
*While the music lasts*[2]

Yet if anyone were to ask at that moment, "*Where* is Beethoven now?" —we could only reply that Beethoven-*qua*-composer was to be found only in the music itself. The music would in some sense be Beethoven's inner musical thought kindled in us and we would genuinely be encountering Beethoven-*qua*-composer. This very closely models, I am suggesting, God's immanence in creation and God's self-communication in and through the processes by means of which God is creating. The processes revealed by the sciences, especially evolutionary biology, are in themselves God-acting-as-Creator. There is no need to look for God as some kind of *additional* factor supplementing the processes of the world. God, to use language usually applied in sacramental theology, is "in, with and under" all-that-is and all-that-goes-on.

### The Mechanism of Biological Evolution

There appear to be no serious biologists who doubt that natural selection is a factor operative in biological evolution—and most would say it is by far the most significant one. At one end of the spectrum authors like Richard Dawkins argue cogently for the all-sufficiency of natural selection in explaining the course of biological evolution. However, other biologists are convinced that, even when all the subtleties of natural selection are taken into account, it is not the whole story and some even go so far as to say that natural selection alone cannot account for speciation, the formation of distinctly new species. Some other considerations which, it is claimed, are needed to be taken into account are said to be: the "evolution of evolvability"; the constraints and selectivity effected by self-organizational principles; how an organism might evolve is a consequences of itself; "genetic assimilation"; the innovative behaviour of an individual living creature in a particular environment; "top-down causation" (or "whole-part constraint") in evolution through a flow of information; much molecular evolutionary change is immune to natural selection; *group* selection (after all!); long-term changes resulting from "molecular drive" (gene-hopping); effects of the context of adapative change and *stasis*.

What is significant about all these proposals is that they are all operating entirely within a naturalistic framework and assume a basically Darwinian process to be operating, even when they disagree about its speed and smoothness. That being so, it has to be recognized that the history of life on earth involves chance in a way unthinkable before Darwin. There is a creative interplay of "chance" and law apparent in the evolution of living matter by natural selection.

The original mutational events are random with respect to the future of the biological organism, even its future survival, but these changes have their consequences in a milieu that has regular and lawlike features. For the biological niche in which the organism exists then filters out, by the processes of natural selection, those changes in the DNA that enable the organisms possessing them to produce more progeny.

The interplay between "chance," at the molecular level of the DNA, and "law" or "necessity" at the statistical level of the population of organisms tempted Jacques Monod, in *Chance and Necessity*[3] to elevate "chance" to the level almost of a metaphysical principle whereby the universe might be interpreted. As is well known, he concluded that the "stupendous edifice of evolution" is, in this sense, rooted in "pure chance" and that *therefore* all inferences of direction or purpose in the development of the biological world, in particular, and of the universe, in general, must be false.

The responses to this thesis and attack on theism came mainly, it is interesting to note, from theologically informed scientists, and some philosophers, rather than from theologians. They have been well surveyed by D. J. Bartholomew[4] and their relative strengths and weaknesses analysed. I shall here pursue what I consider to be the most fruitful line of theological reflection on the processes that Monod so effectively brought to the attention of the twentieth century—a direction that I took[5] in response to Monod and which has been further developed by the statistically informed treatment of Bartholomew.

There is no reason why the randomness of molecular event in relation to biological consequence has to be given the significant metaphysical status that Monod attributed to it. The involvement of what we call "chance" at the level of mutation in the DNA does not, of itself, preclude these events from displaying regular trends and manifesting inbuilt propensities at the higher levels of organisms, populations and eco-systems. To call the mutation of the DNA a "chance" event serves simply to stress its randomness with respect to biological consequence. As I have earlier put it (in a response later supported and amplified by others):

Instead of being daunted by the role of chance in genetic mutations as being the manifestation of irrationality in the universe, it would be more consistent with the observations to assert that the full gamut of the potentialities of living matter could be explored only through the agency of the rapid and frequent randomization which is possible at the molecular level of the DNA.[6]

This role of "chance," or rather randomness (or "free experiment") at the micro-level is what one would expect if the universe were so constituted that all the potential forms of organizations of matter (both living and non-living) which it contains might be thoroughly explored. This interplay of chance and law is in fact creative within time, for it is the combination of the two which allows new forms to emerge and evolve—so that natural selection appears to be opportunistic. As in many games, the consequences of the fall of the dice depend very much on the rules of the game.[7] It has become increasingly apparent that it is chance operating within a law-like framework that is the basis of the inherent creativity of the natural order, its ability to generate new forms, patterns and organizations of matter and energy. If all were governed by rigid law, a repetitive and uncreative order would prevail: if chance alone ruled, no forms, patterns or organizations would persist long enough for them to have any identity or real existence and the universe could never have been a cosmos and susceptible to rational inquiry. It is the combination of the two which makes possible an ordered universe capable of developing within itself new modes of existence.

This combination for a theist, can only be regarded as an aspect of the God-endowed features of the world. The way in which what we call "chance" operates within this "given" framework to produce new structures, entities and processes can then properly be seen as an eliciting of the potentialities that the physical cosmos possessed *ab initio*. One might say that the potential of the "being" of the world is made manifest in the "becoming" that the operation of chance makes actual. God is the ultimate ground and source of both law ("necessity") and "chance."

To return to our musical model—for a theist, God must now be seen as acting rather like a composer extemporising a fugue to create in the world *through* what we call "chance" operating within the created order, each stage of which constitutes the launching pad for the next. The Creator, it now seems, is unfolding the divinely endowed potentialities of the universe, in and through a process in which these creative possibilities and propensities (see next section), inherent by God's own intention within the fundamental entities of that universe and their interrelations, become actualised within a created temporal development shaped and determined by those selfsame God-given potentialities.[8]

### Trends in Evolution?

Given an immanentist understanding of God's presence "in, with and under" the processes of biological evolution adopted up to this point, can God be said to be implementing any purpose in biological evolution? Or is the whole process so haphazard, such a matter of happenstance, such a matter of what Monod and Jacob called *bricolage* (tinkering), that no meaning, least of all a divinely intended one, can be discerned in the process? However Popper[9] has pointed out that the realisation of possibilities, which may be random, depends on the total situation within which the possibilities are being actualised so that "there exist weighted possibilities which are *more than mere possibilities*, but tendencies or propensities to become real"[10] and that these "propensities in physics are properties of *the whole situation* and sometimes even of the particular way in which a situation changes. And the same holds of the propensities in chemistry, in biochemistry, and in biology."[11] I suggest[12] that the evolutionary process is characterized by propensities towards increase in complexity, information-processing and -storage, consciousness, sensitivity to pain, and even self-consciousness (a necessary prerequisite for social and development and the cultural transmission of knowledge down the generations). *Some* successive forms, along *some* branch or "twig" (à la Gould), have a distinct probability of manifesting more and more of these characteristics. However, the actual physical form of the organisms in which these propensities are actualized and instantiated is contingent on the history of the confluence of disparate chains of events, including the survival of the mass extinctions that have occurred (96% of all species in the Permo-Triassic one).[13] So it is not surprising that recent reinterpretation of the fossils of very early (*ca.* 530 million years ago) soft-bodied fauna found in the Burgess shale of Canada show that, had any a larger proportion of these survived and prevailed, the actual forms of contemporary, evolved creatures would have been very much more disparate in anatomical *plans* than those now observed to exist—albeit with a very great diversity in the few surviving designs.[14] But even had these particular organisms, unique to the Burgess shale, been the progenitors of subsequent living organisms, the same propensities towards complexity, etc., would also have been manifest in their subsequent evolution, for these "propensities" simply reflect the advantages conferred in natural selection by these features. The same considerations apply to the arbtrariness and contingency of the mass extinctions, which Gould also strongly emphasizes. So that, providing there had been enough time, a complex organism with consciousness, self-consciousness, social and cultural organization (that is, the basis for the existence of "persons") would have been likely eventually to have evolved and appeared on the Earth (or on some other planet amenable to the emergence of living organisms), though no doubt with a physical form very different from *homo sapiens*. There can, it seems to me (*pace* Stephen Gould)[15] be overall direction and implementation of divine purpose through the interplay of chance and law without a deterministic plan fixing all the details of the structure(s) of what emerges possessing personal qualities. Hence the emergence of self-conscious persons capable of relating personally to God can still be regarded as an intention of God continuously creating through the processes of that to which God has given an existence of this contingent kind and not some other. It certainly must have been possible since it actually happened—with us!

I see no need to postulate any *special* action of God—along the lines, say, of some divine manipulation of mutations at the quantum level—to ensure that persons emerge in the universe, and in particular on Earth. Not to coin a phrase, "I have no need of that hypothesis!"[16]

### The Diversity of Life

The natural world is immensely variegated in its hierarchies of levels of entities. Structures and processes, in its "being"; and abundantly diversifies with a cornucopian fecundity in its "becoming" in time. From the unity in this diversity and the richness of the diversity itself, one may adduce[17] respectively, both the essential oneness of its source of being, namely the one God the Creator, and the unfathomable richness of the unitive Being of that Creator God. But now we must reckon more directly with the diversity itself. The forms even of non-living matter throughout the cosmos as it appears to us is even more diverse than what we can now observe immediately on the Earth. Furthermore the multiply branching bush (as Gould appropriately calls it) of terrestrial biological evolution appears to be primarily opportunist in the direction it follows and, in so doing, produces the enormous variety of biological life on this planet.

We can only conclude that, if there is a Creator, least misleadingly described in terms of "personal" attributes, then that Creator intended this rich multiformity of entities, structures, and processes in the natural world and, if so, that such a Creator God takes what, in the personal world of human experience, could only be called "delight" in this multiformity of what has been created. The existence of the *whole* tapestry of the created order, in its warp and woof, and in the very heterogeneity and multiplicity of its forms must be taken to be the Creator's intention. We can only make sense of that, utilising our resources of personal language, if we say that God may be said to have something akin to "joy" and "delight" in creation. We have a hint of this in the satisfaction attributed to God as Creator in the first chapter of *Genesis*: "And God saw everything he had made, and behold, it was very good."[18] This naturally leads to the idea of the "play" of God in creation, on which I have expanded elsewhere,[19] in relation to Hindu thought as well as to that of Judaism and Christianity. But now the darker side of evolution.

**The Ubiquity of Pain, Suffering and Death**    The ability for information-processing and storage is indeed the necessary, if not sufficient, condition for the emergence of consciousness. This sensitivity to, this sentience of, its surroundings inevitably involves an increase in its ability to experience pain, which constitutes the necessary biological warning signals of danger and disease. So that it is impossible readily to envisage an increase of information-processing ability without an increase in the sensitivity of the signal system of the organism to its environment. In other words, an increase in "information-processing" capacity, with the advantages it confers in natural selection, cannot but have as its corollary an increase, not only in the level of consciousness, but also in the experience of pain. Insulation from the surrounding world in the biological equivalent of three-inch nickel steel would be a sure recipe for preventing the development of consciousness.

New patterns can only come into existence in a finite universe ("finite" in the sense of the conservation of matter-energy) if old patterns dissolve to make place for them. This is a condition of the creativity of the process—that is, of its ability to produce the new—which at the biological level we observe as new forms of life only through death of the old. For the death of individuals is essential for release of food resources for new arrivals, and species simply die out by being ousted from biological "niches" by new ones better adapted to survive and reproduce in them. So there is a kind of *structural* logic about the inevitability of living organisms dying and of preying on each other—for we cannot conceive, in a lawful, non-magical universe, of any way whereby the immense variety of developing, biological, structural complexity might appear, except by utilizing structures already existing, either by way of modification (as in biological evolution) or of incorporation (as in feeding).[20] The statistical logic is inescapable: new forms of matter arise only through the dissolution of the old; new life only through death of the old. So that biological death of the individual is the prerequisite of the creativity of the biological order, that creativity which eventually led to the emergence of human beings.

But death not only of individuals but of whole species has also occurred on the Earth during the periods of mass extinctions which are now widely attributed to chance extreterrestrial collisions of the planet with comet showers, asteroids or other bodies. These could be cataclysmic and global in their effects and have been far more frequent than previously imagined. This adds a further element of sheer contingency to the history of life on the Earth.

Hence pain, suffering and death, which have been called "natural evil"—the features of existence inimical to biological life, in general, and human flourishing, in particular—appear to be inevitable concomitants of a universe that is going to be creative of new forms, some of which are going to be conscious and self-conscious.

Even so, the theist cannot lightly set aside these features of the created order. The spontaneity and fecundity of the biological world is gained at the enormous price of universal death and of pain and suffering during life,[21] even though individual living creatures, other than humans, scarcely ever commit suicide in any way that might be called intentional.

For any concept of God to be morally acceptable and coherent, the ubiquity of pain, suffering and death as the means of creation through biological evolution entails, that if God is also immanently present in and to natural processes, in particular those that generate conscious and self-conscious life, then we cannot but infer that—in some sense hard to define—*God*, like any human creator, suffers in, with and under the creative processes of the world with their costly unfolding in time.

Rejection of the notion of the impassibility of God has, in fact, been a feature of the Christian theology of recent decades. There has been an increasing assent to the idea that it is possible "to speak consistently of *a God who suffers eminently and yet is still God, and a God who suffers universally....*"[22]

As Paul Fiddes points out in his survey and analysis of this change in theological perspective, among the factors that have promoted the view that God suffers are new assessments of "the meaning of love [especially, the love of God], the implications of the cross of Jesus, the problem of [human] suffering, and the structure of the world."[23] It is this last-mentioned—the "structure of the world"—on which the new perspectives of the biological sciences bear by revealing the world processes to be such, as described above, that involvement in them by the immanent Creator has to be regarded as involving suffering on the Creator's part. God, we find ourselves having tentatively to conjecture suffers the "natural" evils of the world along with ourselves because—again we can only hint at this stage—God purposes to bring about a greater good thereby, namely, the kingdom of free-willing, loving persons in communion with God and with each other.[24]

**Other Aspects of Human Evolution**    That *homo sapiens* represents only an extremely small, and very recently arrived, fraction of living organisms that have populated the Earth raises the question of God's purposes in creating such a labrynth of life. To assume it was all there simply to lead to us clearly will not do. Hence the attribution to God of a sheer exuberance in creativity for its own sake, to which we have already referred. Moreover, since biological death was present on the Earth long before human beings arrived on the scene and was the pre-requisite of our coming into existence through the processes of biological evolution, when St Paul says that "sin pays a wage, and the wage is death"[25] that cannot possibly mean for us now *biological* death. It can only mean "death" in some other sense, such as the death of our relation to God consequent upon sin. I can see no sense in regarding biological death as the consequence of that very real alienation from God that is sin, because God had already used biological death as the means for creating new forms of life, including ourselves, long before human beings appeared on the Earth. This means those classical Christian formulations of the theology of the redemptive work of Christ that assume a causal connection between biological death and sin urgently need reconsidering.

Moreover, the biological-historical evidence is that human nature has emerged only gradually by a continuous process from other forms of primates and that there are no sudden breaks of any substantial kind in the sequences noted by palaeontologists and anthropologists. Moreover there is *no* past period for which there is reason to affirm that human beings possessed moral perfection existing in a paradisal situation from which there has been only a subsequent decline. We appear to be rising beasts[26] rather than fallen angels! Although there was no golden age, no perfect past, no original perfect, individual "Adam" from whom all human beings have now declined, what *is* true is that humanity manifests aspirations to a perfection not yet attained, a potentiality not yet actualized, but no "original righteousness." Sin as alienation from God, humanity and nature is real and appears as the consequence of our very possession of that *self*-consciousness which always places ourselves at the egotistical centre of the "universe" created by our consciousness. Sin is about a falling short from, about not having realised, what God intends us to be and is part and parcel of our having evolved into self-consciousness, freedom and intellectual curiosity. The domination of Christian theologies of redemption by classical conceptions of the "Fall" urgently needs, it seems to me, to be rescinded and what we mean by redemption to be rethought if it is to make any sense to our contemporaries.

Now, we all have an awareness of the tragedy of our failure to fulfil our highest aspirations; of our failure to come to terms with finitude, death and suffering; of our failure to realise our potentialities and to steer our path through life. Freedom allows us to make the wrong choices, so that sin and alienation from God, from our fellow human beings and from nature are real features of our existence. So the questions of not only "Who are we?" but, even more acutely, "What should we be becoming—where should we be going?" remain acute for us. Christians find the clue to the answers to these questions in the person of Jesus of Nazareth and what he manifested of God's perennial expression in creation (as the *Logos* of the cosmos, as "God incarnate"). This leads us to consider specifically Christian affirmations in the light of the foregoing re-considerations of God's creating through an evolutionary process.

### The Significance of Jesus the Christ in a Possible Christian Evolutionary Perspective

... [I]n scientific language, the Incarnation may be said to have introduced a new species into the world—the Divine man transcending past humanity, as humanity transcended the rest of the animal creation, and communicating His vital energy by a spiritual process to subsequent generations ...
—(J. R. Illingworth, in the 12th edition of Lux Mundi, 1891, p. 132.)

Jesus' resurrection convinced the disciples, notably Paul, that it is the union of *his* kind of life with God which is not broken by death and capable of being taken up into God. For he manifested the kind of human life which, it was believed, can become fully life with God, not only here and now, but eternally beyond the threshold of death. Hence his imperative "Follow me" constitutes a call for the transformation of humanity into a new kind of human being and becoming. What happened to Jesus, it was thought, *could* happen to all.

In this perspective, Jesus the Christ, the whole Christ even, has, I would suggest, shown us what is possible for humanity. The actualization of this potentiality can properly be regarded as the consummation of the purposes of God already incompletely manifested in evolving humanity. In Jesus there was a *divine* act of new creation because the initiative was *from God* within human history, within the responsive human will of Jesus inspired by that outreach of God into humanity designated as "God the Holy Spirit." Jesus the Christ is thereby seen, in the context of the whole complex of events in which he participated as the paradigm of what God intends for all human beings, now revealed as having the potentiality of responding to, of being open to, of becoming united with, God. In this perspective, he represents the consummation of the evolutionary creative process which God has been effecting in and through the world.

Darwinism and Divinity may join hands—if only rather tentatively!

## Notes

1. E.g., J. R. Moore, *The Post-Darwinian Controversies: a study of the Protestant struggle to come to terms with Darwin in Great Britain and America* (Cambridge: Cambridge University Press, 1979); J. R. Lucas, "Wilberforce and Huxley: a legendary encounter," *The Historical Journal* 22, 2 (1979): 313–330; J. V. Jensen, "Return to the Huxley-Wilberforce Debate," *Brit. J. Hist. Sci.* 21 (1988): 161–179.

2. T. S. Eliot, "The Dry Salvages," *The Four Quartets*, ll. 210–12.

3. Jacques Monod, *Chance and Necessity* (London: Collins, 1972).

4. David J. Bartholomew, *God of Chance* (London: SCM Press, 1984).

5. A. R. Peacocke, "Chance, Potentiality and God," *The Modern Churchman*, 17 (New Series 1973): 13–23; and in *Beyond Chance and Necessity*, ed. J. Lewis (London: Garnstone Press, 1974,), 13–25; "Chaos or Cosmos," *New Scientist*, 63 (1974): 386–389; A. R. Peacocke, *Creation and the World of Science* [henceforth "CWS"] (Oxford: Clarendon Press, 1979), chap. 3.

6. *CWS*, p. 94; see also by this author, *Theology for a Scientific Age: Being and Becoming-Natural, Divine and Human* [henceforth *TSA*] (SCM Press, London, 1993, 2nd enlarged edit.; repr. 1995), pp. 115–121.

7. R. Winkler and M. Eigen, *Laws of the Game* (New York: Knopf, 1981. London: Allen Lane, 1982).

8. Cf., *CWS*, pp. 105–6 and *TSA*, pp. 115–121.

9. Karl R. Popper, *A World of Propensities* (Thoemmes, Bristol, 1990).

10. Op. cit., p. 12.

11. Op. cit., p. 17.

12. It is interesting to note that Richard Dawkins, too, in his *River Out of Eden* (London: Wiedenfeld and Nicolson, 1995) includes (p. 151ff.) amongst the thresholds that will be crossed *naturally* in "a general chronology of a life explosion on any planet, anywhere in the universe ... thresholds that any planetary replication bomb can be expected to pass, those for:—"

high-speed information-processing (no. 5), achieved by possession of a nervous system; consciousness (no. 6, concurrent with brains); and language (no. 7).

13. Gould, *Wonderful Life: The Burgess shale and the nature of history* (Penguin Books, London, 1989), p. 306, citing David M. Raup.

14. Gould, *op. cit.*, p. 49.

15. Gould, op. cit., p. 51 and *passim*. Since this lecture was given S. Conway Morris, an evolutionary palaeobiologist who has devoted his research life to the study of the Burgess Shale and related formations, has affirmed that Gould has in fact over-emphasised the role of contingency and that his argument is based on a

basic confusion concerning the destiny of a given lineage.... Nearly all biologists agree that convergence is a ubiquitous feature of life. (*The Crucible of Creation: The Burgess Shale and the Rise of the Animals*, Oxford University press, Oxford, 1998, pp. 201, 202)

Again and again we have evidence of biological form stumbling on the same solution to a problem. (*idem*, p. 204)

The reality of convergence suggests that the tape of life, to use Gould's metaphor, can be run as many times as we like and in principle intelligence will surely emerge. (*idem*, p. 14)

16. If there are any such influences by God shaping the direction of evolutionary processes at specific points—for which I see no evidence (how could we know?) and no theological need—I myself could only envisage them as being through God's whole-part constraint on all-that-is affecting the confluence of what, to us, would be independent causal chains. Such specifically directed constraints I would envisage as possible by being exerted upon the whole interconnected and interdependent system of the whole Earth in the whole cosmos which is in and present to God, who is therefore its ultimate Boundary Condition and therefore capable of shaping the occurrence of particular patterns of events, if God chooses to do so. For a fuller exposition of this approach, see *TSA*, pp. 157–165; and, more particularly and recently, "God's Interaction with the World—The Implications of Deterministic "Chaos" and of Interconnected and Interdependent Complexity," in *Chaos and Complexity*, eds. R. J. Russell, N. Murphy and A. R. Peacocke (Vatican City State: Vatican Observatory Publications. Berkeley, California, USA: Centre for Theology and the Natural Sciences. 1995—distrubuted by Univ. of Notre Dame Press.) 263–287.

17. *TSA*, Chap. 8, section 1.

18. *Genesis* 1, v. 31.

19. *CWS*, pp. 108–111.

20. The depiction of this process as "nature, red in tooth and claw" (a phrase of Tennyson's that actually pre-dates Darwin's proposal of evolution through natural selection) is a caricature for, as many biologists have pointed out (e.g., G. G. Simpson in *The Meaning of Evolution* (New Haven: Bantam Books, Yale University Press, 1971 edn., 201), natural selection is not even in a figurative sense the outcome of struggle, as such. Natural selection involves many factors that include better integration with the ecological environment, more efficient utilization of available food, better care of the young, more cooperative social organization—and better capacity of surviving such "struggles" as do occur (remembering that it is in the interest of any predator that their prey survive as a species!).

21. Cf., Dawkins' epithet, "DNA neither cares nor knows. DNA just is. And we dance to its music" (*River out of Eden*, p. 133).

22. Paul S. Fiddes, *The Creative Suffering of God* (Oxford: Clarendon Press, 1988) p. 3 (emphasis in the original).

23. Ibid., p. 45 (see also all of chap. 2).

24. I hint here at my broad acceptance of John Hick's "Irenaean" theodicy in relation to "natural" evil (q.v., "An Irenaean Theodicy," in *Encountering Evil*, ed. Stephen T. Davis [Edinburgh: T. & T. Clark, 1981], pp. 39–52; and his earlier *Evil and the God of Love* [London: Macmillan, 1996], especially chapters 15 and 16) and the position outlined by Brain Hebblethwaite in Chapter 5 ("Physical suffering and the nature of the physical world") in his *Evil, Suffering and Religion* (London: Sheldon Press, 1976).

25. *Rom.* 6 v. 23, REB.

26. Rising from an amoral (and in that sense) innocent state to the capability of moral, and immoral, action.

# 20

## The Creation: Intelligently Designed or Optimally Equipped?

Howard J. Van Till

### The Shaping of Questions and Strategies

#### Matters of Heritage and Context

Each of the contributions to this issue of *Theology Today* has a focus of concern that has been influenced by the author's particular intellectual and spiritual odyssey. My own concerns and priorities have developed in the North American context, in which an attitude of antagonism and mutual mistrust has developed between two important portions of the population: the intellectual/scientific community and the more conservative portion of the Christian community.[1] On the one hand, a large portion of the Christian community has come to adopt a very skeptical attitude toward the results of scientific theorizing regarding the formational history of the universe—theorizing that appears to contradict certain beliefs derived from a particular reading of the biblical text. On the other hand, a large portion of the scientific community has assumed an equally skeptical attitude toward Christianity, in part because the Christian faith is so frequently identified with some form of episodic creationism—a concept of the universe's formational history that is at odds with natural science's reading of the empirical evidence.

Within this North American context, my own theological roots extend deeply into the soil of the Calvinist heritage, especially as it has been expressed in the Netherlands. The names of Abraham Kuyper, Herman Bavinck, and Louis Berkhof can still be heard in our denominational

Originally published in *Theology Today* (1998, vol. 55, no. 3, Oct. 1998, pp. 344–364), discourse. From childhood on, most of us were taught to place high value on thoughtfully articulated theological principles and well-examined philosophical foundations. We were taught by example to look beneath the superficial details of an issue and to examine the presuppositions on which a particular position was founded.

Given that background, as soon as I became familiar with the resurgent creation/evolution debate I had the sense that there was something radically wrong about it and I began to puzzle over just what its shortcomings might be. That puzzlement, along with an intense interest in the natural sciences, ultimately led to my becoming deeply involved, as a Christian trained in physics, in reflections on the character of the creation and of God's creative action.

One of the more obvious shortcomings of the common creation/evolution debate is evident when it is framed as a simple either/or choice between only two comprehensive positions: "episodic creationist theism" and "evolutionary naturalism." Before dealing with the substance of the issues, let me here clarify my use of some important terminology. By *episodic creationist theism* (often called *special creationism* and often presented as entailing a commitment to an Ussher-style, young-earth chronology), I mean to denote the belief that the formation of certain physical structures and life forms now found in the creation was accomplished by occasional episodes of extraordinary divine action in which God imposed those structures and forms on matter. The term *naturalism* will be used in this essay to represent the comprehensive, atheistic worldview based on the presumption that the natural world is all there is. In this worldview, nature is presumed to be a self-existent and closed system that requires no transcendent Creator to act as the source of its being and leaves no opportunity for divine action of any sort within it. More specifically, the term *evolutionary naturalism* here denotes a naturalistic worldview in which the scientific concept of evolutionary development in the formational history of the universe is taken to play a major role in warranting its credibility over against episodic creationist theism.

Presented as a simple either/or choice, the creation/evolution debate suffers the fatal flaw commonly called "the fallacy of many questions." The issue under debate is very broad in scope and includes a lengthy list of questions that span a diversity of categories—scientific methodology, empirical evidence, interpretive strategies, metaphysical presuppositions, theological principles, faith commitments, and the like. However, each of the two contestants in the debate—episodic creationism and evolutionary naturalism—brings but one package of answers to this long list of categorially diverse questions and demands that its package be adopted in its entirety at the exclusion of the other. Such a demand is, of course, grossly unfair. Answers to each of the many questions at issue deserve to be evaluated on their own terms, not to be presented as non-negotiable components of some "no-options package deal."

### The Return of Paley's Argument from Design

Recognizing the shortcomings of young-earth episodic creationism and yet desiring to offer a theistic perspective that would be apologetically effective in theism's engagement with evolutionary naturalism, some Christians have become proponents of an approach they wish to call "intelligent design theory."[2] The basic strategy of the intelligent design movement is as follows: Select and consider, in the light of information drawn from the natural sciences, specific life forms and biotic subsystems. Ask the question, "Can one now, with the science of the day, construct a complete and credible account of how that particular life form or biotic subsystem first came to be actualized in a Darwinian gradualist fashion?[3] If not, the intelligent design theorists argue, then it must be the outcome, not of mindless, purposeless, naturalistic, evolutionary processes, but of "intelligent design." The precise meaning of "intelligent design" is not always apparent, but it most often entails the combination of *both* thoughtful conceptualization *and* the first assembly of a new form by extra-natural means.

In contrast to the young-earth episodic creationist movement with its transparently inadequate treatment of both the biblical text and empirical data, this particular form of skepticism toward evolution demands a more careful critique, some of which will appear later in this essay. As I engage in discussion, both formal and informal, with Christians in search of a defensible position somewhere in the vast conceptual space between young-earth creationism and evolutionary naturalism, I find that many Christians, some scholars included, find the argumentation of intelligent design proponents to be very attractive. My own measure of that position is mixed  positive in regard to some of its elements, negative in regard to others. With them, I reject young-earth creationism for numerous biblical, theological, and scientific reasons. And with them, I reject evolutionary naturalism because of its denial of a Creator-God. However, in contrast to the proponents of intelligent design, I find no warrant for rejecting the possibility (or, stated more strongly, the likelihood) that the creation has been gifted with all of the self-organizational and transformational capabilities needed to make something like the macroevolutionary scenario viable. In fact, I have argued elsewhere that I find a great deal of encouragement for envisioning such a robustly gifted creation in early Christian writings, especially in the reflections of Augustine on the early chapters of Genesis.[4]

## The Goal of Articulating a Well-informed Faith

In all of my reflections on the character of the creation and on divine action within it, my goal is to articulate a perspective that is at once faithful to historic Christian doctrine and well-informed by the natural sciences. Given the way in which the North American community continues to puzzle over both the credibility and the theological relevance of evolutionary theorizing, I continue to search for ways in which faithful Christians might perceive the fruits of divine action expressed in the course of creation's formational history. In so doing, my strategy is twofold: (1) to encourage the conservative Christian community to re-examine its concept of divine creative action and to develop a more appreciative attitude toward scientific investigation regarding the creation's gifts for self-organization and transformation, and (2) to encourage persons outside of the Christian faith to consider a contemporary articulation of historic Christianity that welcomes the fruits of scientific theorizing performed with procedural competence, professional integrity, and intellectual humility.

One of the hazards of this approach is that each of the two communities being addressed is inclined to perceive the message as a threat to its own apologetic strategy. If episodic creationist theism, for instance, perceives its defense of the Christian faith to depend on demonstrating the need for episodes of special creation in the formational history of the universe, then any encouragement to consider evolutionary continuity as an acceptable possibility will appear to constitute a call to "give away the store." And if the proponents of evolutionary naturalism take comfort in their belief that the defeat of episodic creationism constitutes a lasting victory over all forms of Christian theism, then any encouragement to consider the viability of an evolving creation perspective will be viewed as an unwelcome threat to the simple either/or format of the creation/ evolution debate. Herein lies a deep irony of the debate: The two diametrically opposed parties—episodic creationism and evolutionary naturalism —agree to promote the idea that a simple either/or debate is meaningful.

## The Formational Economy of the Universe

### The Need for a Reexamination of Fundamental Questions

In spite of the numerous shortcomings of the creation/evolution debate, it does stimulate us to re-examine some profoundly important questions. For example: In the awareness of what has been learned through modern scientific investigation regarding the manner and history of creaturely action, how might we now speak about the character of divine creative activity and about our apprehension of it? What does God, specifically in the role of Creator, do? And how do we come to be aware of the Creator's action? What are its distinguishing marks? Focusing on the diverse phenomena that comprise the formational history of the Creation, how might we distinguish between the Creator's action and the action of creatures in the actualization of novel structures and life forms in the course of time?

Although there is within the Christian community a diversity of concepts concerning divine action, I presume that there would be general agreement on at least the following propositions regarding God's action as Creator: (1) that as Creator, God gave being to the creation "in the beginning"—often taken to be the beginning of time as we know it; and (2) that God continues to sustain the creation in its being from moment to moment. Were God to cease acting as Creator, the creation would not merely decline in some quality, but it would cease to be. In other words, historic Christian theology sees the existence of the universe to be radically dependent on God's creative action at all times, at this moment no less than at the first moment of its existence. What has traditionally been taken to be the evidence for this creative action? Quite simply, the existence of ourselves and the creation of which we are members. If no Creator, then no creation. In the context of a theistic worldview, the evidence of divine creative action is both obvious and undeniable—we are here.

But there are, of course, alternatives to this line of thought. Naturalistic worldviews, for instance, presume that the universe is self-existent and needs no Creator to serve as the divine Source of its existence. In the context of a commitment to naturalism, then, the existence of the universe is evidence only for the existence of the universe—nothing more to be said. Fortunately for us, it might be argued, the universe happens to exist and we are in it. Furthermore, in the more specific context of *evolutionary* naturalism, the universe that happens to exist also happens to satisfy all of the requisite conditions and to possess all of the requisite capabilities—a truly astounding list, we are beginning to realize—to make possible our formation in time from the more elementary units of matter present in the early universe. We are here, says naturalism, not as the outcome of a Creator's intention, but by a remarkable fortuity.

What rejoinder does Christian theism have against these bold assertions of naturalism? What apologetic strategy, for instance, would effectively engage evolutionary naturalism and, presumably, defeat its claims that we are nothing more than amusing artifacts of the self-organizational powers that the universe just happens to possess? One of the most commonly employed strategies among conservative North American Christians is to counter by asserting that the type of evolutionary development of life forms that evolutionary naturalism presumes to be possible did not, in fact, occur. Why not? Because, it is argued, such continuous formational development is, in actuality, impossible. And if evolution is impossible, then evolutionary naturalism is false and episodic creationist theism, the only other possibility offered in the creation/evolution debate, must be true.

From this formulation (I would classify it as a tragic misformulation) of the issue proceeds the familiar shouting match between episodic creationist theism and evolutionary naturalism. A considerable portion of the debate has been carried out in the court of scientific theorizing and has focused on whether the empirical evidence favors (1) the concept of evolutionary continuity in the formational history of all inanimate structures and life forms, or (2) the concept of radical discontinuities of the sort that could be bridged only by episodes of special creation or of intelligent design. Once the issue is framed in this way, a peculiar—and, I would argue, inverted—scoring system falls into place: Evidence for the functioning of natural processes that make unbroken evolutionary continuity appear plausible is credited to evolutionary naturalism, while the credibility of Christian theism, on the other hand, appears to depend on the absence or inadequacy of certain natural form-producing processes. This apologetic strategy places some Christian scientists in the awkward position of looking for evidence that God has withheld from the creation certain crucial capabilities for self-organization or transformation so that macroevolutionary continuity would be impossible. Searching for evidence of gifts withheld from the creation strikes me as precisely the opposite of what the enterprise of Christian scientific scholarship ought to be doing.

## An Important Definition

In order to see more clearly what the underlying issues are, I have found it helpful to define a concept that I call the "formational economy" of the universe. By the *formational economy of the universe* I mean the set of all of the dynamic capabilities of matter and material, physical, and biotic systems that contribute to the actualization of both inanimate structures and biotic forms in the course of the universe's formational history. Special attention would be drawn to capabilities for self-organization and transformation.

Elementary particles called quarks, for instance, possess the capabilities to interact in such a way as to form nucleons (protons and neutrons). Nucleons, in turn, have the capacities to interact and organize, by such processes as thermonuclear fusion, into progressively larger atomic nuclei. Nuclei and electrons have the dynamic capability to interact and organize into atoms. On the macroscopic scale, vast collections of atoms interact to form the inanimate structures of interest to astronomy —galaxies, stars, and planets. On the microscopic scale, atoms interact chemically to form molecules; molecules interact to form more complex molecules. Some molecular ensembles are presumed to possess the capabilities to organize into the fundamental units that constitute living cells and organisms. Organisms and environments interact and organize into ecosystems. All of these organizational and transformational capabilities together comprise the *formational economy* of the universe.

## The Robust Formational Economy Principle

With this concept defined and named, I believe that we can now approach questions regarding the formational history of the creation in a way that will allow us to rise above the self-perpetuating din of the usual creation/evolution debate. I would begin by posing the question, "What is the character or scope of the creation's formational economy?" More specifically, is the formational economy of the universe sufficiently robust (that is, does it possess all of the requisite capabilities) to make possible the actualization in time of all of the inanimate structures and biotic forms that have ever existed? The natural sciences, as now practiced, presume the answer to be, "Yes." Furthermore, the manner of historical actualization is judged to entail an unbroken continuity of increasingly complex and diverse life forms.

On the other hand, a substantial portion of the North American Christian community presumes the answer to our question to be, "No." Episodic creationism, for instance, is well known for its insistence not only that its concept of special creation episodes is a "clear teaching" of the Bible (conveyed unambiguously by its inerrant text) and one of the fundamental "deliverances of the faith," but also that unbroken evolutionary continuity is physically impossible. Strategies for warranting such beliefs vary considerably, but the claim that evolutionary continuity is impossible is ordinarily grounded in an appeal (by what standards is another question) to the empirical evidence.

Typical argumentation purporting to take a person from empirical evidence to an episodic creationist conclusion would include the following: (1) The fossil record fails to support the idea of Darwinian gradual development of new forms. On the contrary, it is argued, it supports the concept of the "sudden appearance" of new "kinds" of creatures, just as the Bible would lead a faithful reader to expect. (2) The second law of thermodynamics precludes the development of more complex organisms from simpler forms of life or the development of any life form from non-living matter; a variant on this argument would be to assert that the second law precludes the spontaneous generation of the "new genetic information" that any novel life form would necessarily require. (3) The only kind of genetic transformation that can be demonstrated empirically is variation within biblical "kinds," and there is no empirical warrant for extrapolating from these microevolutionary variations to the full-scale macroevolutionary transformation presumed possible by contemporary biology. (4) Finally, according to some proponents of intelligent design theory, it is now possible to point to specific life forms and biotic subsystems that could not possibly have come to be assembled by "natural" means.[5]

One common theme in these and other episodic creationist or intelligent design theorist appeals to empirical science is that it is possible to identify notable gaps in our knowledge about the formational history of the universe. On that point, we must agree. There are indeed *epistemological* gaps to which the episodic creationists and intelligent design theorists can call attention. We do *not* now have the scientific competence to say in full detail and with certainty just how each form of life came to be actualized in the course of time. We do *not* know precisely what role each creaturely capability for self-organization or transformation has played in actualizing the diversity of life forms and biotic subsystems that we now see. Even more seriously, we *cannot* now demonstrate the sufficiency of known creaturely capabilities to make the full macroevolutionary scenario of unbroken genealogical continuity possible. Thus, the episodic creationist is technically correct in saying that there are epistemological gaps of the sort that stand in the way of natural science saying that a detailed theory of biotic evolution has been "proved" in the strict, logical sense of the term.

But the "provability" of a particular evolutionary scenario is not, in fact, the issue. The truth is, as philosophers of science have long been reminding us, that *all* scientific theories are necessarily underdetermined by the empirical data. The word, "proof," in the strict, logical sense, has no place in the world of scientific theory evaluation. There will always be more than one possible scientific account for any natural phenomenon. Epistemological gaps and the unprovability of scientific theories are permanent features of the scientific landscape. The goal of scientific theorizing is not to prove, by appeal to empirical data and unassailable logic alone, one theory correct and all others false. Rather, the goal is to construct a theoretical account that is, in the context of all relevant empirical data at hand and within the bounds of certain presuppositions regarding the character of the universe and its formational history, the most adequate account conceivable at a particular time.

One of the questions of the moment, then, concerns the status of those epistemological gaps. In the context of theorizing about the formational history of the universe, contemporary natural science ordinarily presumes that these gaps in our knowledge could, in principle, be filled at some time in the future. The scientific community fully expects that further research will provide the basis for more adequate and comprehensive theories regarding the formational history of the universe and the life forms that inhabit it. One of the most basic—but seldom explicitly stated—presuppositions of the natural sciences, especially relevant to the formulation of theories regarding the formational history of the universe, is that *the formational economy of the universe is sufficiently robust to make possible the actualization of all inanimate structures and all life forms that have ever appeared in the course of time.* I call this proposition the *robust formational economy principle.* In my judgment, it is not only one of the most fundamental presuppositions of the natural sciences but also the fundamental "sticking point" for a large portion of North American Christians in their assessment of evolutionary theorizing.

## The Role of Apologetics

Why would this principle be seen as a sticking point? Why would it be perceived by millions of North American Christians as being incompatible, say, with Christian belief regarding God's creative action? Is it the case, for instance, that God might be incapable of giving being to a creation so richly gifted with formational capabilities? I would presume that the answer of all Christians would be, "Surely not!" We creatures would have to be arrogant beyond measure (an ever present danger) to declare God incapable of so gifting creation. What, then, is the basis for the widespread Christian rejection of the robust formational economy principle? One major reason for its rejection, as noted above, is the belief that the concept of episodic creation (which presumes the presence of substantial gaps in the creation's formational economy) is both a "clear teaching of the Bible" and a fundamental "deliverance of the faith."

But suppose a person were to set those "in house" concerns aside for the moment (important as they might be to many Christians). Suppose one wished to eschew any appeal either to Scripture or to widely-held Christian beliefs and sought instead to develop an apologetic strategy that appealed only to empirical evidence and sound reasoning, so that even a vocal proponent of naturalism would have to pay attention? This is, I believe, one of the principal goals of the contemporary intelligent design movement. Consequently, to ignore the role of apologetic considerations would be to close one's eyes to a major driving force for the movement.

Now, the goal of effective apologetic engagement with the preachers of naturalism is itself a noble one. However, it is imperative that the apologetic strategy employed in defense of the Christian faith be built on a foundation formed not by the preconceptions of naturalism but by the historic Christian theological heritage. Christian apologetic strategy must be shaped by foundational Christian theological commitments, not by the presuppositions of the opposition. The question is, then, what are the foundational theological propositions on which the intelligent design movement builds its case? Or, to entertain an even more problematic possibility, is their case built not on a consciously-examined theological foundation but simply as a reaction to the offensive naturalistic rhetoric of the day?

## Intelligent Design Theory and Its Apologetic Engagement with Evolutionary Naturalism

### The Naturalistic Challenge

If contemporary intelligent design theory does function primarily as an apologetic reaction to naturalism, what is the nature of the challenge to which it is responding? Specifically, what is the most common form in which the naturalistic challenge to belief in a Creator is presented, and what is the particular concept of divine creative action that it presumes to discredit?

Perhaps the best way to answer this question is to look at an example of the rhetoric of evolutionary naturalism—rhetoric that elicits from many Christians not only an intensified antagonism toward naturalism but also the presumption that ownership of the robust formational economy principle should be ceded to that atheistic worldview. The recent book, *Darwin's Dangerous Idea*,[6] by philosopher Daniel C. Dennett, provides a more than ample supply of such rhetoric.

Dennett focuses much of his attention on the matter of "design." How do things that we see in the world around us, especially living things, he asks, come to exhibit *design*, whether in their internal workings or in their adaptation to the peculiarities of some natural environment? The term "design," as Dennett employs it, functions most commonly as the generic category for any feature of the world, especially of its life forms, that is likely to give an observer the impression (by Dennett's measure, a false impression) of being the outcome of an intentional action by an intelligent agent, such as one would recognize in "a cleverly designed artifact" of human craftsmanship.

Dennett is especially critical of the eighteenth-century style of natural theology with its apologetic strategy of arguing from claims (stated in the conceptual vocabulary of the natural sciences) for the empirical detection of "design" in the universe to the (religiously significant) conclusion of the existence of a Designer.

The overwhelming favorite among purportedly scientific arguments for religious conclusions, then and now, was one version or another of the Argument from Design: among the effects we can objectively observe in the world, there are many that are not (cannot be, for various reasons) mere accidents; they must have been designed to be as they are, and there cannot be design without a Designer; therefore, a Designer, God, must exist (or have existed), as the source of all these wonderful effects.[7]

It is, I believe, important to note here that the concept of "Designer," as it was most commonly employed in the eighteenth century by clergyman William Paley and others, was based on the artisan metaphor. One person, the artisan, did both the conceptualization and the construction of what was intended. Paley's watchmaker, for instance, did both the planning and the fabrication of the watch. Paley's Designer (like his watchmaker) was taken to possess both a mind (to conceptualize, or intend) and the divine equivalent of "hands' (the power to manipulate raw materials into the intended form).

The design concept under Dennett's critical scrutiny entails not only the claim that thoughtful conceptualization would be required for a particular sort of outcome but also the presumption that the actualization of what was first conceptualized would require the action of an "intelligent" agent capable of imposing structure or form on relatively inert materials—at least on materials not equipped with the requisite capabilities for self-organization or transformation into the conceptualized form. Typical of the rhetoric offered by proponents of triumphalist naturalism, Dennett rejects the need for either of these two elements. What appears in nature to be designed, according to Dennett, requires neither thoughtful conceptualization nor extra-natural assembly.

The need for thoughtful conceptualization is dismissed by Dennett with facile ease simply by presuming that one can take for granted the self-existence of a universe complete with a robust economy of formational capabilities. Questions regarding the need for extra-natural assembly are handled in a somewhat more reasoned manner. Here Dennett calls upon the natural sciences and their growing awareness of the powers for self-organization and transformation that contribute to the formational economy of the universe. Presuming the robust formational economy principle to be warranted, Dennett forcefully rejects the "Handicrafter-God" of both episodic creationism and the argument from design. He characterizes such approaches as ill-conceived attempts to inject supernatural explanations into circumstances where natural explanations would suffice. In Dennett's colorful metaphor, he sees no need to appeal to a "skyhook" (the top-down action of some higher power) when a "crane" (the bottom-up action of some extant natural mechanism) is able to do the job of lifting a biotic system to new heights of configurational complexity:

The skyhook concept is perhaps a descendant of the *deus ex machina* of ancient Greek dramaturgy: when second-rate playwrights found their plots leading their heroes into inescapable difficulties, they were often tempted to crank down a god onto the scene, like Superman, to save the situation supernaturally.... [A] *skyhook* is a "mind-first" force or power or process, an exception to the principle that all design, and apparent design, is ultimately the result of mindless, motiveless mechanicity. A *crane*, in contrast, is a subprocess or special feature of a design process that can be demonstrated to permit the local speeding up of the basic, slow process of natural selection, *and* that can be demonstrated to be itself the predictable (or retrospectively explicable) product of the basic process.[8]

What, then, is the essence of the naturalistic challenge? As I read Dennett, whose perspective I find to be representative of modern evolutionary naturalism, the thrust of his challenge to theism is this: *If there are no gaps in the formational economy of the universe, then what need is there for a Creator?* Implicit in this challenge is the presumption that, in order to establish the need for a Creator, one would have to demonstrate that the actualization (assembly from constituent parts) of some particular structures or life forms could have been accomplished only by an irruptive divine act. What is being challenged most specifically here is the interventionist concept of divine creative action in the formational history of the universe—the idea that certain novel life forms or "irreducibly complex" biotic subsystems could have come to be actualized only as the outcome of a form-imposing divine intervention.[9]

### Is Divine Intervention Necessary for the Universe's Formational Development?

Dennett's "hit list"—composed of those belief systems that he presumes will collapse under the weight of his attack—includes episodic creationism and intelligent design theory, both of which presume that the robust formational economy principle is not warranted and that material, physical, and biotic systems do not have the requisite capabilities to make macroevolutionary continuity possible. His strategy is to argue that what is taken by theists to be evidence of design (especially in the sense of manifesting some quality of form that could have been assembled only by extra-natural means) has been grossly misread. From Dennett's perspective, those qualities of form are the product, not of divine imposition, but of natural *algorithmic processes*—material, physical, and biotic processes whose outcome, no matter how complex in appearance, proceeds from the actions of basic material units (atoms, molecules, cells) behaving in accordance with relatively simple rules.

Here, then, is Darwin's dangerous idea: the algorithmic level is the level that best accounts for the speed of the antelope, the wing of the eagle, the shape of the orchid, the diversity of species, and all the other occasions for wonder in the world of nature.... No matter how impressive the products of an algorithm, the underlying process always consists of nothing but a set of individually mindless steps succeeding each other without the help of any intelligent supervision....[10]

No "intelligent supervision" is necessary, says Dennett. No episodes of miraculous special creation in the course of time are needed in order to actualize novel forms. No imposition of form by an act of supernatural assembly is needed. Why not? Because the universe has all of the requisite capabilities for self-organization and transformation. There are no gaps in the formational economy of the universe. And, if no gaps, then what need for a Creator?

Why the emphasis on a gapless formational economy? Because what Dennett takes to be the "creationist" position is one that insists on the need for episodes of *special* divine creative action in order to bridge those presumed gaps. Dennett's attack is directed toward episodic creationists, well known for their resistance to the concept of a gapless formational economy and of evolutionary continuity.

The resistance comes from those who think there must be some discontinuities somewhere, some skyhooks, or moments of Special Creation, or some other sort of miracles, between the prokaryotes and the finest treasures in our libraries.[11]

For over a century, skeptics have been trying to find a proof that Darwin's idea just can't work, at least not *all the way*. They have been hoping for, hunting for, praying for skyhooks, as exceptions to what they see as the bleak vision of

Darwin's algorithm churning away. And time and again, they have come up with truly interesting challenges—leaps and gaps and other marvels that do seem, at first, to need skyhooks. But then along have come the cranes, discovered in many cases by the very skeptics who were hoping to find a skyhook.[12]

### Is Divine Intervention Necessary for the Universe's Daily Functioning?

There is something puzzling to me about the way in which empirical support functions in the strategy for warranting belief in the existence of gaps in the universe's formational economy. The bottom line in all appeals for empirical support for this belief is necessarily an argument of the following form: The first appearance of form X cannot at this moment be fully accounted for in terms of what we now know regarding the self-organizational or transformational capacities of material, physical, and biotic systems. This *epistemological* gap is then taken to be sufficient warrant for believing that there exists a corresponding *ontological* gap—the requisite formational capability is missing in the formational economy of the universe. There is, I believe, no way to escape the recognition that this is an appeal to ignorance in which one begins with the statement, "Given our present state of knowledge regarding natural processes, we do not know with certainty how form X could have been assembled by natural means," and then moves to the conclusion, "Therefore form X must have been assembled by extra-natural means."

Why am I puzzled by this strategy of warranting belief in ontological gaps by appeal to epistemological gaps? Mostly because of the inconsistency with which it is employed. To illustrate the inconsistency, suppose we were to define the *functional economy* of the universe to be the set of all active capabilities of material, physical, and biotic systems that contribute to the normal functioning of the universe at any time. Needless to say, this functional economy is no less impressive than the universe's formational economy. Just try to consider for a moment all of the properties and capabilities that the universe must possess in order for us to experience only one day of our life.

With that attempt under way, think of the following questions: Do we at this time know all of the elements of this functional economy? Can we now give a complete natural account of everything that material, physical, and biotic systems presently do? Given the obvious fact that we must confess some degree of ignorance in the face of both of these questions, are we then, as Christians, inclined to claim that these epistemological gaps warrant the presumption of corresponding ontological gaps? Do we, as Christians, judge that we are warranted in presuming that God is daily bridging gaps in the functional economy of the creation with extraordinary divine action in order to effect outcomes that we do not yet fully understand in terms of creaturely action? Setting aside for the moment the question of occasional miraculous acts that God might choose to perform for special purposes, is extraordinary divine action also necessary for the daily operation of the creation? In other words, is extraordinary divine action really quite ordinary?

To the best of my knowledge, most Christians—including episodic creationists—would answer these questions in the negative. As far as the daily functioning of the universe is concerned, we see no need to jump from the recognition of our ignorance regarding numerous particular elements in the creation's functional economy to the conclusion that divine interventions are a necessary supplement to an inadequate set of creaturely actions. Or, to say it more formally, when we consider the creation's functional economy, we see no need to take epistemological gaps as conclusive evidence for the existence of corresponding ontological gaps. This leaves the question: If this move from epistemological to ontological gaps is not done in regard to the creation's *functional* economy, then why would one proceed in this manner in regard to the creation's *formational* economy? Why the inconsistency here?

Numerous reasons could be suggested, but it is difficult to assess their relative importance. Nonetheless, my personal judgment would be that a major contributing factor is the widespread belief that an episodic creationist picture of God's creative action is both a "clear teaching of the Bible" and one of the fundamental "deliverances of the faith." Beneath the surface of appeals to an empirical basis for belief in either episodic creation or intelligent design lies a set of beliefs (relatively unexamined, I suspect) regarding both biblical and traditional support for an interventionist concept of divine creative action. From this underlying concept of divine creative action proceeds the evidentialist apologetic strategy commonly employed in the debate with naturalism. It would seem, therefore, that the interventionist concept of divine creative action that prevails among the proponents both of episodic creationist theism and of evolutionary naturalism has set the unfruitful agenda of the contemporary creation/evolution debate. Hence, the importance of encouraging a theological reexamination of the concept of divine action in the physical world that is also the object of scientific scrutiny.

## The Optimally Gifted Creation Perspective

### Distinguishing Conceptualization from Actualization

As I reflect upon the ongoing shouting match between proponents of evolutionary naturalism and of episodic creationist theism (or of intelligent design theory), I have come to the conclusion that any advancement in understanding, especially of the historic Christian theological perspective, is unlikely unless both parties agree to distinguish between the issues of (1) thoughtful conceptualization and (2) mode of actualization. The two substantive questions that need to be distinguished from one another are these: (1) Does the universe, in the totality of its properties and dynamic capabilities (in other words, its formational and functional economies), display the marks of having been thoughtfully conceptualized or does it bear the marks of being the sort of unconceptualized entity that just happens to exist? (2) Whether thoughtfully conceptualized or not, is the formational economy of the universe sufficiently robust to make possible the actualization (in this context "actualization" means "assembly from the requisite elementary components") of all physical structures and life forms by means of self-organization and transformation in the course of time?

Both the either/or format of the creation/evolution debate and the apologetic strategy modeled by proponents of intelligent design theory have been built on the presumption that the answers to these two questions are inextricably coupled in one particularly way. According to Phillip Johnson, for instance, and authentically theistic perspective on evolution is impossible because evolution and metaphysical naturalism cannot be isolated from one another. In Johnson's own words,

I think that most theistic evolutionists accept as scientific the claim that natural selection performed the creating, but would like to reject the accompanying metaphysical doctrine that the scientific understanding of evolution excludes design and purpose. The problem with this way of dividing things is that the metaphysical statement is no mere embellishment but the essential foundation for the scientific claim.[13]

William Provine, a vocal proponent of naturalism, heartily agrees with Johnson's claim regarding the incompatability of evolution and theistic religion. Johnson takes this agreement as confirmatory of his rhetorical strategy. "Provine and I have become very friendly adversaries, because our agreement about how to define the question is more important than our disagreement on how to answer it."[14] (I would have expected their religious differences to be far more important than their agreement on rhetorical strategy.)

Proponents of evolutionary naturalism (like Johnson's sparring partner Provine, for instance) presume that their case is to be won by amassing evidence for the robustness of the universe's formational economy. As we noted earlier, the essence of the naturalistic challenge is this: If there are no gaps in the universe's formational economy, then what need is there for a Creator? The response of episodic creationists and intelligent design theorists to this challenge is to say, in effect, "Then there must be demonstrable gaps in the creation's formational economy. We think we have found some of them. Therefore, some form of extra-natural assembly (such as fiat creation or its more subtle intelligent design variant) is essential for the formation of at least some life forms or biotic subsystems, and naturalism is thereby discredited."

The two parties in the popular debate categorically disagree, of course, on whether or not the robust formational economy principle is true, but they nonetheless appear to agree that if it is true, then naturalism wins, or at least appears more likely to be true than does Christian theism. As I see it, that agreement constitutes a tragedy of major proportions for the Christian witness to a scientifically literate world. It implies that the apologetic contest between theism and naturalism is to be settled on the basis of the mode by which life-forms came to be actualized in time. The credibility of theism is presumed by both parties to be closely linked with the possibility of demonstrating a need for occasional episodes of divine creative action of the extra-natural assembly variety. If that need could be conclusively demonstrated, it would falsify both the robust formational economy principle and evolutionary naturalism. In fact, as I have already intimated, some proponents of intelligent design theory see little distinction between these two concepts, presuming that the robust formational economy principle and evolutionary naturalism are, for all practical purposes, equivalent.

To illustrate this last point, consider the following sample of the rhetoric of Phillip Johnson. In a published exchange of views regarding the place of divine intervention in the course of creation's formational history, I challenged Johnson to articulate his conception of "just what biological history would have been like if left to natural phenomena without 'supernatural assistance.'" His candid and very telling reply was,

> If God had created a lifeless world, even with oceans rich in amino acids and other organic molecules, and thereafter had left matters alone, life would not have come into existence. If God had done nothing but create a world of bacteria and protozoa, it world still be a world of bacteria and protozoa. Whatever may have been the case in the remote past, the chemicals we see today have no observable tendency or ability to form complex plants and animals. Persons who believe that chemicals unassisted by intelligence can combine to create life, or that bacteria can evolve by natural processes into complex animals, are making an a priori assumption that nature has the resources to do its own creating. I call such persons metaphysical *naturalists*.[15]

I can only take this to mean that, from Johnson's perspective, only a metaphysical naturalist would presume the truth of the robust formational economy principle. If that were so, then the credibility of theism could be convincingly established by demonstrating the existence of gaps in the formational economy of the universe.

The credibility of naturalism, on the other hand, is often presumed to be established by demonstrating that the actualization of all life forms can be accomplished "naturally," that is, without episodes of special creation in time. Implicit in this line of argumentation is the astounding presumption that the truth of the robust formational economy principle warrants the rejection of the idea that the being of the universe bears the marks of having been thoughtfully conceptualized. One can readily see why proponents of evolutionary naturalism would be eager to grant this presumption, but it is exceptionally difficult for me to see why it would be attractive to a Christian. Why would a Christian be inclined to reject the possibility that a creation thoughtfully conceptualized by God could well also be a creation generously gifted by God with a robust and gapless formational economy?

### Should Not Creationists Have High Expectations for the Creation?

It is clear, then, that the status of the robust formational economy principle in relationship to both evolutionary naturalism and Christian theism must be reexamined. The popular debate is structured around the presumption that this principle is the offspring of naturalism and that if

Christian theism is to survive, it must now slay the dragon named "the robust formational economy principle."

But is that actually the case? Suppose that we, as scientifically-informed Christians, were to address anew, in the contemporary context, the question regarding the mode in which the physical structures and life-forms of the creation have come to be actualized in time. Suppose, furthermore, that we were to adopt the position generally held in the scientific community that the answer is, "Yes, the robust formational economy principle is likely to be true." Would the adoption of such a presupposition place the Christian apologist in a position of weakness or disadvantage, as is commonly believed? Emphatically not, I would argue. Recognizing that such an approach might take many persons, both within and outside of the Christian community, by surprise, let me explain how I have come to adopt this stance.

All Christians are authentic "creationists" in the full *theological* sense of that term. We are all committed to the biblically-informed and historic Christian doctrine of creation that affirms that everything that is not God is part of a creation that has being only because God has given it being and continues to sustain it. As a creation, the universe is neither a divine being nor a self-existent entity that has its being independent of divine creative action. This theological core of the doctrine of creation sets Judeo-Christian theism in bold distinction from both pantheism (all is God) and naturalism (all is nature).

It is important here, I believe, to remind ourselves that the being of every creature—that is, every member of the creation, whether animate or inanimate—is defined not only by its "creaturely properties" but also by its characteristic array of "creaturely capabilities" to act and interact in particular ways, often in accordance with patterns—whether deterministic or probabilistic—that are empirically accessible to the natural sciences. Christians committed to the doctrine of creation recognize all of these "creaturely capabilities" (all of the remarkable things, for instance, that fundamental particles, atoms, molecules, and cells are capable of doing) as God's "gifts of dynamic being" to the several members of creation. A creature can do no more than what God has gifted it with the capacities to do.

From this creationist perspective, then, each discovery of a creaturely capability—including every discovery contributed by the natural sciences—

provides the theist with an occasion for expressing awe regarding the Creator's unfathomable creativity (in thoughtfully conceptualizing the gifts to be given) and unlimited generosity (in actually granting this rich array of gifts of being to the creation). Furthermore, given the creationist orientation here described, I would argue that the Judeo-Christian theist should be inclined to have exceedingly high expectations regarding the character of creation's formational and functional economies. Since the richness of these creaturely economies is to be seen as a manifestation of the creativity and the generosity of the Creator, we have every reason to have high expectations for the fullness of being that is resident in the integrated set of creaturely capabilities with which God has gifted creation for the purposes of both formational development and moment-by-moment functioning. This high expectation is affirmed each time the empirical sciences come to an awareness of another member of these dynamic economies.

Note carefully what I have just said. All Christians hold that the creation is "designed" in the fundamental sense of having been *thoughtfully conceptualized* by a Creator possessing unimaginable creativity. But if that creativity is beyond human comprehension, as we should expect, then the mode by which particular structures and forms are to be actualized in the course of time is also likely to be far more wondrous than we could imagine. Therefore, those epistemological gaps of which I spoke earlier provide no sufficient basis whatsoever for presuming the existence of corresponding ontological gaps in the formational economy of the creation.

### The Fully Gifted Creation: Both Thoughtfully Conceptualized and Optimally Equipped

Drawing from a number of biblical and theological considerations, I envision a creation brought into being in nascent form, brimming not only with awesome potentialities—for being organized into an astounding array of both physical structures and biotic forms—but also with a robust set of dynamic pathways for achieving them by the exercise of their creaturely capacities.

Drawing also from the vocabulary of the natural sciences, I envision a creation brought into being by God and gifted not only with a rich "potentiality space" of possible structures and life forms but also with the capabilities for realizing these potentialities by means of self-organization into nucleons, atoms, molecules, galaxies, nebulae, stars, planets, plants, animals, and the like. To say it in another way, I believe that the universe in its present form is to be seen as a potentiality of the creation that has been actualized by the exercise of its God-given creaturely capabilities.

For this to be possible, however, the creation's formational economy must be astoundingly robust and gapless—lacking none of the resources or capabilities necessary to make possible the sort of continuous actualization of new structures and life forms as now envisioned by the natural sciences. The optimally-equipped character of the universe's formational economy is, I believe, a vivid manifestation of the fact that it is the product, not of mere accident or happenstance, as the worldview of naturalism would have it, but of *intention*. In other words, the universe bears the marks of being the *product of thoughtful conceptualization for the accomplishment of some purpose*. From the Christian perspective, this comes as no surprise whatsoever because the formational economy of the universe—every creaturely capability that contributes to it—is a symbol both of God's creativity and of God's generosity.

### How Is Divine Creative Action Manifested?

Among Christians in North America, there appears to be a strong desire for conceiving of divine creative action in a way that would provide a basis for expecting it to be empirically distinguishable from "natural" action. In the minds of many, an appeal to divine action becomes convincing only if it can be demonstrated to have made a difference of the empirically discernible sort. In essence, evidentialist apologetics is presumed to provide the most substantive and convincing reasons to believe.

The question before us, then, is, *What is the character of that creative activity and how does it become manifest to those who have eyes to see it?* When I speak of divine creative action, however, I am inclined to speak of it not in the Aristotelian vocabulary of cause and effect, but in the royal metaphor (frequently employed in Scripture) of creative word and creaturely response. Cause and effect language seems to encourage images of God acting like creatures, only more powerfully. Matter is coerced to assume new forms. Word and response language focuses more attention on matters of authority and accountability. The king speaks, the king's subjects carry out his wishes.

I believe that God acts by calling upon the creation to employ its creaturely capabilities to bring about a fruitful outcome, and that *the fruitful character of creation's formational history is the manifestation of that divine calling.* This is, I believe, the same kind of divine action that we ask for when we pray for God's "blessing" on the work of the surgical team as we prepare for a journey into the operating room—we ask that God act in such a way that the actions of God's creatures (from the medical staff in the hospital to the molecules in our cells) will lead to a fruitful outcome.

In the contemporary discussion of issues regarding natural science and Christian faith, the question of empirical detectability often arises. Is God's action of blessing, for instance, empirically detectable as the "effect" of some non-creaturely "cause" that overpowers creatures in such a way that the outcome is clearly beyond the realm of creaturely possibility? I think not. If that were the sort of divine action that we were expecting in response to our pre-surgery prayer, why not skip the surgery and avoid both the pain and the expense? The kind of divine action we pray for is discernible only by those who have eyes (of faith) to see it. The natural sciences have no instruments with which to measure the level or effectiveness of God's blessing.

Reflect for a moment on the way modern astrophysics and cosmology describe (within the limits of a very restricted conceptual vocabulary) the processes from time $t = 0$ until now. Some nascent, non-material form of energy (whose ultimate source of being lies beyond the competence of science to identify) employs its capabilities for self-organization (capabilities that are in no way self-explanatory) to form the fundamental particles and their four distinct forces of interaction, from which also proceed such macroscopic forms as galaxies, stars, and planets. How could the universe's formational economy be so robust as to make this astrophysical drama a possibility? Only, I believe, as an outcome of God's thoughtful conceptualization and effective will to give being to the creation first conceptualized.

And how could it be that the outcome of this exercise of creaturely capabilities has been so astonishingly fruitful? Even if we could comprehend all of the things that atoms and molecules could do, how could it be that the outcome of their actions could lead to the vast array of astronomical and biotic forms that now comprise the universe? Only, I believe, as the outcome of God's continuing blessing on those creaturely capabilities.

Do I expect to find particular instances in which God's action in the course of cosmic formative history is empirically discernible? Do I expect to catch God in the act of coercing atoms and molecules into doing things differently from what they might otherwise have done (as if I could even know that)? No, I do not. I can observe what creatures have done, but God's act of calling for that particular creaturely action is beyond my empirical grasp. Though I can empirically detect the creaturely response, I cannot record the divine creative word that called for it. Furthermore, what I observe creatures to be doing in response to that creative word is not something of which they were never capable, but rather it is a fruitful exercise of the very God-given capabilities that constitute their being.

**Intelligently Designed or Optimally Equipped?**
The question posed in the title of this essay is an invitation to choose between two differing visions regarding the character of the creation, the nature of divine creative action, and the effective reasons for the fruitful outcome of the creation's formational history. Some have chosen a perspective that presumes the existence of gaps in the creation's formational economy—gaps to be bridged by occasional episodes of form-imposing divine intervention. My own choice strongly favors the concept of a creation optimally gifted by the creator with a robust and gapless formational economy—yes, even robust enough to make possible the evolutionary continuity envisioned by cosmologists and biologists.

Nearly every time that I have presented this perspective to a Christian audience someone expresses the fear that it represents a form of deism, with its concept of a distant and inactive God. I find the frequency of this concern very intriguing. Is it telling us something about how we Christians today are inclined to think about divine action? Has our concept of divine creative action been unduly affected by the "special effects" industry? Perhaps so.

But the "optimally gifted creation perspective" is not at all inclined toward deism. I think the quickest way to dispel that fear is to ask the following question: Has orthodox Christian theology ever suggested that God is able and/or willing to act in the world only within gaps in either the formational economy or the operational economy of the Creation?

To the best of my knowledge the answer is a resounding no. Therefore, if the presence of such gaps is not required to "make room" for divine action, then the absence of such gaps is no loss whatsoever. End of story.

From the vantage point of believing that God gave being to a creation in which the robust formational economy principle is true, God is still as free as ever to act in any way that is consistent with God's nature and will. The optimally gifted creation, complete with a gapless formational economy, does not in any way hinder God from acting as God wills to act. As I have said on numerous occasions, the question at issue is not, "Does God act in or interact with the creation," but rather, "What is the character of the creation in which God acts and with which God interacts?" I believe that it is an optimally gifted creation.

Does this perspective crowd God or divine action out of the picture? Does this perspective entail too high a view of the creation's dynamic capabilities or too lofty a view of the Creator's creativity and generosity? I, for one, think not.

## Notes

1. The term *conservative* has many meanings. As employed here, the term is meant to call attention to that substantial portion of the North American Christian community that places great emphasis on the role of the biblical text in providing clear, fixed, and normative answers to a broad spectrum of questions, including questions regarding the character and timetable of the creation's formational history. The early chapters of Genesis, for instance, would be viewed as a concise chronicle of particular divine acts by which God brought into being new physical, astronomical, geological, and biotic forms—a faithful reading of the text that must be conserved over against the challenges of modern evolutionary science.

2. Representative literature written from this perspective includes books by law professor Philip F. Johnson, *Darwin on Trial* (Downers Grove: InterVarsity, 1991) and *Reason in the Balance* (Downers Grove: InterVarsity, 1995); an essay collection edited by philosophy professor J. P. Moreland, *The Creation Hypothesis* (Downers Grove: InterVarsity, 1994); and the book, *Darwin's Black Box* (New York: The Free Press, 1996) by biochemist Michael J. Behe. A list of active intelligent design proponents would also include Stephen Meyer, Paul Nelson, and William Dembski.

3. In *Darwin's Black Box*, Behe further restricts the science of the day to the conceptual vocabulary of biochemistry.

4. Howard J. Van Till, "Basil, Augustine, and the Doctrine of Creation's Functional Integrity." in *Science and Christian Belief* 8 (1996) 21–38.

5. For an example of this type of argumentation see Behe's book, *Darwin's Black Box*.

6. Daniel C. Dennett, *Darwin's Dangerous Idea* (New York: Touchstone, 1995). For a more extensive critique of this work see my essay, "No Place for a Small God," in *How Large is God?*, ed. John Marks Templeton (Philadelphia: Templeton Foundation, 1997).

7. Dennett, *Darwin's Dangerous Idea*, 28.

8. Ibid., 74, 76.

9. Divine *intervention* is another term that has numerous meanings. In this context, it denotes the concept of an extraordinary divine act in which God directly causes members of the creation (atoms, molecules, cells, or organisms, for instance) to assume a configuration that they would not otherwise have been able to achieve by the employment of their ordinary capabilities for self-organization or transformation. Thus, divine intervention of this character is presumed to "make a difference" (a phrase commonly employed in the books by Phillip Johnson) of the empirically detectable sort.

10. Dennett, *Darwin's Dangerous Idea*, 59.

11. Ibid., 136.

12. Ibid., 75–76.

13. Phillip E. Johnson, *Darwin on Trial*, 2d ed. (Downers Grove: InterVarsity, 1993), 168.

14. Ibid., 165.

15. *First Things* (June/July 1993), 38.

# 21

## Is Theism Compatible with Evolution?

Roy Clouser

### 0. Introduction

Ever since the publication of *The Origin of Species*, the majority of
theists have taken the position that there is no real conflict between the
Genesis account of human origins and the theory that all life forms
including humans evolved gradually. Nevertheless, a vocal minority of
theists have always dissented from that judgment, insisting that one
cannot take seriously the scriptural account accepted by Jew, Christian,
and Muslim alike, and also accept the evolutionary hypothesis. Moreover, there have always been thinkers hostile to theism who have been
only too happy to concur with the alleged inconsistency. They have been
as happy to dismiss theism on the ground that it's incompatible with
science as those theists have been to dismiss evolution on the ground that
it's incompatible with theism.

It seems to me that a good part of the blame for the persistence of the
inconsistency charge is the fault of those who, like myself, take the compatibilist view. I say this because for the most part it's been left dis-
concertingly vague just how a proper understanding of Genesis can be
complimentary rather than inconsistent with an evolutionary account of
human origins. Of course, there have been thinkers who've defended a
compatibilist view simply by not taking the early chapters of Genesis
seriously. Some, for example, have dismissed them as so poetic as to
teach nothing about human origins. This seems wrong in several directions at once. For one thing, while these chapters surely contain symbols,
figurative speech, and poetic turns of phrase, they are just as surely not a
poem. There are parts of the Bible that do speak of God's creating within
a poem (Job 38:7, Psalm 104: 5 9, e.g.), and they don't sound like

Genesis at all. Moreover, the presence of such poetic material in the early chapters of Genesis—indeed, even if they did constitute a poem proper—would not prevent them either from making factual assertions or having cognitive content.[1]

My first order of business, therefore, will be to examine the text in order to give a detailed account of what it teaches concerning human origins. In doing this I will be at pains not to import far-fetched hypotheses in order to make Genesis sound acceptable to modern science, but to take it on its own terms. That is, I will be attempting to hear *its* purpose and focus rather than impose what a twenty-first-century reader might expect from a scientific discussion of origins. In that sense, I will be seeking a "literal" (but not literal*ist*) interpretation of the text. After all, the literal meaning of a text must be ascertained from its grammar, form, historical setting, internal structure, and universe of discourse. It's not whatever any reader thinks of on first reading it. So I'll begin with the reasons why I find that when the text is taken on its own terms there is no conflict between what it teaches and such ideas as an old age for the earth and the gradual development of its life-forms.

My second order of business will be to tackle how religious belief should be seen to relate to science *generally*. This will require identifying features that characterize all religious beliefs, not just theism. On this score, I will end up disagreeing both with fundamentalism and with those theists who, while rejecting the fundamentalist program of reading science out of Scripture, nevertheless wish to infer religion from science. The latter think that being a theist in science requires showing that the findings of science either entail God's existence or make it probable. Over against both these views, I find that clarifying the general nature of religious belief allows us to discover that there is a deeper epistemic relation between it and science that is at once more basic and more pervasive than mere logical consistency or inconsistency between specific theories and specific religious tenets. Moreover, this relation is of great importance to both scientific beliefs and belief in God despite the fact that neither one is inferred from the other.

## 1.   Genesis and the Alleged Points of Incompatibility

The items of the Genesis text that are most often alleged to be incompatible with an evolutionary account are: (1) the teaching that the entire universe was created in six days; (2) the teaching that humans appeared suddenly; and (3) the teaching that the first woman was formed from the first man. I will also comment briefly on a few other alleged points of conflict after these three are treated.

## i Background Assumptions

Crucial to the interpretation of any literary work is what we assume to be its central theme and purpose. Mistake that, and everything will be misunderstood in detail. So what is the central theme of the biblical corpus? Surely it can be nothing other than the self-revelation of God to humans. Moreover, it is equally important to recognize that the form most of that revelation took, and thus the form in which it has been recorded, is that of *covenant*. The collected works comprising the theistic scriptures claim to be the divinely superintended record of the progressively revealed editions of the covenant God has offered to the human race. There were editions of the covenant revealed to Noah, Abraham, Moses, and David, for example. (There are even what were later to become covenantal elements present in God's pre-redemptive dealings with Adam and Eve.)[2] For Christians, these scriptures include the New Testament, so there is another—final—edition of the covenant forged and fulfilled by Jesus. For Muslims, they include the Koran, so there is also the covenant offered to and through Mohammed. But in every case the central unifying theme is still the covenantal offer of God's love, forgiveness, and everlasting life to all who respond by loving him with all their heart, soul, mind, and strength and loving their neighbors as themselves. For this reason, these scriptures must be understood as having an essentially *religious* character. They are intended to teach the truth about the covenant maker, God, the covenant receivers, humans, and always have their focus on how humans are to stand in proper covenant relation to God. There is thus a religious slant to everything so much as mentioned in them. Whether they record historical events, state genealogies, engage in poetry, or speak of the end of the world, their governing purpose is to teach us how to stand in proper relation to God.

The importance of this point becomes especially clear in the light of what happens when it is forgotten. Some theists have been tempted to look past the religious focus of Scripture when investigating various topics, and focus instead on the fact that God has superintended its writing and preservation. The temptation goes like this: since God has inspired and preserved these writings they must be trustworthy, so why not use them as a short-cut to find out other things we'd like to know? We have questions about prehistory, biology, geology, astronomy, etc., and there's no way (or no easy way) to find answers to them. But suppose there are statements or hints about these things in Scripture! Since it's inspired by God, wouldn't everything even alluded to have to be infallibly true? In fact, even if we have ways of investigating certain topics, shouldn't we at least start by canvassing the Scripture to see what it says about them? This attitude, now associated with fundamentalism, was actually fostered by such pioneers of modern science as Newton, Boyle, Locke, and others. I call it the "encyclopedic assumption." It leads to regarding the Scriptures as an encyclopedia to be consulted on virtually any topic whatever, rather than seeing their inspired character as pertaining to their own *religious* purpose. Avoiding this mistake is especially important for reading Genesis, as it is this assumption that leads to reading its opening chapters as a short scientific treatise rather than what it really is, namely, a *prologue to the covenant given through Moses*. If we see it as covenant, there is no excuse for missing the thoroughly religious focus of every part of it. It starts by distinguishing God the Creator from pagan deities, which were all deifications of various aspects of the natural world. Rather than deifying anything in the universe, Genesis proclaims God to have created everything other than himself.[3] That it does not intend to describe what we would have seen had we been there to observe the early stages of the universe is evident from the way it stresses God's total control, repeating over and again that everything comes to be at his command. Before each creative episode we find "And God said, 'Let there be....'" There is no concern for the processes set in motion or the time they may have taken. All the text says about what an observer might have seen is: "And it was so." Finally, not only its content, but the very reason for this prologue is religious; it was intended to identify the Mosaic covenant as the latest installment in the dealings God had initiated with humans ever since their creation.

## ii   The Days of Creation

If we keep the text's religious focus in mind while examining its organization, it then looks even less like an encyclopedic source of scientific information. It speaks of God's creating as taking place in six days, and reports the days as follows: Day 1, God separates light from darkness; Day 2, God separates sea from atmosphere; Day 3, God separates land from sea and creates plants; Day 4, God creates the sun, moon, and stars; Day 5, God creates sea life and birds; Day 6, God creates plants, animals, and humans. There is an obvious correspondence here between days 1, 2, 3 and days 4, 5, 6. Day 1 speaks of the difference between light and darkness as the planned precondition for the appearance of the sun, moon, and stars on Day 4. Day 2 offers the separation of atmosphere from sea as the precondition for the appearance of sea life and birds on Day 5. And the appearance of dry land and plants on Day 3 is the precondition for the creation of animals and humans on Day 6. (Here we may recall St. Augustine's comment that the "days" of Genesis were not intended to be taken temporally since they have God creating light on the first day while the sun, moon, and stars don't appear until the fourth.) This match-up of the first three days with the second three is too striking and essential a feature of the account to be accidental. But if it's not accidental, it shows two very important things. The first is that the intent of the text is to supply a *teleological,* not chronological, account of God's ordering of the earth and its life forms. The order here is an order of purpose, not time. Its main burden is to stress that the purposes behind creation were God's, as was the accomplishment of those purposes. There was no blind chance or fate involved, nor was God limited by preexisting material he had to work with. What it presents is thus more like a religious birth announcement of the universe and humans than it is like a scientific description. And the main point of that announcement is focused on its Father's loving and redeeming purposes, rather than on revealing answers to scientific questions about the early stages of the universe.

The second thing the correspondence supports is that the "days" are to be taken as a literary framework for speaking of God's creative activity.[4] The reason is both obvious and simple when the connection of this account to the Sinaitic covenant is kept in mind: since that covenant requires that humans work six days and rest on the Sabbath, God's work is represented as done the same way. God's work thus becomes the model for how humans are to work. The stress of the text, then, should not be seen so much on the word "day" as on the ordinals designating the covenantal arrangement for the work week. What is religiously important is "first," "second," etc. as contrasted with the "seventh," not whether the "days" can be construed as a geological eras or taken to be twenty-four-hour periods in which the entire universe appeared. There are a number of other reasons for regarding this as the right interpretation that space will not permit, but one is worth mentioning at this point. Traditional theism has always understood Scripture to teach that God brought into existence not only space, matter, and the laws that govern the universe, but also time. In that case no literal account *could* be given in terms drawn from human experience as to just how God brought all into existence and ordered it. Absent time, space, matter, and all natural laws, we have only the limiting idea that God is the reality without which nothing else could be, and any account of God's bringing about the existence of the universe would necessarily be anthropomorphic. So the text draws its language *from* the human experience of a workweek, and does it in a way that sets up a covenantal rule *for* the human workweek in relation to the Sabbath rest. When viewed from the standpoint of this religious focus, then, there is no excuse for treating Genesis as a source of scientific information. It is not good science or bad science because it is not science at all. It is not concerned with how old the earth is, when and how life forms first appeared, or what role the processes of created nature may have had in bringing them about. It is always about God's purposes and chiefly his purposes concerning how humans can stand in proper relation to himself.

### iii   The Formation of Humans

This same point holds true for its account of the first appearance of humans. That, too, is phrased in anthropomorphic language. But to see what the text intends to teach we must keep in mind the importance of the question: what is a human? This is actually a crucial issue for *any* account of human origins though it has rarely received the attention it deserves. Are we asking when tool-making creatures first appeared? Are we asking when rational beings or language users emerged? Are we asking when creatures with an ethical sense of right and wrong made their debut?[5] Here we confront a philosophical assumption that regulates how scientific findings are to be interpreted. For what we really want to know about any discovered remains from the genus Homo is: how like us were they? Answering that depends on what we take to be essential to ourselves and, obviously, whatever definition is accepted will determine how we answer the question of when humans first appeared. Now it is precisely on this point that Genesis is as clear as one could want. From the point of view of the text, a human is a creature in the image of God created for fellowship with God. *The question as to when the first humans appeared is therefore the same as that of the appearance of beings with religious consciousness.*

Taking this definition of humans as our point of departure, the account of the making of Adam and Eve should be read as partly figurative. That is, the text teaches its view of the *nature* of humans by means of a story about God "making" them. Thus the remark that God created Adam "of the dust of the ground" is not intended so much a description of an event as it is a comment on Adam's nature. To be sure there is activity of God involved here (I will explain that presently), but the point of the expression is not to *describe* God's act but to convey that Adam is made of the same stuff that everything else is made of. Humans are not and never can be anything more than creatures of God. They are not little bits of divinity stuffed into earthly bodies, which are therefore merely "the prison house of the soul." They can never have any existence but what is granted them by God. In short, they are mortal. This is born out by the way not only Genesis but also the rest of the Hebrew Scriptures use the expression "the dust of the earth." It always connotes human mortality (e.g., Ps. 22: 15 and 29, 30:9, 40:25, 103:14, 104:29; Ecc. 3:20, 12:7; Isaiah 26:19; and Daniel 12:2). That is why the sentence of death as punishment for disobedience to God is expressed "from dust you came and to dust you will return." The point is that standing in right relation to God is not an extra added to human life, but the most basic condition for it. For sure, humans depend on the sun, air, water, food, etc., but those are penultimate dependencies. The ultimate condition for human life is to be in the right relation to the One on whom all the penultimate conditions depend, and who directly offers humans a loving, life-guaranteeing relationship.

The same figurative language is employed in the account of the formation of Eve by a rib transplant from Adam. Here too, her nature is expressed in a "making" story intended to convey that she was of the same nature as the man. That this is the intent of the account is shown from the context. After commenting that there was no proper mate for the man in the animal world, Adam's reaction to her is that "she is bone of my bone and flesh of my flesh"—that is, the same *kind* of creature as he. So the surrounding context includes the comments "For this reason a man leaves his father and his mother and unites with his wife ...," and "her desire shall be toward her husband...." Moreover, chapter two sees both the man and the woman as created for fellowship with God since both are held responsible for breaking that fellowship. Thus it seems there is good textual reason to reject any reading of chapter two that takes the remarks about God forming Adam of the dust of the ground to intend that God made a mud model of a human, blew on it, and it hopped up and walked around. Yet it is something almost that crude that some theists seem to place in opposition to evolutionary theory.

What then is the right understanding of the action of God that caused humans to appear? Here the text is, I think, even clearer. That act was *God's speaking to Adam and making himself known to him.* God had brought into existence beings "from the dust of the ground," seen to it that they acquired the capacity for religious consciousness, and then actualized that capacity in the first being in whom it appeared by offering him a gracious, personal relationship. That was the last step in making that being fully human.

In this way Genesis employs what I'll call the "commonsense" view of causality. In ordinary speech we often say one event is the cause of another not because it includes all the necessary conditions for the other but because it was the last thing that needed to be added to all the other necessary conditions so as to produce the effect. Just so, the focus of the text is on that last step. Without a religious consciousness, no being is human in the full sense of the term. When God actualizes that capacity and Adam responds, he becomes the first human; whereas a moment before there was no human, now there is. The action by which God did this is described in the words "God breathed on him the breath of life and man became a living soul." Now the word for breath here is the same in Hebrew as the word for "spirit," so there is a pun here intended to convey several things at once: it is by the spirit (breath, command) of God that man has appeared from the dust of the earth, and it is by the breath (spirit, word) of God that man hears the offer of God's love and can stand in right relation to him. One has produced a biologically living creature; the other has produced human life in its fullest sense—the sense that never ends so long as the covenantal relation to God remains intact.

So I find that the text does indeed teach that there was a single act of God in time and space that brought it about that at one moment there were no humans but at the next moment the first human was produced.[6] For while Adam's formation out of "the dust of the ground" could have been a long process, God's gracious word (breath) in making Adam his covenant partner was not. Once again the surrounding context supports this reading. Notice that in the brief synopsis of the origin of humans given in chapter 1, the account of the formation of humans is immediately followed by the statement that God blessed them and gave them responsibilities (later these are both covenant relations). And in the recapitulation of this story in chapter 2, God's breathing on Adam the breath of life is followed by a more detailed repetition of those responsibilities: he must "cultivate the garden and keep it" (i.e., take care of the specially protected environment that is the setting of his probation), and he must not "eat the fruit of that tree."

Thus the account of human origins has a clearly religious focus. It does not regard "human" as synonymous with any strictly *biological* structure or capacity, and it is clueless with regard to any biological processes or the time they may have taken to contribute to the appearance of religious beings. In fact, I'll go further than that: Genesis is not merely unaware of the physical and biological conditions for the appearance of humans, it is *uninterested* in them. I have no doubt that if we could speak to Moses (and whoever else authored it), and were to ask about the processes and time span involved, his answer would be: "You want to know *WHAT?* I've just given you the greatest news any human could ever have! I've reported to you that the Creator of the universe offers us his love, forgiveness, and everlasting life, and what do you say? You want to know how *long* it took God to make us? Have you been listening at all?"

Now it might be objected at this point that even if my construal of Genesis thus far is right, there are still other possible incompatibilities left unresolved. What of the origin of Eve, for example? Even if the rib-transplant is figurative, are we to take seriously the suggestion that her humanity was somehow derived from Adam's? And if the evolutionary account is right, shouldn't we expect that a number of beings would acquire the necessary and sufficient qualifications for humanity (whatever those are taken to be) at about the same time? So even if there were *a* first human, wouldn't there be many others in short order?

Let's take Eve's formation first. If the interpretation offered above is correct then the transplant of Adam's rib not only conveys their common human nature, but also has further religious import. For since it is contact with God's self-revelation that is the last step in becoming human, Eve's humanity does derive from Adam's in the sense that whereas he received the covenant directly from God she received it from Adam. As for there being other humans at about the same time, Genesis agrees. From chapter four onward, the children of Adam and Eve are represented as traveling among, in fear of, and marrying from, other people. Virtually everyone upon first reading those chapters has wondered, "Where did those other people come from?" But although the question is obvious, it's not one the text is interested in answering. It is narrowly *covenant* history, not broadly human history. It is worth noting, however, that if there were an evolutionary process that produced many nearly human beings at about the same time, and if coming into contact with God's offer of the covenant was the last step in their becoming fully human, then the existence of other newly human beings (owing to the spread of religious belief) would be exactly what we would expect. Adam and Eve would still be the progenitors of all humanity in the religious sense; they were the first to know God, the ones who represented all humanity in being placed on religious probation, and were likewise the first recipients of God's offer of a covenant of redemption following their fall.[7]

I conclude, therefore, that on religious grounds there is no need for a theist to reject the evidence for an old age of the universe and the earth, or the theory of a gradual diversification of living things.[8] The acceptance or rejection of such ideas should be decided on the scientific evidence alone. Nor is it a good objection to point out that evolutionary theory gives aid and comfort to nontheists who deny that the processes and laws of nature are the creations of God. Of course there's an atheistic version of evolution, just as there is a theistic version; for a materialist the natural processes are the ultimate explanations of everything, just as for a theist they are but penultimate explanations. But that same difference holds true for the interpretation of other theories. Atomic theory, for example, was invented and advocated by thinkers (Democratus, Leucippus, Epicurus, et al.) who took atoms to be the ultimate realities that explain the existence of everything else. That is not, however, good reason for a theist to reject belief in atoms any more than it would be a good reason for nontheists to reject atomic theory because Newton and Plank believed them to be created by God. Evolution is a parallel case. A theist is as much within his intellectual rights to accept evolutionary theory while believing those processes to have been created by God as he is to believe atomic theory while holding atoms to have been created by God. Physics can't show that matter and space are or are not God's creatures, any more than biology can show life forms are or are not created by God. Both beliefs are *brought to* science not *derived from* it. And both are *religious* beliefs.

## 2. The Nature of Religious Belief

Since there is not the space here to defend that last sentence in detail, I can now only summarize what I have defended elsewhere about the nature of religious belief.[9] Suppose we start by observing that every religious tradition regards something or other as divine. The question is whether there is anything true of all putative divinities that could serve to define "divine"? If we look for a common element shared by the natures of every putative divinity then the answer is surely, "no." But if we look instead at the *status* of divinity we find a very different state of affairs. The difference between these two approaches is like the difference between two ways we could answer the question, "Who is the President of the United States?" One way would be to describe the person holding the office, the other would be to describe the office itself. The difference is important. If an election were in dispute as to who had won, different people would describe the President differently, but everyone's idea of the office would remain the same. This is an exact parallel to what I've found to be the case for religious beliefs. While there are a host of contrary descriptions of who or what is divine, there is a startling unanimity about what it means to be divine. For no matter how differently the putative divinities are described, the *status* of divinity is always *that of having unconditional reality*. The divine is therefore that on which all nondivine reality depends (if it is also believed there is any nondivine reality). The status of divinity, then, is accorded to whatever is believed to be un-caused and unpreventable, or self-existent, or just there. The terms for the status differ; and some traditions don't bother with any of them. They simply trace everything back to a source that has no explanation and drop the subject. In that case the status of divinity has been conferred tacitly by default, but it's still the case that the divine is being regarded as having independent reality. So the crucial question is whether the converse is also true. Are all beliefs in anything as having independent reality religious beliefs? I think an affirmative answer is unavoidable.

Many people have a difficult time with this point because their idea of religious belief is narrowly culture-bound. They have a prototype in mind—the tradition they're most familiar with—and they judge all beliefs as religious or not to the extent they resemble that prototype. So it is necessary to review briefly why such prototypes, along with some of the most widespread ideas of religious belief they generate, are misleading. Let's start with the idea that it's necessary that a belief be embedded in a cultic tradition and associated with worship for it to count as religious. Is this the case? Surely not. Aristotle argued for and believed in a being he called "god," but whom he didn't worship. Likewise the Epicureans, who believed in many gods. Moreover, there are forms of Hinduism and Buddhism that include no worship. Is it necessary for a religion to believe in gods or a God at all? Once again, the answer is "no"; there are forms of Hinduism and Buddhism that believe in no gods whatever. (Being an atheist is like being a vegetarian: the terms convey what is not believed to be divine or not to be eaten, while leaving in the dark what the person does believe to be divine or does like to eat.) Is it necessary for a belief to be associated with ethical teaching to count as religious? There are counterexamples to this proposal too. Ancient Roman religion is one example; another is the Shinto tradition of Japan. Is what makes a belief religious whether it is taken on faith rather than being proven on the basis of argument and evidence? That can't be right either. Many theists have accepted arguments for the existence of God but it would be absurd to say their belief in God was therefore not religious. Besides, there are many principles necessary to science that also have no proof—the rules of proof, for instance. But, you say, what if a theory assumes *matter* to have that status? Isn't materialism the reverse of religion? Again, that is a narrowly culture-bound notion. The ancient

Greek mystery religions worshipped what they called "the everflowing stream of life and matter," and one form of Hinduism presently regards matter as divine (along with souls). The only thing these beliefs have in common with other religious beliefs is that they take matter to have the same status of ultimate reality that the other religions assign to God, Brahman, the Tao, Wakan, Zurvan, Mana, and so on.

To sum up: no matter how hard we try to find other common characteristics to religious beliefs, they always turn out to have exceptions. The only thing they all have in common is being about what has the *status* of unconditional reality, so that's the only thing that could define them. To be sure, there is a difference in the *employment* of such beliefs when they occur in religious traditions from when they occur in theories. In a cultic tradition the main point of knowing what we and everything else depends on is to acquire the right personal relationship to the divine. In a theory it is sought mainly to explain, rather than acquire a personal relation. But even this difference is one of emphasis rather than exclusion. The divinity beliefs involved in theories can't fail to carry some personal implications for life, just as divinity beliefs embedded in cultic traditions serve also to explain. The upshot is that there is no good reason at all for denying the religious character of any divinity belief no matter in what context it occurs. Whatever is given the status of unconditionally independent reality is metaphysically ultimate, and thus divine, no matter how it is conceived. To put the same point from another angle, no idea of what is metaphysically ultimate could fail to have religious import if for no other reason than it is either God or a surrogate for God (or for Brahman-Atman, or for the Tao, etc.).

Stated formally, then, this definition of religious belief is as follows.

A belief, B, is a religious belief iff:
1. it is a belief in something or other as divine, or
2. it is a belief in how to stand in proper relation to the divine, where
3. "divine" means having utterly independent reality.[10]

Taking this definition as point of departure, I will now argue that: (1) any theory of reality is bound to contain or presuppose a divinity belief, and (2) any theory of science is bound to contain or presuppose some view of reality. For this reason I contend there are no religiously neutral theories; all at least presuppose some divinity belief by conceiving of the natures of their postulates in ways that differ relative to what they take to be divine. Please notice how sharply this differs from the fundamentalist program of importing theistic beliefs such as the special actions of God (miracles) into scientific explanations. The claim is that *whatever* is regarded as metaphysically ultimate *regulates* how a theory conceptualizes the nature of its postulates. This is what I see as the most pervasive and important (and neglected!) of all the relations divinity beliefs have to science. I will now try to clarify this position. Then I will end with a brief statement of why theists should be explicating this relation of belief in God to the theories of the sciences, rather than trying to infer support for belief in God from features of the universe.

## 3.   Religious Belief, Metaphysics, and Science

The central issue for metaphysics is to offer a theory of the ultimate nature of reality. It therefore seeks to answer the question: what kind(s) of things are there? This has traditionally been done by picking a particular *kind* of properties and laws the world exhibits to our experience, and enthroning that kind as the nature of the entities that either (1) comprise everything, or (2) are what everything else depends on. In the brief list of sample theories that follows, I'll use italicized adjectives to designate the kinds and unitalicized nouns to designate the entities proposed as having independent existence: *mathematical* numbers, *physical* matter, *sensory* perceptions, *logical* Forms. There have also been dualist theories combining two metaphysical ultimates: *logical* Forms and *physical* matter, *sensory* perceptions and *logical* categories, *logical* minds and *physical* bodies, and so on. Some of the names that go with these theories include Pythagoras who held that all is comprised of numbers and their relations; Plato and Aristotle who said the world or our experience is produced by the relation of Forms to matter; Hobbes, Smart, the Churchlands, and other materialists who have said all is physical; Hume, Mill, Mach, who held that all is sensory; and Kant and the logical positivists who held that all we experience is sensory/logical. In every case the key belief driving the elaboration of these theories is what they regarded as divine.

But even if one or another divinity belief functions as a regulative presupposition to any theory of reality, is it really the case that scientific theories are in turn regulated by some metaphysics? Can't the natural sciences finally declare their independence from philosophy? To see why this is impossible we need only ask this question: is it possible for any science to explain without specifying the nature(s) of its postulates? If not, then the nature ascribed to any postulated entity or process would (along with specifying the range of its explanatory power) reflect its metaphysical underpinnings. A postulated entity would have to be of the same nature as reality *generally*, or have its explanatory power relativized to something else which is of that same nature. Consider, for example, how this works out in physics. Ernst Mach held reality to have an independent *sensory* nature and for that reason held atoms to be "useful fictions".[11] Einstein held external reality to be a combination of physical matter and logical/mathematical order, so he took atoms to be purely physical entities postulated by purely rational thought.[12] Over against these views, Heisenberg held atoms themselves to be not only physical but also essentially mathematical in their own nature (a view he himself said "fits with the Pythagorean religion").[13] This is why he took the "Copenhagen interpretation" of the uncertainty relations: whatever isn't mathematically explicable can't be real. Each of these views takes atomic theory to be something significantly different from the others, and their differences stem precisely from how the nature of an atom is conceived in each case. In turn, the concept of an atom's nature is regulated by what is presupposed about the nature of reality generally, and the general nature of reality is regulated by what is presupposed to have divine status.

In this sense, the theories of science can never be religiously neutral. They may or may not *contain* a declaration of what is being regarded as divine, but they cannot avoid at least presupposing something in that role.[14] For that reason the right understanding of the relation of divinity beliefs to science can never be that of harmonizing them, as though divinity beliefs and science have two independent but equally reliable sources of information. No interpretation of any theory can fail to be consistent with its own presuppositions, just as it cannot fail to be inconsistent with all contrary presuppositions. Harmonization is thus either unnecessary or impossible. Here Calvin had it right rather than Augustine; it's not that there are two books from God to be read independently, the book of nature and the book of Scripture. Rather, it's that in proclaiming the transcendent Creator, Scripture supplies the "spectacles" through which the book of nature must be read. Thus the same hypothesis (e.g., there are atoms) can have as many interpretations as there are metaphysical ideas of the nature of reality, and there will be at least as many ideas of the nature of reality as there are divinity beliefs. The understanding of nature is not neutral, unbiased, and the same for all. It is religiously controlled, not only for the theist but for everyone.

For these reasons, and in this sense, I maintain that there is a distinctly theistic perspective for the interpretation of all theories. The central point of the perspective is the full force of the Creator/creature distinction: *no kind of things in the universe has independent reality*. Nothing in creation is what everything else depends on; that status belongs only to God. The perspective is therefore radically *nonreductionist*.[15] I presume that by this time it's clear I mean metaphysical reduction here rather than theory reduction, or convenience reduction, or whatever. But let me try to be more precise. Reduction explanations come in two main flavors, eliminative and causal dependency. On the eliminative version a particular kind of properties and laws is asserted to be the nature of reality on the ground that all the other kinds of properties and laws exhibited to our ordinary experience are either identical to the kind favored as the nature of reality or are not real at all. The causal version, on the contrary, claims that the one or two kinds the theory defends as the nature of reality generate all the other kinds of properties and laws we experience. The reason for calling the first explanatory strategy "reduction" is obvious: all but one of the kinds of properties and laws exhibited to our pretheoretical experience are denied, so that the kinds of properties things actually have is drastically reduced in number. (As a result the number of types of things and events considered real is inevitably reduced as well.) The causal version is also reductionist, but in a subtler manner. This version allows that things really have as many distinct kinds of properties as they can plausibly be said to have, but it reduces the status and importance of the dependent kinds relative to the one(s) supposed to produce them. Along with most theists, I contend that the first version should be rejected as flat-out incompatible with both our divinity belief and our experience of the world around us, and as philosophically indefensible. But unlike most theists, I call for the rejection of the second version as well, because it gives some part or aspect of creation the status of being that on which all (the rest of) creation depends—the status of divinity that belongs only to God.[16]

Of course, by insisting that only God be accorded that status we do not thereby avoid making creation dependent. But it is not the *dependency* of things or the kinds of properties they exhibit that we are rejecting, but only their *reduced status relative to one another*. While a theistic view regards creation as dependent on God and thus less real than God, nevertheless it does not elevate any kind(s) of properties and laws found in creation above the others; it does not overestimate any one or two kinds and thus correspondingly underestimate (or eliminate) all the others. Instead, by regarding all aspects of creation as equally real because equally and directly dependent on God, theism can affirm the dependency of created reality without reducing the status of any of its kinds of properties and laws relative to any other.[17] And it is exactly the reduced status of these kinds relative to one another that makes for serious theory differences in science, since it is precisely the deification of some kind of properties and laws, and the consequent reduction of all other kinds to it, that results in the competing metaphysical "-isms" listed above (think of the three views of an atom). By offering the alternative of an irreducibly pluralistic metaphysics, theism can present a salutary gift to the scientific enterprise and it is to this more pervasive and constructive project I would urge theists to turn their attention. The project should include both the development of new nonreductionist theories in the sciences and the elaboration of nonreductionist interpretations of existing theories—including the theory of biological evolution.

## 4.    What about Evidence for Intelligent Design?

What about evidence for intelligent design? This question raises the complex debate concerning what is traditionally called "natural theology." As with most of what I've touched on so far, it too is a topic there is not the space to treat adequately. So once again I'm forced to summarize points that deserve lengthy supporting arguments without being able to state those arguments here.

One of the most disturbing things about the attempts to argue from features of the world to God's existence is the way many Christians justify that project by appeal to Scripture, especially to Romans 1: 18–20. This text says that ever since God created the world he's been making himself known to humans, and that in addition to his covenantal self-disclosures the creation itself (somehow) reflects his "power and divinity." But the fact that creation, if viewed rightly, will be seen not to be divine and thus to witness to God's divinity, is no license to suppose God's existence can be *inferred* as certain or probable from specific features of the natural world. In fact, the context of this remark emphasizes (v.25) that most of humanity "changed the truth about God into a lie and worshipped and served something created instead of the Creator." And that, the same author says elsewhere (I Cor. 1: 20–21; II Cor. 4:4; Eph. 4:18), has resulted in spiritual blindness on the part of those who deify some aspect of the created universe. Taken together, these texts amount to saying that although the universe may exhibit dependency on God, those who regard some part of it as divine are thereby blinded to the exhibition. Their false divinity belief blocks the acquisition of the true one.

So should theists engage in the project of *arguing* that the world is in fact designed? Even the great skeptic David Hume thought the world surely gives that appearance, and called anyone to whom it didn't appear that way "stupid." But Hume found that fact insufficient to infer God's existence, and instead took the appearance of design as misleading—like the appearance that the sun rises and sets. And anyone who takes part (or all) of the universe to be divine will no doubt agree with him. No matter what the evidences for the inference to intelligent design, the fact that something else is believed to be divine will prove decisive for the nontheist. A crucial point to keep in mind here is that religious beliefs are not hypotheses, so pointing to confirming consequences of them cannot provide an argument to the best explanation as it can for theories. Divinity beliefs are always prior convictions brought to science, and advocates of every divinity belief can point to features of our experience that are what would be expected if their belief were true. For that very reason none of those features will appear convincing to people holding contrary religious beliefs. In every case thinkers will see the truly important factors to be those that confirm whatever divinity belief they themselves hold. This is not only the case between theism and nontheism, but between contrary nontheisms. Have advocates of the major -isms of metaphysics ever succeeded in converting their opponents purely by argument and evidence? No. Ditto for the competing perspectives for science these -isms provide. As Michael Polanyi put it:

All formal rules for scientific procedure must prove ambiguous, for they will be interpreted quite differently according to the particular conceptions about the nature of things by which the scientist is guided.... [In cases of theory disputes] it appears that the two sides do not accept the same "facts" as facts, and still less the same "evidence" as evidence.... For within two different conceptual frameworks the same range of experience takes the shape of different facts and different evidence.[18]

So it's true that for the theist the order exhibited in the universe will, indeed, count as confirming evidence; reading nature through the spectacles of Scripture, it will be plain that "the heavens declare the glory of God and the firmament shows his handiwork." But it is a horse of a different paint job to try to make what counts as *confirmation of* belief in God into a compelling *argument for* that belief. In one sense, it's understandable why theists have been drawn into such a project. Many naturalist and materialist defenders of evolution have pointed to the random and chaotic history of life forms as counting against their having been created by God. It is tempting, then, to argue the contrary. So some theists have shifted their gaze. Instead of looking at the apparent randomness of the processes, they look instead at the end products of those processes and stress their orderliness. But is there a religiously neutral argument to show which of these is the more important evidence? No. The decisive factor in each case is what is already believed to be divine.

A second objection to this project concerns an assumption shared by many in both the atheist and theist camps of the debate, namely, that if God exists, God would think and plan in basically the same way humans do. Each side looks for what would be signs of *human* rationality or design, and proceeds either to deny their presence in the evolutionary process or assert their presence in its products. Meanwhile, Scripture has all along warned that what looks random to us is still under the Lord's control (Prov. 16:33). The proper theistic response to the charge that there doesn't appear to be a plan behind evolutionary history is not to try to show one, but to say that knowing there is such a plan depends on first knowing God. For us, as for Adam, it is only by encountering God's covenantal word that we can come to know him and, in consequence, to read nature rightly.[19] Here again I want to urge that essential to that right view of nature is not only the insistence that all is the product of God's plan and power, but also a nonreductionist interpretation of all creation. This program cannot be objected to on the ground that it imports religion into science. In bringing a nonreductionist view to bear in every scientific concept, a theist would be doing nothing different from what reductionists are doing on behalf of whatever they regard as ultimate (divine) reality in their theories. All alike are allowing their view of ultimate reality to regulate their concepts of dependent reality.[20] The theist simply does this on behalf of a different idea of ultimate reality, the idea that it is nothing less than the same transcendent Creator who has offered a covenant of redemption to the whole human race. This is why I think Tillich was right when he said:

The famous struggle between the theory of evolution and the theology of some Christian groups was not a struggle between science and faith, but between a science whose faith deprived man of his humanity and a faith whose expression was distorted by biblical literalism.... A theory of evolution which interprets man's descent from older forms of life in a way that removes the infinite, qualitative difference between man and animal is faith not science.[21]

We need, therefore, to challenge *every* metaphysical reduction: that consciousness can be reduced to biology, that biology reduces to physics, and especially that the physical aspect of the world is metaphysically ultimate. The challenge to that religious belief is more than merely respectable, for the materialist divinity-belief has serious difficulties. For openers, we may pose the question as to whether it is possible so much as to conceive of the physical side of reality as really independent of all the nonphysical properties and laws we experience things to have. What is left of the very idea of "physical" if we strip from it every connection to, say, the quantitative, biotic, sensory, logical, and linguistic properties that are also exhibited to our experience? Try to do so as a thought experiment. Is anything left? Isn't trying to combine "physical" with "independently existing" like trying to combine "triangle" with "four-sided"? We know what each term designates when taken alone but when we try to combine them, they both evaporate. Notice that if correct, this undermines both the eliminative and causal versions of the reductionist strategy for explanation. I believe it is along this and other lines of criticism that naturalism and materialism can best be challenged, rather than by trying to infer God from nature.

But in any case, theists need not throw out evolutionary theory on religious grounds and still less should they be tempted to do so because of any materialist confession of faith that may be expressed in conjunction with it. We should never let a scientist who is not a theist get away with passing off his naturalist faith *as* science. Richard Lewontin, for one, admits this difference:

> It is not that the methods and institutions of science somehow compel us to accept a material explanation of the world, but on the contrary, that we are forced by our prior adherence to material causes to create an apparatus of investigation and a set of concepts that produce material explanations, no matter how counterintuitive. . . . Moreover, that materialism is absolute, for we cannot allow a divine foot in the door. . . . (*New York Review of Books*, January 7, 1997, p. 31)

The proper response to that is to point to the benefits for science of the nonreductionist strategy for explanation resulting from our prior adherence to our belief in God. From that perspective a theist may appropriate every element of truth that evolution or any other scientific theory may uncover, so long as it is regarded as the creation of God and for that reason is interpreted within a nonreductionist conceptual framework.[22]

## Notes

1. See G. Wright and R. Fuller, *The Book of the Acts of God* (Garden City: Anchor Books, 1957), pp. 17–29.

2. Cmp. G. Vos, *Biblical Theology* (Grand Rapids: Eerdmans, 1948), pp. 19–44; and Wright and Fuller, *The Book of the Acts of God*, pp. 47–59.

3. This distinguishes the biblical creation story from all pagan accounts in which some part or force in the natural world is regarded as divine. It is worth noting, however, that after God calls the universe into existence from nothing, the formation of life-forms is not described that same way. Instead, the wording suggests some natural process was involved in that God commanded that the *waters* bring forth life.

4. For a detailed defense of this view see N. H. Ridderbos, *Is There a Conflict between Genesis 1 and Natural Science?* (Grand Rapids: Eerdmans, 1957).

5. Roger Lewin has noticed the importance of how "human" is defined, and reports that the currently popular idea is to regard the capacity for ethics as the defining difference. In explaining this he includes the ability to ask the question "why are we here?" as a prime example of ethical consciousness. But as we shall see shortly, this question goes beyond ethics alone and is religious. See *Human Evolution* (New York: Freeman Co., 1984), pp. 24, 25, 98 ff.

6. This strikes some people as odd, but wouldn't it have to be true on virtually any account of human origins? Some individual would have to have been the first to qualify as human no matter how "human" is defined. It will not do to object that perhaps the defining qualities are relational, as Aristotle and Marx held, for prehumans already lived in social groups. Whatever distinguishes humans would

therefore have to be new capacities in individuals even if those could only be actualized socially.

7. Although there is a long theological tradition taking Adam and Eve to be the biological ancestors of all humans, I can find no warrant for it anywhere in the Hebrew Scriptures or the New Testament. The closest thing to any assertion on the topic is Adam's remark calling Eve "the mother of all living"; but that is in connection with her being promised that one of her descendants will be the Messiah. So it appears that this refers to the full sense of "life" that includes the right relationship to God, rather than to merely biological descent. The point is even clearer in the New Testament where Jesus is said to be the Messiah and so to be the "new Adam." His headship of the human race is explicitly and exclusively religious, since he was not the ancestor of anyone.

8. Other alleged incompatibilities are that Genesis teaches there were no such things as death, weeds, or pain in childbirth until Adam and Eve disobeyed God. That is a patent misreading. As the first to be put on probation relative to God's covenant, Adam and Eve were placed in a specially protected environment Genesis calls "the garden of God." Once exiled from that place of special protection, they were subject to all the vicissitudes of life from which they would otherwise have been shielded. That this is the viewpoint of the text can be seen by comparing Gen. 3:24 with Joshua 5: 13–15. In the first, an armed angel of God blocks Adam and Eve from the tree of life while in the second an armed angel leads God's people into the Promised Land—another area of special protection. See esp. Joshua's remark at Numbers 14:9.

9. See chapter 1 of *Knowing with the Heart: Religious Experience and Belief in God,* (Downer's Grove: InterVarsity Press, 1999) and chapters 1 and 2 of *The Myth of Religious Neutrality* (Notre Dame: University of Notre Dame Press, 1991).

10. This definition was held by virtually all the presocratic thinkers; see Werner Jaeger's *The Theology of the Early Greek Philosophers* (New York: Oxford University Press, 1960). It was held by both Plato (*Tim.* 50 ff., *Phil.* 53–54) and Aristotle (*Meta.* 1064a34, 1074b1–13), and has been rediscovered often since. In the past century alone it has been held by: William James, *The Varieties of Religious Experience* (New York: Longmans, Green, 1927), pp. 31, 33; A. C. Bouquet, *Comparative Religion* (Baltimore: Penguin, 1973), pp. 21, 38, 45, 48; Mircea Eliade, *Patterns in Comparative Religion* (New York: New American Library, 1974), pp. 24–30; Herman Dooyeweerd, *A New Critique of Theoretical Thought* (Philadelphia: Presbyterian & Reformed Pub. Co., 1955), I: p. 57; N. K. Smith, *The Credibility of Divine Existence* (New York: St Martin's, 1967), p. 396; Paul Tillich, *Systematic Theology* (Chicago: University of Chicago Press, 1951), I: pp. 9, 13; Hans Kung, *Christianity and the World Religions* (Garden City: Doubleday, 1986), p. xiv; and C. S. Lewis, *Miracles* (New York: Macmillan, 1948), pp. 16–20, 99–107.

11. "If ordinary 'matter' must be regarded merely as a highly natural, unconsciously constructed mental symbol for a . . . complex of [sensations], much more must this be the case with the artificial hypothetical atoms and molecules of

physics and chemistry." From "The Conservation of Energy" in Ernst Blackmore's *Ernst Mach* (Berkeley: University of California Press, 1972), p. 49. "One must not attempt to explain sense-perception. It is something so simple and fundamental, that the attempt to trace it back to something simpler, at least at the present time, can never succeed." From *Knowledge and Error* (Dordrecht: Reidel, 1975), 441.

12. "Our psychological experience contains ... sense experiences ... even the concept of the 'real external world' of everyday thinking rests exclusively on sense impressions ... what we mean when we attribute to the bodily object 'a real existence' ... [is] that, by means of such concepts ... we are able to orient ourselves in the labyrinth of sense perceptions." From *Ideas and Opinions* (New York: Bonanza Books, 1954), pp. 290–291. "I cannot conceive of a God who rewards and punishes his creatures, or who has a will of the kind we experience in ourselves ... I am satisfied with the mystery of the eternity of life and with ... a glimpse of the marvelous structure of the existing world together with ... the Reason that manifests itself in nature" (ibid., p. 11).

13. *Physics and Philosophy* (New York: Harper, 1958), pp. 72–73.

14. By "presuppose" I mean (roughly): a belief, *q*, presupposes another belief, *p*, iff in order to know or believe that *q* one must know or believe that *p* on grounds other than *q*. In *The Myth of Religious Neutrality* I have defined "presuppose" more precisely (pp. 101–107) and have demonstrated the regulative role of divinity-presuppositions for major theories in mathematics, physics, psychology, sociology, and politics.

15. Such nonreductionist metaphysics is not merely a future hope. The Dutch philosopher Herman Dooyeweerd constructed just such a theory and applied it to a number of scientific problems and theories. See *A New Critique of Theoretical Thought*, (Philadelphia: Presbyterian & Reformed Pub. Co., 1955, 4 vols.). I have summarized this theory's main points in the last four chapters of *The Myth of Religious Neutrality*.

16. Many theists have thought to neutralize this point by maintaining a reductionist theory but adding that whatever it is that all created reality reduces to, this in turn depends on God. Two objections: (1) From the theistic standpoint, this ploy runs afoul of the repeated biblical claims that belief in God affects every sort of knowledge and all truth (Ps. 111:10; Prov. 1:7; Luke 11:52; I Cor. 1:5; Eph. 5:9). In violation of this teaching, the ploy renders belief in God a fifth wheel making no difference to the content of a theory. (2) From the standpoint of philosophy, there are a host of insurmountable difficulties in bringing off any reduction and I will shortly add another to the list. Reduction remains a vain hope, not a reality. See my "On the General Relation of Religion, Metaphysics, and Science" in *Facets of Faith and Science*, ed. J. M. van der Meer (Lanham: University of America Press, 1996), pp. 57–79.

17. The expression "equally real" in this sentence does not mean to deny that some kinds of properties are preconditions for the appearance of others. So, for example, it is not denied that a thing has to have physical properties in order to have biotic properties. The nonreductionist view I have in mind allows this to be

true while still maintaining that there exist biotic laws and objective biotic potentialities in nonliving things that are necessary (but not sufficient) for the appearance of distinctly biotic properties in lining things. Important distinctions need to be made in this connection between the active and passive ways things can possess properties, as well as between the actual and potential ways. See *The Myth of Religious Neutrality*, pp. 212–216.

18. *Personal Knowledge* (New York: Harper & Row, 1962), p. 167.

19. Cf. Calvin: "As to the question, How shall we be persuaded that [Scripture] came from God? . . . It is just the same as if we were asked, How shall we learn to distinguish light from darkness, white from black, sweet from bitter? Scripture bears on the face of it such evidence of its truth as white and black do of their color, sweet and bitter of their taste" (*Inst*. I, 7, 2). "[Thus] they who strive to build up a firm faith in scripture by disputation are doing things backwards ..." (I, 7, 4). "Scripture, carrying its own evidence along with it, deigns not to submit to proofs and arguments, but owes the full conviction with which we ought to receive it to the testimony of the Spirit of God" (I, 7, 5).

20. This point also serves to answer another objection to the claim that scientific theories are religiously regulated. The objection is that a scientist may very well work with a theory as a skeptic, without actually believing it to be true. In that case, the objection goes, wouldn't the theory fail to be religiously regulated? The answer is that so long as the theory employs reductionist *concepts*, it still exhibits the sort of regulation I've been pointing to. A scientist may, indeed, not believe the particular hypothesis being tested. But so long as its initial conditions, background assumptions, or the nature of the postulated entities are conceived reductionistically, the regulative affect of some divinity belief remains in force.

21. *The Dynamics of Faith* (New York: Harper & Row, 1957), 83.

22. This position therefore provides a clear stand respecting the debate as to whether theists should join with naturalists in employing a common "naturalist methodology" in science. If this meant only that theists should not offer miracles or other special interventions by God as scientific explanations, I would agree entirely. But that is not all that's at stake. We've now seen why there is a theistic (nonreductionist) perspective for science that is at least as different from all the various reductionist perspectives (materialism, dualism, positivism, rationalism, pragmatism, historicism, etc.) as those are from one another.

# VII

## Intelligent Design and Information

Next to the Paley-inspired creationist challenge that evolutionists cannot explain irreducible complexity (section IV), the second most common challenge is that they cannot explain the "information" in the genetic material. This section takes up the two main aspects of that second challenge as it is now put forward by intelligent design creationists.

The first three essays deal with Phillip Johnson's take on the significance of genetic information. He argues that it calls into question not just Darwinian evolution, but materialism itself. Challenging Oxford biologist Richard Dawkins's claim that the discovery of the structure of DNA and the genetic code dealt the final blow to vitalism and vindicated the mechanistic view of life, Johnson quotes evolutionary biologist George C. Williams to the effect that information and matter are "more or less incommensurable domains" that have to be considered separately. Johnson says that except for a couple of unexplained phrases, Williams sounds like he is endorsing "metaphysical dualism." He claims that biologists have been slow to recognize the significance of genetic information, and writes "Our everyday experience is that linguistic communications . . . cannot be reduced to the material medium in which they are encoded and emerge only from pre-existing intelligence." He concludes that if information and matter belong to separate domains, "then no materialistic evolutionary theory can explain the most important element in life."

Following Johnson's article are Dawkins's and Williams's pithy replies. Dawkins writes that if Johnson were correct he would be hailed as the greatest philosopher of the age, but that, in fact, his argument is specious and misleading. He concludes that the only interesting question that the paper raises is "whether Phillip Johnson really believes the stuff he writes or whether he is just a good lawyer." Williams takes less than a page to explain the "unexplained phrases" that Johnson had noted and says that Johnson's argument is "based on some obvious fallacies, such as information requiring an intelligent author."

The next three essays deal with William Dembski's attempt to show that the kind of information found in DNA does require an intelligent author. He argues that natural causes are incapable of generating complex specified information, which he calls the "law of the conservation of information." Since biological organisms contain complex specified information, he claims, they can only have been produced by nonnatural, intelligent design. In the ID movement, Dembski is the chief advocate of this position, and others regularly cite him as their authority on it. The article reproduced here gives the key elements of his view, explaining the conceptual framework of what he calls "the design inference." Dembski credits University of Texas philosopher Rob Koons for spurring him to write this paper to deliver at the conference showcasing intelligent design (Naturalism, Theism, and the Scientific Enterprise) that he organized in 1997. Koons picked this paper to publish in the issue that he guest-edited of *Perspectives on Science and Christian Faith* (1997, vol. 49, no. 3) devoted to the conference. Dembski expanded the paper and included it as the key chapter in his "pivotal, synthesizing" book, *Intelligent Design: The Bridge between Science and Theology* (InterVarsity, 1999). Making special reference to this chapter in his endorsement of the book, Koons writes that "William Dembski is the Isaac Newton of information theory, and since this is the Age of Information, that makes Dembski one of the most important thinkers of our time. His 'law of conservation of information' represents a revolutionary breakthrough."

Peter Godfrey-Smith, a philosopher of biology at Stanford University, differs in his assessment of Dembski's position. He argues in his essay, "Information and the Argument from Design," written for this volume, that Dembski's argument achieves only "an apparent sophistication" because of the information-theoretic terms in which it is expressed, and that when analyzed carefully it is actually "one of the least plausible versions of the design argument," in part because of his overly sweeping proposed conservation law.

Dembski usually responds to criticisms of his informal presentations of his critique of Darwinism and naturalism by referring critics to his book *The Design Inference* (Cambridge, 1998), where he gives the full, formal treatment of his design inference. The third essay in this section, by philosophers Branden Fitelson, Christopher Stephens, and Elliott Sober, is a critical evaluation of Dembski's technical argument. Dembski's book actually devotes only a few pages to either Darwinism or naturalism and does not explicitly dispute either, so Fitelson, Stephens, and Sober make little reference to those issues and devote most of their essay to showing errors in Dembski's formal account, which they say is "deeply flawed." They conclude that "[n]either creationists nor evolutionists nor people who are trying to detect design in nontheological contexts should adopt Dembski's framework."

Of course, the objection that science cannot explain biological information is neither unique nor original to the intelligent design movement, but has been regularly made by creationists over the years. In the final essay, "The 'Information Challenge,'" Richard Dawkins addresses a key aspect of the creationists' information challenge directly. One problem with it is that creationists (including intelligent design creationists) do not use the term in a careful and consistent manner. Dawkins begins by explaining a technical notion of information as defined by Claude Shannon, one of the founders of information theory. He sorts out how this notion of information applies to information in the genome and in the body in general. He then explains how the question of biological information is connected to the question of biological complexity, showing that the information challenge is simply a variation of the old question, asked by William Paley and answered by Darwin, of how something like the complex human eye could have formed. Dawkins briefly outlines how evolution can increase biological information and concludes, "We need only a little poetic license to say that the information fed into modern genomes by natural selection is actually information about ancient environments in which ancestors survived."

# 22

# Is Genetic Information Irreducible?

Phillip E. Johnson

In his 1992 book *Natural Selection: Domains, Levels, and Challenges*, George C. Williams stated that organisms consist of two separate domains, which Williams called the "material" and the "codical." The former is the domain of chemistry, including particularly the nucleotides of DNA and their chemical components. The codical domain consists of non-material *information*. The information encoded in DNA is fundamentally distinct from the chemical medium in which the information is recorded, just as the information conveyed in a book is fundamentally distinct from the ink and paper on which it is printed.

Williams explicitly invoked this analogy between biology and literature:

> *Don Quixote* is information, most often coded as a pattern of ink on paper, but it can exist in many other media. It is often transmitted visually and stored ephemerally in human brains. It has no doubt been recorded on disks and magnetic tapes for transduction into sound for transmission to brains of people unable to read ink on paper. In any medium, *Don Quixote* can form an archive from which copies can be made to any other medium, but no matter what the medium, it is always the same book [Williams, 1992, p. 11].

In an interview published in 1995, Williams made the consequences of the analogy for biology still more explicit:

> Evolutionary biologists have failed to realize that they work with two more or less incommensurable domains: that of information and that of matter.... These two domains can never be brought together in any kind of the sense usually implied by the term "reductionism." You can speak of galaxies and particles of dust in the same terms, because they both have mass and charge and length and width. You can't do that with information and matter. Information doesn't have

Originally published in *Biology & Philosophy* (1996, vol. 11, no. 4, pp. 535–538).

mass or charge or length in millimeters. Likewise, matter doesn't have bytes.... This dearth of shared descriptors makes matter and information two separate domains of existence, which have to be discussed separately, in their own terms.

The gene is a package of information, not an object. The pattern of base pairs in a DNA molecule specifies the gene. But the DNA molecule is the medium, it's not the message. Maintaining this distinction between the medium and the message is absolutely indispensable to clarity of thought about evolution.... In biology, when you're talking about things like genes and genotypes and gene pools, you're talking about information, not physical objective reality [Brockman, 1995, p. 43].

That sounds like straightforward metaphysical dualism—except insofar as it is weakly qualified by two unexplained phrases, one near the beginning of the passage and the other at the end. If genetic information and matter are only "more or less" incommensurable, then they are not incommensurable at all. Incommensurability, like omnipotence or uniqueness, is an all-or-nothing quality. If the closing phrase implies that only physical entities can be objectively real, it denies what the rest of the passage seems to assert. Information is a separate, non-physical domain of objective reality. The information contained in *Don Quixote*, or the Chicago telephone directory, is no less objectively real than the paper on which it is printed.

If matter and information are truly incommensurable, then Darwinism cannot be true as a theory of information creation. Darwinism is based upon materialism, and posits the creation of genetic information by the material forces of random mutation and natural selection. If the analogy to literature is valid, this is like trying to explain the origin of *Don Quixote* by invoking only the chemistry of ink and paper and omitting the contribution of the author.

No one denies that a book requires an author, but scientific materialists vehemently deny that genetic information comes from an intelligent source. One of these materialists is Richard Dawkins, who recently wrote that the discovery of the structure of DNA and the genetic code

has dealt the final, killing blow to vitalism—the belief that living material is deeply distinct from nonliving material. Up until 1953 it was still possible to believe that there was something fundamentally and irreducibly mysterious in living protoplasm. No longer. Even those philosophers who had been predisposed to a mechanistic view of life would not have dared hope for such total fulfillment of their wildest dreams [Dawkins 1995, p. 17].

The quoted passages from Williams imply that, on the contrary, there is something fundamentally distinct in living organisms, and it is the genetic information.[1] The discovery that some or all of this information is recorded in DNA, as *Don Quixote* and the telephone directory are recorded on paper, in no way implies that the message is reducible to the medium. That the genetic information is irreducible does not mean, however, that it is "mysterious" in the sense of being inaccessible to normal experience. Information is as much a part of everyday life as matter itself. Perhaps that is part of the reason that biologists have been slow to appreciate the significance of the irreducibility of genetic information. Their philosophy assumes that the alternative to strict materialism in biology is a mysterious "elan vital," a term that suggests something ghostlike or uncanny.

The materialist rejoinder to a matter/information dualism must be that the special qualities of information are an instance of "emergence," meaning the tendency of unpredictable properties to appear when different kinds of matter are combined in certain ways. Just as water has qualities not present in the particles that make up hydrogen and oxygen, DNA has properties not present in the elements and compounds of which it consists. According to materialism, genetic information is not a distinct metaphysical entity but an emergent property of organic chemistry—and so, by a more indirect route, is the mind of Cervantes, the author of *Don Quixote*. To a materialist reductionist, clear thinking about biology requires not maintaining the distinction between the medium and the message, but, on the contrary, attributing all seeming non-material entities to the material base from which they emerged.

Is genetic information a distinct non-material substance, or an emergent property of DNA chemistry? If there is disagreement, how can the dispute be resolved by scientific testing—meaning testing that does not assume the very materialist reductionism which is in question? If we judge the dispute in terms of the evidence available today, the dualists have the stronger case. The arrangement of nucleotides directing protein synthesis is no more determined by the chemistry of amino acids than the arrangement of letters on a page of *Don Quixote* is determined by the chemistry of ink and paper. If chemical laws dictated the arrangement of letters and words, then books would contain no information not already contained in the laws of chemistry.

Darwinism attributes information creation to "mutation," but this is to give the phenomenon a label rather than an explanation. (Additional terms like "variation" or "recombination" are even less informative.) Our everyday experience is that linguistic communications, including computer software like Dawkins' own blind watchmaker program, cannot be reduced to the material medium in which they are encoded and emerge only from preexisting intelligence. Materialists may be satisfied to *assume* that intelligence is the product of unintelligent causes, but non-materialists will want to see proof.

The computer analogy answers the classic argument against mind-body dualism, which is that for non-material "mind stuff" to influence matter would violate the known laws of physics. There is no violation of physical law when information is encoded in matter in order to permit matter to be organized in new and surprising ways. The information in the computer software (and the hardware too, for that matter) none-theless comes ultimately from an intelligent source outside the computer, and is not merely an emergent property of silicon and electricity in com-bination. Why should this not also be true of the vastly more complex genetic information?

Understanding the irreducibility of the genetic information puts the so-called evolution/creation debate in a completely new light. Much of what biologists call "evolution"—e.g., selective breeding, industrial melanism in the peppered moth, and the famous variations among the finches of the Galapagos—fits comfortably within a creationist para-digm. The point of creationism is to assert that God is the creator, not that God created types of organisms in so inflexible a form that they cannot vary in response to environmental changes. If the term "evolution" had not become loaded with naturalistic philosophical con-notations, and if creation had not become misleadingly identified with what Darwin called the "fixity of species," one could say that "special creation" and "polyphyletic evolution" are two ways of describing what observation has revealed: a substantial amount of variation occurs in nature, but it occurs within boundaries.

The fundamental issue in dispute between creationists and scientific materialists is not whether "evolution occurs," but whether unintelli-gent material processes created the genetic information that manifestly exists. If George Williams is correct that information and matter belong to separate domains, and if the domain of information is therefore not a product of the domain of matter, then no materialistic evolutionary theory can explain the most important element in life.

## Note

1. Williams singled out Dawkins as one who "defines a replicator in a way that makes it a physical entity duplicating itself in a reproductive process," adding that Dawkins "was misled by the fact that genes are always identified with DNA" [Brockman, 1995, p. 42].

## References

Brockman, J.: 1995, *The Third Culture: Beyond the Scientific Revolution*, Simon & Schuster.

Dawkins, R.: 1995, *River Out of Eden: A Darwinian View of Life* Basic Books.

Williams, G.: 1992, *Natural Selection: Domains, Levels, and Challenges* Oxford University Press.

# 23

# Reply to Phillip Johnson

Richard Dawkins

When I open a page of Darwin I immediately sense that I have been ushered into the presence of a great mind. I have the same feeling with RA Fisher and with GC Williams. When I read Phillip Johnson, I feel that I have been ushered into the presence of a lawyer. A few years ago there was one called Norman MacBeth who wrote a book called *Darwin Retried*, in which he made a great song and dance about bringing the acumen and training of the "legal mind" to bear upon Darwinism. His stated conclusion was that Darwinism would not stand up in a court of law, but the only thing he demonstrated with any enduring success was the limitations of the legal mind. His book is now happily forgotten, but Phillip Johnson seems to have the same style of argument.

It is in the nature of the adversarial system for settling disputes that advocates must on average be right on 50% of occasions and wrong on 50% of occasions. The advocate is paid to argue with the same conviction when wrong as when right, and he is paid to seize upon any point, however specious or even stupid, which may influence a jury in the desired direction. When a lawyer seizes upon a stupid argument we may pay him—along with his fat fee—the compliment of assuming him to be a blackguard not a fool, knowingly cashing in on the limitations of the jury. But a case can be made that really successful advocates do a better job of misleading others if they have first misled themselves.

The point George Williams is making in his Don Quixote analogy is an extremely simple one, but it is just tricky enough to have jury-misleading potential if used aright. As Peter Medawar wittily re-expressed the

Originally published in *Biology & Philosophy* (1996, vol. 11, no. 4, pp. 539–540).

dangers of a little learning, "... the spread of secondary and latterly of tertiary education has created a large population of people, often with well-developed literary and scholarly tastes, who have been educated far beyond their capacity to undertake analytical thought." Williams was saying, correctly, that information is not reducible to the material medium in which it is encoded, for the simple reason that the medium may change from one material (say paper) to another (say magnetic tape). Johnson, almost incredibly, tries to turn this into an argument against materialism and in favour of dualism. If it were that easy, there would be nothing left for philosophers to do. Johnson would not only have triumphed over Darwinism. He would also be hailed as the greatest philosopher of the age, for his brilliant demonstration that dualism is finally proved by the existence of the humble tape recorder.

See how the trick is done. Don Quixote was written by a creative, intelligent human author, Cervantes. It is also a stream of alphabetic characters on paper which can be translated faithfully into a different medium, say a magnetic tape, a CD-ROM or a punched paper tape. Both these points—the creative author and the independence of data from its medium—are true and both are important, but they are not the same as each other. Williams would not deny the first but, as it happens, it was the second point that he was making. To see the distinction clearly, simply replace Don Quixote in his paragraph by a hypothetical book of nonsense character strings. Any book is not the same as the ink it is written in, because it could be etched on a CD-ROM and would still be the same book. The point has nothing to do with emergence, nor— though Williams and I would accept both as valid, separate points—does it have any connection with Cervantes's creative intelligence. Johnson, with a true lawyer's instinct for the misleading, has pounced unerringly on the wrong point. The only interesting question raised by his paper— and it is not a VERY interesting question—is whether Phillip Johnson really believes the stuff he writes or whether he is just a good lawyer.

# 24

# Reply to Johnson

George C. Williams

I can easily explain the "unexplained phrases" that Johnson noted in my writings. The material and codical domains are only "more or less" in-commensurable because they do in fact share one descriptor, *time*. As noted in the 1992 work cited, the conversion of the material library at Alexandria into $CO_2$ and $H_2O$ coincided in time with the destruction of the information in that library. My "physical objective reality" merely meant the material domain as opposed to that of information. I neither doubt that information can be treated objectively nor deny Johnson's point that it is part of everyday life.

Johnson's argument is based on some obvious fallacies, such as infor-mation requiring an intelligent author. The pattern of slow moving waves in sand dunes records information about what the wind has been doing lately. Their shadow pattern observed late in the day is informa-tion about the structure of the dunes and less directly about the wind. The only author recognizable here is the wind. Similar patternings must arise in any complex molecular mixture, including the prebiotic. If one kind of molecular pattern influences others in ways that increase the in-cidence of that pattern, a hypercycle subject to natural selection has arisen. This would be analogous to some pattern of dune shadows mak-ing it more likely that the responsible winds would occur more fre-quently. That the author of genetic information is as stupid as the wind is apparent in the functionally stupid historical constraints discussed in Chapter 6 of my 1992 book.

I suppose that evolutionists do claim that natural selection creates genetic information, but I think their intended meaning is that it edits

Originally published in *Biology & Philosophy* (1996, vol. 11, no. 4, pp. 541).

information. The original texts were chemically generated patterns, informative only of conditions prevalent at their points of origin. Subsequent editing included such things as the selective saving of transposed elements and tandem duplications, and the fusion of the genetic texts of previously separate lineages. An example of such fusion is the origin of eukaryote cells perhaps a billion years ago. I see no reason to doubt that this sort of editing could produce books much bigger than *Don Quixote* in a billion years.

# 25

# Intelligent Design as a Theory of Information

William A. Dembski

For the scientific community, intelligent design represents creationism's latest grasp at scientific legitimacy. Accordingly, intelligent design is viewed as yet another ill-conceived attempt by creationists to straightjacket science within a religious ideology. But, in fact, intelligent design can be formulated as a scientific theory having empirical consequences and devoid of religious commitments. Intelligent design can be unpacked as a theory of information. Within such a theory, information becomes a reliable indicator of design as well as a proper object for scientific investigation. In my essay, I shall (1) show how information can be reliably detected and measured, and (2) formulate a conservation law that governs the origin and flow of information. My broad conclusion is that information is not reducible to natural causes, and that the origin of information is best sought in intelligent causes. Intelligent design, thereby, becomes a theory for detecting and measuring information, explaining its origin, and tracing its flow.

## Information

In *Steps Towards Life*, Manfred Eigen identifies what he regards as the central problem facing origins-of-life research: "Our task is to find an algorithm, a natural law that leads to the origin of information."[1] Eigen is only half right. To determine how life began, it is indeed necessary to understand the origin of information. Even so, neither algorithms nor natural laws can produce information. The great myth of modern evolu-

Originally published in *Perspectives on Science & Christian Faith* (1997, vol. 49, no. 3, pp. 180–190).

tionary biology is that information can be gotten on the cheap without recourse to intelligence. It is this myth I seek to dispel, but to do so I shall need to give an account of information. No one disputes that there is such a thing as information. As Keith Devlin remarks:

> Our very lives depend upon it, upon its gathering, storage, manipulation, transmission, security, and so on. Huge amounts of money change hands in exchange for information. People talk about it all the time. Lives are lost in its pursuit. Vast commercial empires are created in order to manufacture equipment to handle it.[2]

But what exactly is information? The burden of this paper is to answer this question, presenting an account of information that is relevant to biology.

The fundamental intuition underlying information is not, as is sometimes thought, the transmission of signals across a communication channel, but rather the actualization of one possibility to the exclusion of others. As Fred Dretske puts it:

> Information theory identifies the amount of information associated with, or generated by, the occurrence of an event (or the realization of a state of affairs) with the reduction in uncertainty, the elimination of possibilities, represented by that event or state of affairs.[3]

To be sure, whenever signals are transmitted across a communication channel, one possibility is actualized to the exclusion of others, namely, the signal that was transmitted to the exclusion of those that weren't. But this is only a special case. Information in the first instance presupposes not some medium of communication, but contingency. Robert Stalnaker makes this point clearly: "Content requires contingency. To learn something, to acquire information, is to rule out possibilities. To understand the information conveyed in a communication is to know what possibilities would be excluded by its truth."[4] For there to be information, there must be a multiplicity of distinct possibilities, any one of which might happen. When one of these possibilities does happen and the others are ruled out, information becomes actualized. Indeed, information in its most general sense can be defined as the actualization of one possibility to the exclusion of others (observe that this definition encompasses both syntactic and semantic information).

This way of defining information may seem counterintuitive since we often speak of the information inherent in possibilities that are never actualized. Thus we may speak of the information inherent in flipping one-hundred heads in a row with a fair coin even if this event never happens. There is no difficulty here. In counterfactual situations, the definition of information needs to be applied counterfactually. Thus to consider the information inherent in flipping one-hundred heads in a row with a fair coin, we treat this event/possibility as though it were actualized. Information needs to be referenced not just to the actual world, but also cross-referenced with all possible worlds.

## Complex Information

How does our definition of information apply to biology, and to science more generally? To render information a useful concept for science we need to do two things: first, show how to measure information; second, introduce a crucial distinction—the distinction between *specified* and *unspecified* information. First, let us show how to measure information. In measuring information, it is not enough to count the number of possibilities excluded, and offer this number as the relevant measure of information. The problem is that a simple enumeration of excluded possibilities tells us nothing about how those possibilities were individuated in the first place. Consider, for instance, the following individuation of poker hands:

#1 A royal flush.
#2 Everything else.

To learn that something other than a royal flush was dealt (i.e., possibility #2) is clearly to acquire less information than to learn that a royal flush was dealt (i.e., possibility #1). Yet if our measure of information is simply an enumeration of excluded possibilities, the same numerical value must be assigned in both instances since in both instances a single possibility is excluded.

It follows, therefore, that how we measure information needs to be independent of whatever procedure we use to individuate the possibilities under consideration. The way to do this is not simply to count possibilities, but to assign probabilities to these possibilities. For a thoroughly shuffled deck of cards, the probability of being dealt a royal flush (i.e., possibility #1) is approximately .000002 whereas the probability of being dealt anything other than a royal flush (i.e., possibility #2) is approximately .999998. Probabilities by themselves, however, are not information measures. Although probabilities properly distinguish possibilities according to the information they contain, nonetheless probabilities remain an inconvenient way of measuring information. There are two reasons for this. First, the scaling and directionality of the numbers assigned by probabilities need to be recalibrated. We are clearly acquiring more information when we learn someone was dealt a royal flush than when we learn someone wasn't dealt a royal flush. And yet the probability of being dealt a royal flush (i.e., .000002) is minuscule compared to the probability of being dealt something other than a royal flush (i.e., .999998). Smaller probabilities signify more information, not less.

The second reason probabilities are inconvenient for measuring information is that they are multiplicative rather than additive. If I learn that Alice was dealt a royal flush playing poker at Caesar's Palace and that Bob was dealt a royal flush playing poker at the Mirage, the probability that both Alice and Bob were dealt royal flushes is the product of the individual probabilities. Nonetheless, it is convenient for information to be measured additively so that the measure of information assigned to Alice and Bob jointly being dealt royal flushes equals the measure of information assigned to Alice being dealt a royal flush plus the measure of information assigned to Bob being dealt a royal flush.

An obvious way to transform probabilities that circumvents both these difficulties is to apply a negative logarithm to the probabilities. Applying a negative logarithm assigns the more information to the less probability and transforms multiplicative probability measures into additive information measures, because the logarithm of a product is the sum of the logarithms. What's more, in deference to communication theorists, it is customary to use the logarithm to the base 2. The rationale for this choice of logarithmic base is as follows. The most convenient way for communication theorists to measure information is in bits. Any message sent across a communication channel can be viewed as a string of 0's and 1's. For instance, the ASCII code uses strings of eight 0's and 1's to represent the characters on a typewriter, with whole words and sentences in turn represented as strings of such character strings. In like manner, all communication may be reduced to the transmission of sequences of 0's and 1's. Given this reduction, the obvious way for communication theorists to measure information is in the number of bits transmitted across a communication channel. Since the negative logarithm to the base 2 of a probability corresponds to the average number of bits needed to identify an event of that probability, the logarithm to the base 2 is the canonical logarithm for communication theorists. Thus, we define the measure of information in an event of probability $p$ as $-\log_2 p$.[5]

What about the additivity of this information measure? Recall the example of Alice being dealt a royal flush playing poker at Caesar's Palace and Bob being dealt a royal flush playing poker at the Mirage. Let's call the first event A and the second B. Since randomly dealt poker hands are probabilistically independent, the probability of A and B taken jointly equals the product of the probabilities of A and B taken individually. Symbolically, $P(A\&B) = P(A) \times P(B)$. Given our logarithmic definition of information, we thus define the amount of information in an event E as $I(E) =_{\mathrm{def}} -\log_2 P(E)$. It then follows that $P(A\&B) = P(A) \times P(B)$ if and only if $I(A\&B) = I(A) + I(B)$. Since in the example of Alice and Bob $P(A) = P(B) = .000002$, $I(A) = I(B) = 19$, and $I(A\&B) = I(A) + I(B) = 19 + 19 = 38$. Thus the amount of information inherent in Alice and Bob jointly obtaining royal flushes is 38 bits.

Since many events are probabilistically independent, information measures exhibit much additivity. But since many events are also correlated, information measures exhibit much nonadditivity as well. In the example of Alice and Bob, Alice being dealt a royal flush is probabilistically independent of Bob being dealt a royal flush, and so the amount of information in Alice and Bob both being dealt royal flushes equals the sum of the individual amounts of information.

Now let's consider a different example. Alice and Bob together toss a coin five times. Alice observes the first four tosses but is distracted, and so misses the fifth toss. On the other hand, Bob misses the first toss, but observes the last four tosses. Let's say the actual sequence of tosses is 11001 (1 = heads, 0 = tails). Thus Alice observes 1100* and Bob observes *1001. Let A denote the first observation, B the second. It follows that the amount of information in A&B is the amount of information in the completed sequence 11001, namely, 5 bits. On the other hand, the amount of information in A alone is the amount of information in the incomplete sequence 1100*, namely 4 bits. Similarly, the amount of information in B alone is the amount of information in the incomplete sequence *1001, also 4 bits. This time the information doesn't add up: $5 = I(A\&B) \neq I(A) + I(B) = 4 + 4 = 8$.

Here A and B are correlated. Alice knows all but the last bit of information in the completed sequence 11001. Thus when Bob gives her the incomplete sequence *1001, all Alice really learns is the last bit in this sequence. Similarly, Bob knows all but the first bit of information in the completed sequence 11001. Thus when Alice gives him the incomplete sequence 1100*, all Bob really learns is the first bit in this sequence. What appears to be four bits of information actually ends up being only one bit of information once Alice and Bob factor in the prior information they possess about the completed sequence 11001. If we introduce the idea of conditional information, this is just to say that $5 = I(A\&B) = I(A) + I(B|A) = 4 + 1$. $I(B|A)$, the conditional information of B given A, is the amount of information in Bob's observation once Alice's observation is taken into account. This, as we just saw, is 1 bit.

$I(B|A)$, like $I(A\&B)$, $I(A)$, and $I(B)$, can be represented as the negative logarithm to the base two of a probability, only this time the probability under the logarithm is a conditional, as opposed to an unconditional, probability. By definition, $I(B|A) =_{def} -\log_2 P(B|A)$, where $P(B|A)$ is the conditional probability of B given A. But since $P(B|A) =_{def} P(A\&B)/P(A)$ and since the logarithm of a quotient is the difference of the logarithms, $\log_2 P(B|A) = \log_2 P(A\&B) - \log_2 P(A)$, and so $-\log_2 P(B|A) = -\log_2 P(A\&B) + \log_2 P(A)$, which is just $I(B|A) = I(A\&B) - I(A)$. This last equation is equivalent to

$$I(A\&B) = I(A) + I(B|A) \tag{*}$$

Formula (*) holds with full generality, reducing to $I(A\&B) = I(A) + I(B)$ when A and B are probabilistically independent, in which case $P(B|A) = P(B)$ and thus $I(B|A) = I(B)$.

Formula (*) asserts that the information in both A and B jointly is the information in A plus the information in B that is not in A. Its point, therefore, is to spell out how much additional information B contributes to A. As such, this formula places tight constraints on the generation of new information. Does, for instance, a computer program, call the program A, by outputting some data, call the data B, generate new information? Computer programs are fully deterministic, and so B is fully determined by A. It follows that $P(B|A) = 1$, and thus $I(B|A) = 0$ (the logarithm of 1 is always 0). From Formula (*) it therefore follows that I(A&B) = I(A), and that the amount of information in A and B jointly is no more than the amount of information in A by itself.

For an example in the same spirit, consider that there is no more information in two copies of Shakespeare's *Hamlet* than in a single copy. This is patently obvious, and any formal account of information had better agree. To see that our formal account does indeed agree, let A denote the printing of the first copy of *Hamlet*, and B the printing of the second copy. Once A is given, B is entirely determined. Indeed, the correlation between A and B is perfect. Probabilistically this is expressed by saying the conditional probability of B given A is 1, namely, P(B | A) = 1. In information-theoretic terms this is to say that I(B | A) = 0. As a result I(B | A) drops out of Formula (*), and so I(A&B) = I(A). Our information-theoretic formalism, therefore, agrees with our intuition that two copies of *Hamlet* contain no more information than a single copy.

Information is a complexity-theoretic notion. As a purely formal object, the information measure described here is a complexity measure.[6] Complexity measures arise whenever we assign numbers to degrees of complication. A set of possibilities will often admit varying degrees of complication, ranging from extremely simple to extremely complicated. Complexity measures assign non-negative numbers to these possibilities so that 0 corresponds to the most simple and $\infty$ to the most complicated. For instance, computational complexity is always measured in terms of either time (i.e., number of computational steps) or space (i.e., size of memory, usually measured in bits or bytes) or some combination of the two. The more difficult a computational problem, the more time and space are required to run the algorithm that solves the problem. For information measures, the degree of complication is measured in bits. Given an event A of probability P(A), $I(A) = -\log_2 P(A)$ measures the number of bits associated with the probability P(A). We therefore speak of the "complexity of information" and say that the complexity of information increases as I(A) increases (or, correspondingly, as P(A) decreases). We also speak of "simple" and "complex" information according to whether I(A) signifies few or many bits of information. This notion of complexity is important to biology since not just the origin of information stands in question, but also the origin of complex information.

## Complex Specified Information

Given a means of measuring information and determining its complexity, we turn now to the distinction between *specified* and *unspecified* information. This is a vast topic whose full elucidation is beyond the scope of this paper.[7] Nonetheless, in what follows I shall try to make this distinction intelligible, and offer some hints on how to make it rigorous. For an intuitive grasp of the difference between specified and unspecified information, consider the following example. Suppose an archer stands 50 meters from a large blank wall with bow and arrow in hand. The wall, let us say, is sufficiently large that the archer cannot help but hit it. Consider now two alternative scenarios. In the first scenario, the archer simply shoots at the wall. In the second scenario, the archer first paints a target on the wall, and then shoots at the wall, squarely hitting the target's bull's-eye. Let us suppose that in both scenarios the arrow lands in the same spot. In both scenarios, the arrow might have landed anywhere on the wall. What's more, any place where it might land is highly improbable. It follows that in both scenarios highly complex information is actualized. Yet the conclusions we draw from these scenarios are very different. In the first scenario, we can conclude absolutely nothing about the archer's ability as an archer, whereas in the second scenario, we have evidence of the archer's skill.

The obvious difference between the two scenarios is that in the first, the information follows no pattern, whereas in the second, it does. Now the information that tends to interest us as rational inquirers generally, and scientists in particular, is not the actualization of arbitrary possibilities which correspond to no patterns, but the actualization of circumscribed possibilities which *do* correspond to patterns. There's more. Patterned information, though a step in the right direction, still doesn't quite get us specified information. The problem is that patterns can be concocted after the fact so that instead of helping explain information, the patterns are merely read off already actualized information.

To see this, consider a third scenario in which an archer shoots at a wall. As before, we suppose the archer stands 50 meters from a large blank wall with bow and arrow in hand, the wall being so large that the archer cannot help but hit it. As in the first scenario, the archer shoots at the wall while it is still blank. This time suppose that after having shot the arrow, and finding the arrow stuck in the wall, the archer paints a target around the arrow so that the arrow sticks squarely in the bull's-eye. Let us suppose further that the precise place where the arrow lands in this scenario is identical with where it landed in the first two scenarios. Since any place where the arrow might land is highly improbable, highly complex information has been actualized as in the other scenarios. What's more, since the information corresponds to a pattern, we can even say that in this third scenario highly complex patterned information has been actualized. Nevertheless, it would be wrong to say that highly complex specified information has been actualized. Of the three scenarios, only the information in the second scenario is specified. In that scenario, by *first* painting the target and *then* shooting the arrow, the pattern is given independently of the information. On the other hand, in the third scenario, by first shooting the arrow and then painting the target around it, the pattern is merely read off the information.

Specified information is always patterned information, but patterned information is not always specified information. For specified information, not just any pattern will do. Therefore we must distinguish between the "good" patterns and the "bad" patterns. We will call the "good" patterns *specifications*. Specifications are the independently given patterns that are not simply read off information. By contrast, we will call the "bad" patterns *fabrications*. Fabrications are the *post hoc* patterns that are simply read off already existing information.

Unlike specifications, fabrications are wholly unenlightening. We are no better off with a fabrication than without one. This is clear from comparing the first and third scenarios. Whether an arrow lands on a blank wall and the wall stays blank (as in the first scenario), or an arrow lands on a blank wall and a target is then painted around the arrow (as in the third scenario), any conclusions we draw about the arrow's flight remain the same. In either case, chance is as good an explanation as any for the arrow's flight. The fact that the target in the third scenario constitutes a pattern makes no difference since the pattern is constructed entirely in response to where the arrow lands. Only when the pattern is given independently of the arrow's flight does a hypothesis other than chance come into play. Thus only in the second scenario does it make sense to ask whether we are dealing with a skilled archer. Only in the second scenario does the pattern constitute a specification. In the third scenario, the pattern constitutes a mere fabrication.

The distinction between specified and unspecified information may now be defined as follows: the actualization of a possibility (i.e., information) is specified if the possibility's actualization is independently identifiable by means of a pattern. If not, then the information is unspecified. Note that this definition implies an asymmetry between specified and unspecified information: specified information cannot become unspecified information, though unspecified information can become specified information. Unspecified information can become specified as our background knowledge increases. For example, a cryptographic transmission, whose cryptosystem we have yet to break, will constitute unspecified information. However, as soon as we break the cryptosystem, the cryptographic transmission becomes specified information.

What is it for a possibility to be identifiable by means of an independently given pattern? A full exposition of specification requires a detailed answer to this question. Unfortunately, such an exposition is beyond the scope of this paper. The key conceptual difficulty here is to characterize the independence condition between patterns and information. This independence condition breaks into two subsidiary conditions: (1) a condition to stochastic conditional independence between the information in question and particular relevant background knowledge; and (2) a tractability condition by which the pattern in question can be constructed from the aforementioned background knowledge. Though these conditions make good intuitive sense, they are not easily formalized.[8]

If formalizing what it means for a pattern to be given independently of a possibility is difficult, determining in practice whether a pattern is given independently of a possibility is much easier. If the pattern is given prior to the possibility being actualized—as in the second scenario above where the target was painted before the arrow was shot—then the pattern is automatically independent of the possibility, and we are dealing with specified information. Patterns given prior to the actualization of a possibility are just the rejection regions of statistics. There is a well-established statistical theory that describes such patterns and their use in probabilistic reasoning. These are clearly specifications since having been given prior to the actualization of some possibility, they have already been identified, and thus are identifiable independently of the possibility being actualized.[9]

Many interesting cases of specified information, however, are those in which the pattern is given *after* a possibility has been actualized. This is the case with the origin of life: life originates first and only afterwards do pattern-forming, rational agents (like ourselves) enter the scene. It remains the case, however, that a pattern corresponding to a possibility, though formulated after the possibility has been actualized, can constitute a specification. Certainly this was not so in the third scenario above, where the target was painted around the arrow only after it hit the wall. But consider the following example. Alice and Bob are celebrating their fiftieth wedding anniversary. Their six children all show up bearing gifts. Each gift is part of a matching set of china. There is no duplication of gifts, and together the gifts constitute a complete set of china. Suppose Alice and Bob were satisfied with their old set of china, and had no inkling prior to opening their gifts that they might expect a new set of china. Alice and Bob are therefore without a relevant pattern whither to refer their gifts prior to actually receiving the gifts from their children. Nevertheless, the pattern they explicitly formulate only after receiving the gifts could be formed independently of receiving the gifts—we all know about matching sets of china and how to distinguish them from unmatched sets. This pattern therefore constitutes a specification. What's more, there is an obvious inference connected with this specification: Alice and Bob's children were in collusion, and did not present their gifts as random acts of kindness.

But what about the origin of life? Is life specified? If so, to what patterns does life correspond, and how are these patterns given independently of life's origin? Obviously, pattern-forming rational agents like ourselves don't enter the scene till after life originates. Nonetheless, there are functional patterns to which life corresponds, and which are given independently of the actual living systems. An organism is a functional system comprising many functional subsystems. The functionality of organisms can be cashed out in any number of ways. Arno Wouters cashes it out globally in terms of viability of whole organisms.[10] Michael Behe cashes it out in terms of the irreducible complexity and minimal function of biochemical systems.[11] Even the staunch Darwinist Richard Dawkins admits that life is specified functionally, cashing out the functionality of organisms in terms of reproduction of genes. Thus he writes: "Complicated things have some quality, specifiable in advance, that is highly unlikely to have been acquired by random chance alone. In the case of living things, the quality that is specified in advance is ... the ability to propagate genes in reproduction."[12]

Information can be specified, complex, or both complex and specified. Information that is both complex and specified I call "complex specified information," or CSI for short. CSI is what all the fuss over information has been about in recent years, not just in biology, but in science generally. It is CSI that for Manfred Eigen constitutes the great mystery of biology, and one he hopes eventually to unravel in terms of algorithms and natural laws. It is CSI that for cosmologists underlies the fine-tuning of the universe, and which the various anthropic principles attempt to understand.[13] It is CSI that David Bohm's quantum potentials are extracting when they scour the microworld for what Bohm calls "active information."[14] It is CSI that enables Maxwell's demon to outsmart a thermodynamic system tending toward thermal equilibrium.[15] It is CSI on which David Chalmers hopes to base a comprehensive theory of human consciousness.[16] It is CSI that within the Kolmogorov-Chaitin theory of algorithmic information takes the form of highly compressible, nonrandom strings of digits.[17]

CSI is not just confined to science. It is indispensable in our everyday lives. The 16-digit number on your VISA card is an example of CSI. The complexity of this number ensures that a would-be thief cannot randomly pick a number and have it turn out to be a valid VISA card number. What's more, the specification of this number ensures that it is your number, and not anyone else's. Even your telephone number constitutes CSI. As with the VISA card number, the complexity ensures that this number won't be dialed randomly (at least not too often), and the specification ensures that this number is yours and yours only. All the numbers on our bills, credit slips, and purchase orders represent CSI. CSI makes the world go round. It follows that CSI is a rife field for criminality. CSI is what motivated the greedy Michael Douglas character in the movie *Wall Street* to lie, cheat, and steal. CSI's total and absolute control was the objective of the monomaniacal Ben Kingsley character in the movie *Sneakers*. CSI is the artifact of interest in most techno-thrillers. Ours is an information age, and the information that captivates us is CSI.

## Intelligent Design

From where does the origin of complex specified information come? In this section, I shall argue that intelligent causation, or equivalently design, accounts for the origin of complex specified information. My argument focuses on the nature of intelligent causation, and specifically, on what it is about intelligent causes that makes them detectable. To see why CSI is a reliable indicator of design, we need to examine the nature of intelligent causation. The principal characteristic of intelligent causation is *directed contingency*, or what we call *choice*. Whenever an intelligent cause acts, it chooses from a range of competing possibilities. This is true not just of humans, but also of animals and extraterrestrial intelligences. A rat navigating a maze must choose whether to go right or left at various points in the maze. When SETI (Search for Extra-Terrestrial Intelligence) researchers try to discover intelligence in the extraterrestrial radio transmissions they are monitoring, they first assume that an extraterrestrial intelligence could have chosen any number of possible radio transmissions. Then they try to match the transmissions they observe with certain patterns as opposed to others (patterns that presumably are markers of intelligence). Whenever a human being utters meaningful speech, a choice is made from a range of possible sound-combinations that might have been uttered. Intelligent causation always entails discrimination, choosing certain things, ruling out others.

Given this characterization of intelligent causes, the crucial question is how to recognize their operation. Intelligent causes act by making a choice. How then do we recognize that an intelligent cause has made a choice? A bottle of ink spills accidentally onto a sheet of paper; someone takes a fountain pen and writes a message on a sheet of paper. In both instances, ink is applied to paper. In both instances, one among an almost infinite set of possibilities is realized. In both instances, a contingency is actualized and others are ruled out. Yet in one instance we infer design, in the other chance. What is the relevant difference? Not only do we need to observe that a contingency was actualized, but we ourselves need also to be able to specify that contingency. The contingency must conform to an independently given pattern, and we must be able to independently formulate that pattern. A random ink blot is unspecifiable; a message written with ink on paper is specifiable. Wittgenstein made the same point: "We tend to take the speech of a Chinese for inarticulate gurgling. Someone who understands Chinese will recognize *language* in what he hears. Similarly I often cannot discern the *humanity* in man."[18]

In hearing a Chinese utterance, someone who understands Chinese not only recognizes that one from a range of all possible utterances was actualized, but is also able to specify the utterance as coherent Chinese speech. Contrast this with someone who does not understand Chinese. In hearing a Chinese utterance, someone who does not understand Chinese also recognizes that one from a range of possible utterances was actualized, but this time, because lacking the ability to understand Chinese, is unable to specify the utterance as coherent speech. To someone who does not understand Chinese, the utterance will appear gibberish. Gibberish— the utterance of nonsense syllables uninterpretable within any natural language—always actualizes one utterance from the range of possible utterances. Nevertheless, gibberish, by corresponding to nothing we can understand in any language, cannot be specified. As a result, gibberish is never taken for intelligent communication, but always for what Wittgenstein calls "inarticulate gurgling."

The actualization of one among several competing possibilities, the exclusion of the rest, and the specification of the possibility actualized encapsulate how we recognize intelligent causes, or equivalently, how we detect design. The Actualization-Exclusion-Specification triad constitutes a general criterion for detecting intelligence—be it animal, human, or extraterrestrial. Actualization establishes that the possibility in question is the one that actually occurred. Exclusion establishes that there was genuine contingency (i.e., that there were other live possibilities, and that these were ruled out). Specification establishes that the actualized possibility conforms to a pattern given independently of its actualization.

Now where does choice, which we've cited as the principal characteristic of intelligent causation, figure into this criterion? The problem is that we never witness choice directly. Instead, we witness actualizations of contingency which might be the result of choice (i.e., directed contingency) or the result of chance (i.e., blind contingency). Specification is the only means available to us for distinguishing choice from chance, directed contingency from blind contingency. Actualization and exclusion together guarantee that we are dealing with contingency Specification guarantees that we are dealing with a directed contingency. The

Actualization-Exclusion-Specification triad is therefore precisely what we need to identify choice and with it intelligent causation.

Psychologists who study animal learning and behavior have known of the Actualization-Exclusion-Specification triad all along, even if implicitly. For these psychologists—known as learning theorists—learning is discrimination.[19] To learn a task an animal must acquire the ability to actualize behaviors suitable for the task as well as the ability to exclude behaviors unsuitable for the task. Moreover, for a psychologist to recognize that an animal has learned a task, it is necessary not only to observe the animal making the appropriate behavior, but also to specify this behavior. Thus to recognize whether a rat has successfully learned how to traverse a maze, a psychologist must first specify the sequence of right and left turns that conducts the rat out of the maze. No doubt, a rat randomly wandering a maze also discriminates a sequence of right and left turns. But by randomly wandering the maze, the rat gives no indication that it can discriminate the appropriate sequence of right and left turns for exiting the maze. Consequently, the psychologist studying the rat will have no reason to think the rat has learned how to traverse the maze. Only if the rat executes the sequence of right and left turns specified by the psychologist will the psychologist recognize that the rat has learned how to traverse the maze. We regard these learned behaviors as intelligent causes in animals. Thus, it is no surprise that the same scheme for recognizing animal learning recurs for recognizing intelligent causes generally, to wit, actualization, exclusion, and specification.

This general scheme for recognizing intelligent causes coincides precisely with how we recognize complex specified information. First, the basic precondition for information to exist must hold, namely, contingency. Thus one must establish that any one of a multiplicity of distinct possibilities might obtain. Next, one must establish that the possibility which was actualized after the others were excluded was also specified. So far the match between this general scheme for recognizing intelligent causation and how we recognize complex specified information is exact. Only one loose end remains—complexity. Although complexity is essential to CSI (corresponding to the first letter of the acronym), its role in this general scheme for recognizing intelligent causation is not immediately evident. In this scheme, one among several competing possibilities is actualized, the rest are excluded, and the possibility which was actualized is specified. Where in this scheme does complexity figure in?

The answer is that it is there implicitly. To see this, consider again a rat traversing a maze, but now take a very simple maze in which two right turns conduct the rat out of the maze. How will a psychologist studying the rat determine whether it has learned to exit the maze? Just putting the rat in the maze will not be enough. Because the maze is so simple, the rat could by chance just happen to take two right turns, and thereby exit the maze. The psychologist will therefore be uncertain whether the rat actually learned to exit this maze, or whether the rat just got lucky. But contrast this now with a complicated maze in which a rat must take just the right sequence of left and right turns to exit the maze. Suppose the rat must take one hundred appropriate right and left turns, and that any mistake will prevent the rat from exiting the maze. A psychologist who sees the rat take no erroneous turns and in short order exit the maze will be convinced that the rat has indeed learned how to exit the maze, and that this was not dumb luck. With the simple maze, there is a substantial probability that the rat will exit the maze by chance; with the complicated maze, this is exceedingly improbable. The role of complexity in detecting design is now clear, since improbability is precisely what we mean by complexity (cf. section "Complex Information").

We can summarize this argument for showing that CSI is a reliable indicator of design as follows: CSI is a reliable indicator of design because its recognition coincides with how we recognize intelligent causation generally. To recognize intelligent causation, we must establish that one possibility from a range of competing possibilities was actualized, determine which possibilities were excluded, and then specify the actualized possibility. What's more, the competing possibilities that were excluded must be live possibilities, sufficiently numerous so that specifying the actualized possibility cannot be attributed to chance. In terms of probability, this means that the specified possibility is highly improbable. In terms of complexity, this means that the specified possibility is highly complex. All the elements in the general scheme for recognizing intelligent causation (i.e., Actualization-Exclusion-Specification) find their counterpart in complex specified information—CSI. CSI pinpoints what we need to be looking for when we detect design.

As a postscript, I call the reader's attention to the etymology of the word "intelligent." It derives from two Latin words, the preposition *inter*, meaning between, and the verb *lego*, meaning to choose or select.

Thus, according to its etymology, intelligence consists in *choosing between*. It follows that the etymology of the word "intelligent" parallels the formal analysis of intelligent causation just given. Thus, "Intelligent design" is a thoroughly apt phrase, signifying that design is inferred precisely because an intelligent cause has done what only an intelligent cause can do—make a choice.

## The Law of Conservation of Information

Evolutionary biology has steadfastly resisted attributing CSI to intelligent causation. Though Eigen recognizes that the central problem of evolutionary biology is the origin of CSI, he has no thought of attributing CSI to intelligent causation. According to Eigen, natural causes are adequate to explain the origin of CSI. The only question for him is which natural causes explain the origin of CSI. Eigen ignores the logically prior question of whether natural causes can even, in principle, explain the origin of CSI. Yet this is a question that undermines his entire project.[20] Natural causes are, in principle, incapable of explaining the origin of CSI. They can explain the flow of CSI, being ideally suited for transmitting already existing CSI. What they cannot do, however, is originate CSI. This strong proscriptive claim, that natural causes can only transmit CSI but never originate it, I call the Law of Conservation of Information. It is this law that gives definite scientific content to the claim that CSI is intelligently caused. The aim of this last section is briefly to sketch the Law of Conservation of Information.[21]

To see that natural causes cannot account for CSI is straightforward. Natural causes comprise chance and necessity.[22] Because information presupposes contingency, necessity is by definition incapable of producing information, much less complex specified information. For there to be information, there must be a multiplicity of live possibilities, one of which is actualized, and the rest of which are excluded. This is contingency. But if some outcome B is necessary given antecedent conditions A, then the probability of B given A is one, and the information in B given A is zero. If B is necessary given A, Formula (*) reduces to $I(A\&B) = I(A)$, which is to say that B contributes no new information to A. It follows that necessity is incapable of generating new information. Observe that what Eigen calls "algorithms" and "natural laws" fall under necessity.

Since information presupposes contingency, let us take a closer look at contingency. Contingency can assume only one of two forms. Either the contingency is a blind, purposeless contingency—which is chance; or it is a guided, purposeful contingency—which is intelligent causation. Since we already know that intelligent causation is capable of generating CSI (cf. section, "Intelligent Design"), let us next consider whether chance might also be capable of generating CSI. First notice that pure chance, entirely unsupplemented and left to its own devices, is incapable of generating CSI. Chance can generate complex unspecified information, and chance can generate noncomplex specified information. What chance cannot generate is information that is jointly complex and specified.

Biologists by and large do not dispute this claim. Most agree that pure chance—what Hume called the Epicurean hypothesis—does not adequately explain CSI. Jacques Monod is one of the few exceptions, arguing that the origin of life, though vastly improbable, can nonetheless be attributed to chance because of a selection effect.[23] Just as the winner of a lottery is shocked at winning, so we are shocked to have evolved. But the lottery was bound to have a winner, and so too something was bound to have evolved. Something vastly improbable was bound to happen, and so, the fact that it happened to us (i.e., that we were selected—thus the name selection effect) does not preclude chance. This is Monod's argument and it is fallacious. It utterly fails to come to grips with specification. Moreover, it confuses a necessary condition for life's existence with its explanation. Monod's argument has been refuted by the philosophers John Leslie,[24] John Earman,[25] and Richard Swinburne.[26] It has also been refuted by the biologists Francis Crick,[27] Bernd-Olaf Küppers,[28] and Hubert Yockey.[29] Selection effects do nothing to render chance an adequate explanation of CSI.

Most biologists, therefore, reject pure chance as an adequate explanation of CSI. The problem here is not simply one of faulty statistical reasoning. Pure chance as an explanation of CSI is also scientifically unsatisfying. To explain CSI in terms of pure chance is no more instructive than pleading ignorance or proclaiming CSI a mystery. It is one thing to explain the occurrence of heads on a single coin toss by appealing to chance. It is quite another, as Küppers points out, to follow Monod and take the view that "the specific sequence of the nucleotides in the DNA molecule of the first organism came about by a purely random process in the early history of the earth."[30] CSI cries out for an explanation, and pure chance won't do. As Richard Dawkins correctly notes: "We can accept a certain amount of luck in our [scientific] explanations, but not too much."[31]

If chance and necessity left to themselves cannot generate CSI, is it possible that chance and necessity working together might generate CSI? The answer is "No." Whenever chance and necessity work together, the respective contributions of chance and necessity can be arranged sequentially. But by arranging them sequentially, it becomes clear that at no point in the sequence is CSI generated. Consider the case of trial-and-error (trial corresponds to necessity and error to chance). Once considered a crude method of problem solving, trial-and-error has so risen in the estimation of scientists that it is now regarded as the ultimate source of wisdom and creativity in nature. The probabilistic algorithms of computer science all depend on trial-and-error.[32] So too, the Darwinian mechanism of mutation and natural selection is a trial-and-error combination in which mutation supplies the error and selection the trial. An error is committed after which a trial is made. But at no point is CSI generated.

Natural causes are therefore incapable of generating CSI. This broad conclusion I call the Law of Conservation of Information, or LCI for short. LCI has profound implications for science. Among its corollaries are the following: (1) The CSI in a closed system of natural causes remains constant or decreases; (2) CSI cannot be generated spontaneously, originate endogenously, or organize itself (as these terms are used in origins-of-life research); (3) The CSI in a closed system of natural causes either has been in the system eternally or was at some point added exogenously (implying that the system though now closed was not always closed); (4) In particular, any closed system of natural causes that is also of finite duration received whatever CSI it contains before it became a closed system.

This last corollary is especially pertinent to the nature of science for it shows that scientific explanation is not coextensive with reductive explanation. Richard Dawkins, Daniel Dennett, and many scientists are convinced that proper scientific explanations must be reductive, moving from the complex to the simple. Dawkins writes: "The one thing that makes evolution such a neat theory is that it explains how organized complexity can arise out of primeval simplicity."[33] Dennett views any scientific explanation that moves from simple to complex as "question-begging."[34] Thus Dawkins explicitly equates proper scientific explanation with what he calls "hierarchical reductionism," according to which "a complex entity at any particular level in the hierarchy of organization" must properly be explained "in terms of entities only one level down the hierarchy."[35] While no one will deny that reductive explanation is extremely effective within science, it is hardly the only type of explanation available to science. The divide-and-conquer mode of analysis behind reductive explanation has strictly limited applicability within science. In particular, this mode of analysis is utterly incapable of making headway with CSI. CSI demands an intelligent cause. Natural causes will not do.

## Notes

1. Manfred Eigen, *Steps Towards Life: A Perspective on Evolution*, translated by Paul Woolley (Oxford: Oxford University Press, 1992), 12.

2. Keith J. Devlin, *Logic and Information* (New York: Cambridge University Press, 1991), 1.

3. Fred I. Dretske, *Knowledge and the Flow of Information* (Cambridge, MA: MIT Press, 1981), 4.

4. Robert Stalnaker, *Inquiry* (Cambridge, MA: MIT Press, 1984), 85.

5. See Claude E. Shannon and W. Weaver, *The Mathematical Theory of Communication* (Urbana, IL: University of Illinois Press, 1949), 32; R. W. Hamming, *Coding and Information Theory*, 2d edition (Englewood Cliffs, NJ: Prentice-Hall, 1986); or any mathematical introduction to information theory.

6. Cf. William A. Dembski, *the Design Inference: Eliminating Chance through Small Probabilities* (Forthcoming, Cambridge University Press, 1998), Chap. 4.

7. The details can be found in my monograph, *The Design Inference*.

8. For the details refer to my monograph, *The Design Inference*.

9. Cf. Ian Hacking, *Logic of Statistical Inference* (Cambridge: Cambridge University Press, 1965).

10. Arno Wouters, "Viability Explanation," *Biology and Philosophy*, 10 (1995): 435–457

11. Michael Behe, *Darwin's Black Box: The Biochemical Challenge to Evolution.* (New York: The Free Press, 1996).

12. Richard Dawkins, *The Blind Watchmaker* (New York: Norton, 1987), 9.

13. Cf. John D. Barrow and Frank J. Tipler, *The Anthropic Cosmological Principle* (Oxford: Oxford University Press, 1986).

14. Cf. David Bohm, *the Undivided Universe: An Ontological Interpretation of Quantum Theory* (London: Routledge, 1993), 35–38.

15. Cf. Rolf Landauer, "Information is Physical," *Physics Today* (May: 23–29, 1991): 26.

16. Cf. David J. Chalmers, *The Conscious Mind: In Search of a Fundamental Theory* (New York: Oxford University Press, 1996), Chap. 8.

17. Cf. Andrei N. Kolmogorov, "Three Approaches to the Quantitative Definition of Information," *Problemy Peredachi Informatsii* (in translation), 1 (1) (1965): 3–11; Gregory J. Chaitin, "On the Length of Programs for Computing Finite Binary Sequences," *Journal of the ACM*, 13 (1966): 547–569.

18. Ludwig Wittgenstein, *Culture and Value*, edited by G. H. von Wright, translated by P. Winch (Chicago: University of Chicago Press, 1980), 1e.

19. Cf. James. E. Mazur, *Learning and Behavior*, 2d ed. (Englewood Cliffs, NJ: Prentice Hall, 1990); Barry Schwartz, *Psychology of Learning and Behavior*, 2d edition (New York: Norton, 1984).

20. Manfred Eigen, *Steps Towards Life*.

21. A full treatment will be given in *Uncommon Descent*, a book I am jointly authoring with Stephen Meyer and Paul Nelson.

22. Cf. Jacques Monod, *Chance and Necessity* (New York: Vintage, 1972).

23. Ibid.

24. John Leslie, *Universes* (London: Routledge, 1989).

25. John Earman, "The Sap Also Rises: A Critical Examination of the Anthropic Principle," *American Philosophical Quarterly*, 24 (4) (1987): 307–317.

26. Richard Swinburne, *The Existence of God* (Oxford: Oxford University Press, 1979).

27. Francis Crick, *Life Itself: Its Origin and Nature* (New York: Simon and Schuster, 1981), Chap. 7.

28. Bernd-Olaf Küppers, *Information and the Origin of Life* (Cambridge, MA: MIT Press, 1990), Chap. 6.

29. Hubert P. Yockey, *Information Theory and Molecular Biology* (Cambridge: Cambridge University Press, 1992), Chap. 9.

30. Bernd-Olaf Küppers, *Information and the Origin of Life*, 59.

31. Richard Dawkins, *The Blind Watchmaker*, 139.

32. E.g., genetic algorithms—see Stephanie Forrest, "Genetic Algorithms: Principles of Natural Selection Applied to Computation," *Science*, 261 (1993): 872–878.

33. Richard Dawkins, *The Blind Watchmaker*, 316.

34. Daniel C. Dennett, *Darwin's Dangerous Idea: Evolution and the Meanings of Life* (New York: Simon & Schuster, 1995), 153.

35. Richard Dawkins, *The Blind Watchmaker*, 13.

# 26
## Information and the Argument from Design

Peter Godfrey-Smith

## 1. Introduction

William Dembski holds that "the origin of information is best sought in intelligent causes" ("Intelligent Design ...," this volume, p. 553). In particular, Dembski argues that Darwinism is not able to explain the existence of structures that contain a certain kind of information—"complex specified information" (CSI). To explain these informational properties of living systems, we must instead appeal to the choices made by an intelligent designer.

Dembski's version of the argument from design derives an apparent sophistication from the information-theoretic terms in which it is expressed. But in the end, Dembski's version is one of the least plausible versions of the design argument. Partly because of the formal apparatus, Dembski's version of the argument is far too sweeping; it omits qualifications that other opponents of Darwinism sensibly include.

My discussion will have two main sections. One outlines information theory, and Dembski's use of the concepts of information and complexity. The other looks at the alleged consequences of these ideas for Darwinism.

There is so much wrong with Dembski's argument that it is quite difficult to discuss without giving constant and distracting criticisms of relatively minor points. My strategy will be to stick as closely as possible to what I take to be the main thrust of his argument, and mostly resist criticizing details. I will make as many concessions to Dembski as I can, and show that despite all these concessions his argument does not work. But readers should not infer that I agree with Dembski on points that I grant him, or remain silent on, for the purposes of this discussion.[1]

## 2.   Information and Probability

Dembski uses the resources of information theory to describe the phenomena he thinks natural causes cannot explain. At bottom, the only thing information theory does in his argument is provide an alternative way of expressing the idea that some events are *highly unlikely to have arisen by chance*, and unlikely in a way that can be described without hindsight. Events of this kind require a special type of explanation—on that point Dembski and many Darwinians agree. But Dembski argues further that no set of natural causes can provide the required explanation.

Dembski's sketch of information theory in "Intelligent Design" is informal and sometimes imprecise. I will sketch a few key ideas here, but a good place for the reader to turn for a precise and well-organized introduction to these concepts is Fred Dretske's *Knowledge and the Flow of Information* (1981). Dretske's book is also helpful in this context as he treats an issue that is not discussed by Dembski but that has arisen elsewhere in debates about Darwinism and creationism. This is the issue of how informational properties are related to physical properties. (See Johnson 1996 and the accompanying replies, pp. 543–552 in this volume.)

Information theory, as Dembski says, starts with the idea of contingency —with the idea of a situation in which one out of a range of possible outcomes is actualized. The concept of information, Dembski says, does not require the existence of a channel transmitting the information, or anyone receiving it. Anything that has a range of possible states or outcomes, where these states or outcomes each have some definite probability of occurring, has information associated with it. The more different possible states a "source" of this kind has, and the more similar are the probabilities of these states, the more information is associated with the source.

Describing information as something that exists at a source independently of any channel or receiver is not completely uncontroversial within information theory.[2] Some prefer to only use the term "entropy" for the property of the source itself, and to introduce the term "information" when the source is being considered in relation to another variable with which it might be associated (see Cover and Thomas 1991, chap. 2). But this is not important to the argument.

The idea that information is associated with uncertainty or contingency is one basic idea in information theory. Another has to do with the transmission of information. If there are two variables (A and B) that both have some uncertainty associated with them, and where the state of variable A is reliably associated with the state of B, then the state of A carries some information about the state of B. That is, if the reduction of uncertainty at A also reduces uncertainty at B, then A carries information about B. The number of tree rings in a tree stump, for example, carries information about the age of the tree. So information, in this sense, is something that exists wherever there is measurable uncertainty. And information is transmitted by means of physical correlations.[3]

Information theory itself is mostly concerned with the mathematical treatment of relationships between quantities of information. The main point of the theory is to describe the transmission of information over channels, the storage of information in memory, and other kinds of use and manipulation of information. A paradigm situation for the theory is one where we have a source of information, a receiver of information about the source, and a channel that connects them and makes it possible for information to flow from source to receiver, in the form of signals. The theory can be used to describe which kinds of channels have the capacity to transmit messages with given mathematical properties. To pick a very simple example, if a source has four possible states, in order for the receiver to be able to tell from a signal which one of these four states is actual, the signal must have at least four possible states. Otherwise the signal must be ambiguous. Although "flow" of information has an evocative sound, all that is required for information flow is a certain kind of reliable association between what happens at the source and the state of a signal received by the receiver.

Applications like Dembski's make use of just one part of this theoretical structure—the first of the basic ideas listed above. This is the idea that a source of information (or "probabilistic source") is anything with a number of distinct possible states, where the states each have a probability of occurring.

The point about probability is important; in order to use information theory in some domain, we have to be able to ascribe an *objective probability* to each possible state of the source. Ascribing objective probabilities to some events, like the outcomes of lotteries, is often regarded as fairly easy. Other events are much harder to describe in this way, not just because the probabilities are well hidden but because it is hard to know what is even meant by the description. Unique and complex events are hard cases, for example. Ancient Rome was first sacked by Visigoths, in 410 A.D. This was a contingent matter; other tribes might have done the job instead, and it might have happened earlier or later. So we have a space of possibilities with only one being actualized. What was the probability that Rome be first sacked by the Celts, or the Chinese? How about by New Zealand Maoris? In some sense these alternatives seem unlikely (especially the last) but in describing a unique and complex event (as opposed to the outcome of a gambling game, which is repeated over and over) it is very hard to ascribe probabilities to specific outcomes, and many statisticians would refuse to even try. To use information theory, though, you have to have probabilities. So even though we believe that the sacking of Rome was a contingent matter, this is apparently not the kind of case where we can say that there is some definite "amount of information" associated with the fact that the Visigoths first sacked Rome in 410 A.D. Just because some event can be thought about in contrast with a range of possible alternatives, this does not imply that there is a definite and measurable amount of information associated with that event.

Although I think it is very implausible to claim that there is a definite amount of information associated with every contingent event or state of affairs, in this paper I will concede to Dembski all the assumptions he needs about the existence of physical probabilities, and their connection to information. I do this partly for the sake of discussion, but also because many biologists, including Darwinians, are happy to use probabilistic and information-theoretic frameworks in roughly the same way that Dembski does. So it is better to concede these issues to Dembski for now, and reply to him in a way that does not require controversial assumptions about probability.

For a particular event or structure to contain information, there must be a range of alternatives that could have been actual instead. Information theory is usually not much concerned with specific events. But as Dembski says, it is possible in principle to take a single, specific event that has a single probability and assign a measure of information to that event. The more improbable the event, the more information is associated with it. This measure is sometimes referred to as the "surprisal" of the particular event.

When we are talking only about one specific event, the relationship between a description in terms of probability and a description in terms of information is trivial. To assign a measure of information to the event, you just mathematically transform its probability. You find the logarithm to the base 2 of that probability, and take the negative of that logarithm. A probability of 1/4 becomes 2 bits of information, as the logarithm to the base 2 of 1/4 is −2. A probability of 1/32 becomes 5 bits of information, and so on. In saying these things, we are doing *no more* than applying a mathematical transformation to the probabilities. Because the term "information" is now being used, it might seem that we have done something important. But we have just rescaled the probabilities that we already had.

So why do the rescaling at all? Logarithms are convenient to work with for some purposes; if we were going to multiply the probabilities we could add their logarithms instead. But that is just convenience. If we were faced with a problem of the sort that information theory is intended to deal with, such as the problem of working out whether a particular medium of communication is suitable for the messages we want to send, then there is more we could do. But assessing the power of Darwinism does not have much to do with choosing communication media. Despite all the detail that Dembski gives in describing information theory, information is not making any essential contribution to his argument. What is doing the work is just the idea of objective probability. We have objective probabilities associated with events or states of affairs, and we are re-expressing these probabilities with a mathematical transformation.[4]

So far I have discussed Dembski's use of the term "information." Something should be said also about "complexity" and "specification," as Dembski claims that the problem for Darwinism is found in cases of "complex specified information" (CSI). Do these concepts add anything to Dembski's argument? "Complexity" as used by Dembski certainly does not add anything, as by "complex information" in his paper Dembski just means "lots of information." He uses the distinction between "simple" and "complex" as a way of referring to the amount of information associated with an event (p. 559, this vol.).

Dembski motivates his introduction of the loaded term "complex" by linking his discussion to another mathematical framework, "complexity theory." Complexity theory, in the sense Dembski intends, is mostly concerned with the measurement of the complexity of formal problems and computational operations. Computational complexity theory uses a special sense of "complex," which should not be confused with everyday senses of the term, or with biological uses. And as in the case of information, Dembski is making his connection to complexity theory in only a minimal way; he says that there is more complexity as the amount of information associated with an event goes up (or equivalently, as the probability of the event goes down). There is nothing stopping Dembski calling this "complexity," but it is important to realize how little the term contributes to his argument.

When a full house is dealt in a hand of poker, we have a more improbable event than we would have had if two pairs had been dealt. So there is more information generated by the full house. And in Dembski's special sense, the full house is associated with "more complex information" than the two pairs—again, he just means the full house is less likely than two pairs. Dembski is saying, in lots of different ways, that some events are less likely to arise by chance than others.

The term "complexity" is ambiguous and hard to define in everyday usage. I doubt that in any everyday sense, a full house is "more complex" than two pairs. And more important, the various biological uses of the term "complex" are not captured simply by the idea of low probability or high informational content. Admittedly it is hard to work out what some biologists mean by the term "complex." I have argued (1996) that the core meaning of complexity, in biological contexts, is *heterogeneity*. Something is complex when it has many different kinds of parts, or does many different things. For me, there are as many kinds of complexity as there are kinds of heterogeneity. John Bonner, in *The Evolution of Complexity* (1988), measures the biological complexity of an organism by counting the number of different types of cell the organism contains. This is one kind of heterogeneity—heterogeneity in cell types—and Bonner's book shows this to be a clear and useful sense of "complexity." Dan McShea (1991) has discussed a number of other senses used by biologists. Some of these are linked to information-theoretic measures, such as the entropy of a probability distribution. But none of the senses of complexity recognized or used by biologists can be defined simply in terms of the improbability of a particular structure or event. In Dembski's article, however, the three phrases "very improbable," "containing a lot of information," and "containing complex information" all pick out the same property.[5]

The final important term Dembski introduces is "specified," which also applies to information. His treatment of this term is vague but I will not make an issue of this, because he is dealing with a genuinely tricky problem. The motivation for the discussion is the fact that some events can be truly described as improbable, but only in a sense involving hindsight. If you are dealt a worthless poker hand, it is nonetheless extremely improbable for you to have been dealt *that exact* worthless hand. All poker hands are improbable in that sense, but the game of poker singles out some combinations of cards as valuable, and specifies these ahead of time. So, in many judgments about probability we are interested only in events that are improbable *in a sense that does not involve hindsight*. That is, we assess the probabilities according to some preordained scheme for classifying outcomes (full house, three of a kind, flush, etc.).

These terms—information, complexity, and specification—complete the outline of Dembski's information-theoretic framework. Dembski goes on to claim that life contains CSI—complex specified information. This *looks* like an interesting and theoretically rich property, but in fact it is nothing special. Dembski's use of the term "information" should not be taken to suggest that meaning or representation is involved. His use of the term "complex" should not be taken to suggest that things with CSI must be complex in either the everyday sense of the term or a biologist's sense. Anything that is unlikely to have arisen by chance (in a sense that does not involve hindsight) contains CSI, as Dembski has defined it.

So, Dembski's use of information theory provides a roundabout way of talking about probability. Is this probabilistic approach a useful one for tackling general questions about Darwinism? In my view, there is little to recommend casting these questions in probabilistic terms. They are understood well enough without the probabilistic apparatus, and objective probability is controversial, even in contexts where there is some consensus about the actual judgments. We probably all agree that the chance Rome had of being first sacked by Chinese was lower than its chance of being first sacked by Goths. In saying this, we have in mind the proximity of Rome to the Gothic tribes, the history of antagonism between them, and so on. But it has proved very hard to give a precise meaning to probabilistic claims of this kind. And claims about the probability of life appearing, or of elephants evolving by natural selection, are in the same boat. I think the probabilistic language introduces more problems than it solves here.

In the present context though, it would be unwise for me to reject this way of approaching evolutionary questions. Quite a number of Darwinians see some usefulness in using broad claims about probability, and also claims about information, when discussing Darwinism. G. C. Williams, for example, has placed a lot of stress on the role of informational properties in evolution (Williams 1992). Maynard Smith and Szathmáry suggest that various major transitions in evolution resulted from "new ways of transmitting information" across generations (1995, p. 13). Biologists also sometimes use information to discuss the ways in which genes and other developmental factors causally affect biological traits. As DNA sequences make use of a fixed "alphabet" of four elements, the bases C, A, T, and G, it is possible to discuss the informational content of stretches of DNA. As I said earlier, to associate a measure of information with an event or structure is to view it as one of a definite range of possible alternatives, each of which had some probability of being actual. So each actual base in a DNA sequence is thought of as one of four possible alternatives for that position in the stretch of DNA. If some totally random process was selecting which of the four bases goes in each position, then each actual base in a DNA sequence would be associated with 2 bits of information.

The processes that determine the sequence of bases in a particular DNA sequence are in fact far from random, but biologists do sometimes treat a stretch of DNA as containing an amount of information that is a simple function of its length, where each base is treated as in some sense equally likely to be found in each position. A biologist might say: "The information content or determinacy of a complex anatomical structure is orders of magnitude higher than that of the genome...." (Gilbert, Opitz, and Raff 1996, pp. 366–367. An organism's genome is its total genetic sequence). What they mean here is that the space of possible genomes (of a certain size) is much smaller than the space of possible anatomical structures that can be relevantly contrasted with the structure in question. The "gap" between the two amounts of information is taken to tell us something about the causation of the anatomical structures. To assert this while using a standard sense of "information content," and without a great deal more probabilistic knowledge, it is necessary to idealize and treat each genome of a certain length as equally likely and do the same for each possible anatomical structure. Biologists often do not worry about this idealization; the main aim of the description is just to contrast an actual genetic or phenotypic structure with some range of relevant possible alternatives.

So plenty of biologists use the concept of information in describing genes, organisms, and evolution. Most of these discussions make use of more of the structure of information theory than Dembski does. Dembski, as I said, uses information merely as a way of redescribing probabilities. And his use of probabilities is very close to that of Richard Dawkins.

Parts of Dembski's discussion develop in a way that resembles the opening pages of Dawkins's *Blind Watchmaker* (1986). Dawkins says that living things have a special property that makes them very hard to scientifically explain. That property can be described as a certain kind of improbability. And living things are not merely improbable in a sense involving hindsight; Dawkins says the kind of improbability that is relevant here is a kind that must be "specifiable in advance." Dembski quotes Dawkins and notes one aspect of their agreement, though he expresses Dawkins's claim as if it makes a concession to Dembski's position. Both Dembski and Dawkins hold that because living things have features that are highly unlikely to have arisen by pure chance, a special kind of explanation is needed for them.

Whereas the biologists mentioned above tend to use information to express the idea that any particular organism is one of a vast range of possible organisms, and in that sense improbable, Dawkins and Dembski use probability to express the fact that a living organism is one highly organized state for a pile of physical material, material that *could* be combined together in many other ways, most of which would not produce a living system. Life is "improbable" in the sense that there are far more nonliving ways for matter to be arranged.

To summarize so far, then: Dembski uses the idea of information merely to redescribe probabilities. Turning a number into a logarithm does change how you can compute with it, but does not change what (if anything) the number measures. "Information" for Dembski is no more than improbability. "Complex information" is just lots of improbability. An event is associated with "specified" information if the probability associated with it is assigned using a categorization of outcomes that is specifiable in advance and not merely with hindsight.

### 3.   On What Evolutionary Processes Can Bring About

Dembski says that natural causes cannot explain complex specified information. More precisely: an event or structure associated with complex specified information cannot be fully explained in naturalistic terms that do not include some role for intelligent choice. As above, I will ignore quite a lot of Dembski's discussion and focus on what I take to be his primary line of argument. Dembski holds that natural causes cannot create CSI because natural causes are a mixture or combination of "chance" and "necessity." Necessity cannot bring new information into being, and chance can create some information but not complex specified information. So mutation and natural selection, in particular, cannot create CSI because mutation-plus-selection is just a combination of chance and necessity. Dembski expresses his main claim as a law, the law of the "conservation of information" in all systems of wholly natural causes. Dembski really should not call this a conservation law, because he does allow that CSI can *decrease* in a system of natural causes. So CSI is not in fact conserved, according to Dembski. But this is a minor issue.

To assess Dembski's claims from a biological standpoint, we need to know what the phenomenon to be explained is. Dembski says the phenomenon is CSI, but where exactly is this CSI found? And if CSI cannot arise from natural causes, which parts of the standard scientific picture of the world—the picture given in standard college biology textbooks, for example—must be changed?

The standard scientific picture includes, first, some purely historical claims. The first life appeared on earth over 3.5 billion years ago, as there are fossil traces of bacteria from around that time. There must have been even more basic forms of life prior to bacteria, but these are unknown. For something like 2 billion years, bacteria were the only life on earth, until the appearance of eukaryotic cells, at least 1.7 billion years ago. Multicellular organisms date from about 800 million years ago. Later there appear, among other things, vertebrates (over 500 million years ago), flowering plants (over 120 million years ago), and homo sapiens (about 300,000 years ago). That much is pure history. It is based not just on biological data, but on geology, physics, and other sciences.

Along with this historical story we have, in the standard scientific picture, claims about the nature of the processes linking these historical events. These events occurred in a system governed, at bottom, only by physical laws—"natural causes" in Dembski's sense. But these physical laws, operating in certain kinds of conditions, give rise to the processes of replication, variation, and natural selection, and all the other factors that contribute to evolution. These factors are taken to be sufficient to generate, on a planet bathed in solar energy and containing the right raw materials, both simple forms of life and later organisms like people arising eventually from organisms like bacteria. If CSI was really "conserved," which parts of this story would have to be false?

Certainly Dembski claims that the origin of life itself cannot be understood in terms of natural causes, so that part of the standard story must be false. But what about the later events? Dembski also says that "the functionality of organisms" (p. 563, this vol.) is something that natural causes cannot explain. Dembski does not commit himself to a definition of "functionality" but gestures toward the complicated and hierarchical organization of organisms, their ability to maintain their structure and reproduce, and so on. Here we find CSI.

But this property of "functionality" apparently comes in degrees. Bacteria have a certain amount of "functionality" and a certain amount of CSI, but a camel apparently has a lot more. A bacterium is just a single cell, containing perhaps a few thousand genes. Every cell of the trillions of cells within a camel has a more complex organization than a bacterial cell—though the bacterial cell does have many notable properties that a camel cell lacks, such as the ability to survive on its own. And a camel contains many different kinds of cell, working together in a hierarchically organized way. It is fairly standard, though not totally uncontroversial, to say that a camel is a more complex organism than a bacterium. Certainly I would say this—in all relevant senses, a camel is a more heterogeneous object than a bacterium. So if there is a property of "complex specified information" recognizable in a bacterium, there is apparently more of this same property in a camel.

As a consequence, Dembski seems not to be just denying naturalistic views about the origins of life. He must also hold that getting *from* simpler one-celled organisms *to* more complex, integrated, and even intelligent forms of life is not possible with wholly natural causes. So if simpler organisms did appear on earth prior to complex ones, and are connected to them via lines of descent, then the processes by which the more complex came from the less complex must have involved more than natural causes. Intelligent design must have played a role in many later stages, as well as the initial stage, of the history of life.

That is how I interpret Dembski. Possibly he has something else in mind—maybe the earth 3 billion years ago contained both bacteria and a store of hidden CSI that was slowly incorporated, via natural processes, into sharks, camels, and all the rest. As it seems unlikely that camel CSI, orchid CSI, and salamander CSI could all be hidden inside bacteria, the extra CSI must have been distributed somehow in the nonbiological part of the world before being absorbed into organisms. This is a picturesque option. Or perhaps Dembski thinks that God called everything into being less than 10,000 years ago, and denies that species are linked to each other by any kind of "descent with modification." Dembski presents his argument in such abstract terms that it is hard to tell.[6] But it is important to insist that views like Dembski's confront the *specific historical claims* made by the standard scientific view. These claims concern both the order in which organisms appeared on earth, and the lines of descent that connect them. Either bacteria appeared on earth before camels, or they did not. Either oak trees and water lilies have a more recent common ancestor than oak trees have with pine trees, or they do not.

We can take Dembski's claim to be restricted to the problem of the origin of life, or we can take it to apply more generally. In this discussion I will assume that it applies more generally, not just to the origin of life but to subsequent evolutionary processes as well. So bacteria contain CSI, but camels contain more CSI than bacteria do. And I take Dembski to claim that the processes described in neo-Darwinian evolutionary theory cannot explain how we can go from bacteria to camels. When I say "neo-Darwinian theory" I have in mind a fairly inclusive sense of this term. Neo-Darwinism combines the basic explanatory structure outlined by Darwin with modern genetics and molecular biology. Sometimes "neo-Darwinism" is also associated with strong claims about the size of mutations and the constancy of the rate of evolution, but I am assuming a flexible attitude toward these details. So I take moderate unorthodoxies such as Stephen Jay Gould's (1989) view to be neo-Darwinian in this inclusive sense.

We must explain both the origin of life and its subsequent development. I will not discuss the origin of life in detail here. There is not the kind of consensus about the origin of life that we find about the broad outlines of subsequent evolution. The whole area is more speculative, and it is tricky for a Darwinist because we are discussing the appearance of the mechanisms that make later episodes of mutation and natural selection possible. A classic experiment by Miller and Urey in 1953, and its subsequent elaboration, has shown that many of the organic molecules fundamental to living systems, including some seemingly "improbable" ones, can be created from inorganic raw materials spontaneously in laboratory experiments that simulate early conditions on earth. The molecules produced include amino acids, which make up proteins, and also the bases that are key components of DNA and RNA. Short chains of amino acids ("proteinoids") have also been produced from solutions of individual amino acids with the aid of heat and the right kind of surface. Once we leave the laboratory results, however, most issues concerning the origin of life are very controversial. General theories divide roughly into those positing a single type of molecule with special properties of self-replication, possibly a form of RNA, and theories positing self-reproducing chemical networks, in which no single molecule is central. The appearance of cell-like encapsulation of self-replicating structures is also crucial. Discussions of the origins of life rapidly become technical. Here I will mostly discuss the role of natural causes in life's subsequent evolution. But many of the points I will make below apply also the question of the origin of life.[7]

So how can you get from a bacterium to a camel? A large part of the answer is: slowly! Neo-Darwinism explains a process of this kind by breaking it down to a large number of small steps. The exact size of the important steps, and the variation in step-size, are famously controversial issues. But there is consensus about the important *physical* features of these processes, and about the fact that *many* of these steps are needed for large-scale evolutionary transitions. Genes mutate, as a consequence of molecular mishaps. Organisms have their structure and behavior affected by the mutations, usually for ill but occasionally for good. The organisms live, reproduce, and die, and those carrying novel genes either reproduce more or less than other organisms in the population. If they reproduce more, and certain other conditions are realized, the frequency of those genes in the population will tend to increase. Through this process, useful modifications slowly accumulate. Genetic material is duplicated within the genome, and the duplicates acquire new roles, making more complex structures possible. Populations change over time, split, and diverge. The striking features of evolution—elephants from bacteria, roses related (distantly) to tuna, and so on—are a consequence of the accumulation of a great many of these small steps. I talk of "steps" here, but it is important not to have in one's mind a single linear chain of events. The local events that give rise to large-scale evolutionary processes are scattered, as the individual organisms of all the various species are. Organisms live and die locally. But a total family tree connects life on earth—connects all individuals as well as all species.

If there is more "complex specified information" in a camel than in a bacterium, then the natural process described above is able to create this information. There is no metaphysical mystery in such a creation; consider what is required for natural causes to take us from a simple bacterium to a slightly more complex, well-adapted, and "unlikely" bacterium. What this requires is a molecular accident in the replication of the bacterial DNA that happens to result in a change to some protein molecule produced by that DNA, a change that happens to help the bacterium survive and reproduce—perhaps, for example, by enabling it to break down an environmental chemical that was formerly toxic to it. The innovation will proliferate, if circumstances allow, and in time we will have a bacterial population able to do something novel and adaptive.

A case like this is a very low-level case of Darwinian evolution, a far cry from the evolution of human eyes or brains. But the opponents of Darwinism do have to deal with these cases. Anti-Darwinians have two options here. One option is to concede these low-level cases to the Darwinist, but claim that these are not the important cases. The other option is to deny even the low-level cases.

The first option is the smarter choice for the anti-Darwinian: Darwinism, on this view, can give bacteria new adaptations to their environments but cannot bring about "irreducible complexity," and cannot generate the big transitions between whole phyla of organisms. If this option is taken, the battle must be fought over the question of how different the low-level cases are from the big and striking cases. In particular, are the largest and most dramatic evolutionary transitions explainable in terms of long chains of lower-level evolutionary events? This first option is the smarter one because it does not commit the anti-Darwinian to an absurd fight about whether bacteria can acquire antibiotic resistance or moths can get darker wings through Darwinian processes. It is also a smarter option because every time a Darwinian explains how an evolutionary process could create some particular noteworthy structure, and explains why the traces of evolutionary history are in fact visible in that structure, the anti-Darwinian can concede the case (or just stop talking about it) and pick a new one. (For this sort of debate, see the exchange between Behe, Orr, Doolittle, Futuyma, and others in the *Boston Review* issues for December/January and February/March 1997.)

Although this hedged option is the sensible one for the anti-Darwinian, it does not seem to be available to Dembski. This is because Dembski expresses his claims in the form of a "law," the "law of the conservation of information." The law is expressed in sweeping terms. Natural causes cannot create *any* new CSI, not even a little bit; "the CSI in a closed system of natural causes remains constant or decreases" (p. 571, this vol.). So he must insist that natural causes cannot add even a tiny increment of CSI to a population of bacteria.

The exact scope of this "law" is hard to determine because, on Dembski's own view, there is no sharp cut off between "complex" information and information that is not complex. The law does not say that information in general is conserved. But Dembski clearly thinks that something very important cannot be created, even in small amounts, by natural processes; I pass over the problem of what exactly this important thing is.

So Dembski claims that even small additions of CSI via natural causes are impossible. He grounds this claim on an abstract argument about "chance" and "necessity." Dembski holds that neither of these, in principle, can create CSI, and a combination of the two can fare no better. But mutation and selection is, Dembski says, just a combination of chance and necessity.

Dembski misdescribes the structure of neo-Darwinian explanation in a number of ways. I will quickly mention a terminological oddity first. Dembski correctly says that the process of mutation and selection resembles learning by trial and error. But Dembski expresses the analogy by likening mutation to "error" and selection to "trial." This is a strange flipping of the usual way of describing the analogy. Usually people writing about evolution say that mutation is a trial (in the sense of a "trying-out," or an experiment) and selection corresponds to error (feedback from the environment, usually negative). In this way, a Darwinian process is seen as searching a space of possible structures. I suppose the terms "trial" and "error" always had within them the possibility of a switch from their usual roles in the analogy. But Dembski is the first person I have ever seen make this switch. Maybe he meant to do it, and a terminological oddity is certainly not a big problem. But this does illustrate how far Dembski's discussion is from the large literature discussing how natural selection actually works.

Turning to the argument itself, Dembski says we can discover some limitations of Darwinism by looking at the limitations associated with chance and with necessity. It is true that Monod (1971) and others have described Darwinism as a combination of chance and necessity. But this is a metaphorical description of Darwinism that has little value and much potential to mislead, as it does here. Mutation of genes can be treated like a random event for the purposes of much evolutionary discussion and modeling, but mutation is of course a physical process. The main reason to use the term "chance" has been to stress that mutation occurs haphazardly in many directions, and occurs in ways that are not controlled by the needs of the organism. A mutation, however, is a certain type of change to a certain type of molecule within a cell. This process results in one chemical compound being substituted for another within a particular larger molecule. These molecular events are part of the ordinary causal order. Such processes might be irreducibly chancy in the sense associated with quantum mechanics—but only if everything in the world is irreducibly chancy in the same sense.

Referring to mutations as "chance" events is often a useful shorthand, but referring to selection as a kind of "necessity" has no positive use. I think people like to call selection a kind of "necessity" just because necessity is a complement to chance, and the pairings of chance/mutation and necessity/selection look neat. But selection, too, is a type of causal process (Sober 1984). This process involves the interaction of organisms with their environments, and involves in particular the various causes and effects of individual reproduction. The physical processes involved in selection have as much determinism or necessity at the lowest level as basic physics allows—just as mutation does. So the processes of natural selection are not really any more "necessary" than mutation is. Selection processes are in some respects more patterned and regular, but they are not especially "necessary." Consider our evolving bacterial population again. The mutations in bacteria all have their specific molecular causes. The mutations result in different protein molecules being produced, which affect the metabolic processes of each of the mutated bacteria. Most mutations are bad, but every now and then one turns out to be useful.

Monod's metaphorical description of evolution in terms of chance and necessity will not bear any weight, and certainly will not support an argument like Dembski's. There is no way one can deduce any important properties of evolution by working from a discussion of the general properties of chance and the general properties of necessity.[8]

More generally, when working out what evolution can bring about, it is a mistake to try to deduce strong conclusions from the basic properties of information, contingency, chance, necessity, probability, and so on. All these concepts have only slippery applications to the actual processes of evolution as understood by Darwinians. They provide picturesque ways of talking, but they will not support inferences of the kind Dembski engages in. To work out what evolution can and cannot do, the way to proceed is to focus on the biological processes themselves—the variation found in populations, the phenomena of heritability and their underlying genetic mechanisms, and the interaction of organisms with their physical and biological environments. When we look at how these processes in fact work, we find that producing outcomes that would otherwise be unlikely is *exactly* what Darwinian processes *are* able to do. The causal processes of mutation and natural selection provide a way for a population of organisms to accumulate useful bits and pieces of biological machinery—the products of very occasional good luck in mutation's products. The hard problem for Darwinism has always been showing that the biggest transitions in evolution can be explained in these terms. (For a fascinating attempt to tackle these hardest problems, see Maynard Smith and Szathmáry 1995.)

In this section I have argued that his sweeping formulations and abstract claims about chance and necessity make Dembski's argument particularly vulnerable, even by creationist standards. The role played by Dembski's "law of the conservation of information," which is a vague relative of real laws in thermodynamics, exemplifies this most clearly. In these respects Dembski is, ironically, a little bit like Herbert Spencer (1871, 1872). Spencer was an English speculative polymath, influential during the late nineteenth century, who developed a grand theory of evolution that applied to everything, from the cosmos to human society. Spencer had some interesting ideas, but one of his pitfalls was his tendency to express these ideas in the form of grand, general laws, which were sometimes vague analogues of Newtonian laws. Spencer had a particular fondness for conservation principles. This led him into hopelessly sweeping formulations of his claims. None of Spencer's laws is taken seriously today.

## 4.  Conclusion

In section 2 of this paper I outlined Dembski's use of information theory. My main claim in this section was that Dembski's "complex specified information" is just improbability. Something contains CSI if it is very improbable for that thing to be produced via chance processes, where this improbability is not of a kind dependent on hindsight. Dembski sees living organisms as improbable things in this sense, and although there are problems surrounding this kind of application of probability theory, quite a few Darwinian biologists have expressed similar ideas. Dembski's concept of "complex specified information" does not have much to do with complexity as ordinarily understood, or as understood by biologists. Complex structures (in a biologist's sense) might indeed tend to be improbable, but plenty of improbable things are not especially complex (like winning poker hands). Dembski's use of information theory is minimal and raises no new problems for Darwinians.

In section 3 I looked at Dembski's claim that natural processes, especially those described by Darwinians, cannot generate CSI. I argued that Dembski's use of a "conservation law" to express this idea lands him in the uncomfortable position of claiming that even slight increases in complexity or adaptedness cannot be produced by Darwinian processes. I also argued that there is no way to deduce any important properties of evolution from abstract discussions of "chance and necessity." Mutation and selection are causal processes, and they are causal processes that enable populations of organisms to discover and accumulate useful new biological traits, gradually leading to large-scale evolutionary change.

Dembski's paper is part of a recent movement aimed at defending a place for "intelligent design" in the overall structure of the universe. These defenses include erroneous but well-packaged attacks on the explanatory power of Darwinism, and the publicity surrounding these attacks makes it harder to educate people about evolution. Understanding how evolutionary processes work is not just a matter of theoretical, academic interest. For example, in this paper I have sometimes used examples involving bacteria. There are lots of reasons to be interested in bacterial evolution, and one is the fact that the history of human disease is made up in part by the evolutionary history of disease-causing microbes (Ewald 1994; Diamond 1997). Bacteria and viruses evolve much faster than we do; they evolve on the rapid time-scale of human social change. As a consequence, we can study their evolution readily, but we are also affected by their evolution in more mundane ways. Why is the routine overuse of antibiotics a bad thing? It is bad for Darwinian reasons. The more often each antibiotic is used, the more selection pressure is put on bacteria to evolve ways to survive the onslaught from that drug. And evolving resistance is exactly what they tend to do. Over recent decades bacteria have found ways to break down, pump out, or in other ways neutralize a great range of antibiotic drugs. Some have argued that bacteria are *so* good at this sort of thing that Darwinian explanations of bacterial evolution must be supplemented with "directed mutation." (See Jablonka and Lamb 1995, chap. 3, for a survey of these issues). The jury is still out on the issue of whether there is any role for a significantly non-Darwinian form of evolution in bacteria, but even those defending unorthodox views in this area do not doubt that a great deal of bacterial evolution is straightforwardly Darwinian.

Hospitals are where many of these dangerous bacteria are commonly found, both because hospitals are full of sick people and because hospitals are special environments where the only bacteria that can survive are the ones that can beat the chemicals we use against them. When a mutant bacterium appears in a hospital and happens to be able to survive in the presence of available antibiotics, the new strain will often multiply and we will be the worse for it (Garrett 1995). There is no way to avoid altogether this problem with antibiotics; given the nature of bacteria and the nature of evolutionary processes, we must expect resistance to arise eventually in many cases. But knowledge of microbial evolution instructs us to use antibiotics only when there is good reason to do so. The drugs will inevitably be more effective the less they are used. These are Darwinian facts, and facts with immediate practical importance.

In sum: the classical argument from design was answered by Darwin and by the subsequent development of evolutionary biology. Recasting the argument in terms of "information" does not change that situation. And a recasting in terms of a general "law of conservation of information" makes the argument worse rather than better.

### Acknowledgment

Thanks to Lori Gruen, Amir Najmi, Robert Pennock, and an anonymous referee for helpful comments.

### Notes

1. Dembski's paper "Intelligent Design ..." is my main target in this discussion. Dembski has also published a book (1998) that develops his claims about probability in more detail, and which includes a brief discussion of the evolution/creation controversy (pp. 55–62). The discussion of evolution and creation in the book does not take sides. In footnotes to this paper I will indicate a few areas in which Dembski's book and his article differ on points of relevant detail. None of these issues is important to the overall argument in "Intelligent Design ..." or my criticisms of that argument.

2. Thanks to Amir Najmi for assistance with these issues. Shannon (1948) did refer to the source as a "source of information," although he was considering the source in relation to a signal and a receiver. Through the rest of this paper I will

use the term "information" in the way Dembski does, without further comment on the relation between the terms "information," "entropy," and so on.

3. In this paper I use the term "correlation" in a broad, vernacular sense, not in the strict sense measured by a correlation coefficient. I also use the terms "probable" and "likely" interchangeably; "likely" does not refer specifically to likelihoods, in the statistician's sense.

4. In his 1998 book, Dembski focuses more on probabilities and does not dress his argument up in information theoretic terms.

5. In his 1998 book, Dembski discusses complexity theory in more detail. I am not qualified to comment on that discussion of complexity theory. But, again, the only role played by "complexity" in his argument in "Intelligent Design" is as a measure of low probability.

6. Dembski expands on his positive views, though not on biological matters, in a paper called "The Act of Creation: Bridging Transcendence and Immanence," which is available at http://www.arn.org/docs/dembski/wd_actofcreation.htm.

7. See Maynard Smith and Szathmáry 1995 for a critical survey of recent work on the origin of life. A standard college biology textbook, Campbell, Reece, and Mitchell 1999, also has a very good overview, pp. 492–498.

8. Dembski's book makes use of a classification in which there are only three possible kinds of explanation for any event—regularity, chance and design (1998, chap. 2). But many events, and a great many of the ones relevant to these debates about evolution, result from a complicated causal network that includes both likely and unlikely events, interacting over time to yield an outcome. A simple three-way distinction between regularity, chance, and design is entirely inadequate for understanding explanation in such cases.

For other criticisms of Dembski's book, see Fitelson, Stephens, and Sober (1999), chapter 27 in this volume.

## References

Bonner, J. T. 1988. *The Evolution of Complexity*. Princeton, N.J.: Princeton University Press.

Campbell, N. A., J. B. Reece, and L. G. Mitchell. 1999. *Biology*, fifth edition. Menlo Park: Addison Wesley.

Cover, T. M. and J. A. Thomas. 1991. *Elements of Information Theory*. New York: Wiley.

Dawkins, R. 1986. *The Blind Watchmaker*. New York: Norton.

Dembski, W. 1997. Intelligent Design as a Theory of Information. *Perspectives on Science and Christian Faith* 49 (3): 180–190. Reprinted as chap. 25 in this volume.

———. 1998. *The Design Inference*. Cambridge: Cambridge University Press.

Diamond, J. 1997. *Guns, Germs, and Steel: The Fates of Human Societies*. New York: Norton.

Dretske, F. 1981. *Knowledge and the Flow of Information*. Cambridge, Mass.: MIT Press.

Ewald, P. A. 1994. *The Evolution of Infectious Disease*. Oxford: Oxford University Press.

Fitelson, B., C. Stephens, and E. Sober. 1999. How Not to Detect Design—Critical Noble: William A. Dembski, *The Design Inference*. *Philosophy of Science*. Reprinted as chap. 27 in this volume.

Garrett, L. 1995. *The Coming Plague*. New York: Penguin.

Gilbert, S. F., J. M. Opitz, and R. A. Raff. 1996. Resynthesizing Evolutionary and Developmental Biology. *Developmental Biology* 173: 357–372.

Godfrey-Smith, P. 1996. *Complexity and the Function of Mind in Nature*. Cambridge: Cambridge University Press.

Gould, S. J. 1989. *Wonderful Life*. New York: Norton.

Jablonka, E. and M. Lamb. 1995. *Epigenetic Inheritance and Evolution*. Oxford: Oxford University Press.

Johnson. P. 1996. Is Genetic Information Irreducible? *Biology and Philosophy* 11: 535–538.

Maynard Smith, J. and E. Szathmáry. 1995. *The Major Transitions in Evolution*. Oxford: Oxford University Press.

McShea, D. 1991. Complexity and Evolution: What Everybody Knows. *Biology and Philosophy* 6: 303–324.

Monod, J. 1971. *Chance and Necessity*. Translated by Austryn Wainhouse. New York: Alfred A. Knopf.

Shannon, C. E. 1948. A Mathematical Theory of Communication. *Bell System Technical Journal* 27: 379–423, 623–656.

Sober, E. 1984. *The Nature of Selection*. Cambridge, Mass.: MIT Press.

Spencer, H. 1871. *Principles of Psychology*, second edition. 2 volumes. New York: Appleton.

Spencer, H. 1872. *First Principles of a New System of Philosophy*, second edition. New York: Appleton.

Williams, G. C. 1992. *Natural Selection: Domains, Levels, and Challenges*. Oxford: Oxford University Press.

# 27

# How Not to Detect Design—Critical Notice: William A. Dembski, *The Design Inference*

Branden Fitelson, Christopher Stephens, and Elliott Sober

As every philosopher knows, "the design argument" concludes that God exists from premises that cite the adaptive complexity of organisms or the lawfulness and orderliness of the whole universe. Since 1859, it has formed the intellectual heart of creationist opposition to the Darwinian hypothesis that organisms evolved their adaptive features by the mindless process of natural selection. Although the design argument developed as a defense of theism, the logic of the argument in fact encompasses a larger set of issues. William Paley saw clearly that we sometimes have an excellent reason to postulate the existence of an intelligent designer. If we find a watch on the heath, we reasonably infer that it was produced by an intelligent watchmaker. *This* design argument makes perfect sense. Why is it any different to claim that the eye was produced by an intelligent designer? Both critics and defenders of the design argument need to understand what the ground rules are for inferring that an intelligent designer is the unseen cause of an observed effect.

Dembski's book is an attempt to clarify these ground rules. He proposes a procedure for detecting design and discusses how it applies to a number of mundane and nontheological examples, which more or less resemble Paley's watch. Although the book takes no stand on whether creationism is more or less plausible than evolutionary theory, Dembski's epistemology can be evaluated without knowing how he thinks it bears on this highly charged topic. In what follows, we will show that Dembski's account of design inference is deeply flawed. Sometimes he is too hard on hypotheses of intelligent design; at other times he is too lenient.

Originally published in *Philosophy of Science* (Sept 1999, vol. 66, pp. 472–488).

Neither creationists, nor evolutionists, nor people who are trying to detect design in nontheological contexts should adopt Dembski's framework.

## The Explanatory Filter

Dembski's book provides a series of representations of how design inference works. The exposition starts simple and grows increasingly complex. However, the basic pattern of analysis can be summarized as follows. Dembski proposes an "explanatory filter" (37), which is a procedure for deciding how best to explain an observation E:

(1) There are three possible explanations of E—Regularity, Chance, and Design. They are mutually exclusive and collectively exhaustive. The problem is to decide which of these explanations to accept.

(2) The Regularity hypothesis is more parsimonious than Chance, and Chance is more parsimonious than Design. To evaluate these alternatives, begin with the most parsimonious possibility and move down the list until you reach an explanation you can accept.

(3) If E has a high probability, you should accept Regularity; otherwise, reject Regularity and move down the list.

(4) If the Chance hypothesis assigns E a sufficiently low probability and E is "specified," then reject Chance and move down the list; otherwise, accept Chance.

(5) If you have rejected Regularity and Chance, then you should accept Design as the explanation of E.

The entire book is an elaboration of the ideas that comprise the Explanatory Filter.[1] Notice that the filter is *eliminativist*, with the Design hypothesis occupying a special position.

We have interpreted the Filter as sometimes recommending that you should accept Regularity or Chance. This is supported, for example, by Dembski's remark (38) that "if E happens to be an HP [a high probability] event, we stop and attribute E to a regularity." However, some of the circumlocutions that Dembski uses suggest that he doesn't think you should ever "accept" Regularity or Chance.[2] The most you should do is "not reject" them. Under this alternative interpretation, Dembski is saying that if you fail to reject Regularity, you can believe any of the three hypotheses, or remain agnostic about all three. And if you reject Regularity, but fail to reject Chance, you can believe either Chance or Design, or remain agnostic about them both. Only if you have rejected Regularity and Chance must you accept one of the three, namely Design. Construed in this way, a person who believes that every event is the result of Design has nothing to fear from the Explanatory Filter—no evidence can ever dislodge that opinion. This *may* be Dembski's view, but for the sake of charity, we have described the Filter in terms of rejection *and* acceptance.

## The Caputo Example

Before discussing the filter in detail, we want to describe Dembski's treatment of one of the main examples that he uses to motivate his analysis (9–19, 162–166). This is the case of Nicholas Caputo, who was a member of the Democratic party in New Jersey. Caputo's job was to determine whether Democrats or Republicans would be listed first on the ballot. The party listed first in an election has an edge, and this was common knowledge in Caputo's day. Caputo had this job for 41 years and he was supposed to do it fairly. Yet, in 40 out of 41 elections, he listed the Democrats first. Caputo claimed that each year he determined the order by drawing from an urn that gave Democrats and Republicans the same chance of winning. Despite his protestations, Caputo was brought up on charges and the judges found against him. They rejected his claim that the outcome was due to chance, and were persuaded that he had rigged the results. The ordering of names on the ballots was due to Caputo's intelligent design.

In this story, the hypotheses of Chance and Intelligent Design are prominent. But what of the first alternative, that of Regularity? Dembski (11) says that this can be rejected because our background knowledge tells us that Caputo probably did not innocently use a biased process. For example, we can rule out the possibility that Caputo, with the most honest of intentions, spun a roulette wheel in which 00 was labeled "Republican" and all the other numbers were labeled "Democrat." Apparently, we know before we examine Caputo's 41 decisions that there are just two possibilities—he did the equivalent of tossing a fair coin (Chance) or he intentionally gave the edge to his own party (Design).

There is a straightforward reason for thinking that the observed outcomes favor Design over Chance. If Caputo had allowed his political allegiance to guide his arrangement of ballots, you would expect Democrats to be listed first on all or almost all of the ballots. However, if Caputo did the equivalent of tossing a fair coin, the outcome he obtained would be very surprising. This simple analysis also can be used to represent Paley's argument about the watch (Sober 1993). The key concept is *likelihood*. The likelihood of a hypothesis is the probability it confers on the observations; it is not the probability that the observations confer on the hypothesis. The likelihood of H relative to E is $Pr(E|H)$, not $Pr(H|E)$. Chance and Design can be evaluated by comparing their likelihoods, relative to the same set of observations. We do not claim that likelihood is the whole story, but surely it is relevant.

The reader will notice that the Filter does not use this simple likelihood analysis to help decide between Chance and Design. The likelihood of Chance is considered, but the likelihood of Design never is. Instead, the Chance hypothesis is evaluated for properties additional to its likelihood. Dembski thinks it is possible to reject Chance and accept Design without asking what Design predicts. Whether the Filter succeeds in showing that this is possible is something we will have to determine.

## The Three Alternative Explanations

Dembski defines the Regularity hypothesis in different ways. Sometimes it is said to assert that the evidence E is noncontingent and is reducible to law (39, 53); at other times it is taken to claim that E is a deterministic consequence of earlier conditions (65; 146, fn. 5); and at still other times, it is supposed to say that E was highly probable, given some earlier state of the world (38). The Chance Hypothesis is taken to assign to E a lower probability than the Regularity Hypothesis assigns (40). The Design Hypothesis is said to be the complement of the first two alternatives. As a matter of stipulation, the three hypotheses are mutually exclusive and collectively exhaustive (36).

Dembski emphasizes that design need not involve intelligent agency (8–9, 36, 60, 228–229). He regards design as a mark of intelligent agency; intelligent agency can produce design, but he seems to think that there could be other causes as well. On the other hand, Dembski says that "the explanatory filter pinpoints how we recognize intelligent agency (66)" and his Section 2.4 is devoted to showing that design is reliably correlated with intelligent agency. Dembski needs to supply an account of what he means by design and how it can be caused by something other than intelligent agency.[3] His vague remark (228–229) that design is equivalent to "information" is not enough. Dembski quotes Dretske (1981) with approval, as deploying the concept of information that the design hypothesis uses. However, Dretske's notion of information is, as Dembski points out, the Shannon-Weaver account, which describes a probabilistic dependency between two events labeled source and receiver. Hypotheses of mindless chance can be stated in terms of the Shannon-Weaver concept. Dembski (39) also says that the design hypothesis is not "characterized by probability."

Understanding what "regularity," "chance," and "design" mean in Dembski's framework is made more difficult by some of his examples. Dembski discusses a teacher who finds that the essays submitted by two students are nearly identical (46). One hypothesis is that the students produced their work independently; a second hypothesis asserts that there was plagiarism. Dembski treats the hypothesis of independent origination as a Chance hypothesis and the plagiarism hypothesis as an instance of Design. Yet, both describe the matching papers as issuing from intelligent agency, as Dembski points out (47). Dembski says that context influences how a hypothesis gets classified (46). How context induces the classification that Dembski suggests remains a mystery.

The same sort of interpretive problem attaches to Dembski's discussion of the Caputo example. We think that all of the following hypotheses appeal to intelligent agency: (i) Caputo decided to spin a roulette wheel on which 00 was labeled "Republican" and the other numbers were labeled "Democrat"; (ii) Caputo decided to toss a fair coin; (iii) Caputo decided to favor his own party. Since all three hypotheses describe the ballot ordering as issuing from intelligent agency, all, apparently, are instances of Design in Dembski's sense. However, Dembski says that they are examples, respectively, of Regularity, Chance, and Design.

## The Parsimony Ordering

Dembski says that Regularity is a more parsimonious hypothesis than Chance, and that Chance is more parsimonious than Design (38–39). He defends this ordering as follows:

Note that explanations that appeal to regularity are indeed simplest, for they admit no contingency, claiming things always happen that way. Explanations that appeal to chance add a level of complication, for they admit contingency, but one characterized by probability. Most complicated are those explanations that appeal to design, for they admit contingency, but not one characterized by probability. (39)

Here Dembski seems to interpret Regularity to mean that E is nomologically necessary or that E is a deterministic consequence of initial conditions. Still, why does this show that Regularity is simpler than Chance? And why is Chance simpler than Design? Even if design hypotheses were "not characterized by probability," why would that count as a reason? But, in fact, design hypotheses *do* in many instances confer probabilities on the observations. The ordering of Democrats and Republicans on the ballots is highly probable, given the hypothesis that Caputo rigged the ballots to favor his own party. Dembski supplements this general argument for his parsimony ordering with two examples (39). Even if these examples were convincing,[4] they would not establish the general point about the parsimony ordering.

It may be possible to replace Dembski's faulty argument for his parsimony ordering with a different argument that comes close to delivering what he wants. Perhaps determinism can be shown to be more parsimonious than indeterminism (Sober 1999a) and perhaps explanations that appeal to mindless processes can be shown to be simpler than explanations that appeal to intelligent agency (Sober 1998). But even if this can be done, it is important to understand what this parsimony ordering means. When scientists choose between competing curves, the simplicity of the competitors matters, but so does their fit-to-data. You do not reject a simple curve and adopt a complex curve just by seeing how the simple curve fits the data and without asking how well the complex curve does so. You need to ask how well *both* hypotheses fit the data. Fit-to-data is important in curve-fitting because it is a measure of *likelihood*; curves that are closer to the data confer on the data a higher probability than curves that are more distant. Dembski's parsimony ordering, even if correct, makes it puzzling why the Filter treats the likelihood of the Chance hypothesis as relevant, but ignores the likelihoods of Regularity and Design.

## Why Regularity Is Rejected

As just noted, the Explanatory Filter evaluates Regularity and Chance in different ways. The Chance hypothesis is evaluated in part by asking how probable it says the observations are. However, Regularity is not evaluated by asking how probable it says the observations are. The filter starts with the question, "Is E a high probability event?" (38). This does not mean "Is E a high probability event according to the Regularity hypothesis?" Rather, you evaluate the probability of E on its own. Presumably, if you observe that events like E occur frequently, you should say that E has a high probability and so should conclude that E is due to Regularity. If events like E rarely occur, you should reject Regularity and move down the list.[5] However, since a given event can be described in many ways, any event can be made to appear common, and any can be made to appear rare.

Dembski's procedure for evaluating Regularity hypotheses would make no sense if it were intended to apply to *specific* hypotheses of that kind. After all, specific Regularity hypotheses (e.g., Newtonian mechanics) are often confirmed by events that happen rarely—the return of a comet, for example. And specific Regularity hypotheses are often *dis*confirmed by events that happen frequently. This suggests that what gets evaluated under the heading of "Regularity" are not *specific* hypotheses of that kind, but the *general* claim that E is due to some regularity or other. Understood in this way, it makes more sense why the likelihood of the Regularity hypothesis plays no role in the Explanatory Filter. The claim that E is due to some regularity or other, *by definition*, says that E was highly probable, given antecedent conditions.

It is important to recognize that the Explanatory Filter is enormously ambitious. You do not just reject a given Regularity hypothesis; you reject all possible Regularity explanations (53). And the same goes for Chance—you reject the whole category; the Filter "sweeps the field clear" of *all* specific Chance hypotheses (41, 52–53). We doubt that there is any general inferential procedure that can do what Dembski thinks the Filter accomplishes. Of course, you presumably can accept "E is due to some regularity or other" if you accept a specific regularity hypothesis. But suppose you have tested and rejected the various specific regularity hypotheses that your background beliefs suggest. Are you obliged to reject the claim that *there exists* a regularity hypothesis that explains E? Surely it is clear that this does not follow.

The fact that the Filter allows you to accept or reject Regularity without attending to what specific Regularity hypotheses predict has some peculiar consequences. Suppose you have in mind just one specific regularity hypothesis that is a candidate for explaining E; you think that if E has a regularity-style explanation, this has got to be it. If E is a rare type of event, the Filter says to conclude that E is not due to Regularity. This can happen even if the specific hypothesis, when conjoined with initial condition statements, predicts E with perfect precision. Symmetrically, if E is a common kind of event, the Filter says not to reject Regularity, even if your lone specific Regularity hypothesis deductively entails that E is false. The Filter is too hard on Regularity, and too lenient.

### The Specification Condition

To reject Chance, the evidence E must be "specified." This involves four conditions—CINDE, TRACT, DELIM, and the requirement that the description D* used to delimit E must have a low probability on the Chance hypothesis. We consider these in turn.

**CINDE.** Dembski says several times that you cannot reject a Chance hypothesis just because it says that what you observe was improbable. If Jones wins a lottery, you cannot automatically conclude that there is something wrong with the hypothesis that the lottery was fair and that Jones bought just one of the 10,000 tickets sold. To reject Chance, further conditions must be satisfied. CINDE is one of them.

CINDE means conditional independence. This is the requirement that $Pr(E \mid H\&I) = Pr(E \mid H)$, where H is the Chance hypothesis, E is the observations, and I is your background knowledge. H must render E conditionally independent of I. CINDE requires that H capture everything that your background beliefs say is probabilistically relevant to the occurrence of E.

CINDE is too lenient on Chance hypotheses—it says that their violating CINDE suffices for them to be accepted (or not rejected). Suppose you want to explain why Smith has lung cancer (E). It is part of your background knowledge (I) that he smoked cigarettes for thirty years, but you are considering the hypothesis (H) that Smith read the works of Ayn Rand and that this helped bring about his illness. To investigate this question, you do a statistical study and discover that smokers who read Rand have the same chance of lung cancer as smokers who do not. This study allows you to draw a conclusion about Smith—that $Pr(E|H\&I) = Pr(E|\text{not-}H\&I)$. Surely this equality is evidence *against* the claim that E is due to H. However, the filter says that you cannot reject the causal claim, because CINDE is false—$Pr(E|H\&I) \neq Pr(E|H)$.[6]

**TRACT and DELIM.**   The ideas examined so far in the Filter are probabilistic. The TRACT condition introduces concepts from a different branch of mathematics—the theory of computational complexity. TRACT means tractability—to reject the Chance hypothesis, it must be possible for you to use your background information to formulate a description D* of features of the observations E. To construct this description, you needn't have any reason to think that it might be true. For example, you could satisfy TRACT by obtaining the description of E by "brute force"—that is, by producing descriptions of *all* the possible outcomes, one of which happens to cover E (150–151).

Whether you can produce a description depends on the language and computational framework used. For example, the evidence in the Caputo example can be thought of as a specific sequence of 40 Ds and 1 R. TRACT would be satisfied if you have the ability to generate all of the following descriptions: "0 Rs and 41 Ds," "1 R and 40 Ds," "2 Rs and 39 Ds," … "41 Rs and 0 Ds." Whether you can produce these descriptions depends on the character of the language you use (does it contain those symbols or others with the same meaning?) and on the computational procedures you use to generate descriptions (does generating those descriptions require a small number of steps, or too many for you to perform in your lifetime?). Because tractability depends on your choice of language and computational procedures, we think that TRACT has no evidential significance at all. Caputo's 41 decisions count against the hypothesis that he used a fair coin, and in favor of the hypothesis that he cheated, for reasons that have nothing to do with TRACT. The relevant point is simply that $Pr(E|\text{Chance}) \ll Pr(E|\text{Design})$. This fact is not relative to the choice of language or computational framework.

The DELIM condition, as far as we can see, adds nothing to TRACT. A description D*, generated by one's background information, "delimits" the evidence E just in case E entails D*. In the Caputo case, TRACT and DELIM would be satisfied if you were able to write down all possible sequences of D's and R's that are 41 letters long. They also would be satisfied by generating a series of weaker descriptions, like the one just mentioned. In fact, just writing down a tautology satisfies TRACT and DELIM (165). On the assumption that human beings are able to write down tautologies, we conclude that these two conditions are always satisfied and so play no substantive role in the Filter.

## Do CINDE, TRACT, and DELIM "Call the Chance Hypothesis into Question"?

Dembski argues that CINDE, TRACT and DELIM, if true, "call the chance hypothesis H into question." We quote his argument in its entirety:

> The interrelation between CINDE and TRACT is important. Because I is conditionally independent of E given H, any knowledge S has about I ought to give S no knowledge about E so long as—and this is the crucial assumption—E occurred according to the chance hypothesis H. Hence, any pattern formulated on the basis of I ought not give S any knowledge about E either. Yet the fact that it does in case D delimits E means that I is after all giving S knowledge about E. The assumption that E occurred according to the chance hypothesis H, though not quite refuted, is therefore called into question . . . .
>
> To actually refute this assumption, and thereby eliminate chance, S will have to do one more thing, namely, show that the probability $P(D^* | H)$, that is, the probability of the event described by the pattern D, is small enough. (147)

We'll address this claim about the impact of low probability later.

To reconstruct Dembski's argument, we need to clarify how he understands the conjunction TRACT & DELIM. Dembski says that when TRACT and DELIM are satisfied, your background beliefs I provide you with "knowledge" or "information" about E (143, 147). In fact, TRACT and DELIM have nothing to do with informational relevance understood as an evidential concept. When I provides information about E, it is natural to think that $\Pr(E | I) \neq \Pr(E)$; I provides information because taking it into account changes the probability you assign to E. It is easy to see how TRACT & DELIM can both be satisfied by brute force without this evidential condition's being satisfied. Suppose you have no idea how Caputo might have obtained his sequence of D's and R's; still, you are able to generate the sequence of descriptions we mentioned before. The fact that you can generate a description which delimits (or even matches) E does not ensure that your background knowledge provides evidence as to whether E will occur. As noted, generating a tautology satisfies both TRACT and DELIM, but tautologies do not provide information about E.

Even though the conjunction TRACT & DELIM should not be understood evidentially (i.e., as asserting that $Pr[E \mid I] \neq Pr[E]$), we think this *is* how Dembski understands TRACT & DELIM in the argument quoted. This suggests the following reconstruction of Dembski's argument:

(1) CINDE, TRACT, and DELIM are true of the chance hypothesis H and the agent S.

(2) If CINDE is true and S is warranted in accepting H (i.e., that E is due to chance), then S should assign $Pr(E \mid I) = Pr(E)$.

(3) If TRACT and DELIM are true, then S should not assign $Pr(E \mid I) = Pr(E)$.

∴ (4) Therefore, S is not warranted in accepting H.

Thus reconstructed, Dembski's argument is valid. We grant premise (1) for the sake of argument. We have already explained why (3) is false. So is premise (2); it seems to rely on something like the following principle:

(*)   If S should assign $Pr(E \mid H\&I) = p$ and S is warranted in accepting H, then S should assign $Pr(E \mid I) = p$.

If (*) were true, (2) would be true. However, (*) is false. For (*) entails

If S should assign $Pr(H \mid H) = 1.0$ and S is warranted in accepting H, then S should assign $Pr(H) = 1.0$.

Justifiably accepting H does not justify assigning H a probability of unity. Bayesians warn against assigning probabilities of 1 and 0 to any proposition that you might want to consider revising later. Dembski emphasizes that the Chance hypothesis is always subject to revision.

It is worth noting that a weaker version of (2) is true:

(2*)   If CINDE is true and S should assign $Pr(H) = 1$, then S should assign $Pr(E \mid I) = Pr(E)$.

One then can reasonably conclude that

(4*)   S should not assign $Pr(H) = 1$.

However, a fancy argument isn't needed to show that (4*) is true. Moreover, the fact that (4*) is true does nothing to undermine S's confidence that the Chance hypothesis H is the true explanation of E, provided that S has not stumbled into the brash conclusion that H is entirely certain. We conclude that Dembski's argument fails to "call H into question."

It may be objected that our criticism of Dembski's argument depends on our taking the conjunction TRACT & DELIM to have probabilistic consequences. We reply that this is a *charitable* reading of his argument. If the conjunction does not have probabilistic consequences, then the argument is a nonstarter. How can purely non-probabilistic conditions come into conflict with a purely probabilistic condition like CINDE? Moreover, since TRACT and DELIM, *sensu strictu*, are always true (if the agent's side information allows him/her to generate a tautology), how could these trivially satisfied conditions, when coupled with CINDE, possibly show that H is questionable?

### The Improbability Threshold

The Filter says that $\Pr(E \mid \text{Chance})$ must be sufficiently low if Chance is to be rejected. How low is low enough? Dembski's answer is that $\Pr(E(n) \mid \text{Chance}) < 1/2$, where n is the number of times in the history of the universe that an event of kind E actually occurs (209, 214–217). As mentioned earlier, if Jones wins a lottery, it does not follow that we should reject the hypothesis that the lottery was fair and that he bought just one of the 10,000 tickets sold. Dembski thinks the reason this is so is that lots of *other* lotteries have occurred. If p is the probability of Jones's winning the lottery if it is fair and he bought one of the 10,0000 tickets sold, and if there are n such lotteries that ever occur, then the relevant probability to consider is $\Pr(E(n) \mid \text{Chance}) = 1 - (1 - p)^n$. If n is large enough this quantity can be greater than $1/2$, even though p is very small. As long as the probability exceeds $1/2$ that Smith wins lottery L2, or Quackdoodle wins lottery L3, or ... or Snerdley wins lottery Ln, given the hypothesis that each of these lotteries was fair and the individuals named each bought one of the 10,000 tickets sold, we shouldn't reject the Chance hypothesis about Jones.

Why is $1/2$ the relevant threshold? Dembski thinks this follows from the Likelihood Principle (190–198). As noted earlier, that principle states that if two hypotheses confer different probabilities on the same observations, the one that entails the higher probability is the one that is better supported by those observations. Dembski thinks this principle solves the following prediction problem. If the Chance hypothesis predicts that either F or not-F will be true, but says that the latter is more probable, then, if you believe the Chance hypothesis and must predict whether F or not-F will be true, you should predict not-F. We agree that if a gun were put to your head, that you should predict the option that the Chance hypothesis says is more probable if you believe the Chance hypothesis and this exhausts what you know that is relevant. However, this does not follow from the likelihood principle. The likelihood principle tells you how to evaluate different hypotheses by seeing what probabilities they confer on the observations. Dembski's prediction principle describes how you should choose between two predictions, not on the basis of observations, but on the basis of a theory you already accept; the theory says that one prediction is more *probable*, not that it is more *likely*.

Even though Dembski's prediction principle is right, it does not entail that you should reject Chance if $\Pr(E(n) \,|\, \text{Chance}) < 1/2$ and the other specification conditions are satisfied. Dembski thinks that you face a "probabilistic inconsistency" (196) if you believe the Chance hypothesis and the Chance hypothesis leads you to predict not-F rather than F, but you then discover that E is true and that E is an instance of F. However, there is no inconsistency here of any kind. Perfectly sensible hypotheses sometimes entail that not-F is more probable than F; they can remain perfectly sensible even if F has the audacity to occur.

An additional reason to think that there is no "probabilistic inconsistency" here is that H and not-H can *both* confer an (arbitrarily) low probability on E. In such cases, Dembski must say that you are caught in a "probabilistic inconsistency" *no matter what you accept.* Suppose you know that an urn contains either 10% green balls or 1% green balls; perhaps you saw the urn being filled from one of two buckets (you do not know which), whose contents you examined. Suppose you draw 10 balls from the urn and find that 7 are green. From a likelihood point of view, the evidence favors the 10% hypothesis. However, Dembski would point out that the 10% hypothesis predicted that most of the balls in your sample would fail to be green. Your observation contradicts this prediction. Are you therefore forced to reject the 10% hypothesis? If so, you are forced to reject the 1% hypothesis on the same grounds. But you know that one or the other hypothesis is true. Dembski's talk of a "probabilistic inconsistency" suggests that he thinks that improbable events can't really occur—a true theory would *never* lead you to make probabilistic predictions that fail to come true.

Dembski's criterion is simultaneously too hard on the Chance hypothesis, and too lenient. Suppose there is just one lottery in the whole history of the universe. Then the Filter says you should reject the hypothesis that Jones bought one of 10,000 tickets in a fair lottery, just on the basis of observing that Jones won (assuming that CINDE and the other conditions are satisfied). But surely this is too strong a conclusion. Shouldn't your acceptance or rejection of the Chance hypothesis depend on what alternative hypotheses you have available? Why can't you continue to think that the lottery was fair when Jones wins it? The fact that there is just one lottery in the history of the universe hardly seems relevant. Dembski is too hard on Chance in this case. To see that he also is too lenient, let us assume that there have been many lotteries, so that $Pr(E(n) \mid Chance) > 1/2$. The Filter now requires that you not reject Chance, even if you have reason to consider seriously the Design hypothesis that the lottery was rigged by Jones's cousin, Nicholas Caputo. We think you should embrace Design in this case, but the Filter disagrees. The flaw in the Filter's handling of both these examples traces to the same source. Dembski evaluates the Chance hypothesis without considering the likelihood of Design.

We have another objection to Dembski's answer to the question of how low $Pr(E(n) \mid Chance)$ must be to reject Chance. How is one to decide which actual events count as "the same" with respect to what the Chance hypothesis asserts about E? Consider again the case of Jones and his lottery. Must the other events that are relevant to calculating $E(n)$ be lotteries? Must exactly 10,000 tickets have been sold? Must the winners of the other lotteries have bought just one ticket? Must they have the name "Jones"? Dembski's $E(n)$ has no determinate meaning.

Dembski supplements his threshold of $Pr(E(n) \mid Chance) < 1/2$ with a separate calculation (209). He provides generous estimates of the number of particles in the universe ($10^{80}$), of the duration of the universe ($10^{25}$ seconds), and of the number of changes per second that a particle can experience ($10^{45}$). From these he computes that there is a maximum of $10^{150}$ specified events in the whole history of the universe. The reason is that there cannot be more agents than particles, and there cannot be more acts of specifying than changes in particle state.[7] Dembski thinks it follows that if the Chance hypothesis assigns to any event that occurs a probability lower than $1/[(2)10^{150}]$, that you should reject the Chance hypothesis (if CINDE and the other conditions are satisfied). This is a fallacious inference. The fact that there are no more than $10^{150}$ acts of specifying in the whole history of the universe tells you nothing about what the probabilities of those specified events are or should be thought to be. Even if sentient creatures manage to write down only N inscriptions, why can't those creatures develop a well confirmed theory that says that some actual events have probabilities that are less than $1/(2N)$?

## Conjunctive, Disjunctive, and Mixed Explananda

Suppose the Filter says to reject Regularity and that TRACT, CINDE and the other conditions are satisfied, so that accepting or rejecting the Chance hypothesis is said to depend on whether $Pr(E(n) | Chance) < 1/2$. Now suppose that the evidence E is the conjunction E1&E2& ... &Em. It is possible for the conjunction to be sufficiently improbable on the Chance hypothesis that the Filter says to reject Chance, but that each conjunct is sufficiently probable according to the Chance hypothesis that the Filter says that Chance should be accepted. In this case, the Filter concludes that Design explains the conjunction while Chance explains each conjunct. For a second example, suppose that E is the disjunction E1 ∨ E2 ∨ ⋯ ∨ Em. Suppose that the disjunction is sufficiently probable, according to the Chance hypothesis, so that the Filter says not to reject Chance, but that each disjunct is sufficiently improbable that the Filter says to reject Chance. The upshot is that the Filter says that each disjunct is due to Design though the disjunction is due to Chance. For a third example, suppose the Filter says that E1 is due to Chance and that E2 is due to Design. What will the Filter conclude abut the conjunction E1&E2? The Filter makes no room for "mixed explanations"—it cannot say that the explanation of E1&E2 is simply the conjunction of the explanations of E1 and E2.

## Rejecting Chance as a Category Requires a Kind of Omniscience

Although specific chance hypotheses may confer definite probabilities on the observations E, this is not true of the generic hypothesis that E is due to some chance hypothesis or other. Yet, when Dembski talks of "rejecting Chance" he means rejecting the whole category, not just the specific chance hypotheses one happens to formulate. The Filter's treatment of Chance therefore applies only to agents who believe they have a complete list of the chance processes that might explain E. As Dembski (41) says, "before we even begin to send E through the Explanatory Filter, we need to know what probability distribution(s), if any, were operating to produce the event." *Dembski's epistemology never tells you to reject Chance if you do not believe you have considered all possible chance explanations.*

Here Dembski is *much* too hard on Design. Paley reasonably concluded that the watch he found is better explained by postulating a watchmaker than by the hypothesis of random physical processes. This conclusion makes sense even if Paley admits his lack of omniscience about possible Chance hypotheses, but is does not make sense according to the Filter. What Paley did was compare a *specific* chance hypothesis and a *specific* design hypothesis without pretending that he thereby surveyed all possible chance hypotheses. For this reason as well as for others we have mentioned, friends of Design should shun the Filter, not embrace it.

## Concluding Comments

We mentioned at the outset that Dembski does not say in his book how he thinks his epistemology resolves the debate between evolutionary theory and creationism.[8] Still, it is abundantly clear that the overall shape of his epistemology reflects the main pattern of argument used in "the intelligent design movement." Accordingly, it is no surprise that a leading member of this movement has praised Dembski's epistemology for clarifying the logic of design inference (Behe 1996, 285–286). Creationists frequently think they can establish the plausibility of what they believe merely by criticizing the alternatives (Behe 1996; Plantinga 1993, 1994; Phillip Johnson, as quoted in Stafford 1997, 22). This would make sense if two conditions were satisfied. If those alternative theories had deductive consequences about what we observe, one could demnstrate that those theories are false by showing that the predictions they entail are false. If, in addition, the hypothesis of intelligent design were the only alternative to the theories thus refuted, one could conclude that the design hypothesis is correct. However, neither condition obtains. Darwinian theory makes probabilistic, not deductive, predictions. And there is no reason to think that the only alternative to Darwinian theory is intelligent design.

When prediction is probabilistic, a theory cannot be accepted or rejected just by seeing what it predicts (Royall 1997, Ch. 3). The best you can do is compare theories with each other. To test evolutionary theory against the hypothesis of intelligent design, you must know what *both* hypotheses predict about observables (Fitelson and Sober 1998, Sober 1999b). The searchlight therefore must be focused on the design hypothesis itself. What does *it* predict? If defenders of the design hypothesis want their theory to be scientific, they need to do the scientific work of formulating and testing the predictions that creationism makes (Kitcher 1984, Pennock 1999). Dembski's Explanatory Filter encourages creationists to think that this responsibility can be evaded. However, the fact of the matter is that the responsibility must be faced.

## Notes

We thank William Dembski and Philip Kitcher for comments on an earlier draft.

1. Dembski (48) provides a deductively valid argument form in which "E is due to design" is the conclusion. However, Dembski's final formulation of "the design inference" (221–223) deploys an epistemic version of the argument, whose conclusion is "S is warranted in inferring that E is due to design." One of the premises of this latter argument contains two layers of epistemic operators; it says that if certain (epistemic) assumptions are true, then S is warranted in asserting that "S is not warranted in inferring that E did not occur according to the chance hypothesis." Dembski claims (223) that this convoluted epistemic argument is valid, and defends this claim by referring the reader back to the quite different, nonepistemic, argument presented on p. 48. This establishes nothing as to the validity of the (official) epistemic rendition of "the design inference."

2. For example, he says that "to retain chance a subject S must simply lack warrant for inferring that E did not occur according to the chance hypothesis H" (220).

3. Dembski (1998a) apparently abandons the claim that design can occur without intelligent agency; here he says that after regularity and chance are eliminated, what remains is the hypothesis of an intelligent cause.

4. In the first example, Dembski (39) says that Newton's hypothesis that the stability of the solar system is due to God's intervention into natural regularities is less parsimonious than Laplace's hypothesis that the stability is due solely to regularity. In the second, he compares the hypothesis that a pair of dice is fair

with the hypothesis that each is heavily weighted towards coming up 1. He claims that the latter provides the more parsimonious explanation of why snake-eyes occurred on a single roll. We agree with Dembski's simplicity ordering in the first example; the example illustrates the idea that a hypothesis that postulates two causes R and G is less parsimonious than a hypothesis that postulates R alone. However, this is not an example of Regularity versus Design, but an example of Regularity & Design versus Regularity alone; in fact, it is an example of two causes versus one, and the parsimony ordering has nothing to do with the fact that one of those causes involves design. In Dembski's second example, the hypotheses differ in likelihood, relative to the data cited; however, if parsimony is supposed to be a different consideration from fit-to-data, it is questionable whether these hypotheses differ in parsimony.

5. Dembski incorrectly applies his own procedure to the Caputo example when he says (11) that the regularity hypothesis should be rejected on the grounds that background knowledge makes it improbable that Caputo in all honesty used a biased device. Here Dembski is describing the probability of Regularity, not the probability of E.

6. Strictly speaking, CINDE requires that $Pr(E|H\&J) = Pr(E|J)$, for all J such that J can be "generated' by the side information I (145). Without going into details about what Dembski means by "generating," we note that this formulation of CINDE is logically stronger than the one discussed above. This entails that it is even harder to reject chance hypotheses than we suggest in our cancer example.

7. Note the *materialistic* character of Dembski's assumptions here.

8. Dembski has been more forthcoming about his views in other manuscripts. The interested reader should consult Dembski 1998a.

## References

Behe, M. (1996), *Darwin's Black Box*. New York: Free Press.

Dembski, William A. (1998), *The Design Inference—Eliminating Chance Through Small Probabilities*. Cambridge: Cambridge University Press.

———. (1998a), "Intelligent Design as a Theory of Information," unpublished manuscript, reprinted electronically at the following web site: http://www.arn.org/docs/dembski/. Reprinted in this volume, chap. 25.

Dretske, F. (1981), *Knowledge and the Flow of Information*. Cambridge, MA: MIT Press.

Fitelson, B. and E. Sober (1998): "Plantinga's Probability Arguments Against Evolutionary Naturalism," *Pacific Philosophical Quarterly* 79: 115–129.

Kitcher, P. (1984), *Abusing Science—The Case against Creationism*. Cambridge, MA: MIT Press.

Pennock, R. (1999), *Tower of Babel: The Evidence against the New Creationism*. Cambridge, MA: MIT Press.

Plantinga, A. (1993), *Warrant and Proper Function*. Oxford: Oxford University Press.

———. (1994), "Naturalism Defeated," unpublished manuscript.

Royall, R. (1997), *Statistical Evidence—A Likelihood Paradigm*. London: Chapman and Hall.

Sober, E. (1993), *Philosophy of Biology*. Boulder, CO: Westview Press.

———. (1998), "Morgan's Canon," in C. Allen and D. Cummins (eds.), *The Evolution of Mind*. Oxford: Oxford University Press, 224–242.

———. (1999a), "Physicalism from a Probabilistic Point of View," *Philosophical Studies*, 95: 135–174.

———. (1999b), "Testability," *Proceedings and Addresses of the American Philosophical Association*, 73: 47–76.

Stafford, T. (1997), "The Making of a Revolution," *Christianity Today* December 8: 16–22.

# 28

## The "Information Challenge"

Richard Dawkins

In September 1997, I allowed an Australian film crew into my house in Oxford without realizing that their purpose was creationist propaganda. In the course of a suspiciously amateurish interview, they issued a truculent challenge to me to "give an example of a genetic mutation or an evolutionary process which can be seen to increase the information in the genome." It is the kind of question only a creationist would ask in that way, and it was at this point I tumbled to the fact that I had been duped into granting an interview to creationists—a thing I normally don't do, for good reasons. In my anger I refused to discuss the question further, and told them to stop the camera. However, I eventually withdrew my peremptory termination of the interview as a whole. This was solely because they pleaded with me that they had come all the way from Australia specifically in order to interview me. Even if this was a considerable exaggeration, it seemed, on reflection, ungenerous to tear up the legal release form and throw them out. I therefore relented.

My generosity was rewarded in a fashion that anyone familiar with fundamentalist tactics might have predicted. When I eventually saw the film a year later,[1] I found that it had been edited to give the false impression that I was incapable of answering the question about information content.[2] In fairness, this may not have been quite as intentionally deceitful as it sounds. You have to understand that these people really believe that their question cannot be answered! Pathetic as it sounds, their entire journey from Australia seems to have been a quest to film an evolutionist failing to answer it.

Originally published in *The Skeptic* (1998, vol. 18, no. 4, pp. 22–25).

With hindsight—given that I had been suckered into admitting them into my house in the first place—it might have been wiser simply to answer the question. But I like to be understood whenever I open my mouth—I have a horror of blinding people with science—and this was not a question that could be answered in a soundbite. First you have to explain the technical meaning of "information." Then the relevance to evolution, too, is complicated—not really difficult, but it takes time. Rather than engage now in further recriminations and disputes about exactly what happened at the time of the interview (for, to be fair, I should say that the Australian producer's memory of events seems to differ from mine), I shall try to redress the matter now in constructive fashion by answering the original question, the "information challenge," at adequate length— the sort of length you can achieve in a proper article.

## Information

The technical definition of "information" was introduced by the American engineer Claude Shannon in 1948. An employee of the Bell Telephone Company, Shannon was concerned to measure information as an economic commodity. It is costly to send messages along a telephone line. Much of what passes in a message is not information: it is redundant. You could save money by recoding the message to remove the redundancy. Redundancy was a second technical term introduced by Shannon, as the inverse of information. Both definitions were mathematical, but we can convey Shannon's intuitive meaning in words.

Redundancy is any part of a message that is not informative, either because the recipient already knows it (is not surprised by it) or because it duplicates other parts of the message. In the sentence "Rover is a poodle dog," the word "dog" is redundant because "poodle" already tells us that Rover is a dog. An economical telegram would omit it, thereby increasing the informative proportion of the message. "Arr JFK Fri pm pls mt BA Cncrd flt" carries the same information as the much longer, but more redundant, "I'll be arriving at John F. Kennedy airport on Friday evening; please meet the British Airways Concorde flight." Obviously the brief, telegraphic message is cheaper to send (although the recipient may have to work harder to decipher it—redundancy has its virtues if we forget economics). Shannon wanted to find a mathematical way to capture the idea that any message could be broken into the information (which is worth paying for), the redundancy (which can, with economic advantage, be deleted from the message because, in effect, it can be reconstructed by the recipient), and the noise (which is just random rubbish).

"It rained in Oxford every day this week" carries relatively little information, because the receiver is not surprised by it. On the other hand, "It rained in the Sahara desert every day this week" would be a message with high information content, well worth paying extra to send. Shannon wanted to capture this sense of information content as "surprise value." It is related to the other sense—"that which is not duplicated in other parts of the message"—because repetitions lose their power to surprise. Note that Shannon's definition of the quantity of information is independent of whether it is true. The measure he came up with was ingenious and intuitively satisfying. Let's estimate, he suggested, the receiver's ignorance or uncertainty before receiving the message, and then compare it with the receiver's remaining ignorance after receiving the message. The quantity of ignorance-reduction is the information content. Shannon's unit of information is the bit, short for "binary digit." One bit is defined as the amount of information needed to halve the receiver's prior uncertainty, however great that prior uncertainty was (mathematical readers will notice that the bit is, therefore, a logarithmic measure).

In practice, you first have to find a way of measuring the prior uncertainty—that which is reduced by the information when it comes. For particular kinds of simple message, this is easily done in terms of probabilities. An expectant father watches the Caesarian birth of his child through a window into the operating theater. He can't see any details, so a nurse has agreed to hold up a pink card if it is a girl, blue for a boy. How much information is conveyed when, say, the nurse flourishes the pink card to the delighted father? The answer is one bit—the prior uncertainty is halved. The father knows that a baby of some kind has been born, so his uncertainty amounts to just two possibilities—boy or girl—and they are (for purposes of this discussion) equal. The pink card halves the father's prior uncertainty from two possibilities to one (girl). If there'd been no pink card but a doctor had walked out of the operating theater, shook the father's hand and said "Congratulations old chap, I'm delighted to be the first to tell you that you have a daughter," the information conveyed by the seventeen-word message would still be only one bit.

## Computer Information

Computer information is held in a sequence of noughts and ones. There are only two possibilities, so each 0 or 1 can hold one bit. The memory capacity of a computer, or the storage capacity of a disc or tape, is often measured in bits, and this is the total number of 0s or 1s that it can hold. For some purposes, more convenient units of measurement are the byte (8 bits), the kilobyte (1000 bytes or 8000 bits), the megabyte (a million bytes or 8 million bits) or the gigabyte (1000 million bytes or 8000 million bits). Notice that these figures refer to the total available capacity. This is the maximum quantity of information that the device is capable of storing. The actual amount of information stored is something else. The capacity of my hard disc happens to be 4.2 gigabytes. Of this, about 1.4 gigabytes are actually being used to store data at present. But even this is not the true information content of the disc in Shannon's sense. The true information content is smaller, because the information could be more economically stored. You can get some idea of the true information content by using one of those ingenious compression programs like "Stuffit." Stuffit looks for redundancy in the sequence of 0s and 1s, and removes a hefty proportion of it by recoding—stripping out internal predictability. Maximum information content would be achieved (probably never in practice) only if every 1 or 0 surprised us equally. Before data is transmitted in bulk around the internet, it is routinely compressed to reduce redundancy.

That's good economics. But on the other hand it is also a good idea to keep some redundancy in messages, to help correct errors. In a message that is totally free of redundancy, after there's been an error there is no means of reconstructing what was intended. Computer codes often incorporate deliberately redundant "parity bits" to aid in error detection. DNA, too, has various error-correcting procedures that depend on redundancy. When I come on to talk of genomes, I'll return to the three-way distinction between total information capacity, information capacity actually used, and true information content.

It was Shannon's insight that information of any kind, no matter what it means, no matter whether it is true or false, and no matter by what physical medium it is carried, can be measured in bits, and is translatable into any other medium of information. The great biologist J. B. S. Haldane used Shannon's theory to compute the number of bits of information conveyed by a worker bee to her hivemates when she "dances" the location of a food source (about 3 bits to tell about the direction of the food and another 3 bits for the distance of the food). In the same units, I recently calculated that I'd need to set aside 120 megabits of laptop computer memory to store the triumphal opening chords of Richard Strauss's "Also Sprach Zarathustra" (the "2001" theme), which I wanted to play in the middle of a lecture about evolution. Shannon's economics enable you to calculate how much modem time it'll cost you to e-mail the complete text of a book to a publisher in another land. Fifty years after Shannon, the idea of information as a commodity, as measurable and interconvertible as money or energy, has come into its own.

## DNA Information

DNA carries information in a very computer-like way, and we can measure the genome's capacity in bits too, if we wish. DNA doesn't use a binary code, but a quaternary one. Whereas the unit of information in the computer is a 1 or a 0, the unit in DNA can be T, A, C, or G. If I tell you that a particular location in a DNA sequence is a T, how much information is conveyed from me to you? Begin by measuring the prior uncertainty. How many possibilities are open before the message "T" arrives? Four. How many possibilities remain after it has arrived? One. So you might think the information transferred is four bits, but actually it is two. Here's why (assuming that the four letters are equally probable, like the four suits in a pack of cards). Remember that Shannon's metric is concerned with the most economical way of conveying the message. Think of it as the number of yes/no questions that you'd have to ask in order to narrow down to certainty, from an initial uncertainty of four possibilities, assuming that you planned your questions in the most economical way. "Is the mystery letter before D in the alphabet?" No. That narrows it down to T or G, and now we need only one more question to clinch it. So, by this method of measuring, each "letter" of the DNA has an information capacity of 2 bits.

Whenever prior uncertainty of recipient can be expressed as a number of equiprobable alternatives $N$, the information content of a message that narrows those alternatives down to one is log $2N$ (the power to which 2 must be raised in order to yield the number of alternatives $N$). If you pick a card, any card, from a normal pack, a statement of the identity of the card carries log 252 or 5.7 bits of information. In other words, given a large number of guessing games, it would take 5.7 yes/no questions on average to guess the card, provided the questions are asked in the most economical way. The first two questions might establish the suit. (Is it red? Is it a diamond?) the remaining three or four questions would successively divide and conquer the suit (is it a 7 or higher? etc.), finally homing in on the chosen card. When the prior uncertainty is some mixture of alternatives that are not equiprobable, Shannon's formula becomes a slightly more elaborate weighted average, but it is essentially similar. By the way, Shannon's weighted average is the same formula physicists have used, since the nineteenth century, for entropy. The point has interesting implications but I shall not pursue them here.

**Information and Evolution**

That's enough background on information theory. It is a theory that has long held a fascination for me, and I have used it in several of my research papers over the years. Let's now think how we might use it to ask whether the information content of genomes increases in evolution. First, recall the three-way distinction between total information capacity, the capacity that is actually used, and the true information content when stored in the most economical way possible. The total information capacity of the human genome is measured in gigabits. That of the common gut bacterium *Escherichia coli* is measured in megabits. We, like all other animals, are descended from an ancestor that, were it available for our study today, we'd classify as a bacterium. So perhaps, during the billions of years of evolution since that ancestor lived, the information capacity of our genome has gone up about three orders of magnitude (powers of ten)—about a thousandfold. This is satisfyingly plausible and comforting to human dignity. Should human dignity feel wounded, then, by the fact that the crested newt, *Triturus cristatus*, has a genome capacity estimated at 40 gigabits, an order of magnitude larger than the human genome? No, because, in any case, most of the capacity of the genome of any animal is not used to store useful information. There are many nonfunctional pseudogenes (see below) and lots of repetitive nonsense, useful for forensic detectives but not translated into protein in the living cells. The crested newt has a bigger "hard disc" than we have, but since the great bulk of both our hard discs is unused, we needn't feel insulted. Related species of newt have much smaller genomes. Why the Creator should have played fast and loose with the genome sizes of newts in such a capricious way is a problem that creationists might like to ponder. From an evolutionary point of view the explanation is simple (see *The Selfish Gene* pp. 44–45, and p. 275 in the second edition).

### Gene Duplication

Evidently the total information capacity of genomes is very variable across the living kingdoms, and it must have changed greatly in evolution, presumably in both directions. Losses of genetic material are called deletions. New genes arise through various kinds of duplication. This is well illustrated by hemoglobin, the complex protein molecule that transports oxygen in the blood.

Human adult hemoglobin is actually a composite of four protein chains called globins, knotted around each other. Their detailed sequences show that the four globin chains are closely related to each other, but they are not identical. Two of them are called alpha globins (each a chain of 141 amino acids), and two are beta globins (each a chain of 146 amino acids). The genes coding for the alpha globins are on chromosome 11; those coding for the beta globins are on chromosome 16. On each of these chromosomes, there is a cluster of globin genes in a row, interspersed with some junk DNA. The alpha cluster, on chromosome 11, contains seven globin genes. Four of these are pseudogenes, versions of alpha disabled by faults in their sequence and not translated into proteins. Two are true alpha globins, used in the adult. The final one is called zeta and is used only in embryos. Similarly the beta cluster, on chromosome 16, has six genes, some of which are disabled, and one of which is used only in the embryo. Adult hemoglobin, as we've seen, contains two alpha and two beta chains.

Never mind all this complexity. Here's the fascinating point. Careful letter-by-letter analysis shows that these different kinds of globin genes are literally cousins of each other, literally members of a family. These distant cousins still coexist inside our own genome, and that of all vertebrates. On the scale of whole organism, the vertebrates are our cousins too. The tree of vertebrate evolution is the family tree we are all familiar with, its branch-points representing speciation events—the splitting of species into pairs of daughter species. But there is another family tree occupying the same timescale, whose branches represent not speciation events but gene duplication events within genomes.

The dozen or so different globins inside you are descended from an ancient globin gene that duplicated in a remote ancestor who lived about half a billion years ago, after which both copies stayed in the genome. There were then two copies of it, in different parts of the genome of all descendant animals. One copy was destined to give rise to the alpha cluster (on what would eventually become chromosome 11 in our genome), the other to the beta cluster (on chromosome 16). As the aeons passed, there were further duplications (and doubtless some deletions as well). Around 400 million years ago the ancestral alpha gene duplicated again, but this time the two copies remained near neighbors of each other, in a cluster on the same chromosome. One of them was destined to become the zeta of our embryos, the other became the alpha globin genes of adult humans (other branches gave rise to the nonfunctional pseudogenes I mentioned). It was a similar story along the beta branch of the family, but with duplications at other moments in geological history.

Here's an equally fascinating point. Given that the split between the alpha cluster and the beta cluster took place 500 million years ago, it will of course not be just our human genomes that show the split—possess alpha genes in a different part of the genome from beta genes. We should see the same within-genome split if we look at any other mammals, at birds, reptiles, amphibians, and bony fish, for our common ancestor with all of them lived less than 500 million years ago. Wherever it has been investigated, this expectation has proved correct. Our greatest hope of finding a vertebrate that does not share with us the ancient alpha/beta split would be a jawless fish like a lamprey, for they are our most remote cousins among surviving vertebrates; they are the only surviving vertebrates whose common ancestor with the rest of the vertebrates is sufficiently ancient that it could have predated the alpha/beta split. Sure enough, these jawless fishes are the only known vertebrates that lack the alpha/beta divide.

Gene duplication, within the genome, has a similar historic impact on species duplication ("speciation") in phylogeny. It is responsible for gene diversity, in the same way as speciation is responsible for phyletic diversity. Beginning with a single universal ancestor, the magnificent diversity of life has come about through a series of branchings of new species, which eventually gave rise to the major branches of the living kingdoms and the hundreds of millions of separate species that have graced the earth. A similar series of branchings, but this time within genomes—gene duplications—has spawned the large and diverse population of clusters of genes that constitutes the modern genome.

The story of the globins is just one among many. Gene duplications and deletions have occurred from time to time throughout genomes. It is by these, and similar means, that genome sizes can increase in evolution. But remember the distinction between the total capacity of the whole genome, and the capacity of the portion that is actually used. Recall that not all the globin genes are actually used. Some of them, like theta in the alpha cluster of globin genes, are pseudogenes, recognizably kin to functional genes in the same genomes, but never actually translated into the action language of protein. What is true of globins is true of most other genes. Genomes are littered with nonfunctional pseudogenes, faulty duplicates of functional genes that do nothing, while their functional cousins (the word doesn't even need scare quotes) get on with their business in a different part of the same genome. And there's lots more DNA that doesn't even deserve the name pseudogene. It, too, is derived by duplication, but not duplication of functional genes. It consists of multiple copies of junk, "tandem repeats," and other nonsense that may be useful for forensic detectives but which doesn't seem to be used in the body itself.

Once again, creationists might spend some earnest time speculating on why the Creator should bother to litter genomes with untranslated pseudogenes and junk tandem repeat DNA.

## Information in the Genome

Can we measure the information capacity of that portion of the genome that is actually used? We can at least estimate it. In the case of the human genome it is about two percent—considerably less than the proportion of my hard disc that I have used since I bought it. Presumably the equivalent figure for the crested newt is even smaller, but I don't know if it has been measured. In any case, we mustn't run away with a chauvinistic idea that the human genome somehow ought to have the largest DNA database because we are so wonderful. The great evolutionary biologist George C. Williams has pointed out that animals with complicated life cycles need to code for the development of all stages in the life cycle, but they only have one genome with which to do so. A butterfly's genome has to hold the complete information needed for building a caterpillar as well as a butterfly. A sheep liver fluke has six distinct stages in its life cycle, each specialized for a different way of life. We shouldn't feel too insulted if liver flukes turned out to have bigger genomes than we have (actually they don't).

Remember, too, that even the total capacity of genome that is actually used is still not the same thing as the true information content in Shannon's sense. The true information content is what's left when the redundancy has been compressed out of the message, by the theoretical equivalent of Stuffit. There are even some viruses that seem to use a kind of Stuffit-like compression. They make use of the fact that the RNA (not DNA in these viruses, as it happens, but the principle is the same) code is read in triplets. There is a "frame" that moves along the RNA sequence, reading off three letters at a time. Obviously, under normal conditions, if the frame starts reading in the wrong place (as in a so-called frame-shift mutation), it makes total nonsense: the "triplets" that it reads are out of step with the meaningful ones. But these splendid viruses actually exploit frame-shifted reading. They get two messages for the price of one, by having a completely different message embedded in the very same series of letters when read frame-shifted. In principle you could even get three messages for the price of one, but I don't know whether there are any examples.

## Information in the Body

It is one thing to estimate the total information capacity of a genome, and the amount of the genome that is actually used, but it's harder to estimate its true information content in the Shannon sense. The best we can do is probably to forget about the genome itself and look at its product, the "phenotype," the working body of the animal or plant itself. In 1951, J. W. S. Pringle, who later became my professor at Oxford, suggested using a Shannon-type information measure to estimate "complexity." Pringle wanted to express complexity mathematically in bits, but I have long found the following verbal form helpful in explaining his idea to students.

We have an intuitive sense that a lobster, say, is more complex (more "advanced," some might even say more "highly evolved") than another animal, perhaps a millipede. Can we measure something in order to confirm or deny our intuition? Without literally turning it into bits, we can make an approximate estimation of the information contents of the two bodies as follows. Imagine writing a book describing the lobster. Now write another book describing the millipede down to the same level of detail. Divide the word-count in one book by the word-count in the other, and you have an approximate estimate of the relative information content of lobster and millipede. It is important to specify that both books describe their respective animals "down to the same level of detail." Obviously if we describe the millipede down to cellular detail, but stick to gross anatomical features in the case of the lobster, the millipede would come out ahead.

But if we do the test fairly, I'll bet the lobster book would come out longer than the millipede book. It's a simple plausibility argument, as follows. Both animals are made up of segments—modules of bodily architecture that are fundamentally similar to each other, arranged fore-and-aft like the trucks of a train. The millipede's segments are mostly identical to each other. The lobster's segments, though following the same basic plan (each with a nervous ganglion, a pair of appendages, and so on) are mostly different from each other. The millipede book would consist of one chapter describing a typical segment, followed by the phrase "Repeat *N* times" where *N* is the number of segments. The lobster book would need a different chapter for each segment. This isn't quite fair on the millipede, whose front and rear end segments are a bit different from the rest. But I'd still bet that, if anyone bothered to do the experiment, the estimate of lobster information content would come out substantially greater than the estimate of millipede information content.

It's not of direct evolutionary interest to compare a lobster with a millipede in this way, because nobody thinks lobsters evolved from millipedes. Obviously no modern animal evolved from any other modern animal. Instead, any pair of modern animals had a last common ancestor that lived at some (in principle) discoverable moment in geological history. Almost all of evolution happened way back in the past, which makes it hard to study details. But we can use the "length of book" thought-experiment to agree on what it would mean to ask the question of whether information content increases over evolution, if only we had ancestral animals to look at.

The answer in practice is complicated and controversial, all bound up with a vigorous debate over whether evolution is, in general, progressive. I am one of those associated with a limited form of yes answer. My colleague Stephen Jay Gould tends toward a no answer. I don't think anybody would deny that, by any method of measuring—whether bodily information content, total information capacity of genome, capacity of genome actually used, or true ("Stuffit compressed") information content of genome—there has been a broad overall trend toward increased information content during the course of human evolution from our remote bacterial ancestors. People might disagree, however, over two important questions: first, whether such a trend is to be found in all, or a majority of evolutionary lineages (for example parasite evolution often shows a trend toward decreasing bodily complexity, because parasites are better off being simple); second, whether, even in lineages where there is a clear overall trend over the very long term, it is bucked by so many reversals and re-reversals in the shorter term as to undermine the very idea of progress. This is not the place to resolve this interesting controversy. There are distinguished biologists with good arguments on both sides.

Supporters of "intelligent design" guiding evolution, by the way, should be deeply committed to the view that information content increases during evolution. Even if the information comes from God, perhaps especially if it does, it should surely increase, and the increase should presumably show itself in the genome. Unless, of course—for anything goes in such addle-brained theorizing—God works his evolutionary miracles by nongenetic means.

Perhaps the main lesson we should learn from Pringle is that the information content of a biological system is another name for its complexity. Therefore the creationist challenge with which we began is tantamount to the standard challenge to explain how biological complexity can evolve from simpler antecedents, one that I have devoted three books to answering (*The Blind Watchmaker, River Out of Eden, Climbing Mount Improbable*), and I do not propose to repeat their contents here. The "information challenge" turns out to be none other than our old friend: "How could something as complex as an eye evolve?" It is just dressed up in fancy mathematical language—perhaps in an attempt to bamboozle. Or perhaps those who ask it have already bamboozled themselves, and don't realize that it is the same old—and thoroughly answered—question.

## The Genetic Book of the Dead

Let me turn, finally, to another way of looking at whether the information content of genomes increases in evolution. We now switch from the broad sweep of evolutionary history to the minutiae of natural selection. Natural selection itself, when you think about it, is a narrowing down from a wide initial field of possible alternatives, to the narrower field of the alternatives actually chosen. Random genetic error (mutation), sexual recombination, and migratory mixing, all provide a wide field of genetic variation: the available alternatives. Mutation is not an increase in true information content, but rather the reverse, for mutation, in the Shannon analogy, contributes to increasing the prior uncertainty. But now we come to natural selection, which reduces the "prior uncertainty" and therefore, in Shannon's sense, contributes information to the gene pool. In every generation, natural selection removes the less successful genes from the gene pool, so the remaining gene pool is a narrower subset. The narrowing is nonrandom, in the direction of improvement, where improvement is defined, in the Darwinian way, as improvement in fitness to survive and reproduce. Of course the total range of variation is topped up again in every generation by new mutation and other kinds of variation. But it still remains true that natural selection is a narrowing down from an initially wider field of possibilities, including mostly unsuccessful ones, to a narrower field of successful ones. This is analogous to the definition of information with which we began: information is what enables the narrowing down from prior uncertainty (the initial range of possibilities) to later certainty (the "successful" choice among the prior probabilities). According to this analogy, natural selection is by definition a process whereby information is fed into the gene pool of the next generation.

If natural selection feeds information into gene pools, what is the information about? It is about how to survive. Strictly it is about how to survive and reproduce, in the conditions that prevailed when previous generations were alive. To the extent that present-day conditions are different from ancestral conditions, the ancestral genetic advice will be wrong. In extreme cases, the species may then go extinct. To the extent that conditions for the present generation are not too different from conditions for past generations, the information fed into present-day genomes from past generations is helpful information. Information from the ancestral past can be seen as a manual for surviving in the present: a family bible of ancestral "advice" on how to survive today. We need only a little poetic license to say that the information fed into modern genomes by natural selection is actually information about ancient environments in which ancestors survived.

This idea of information fed from ancestral generations into descendant gene pools is one of the themes of my new book, *Unweaving the Rainbow*. It takes a whole chapter, "The Genetic Book of the Dead," to develop the notion, so I won't repeat it here except to say two things. First, it is the entire gene pool of the species as a whole, not the genome of any particular individual, that is best seen as the recipient of the ancestral information about how to survive. The genomes of particular individuals are random samples of the current gene pool, randomized by sexual recombination. Second, we are privileged to "intercept" the information if we wish, and "read" an animal's body, or even its genes, as a coded description of ancestral worlds. To quote from *Unweaving the Rainbow*: "And isn't it an arresting thought? We are digital archives of the African Pliocene, even of Devonian seas; walking repositories of wisdom out of the old days. You could spend a lifetime reading in this ancient library and die unsated by the wonder of it."

## Notes

1. The producers never deigned to send me a copy: I completely forgot about it until an American colleague called it to my attention.

2. See Barry Williams (1998): "Creationist deception exposed," *The Skeptic* 18, 3, pp. 7–10, for an account of how my long pause (trying to decide whether to throw them out) was made to look like hesitant inability to answer the question, followed by an apparently evasive answer to a completely different question.

# VIII

## Intelligent Design Theorists Turn the Tables

In response to the charge that creationism is not science but religion, creationists often try to turn the tables on evolution and argue that it takes more faith to believe in it than in Creation, or that evolution is itself a religion. Phillip Johnson has taken this tack as well, claiming that evolutionary naturalism is the "established religion" of the West. In this section, we look at two more variations of this argument from other intelligent design creationists, one from William Dembski and another from Paul Nelson.

The first article is William Dembski's response to my book *Tower of Babel: The Evidence against the New Creationism*. He writes that I have criticized intelligent design creationism as though the issue between evolution and creationism were a choice between, respectively, mechanism and magic. He argues that I have gotten the relationship backwards. Design, he argues, citing the Stoics, "requires neither creator nor miracles" and so is not guilty of the charge of magic. It is rather evolutionary naturalists who are the magicians, he charges, in the sense that they purport to do the impossible, namely, "to get something from nothing"—he holds that complex biological information cannot arise from purely natural processes.

In "The Wizards of ID: Reply to Dembski," I address Dembski's charges. I had previously shown in *Tower of Babel* that Philip Johnson and other IDCs do want to use supernatural interventions to explain biological phenomena, so in this article I focus on documenting that Dembski holds the same view. I also show that evolutionary theory is not subject to his charge, and that it does have the resources to explain how biological complexity arises. Despite Dembski's protests to the contrary, the ID movement is indeed promoting a form of creationism, though they try to disguise it by verbal smoke and mirrors.

Creationists regularly claim that evolutionary biologists aim to undermine religion in general and Christianity in particular; Henry Morris calls evolution the main weapon in "the long war against God." Creationists name paleontologist Stephen Jay Gould as one of the worst offenders, and they often cite the essay included here as one of the main cases in point. Gould wrote "The Panda's Peculiar Thumb" as one of his ongoing series of articles in *Natural History* (Nov. 1978), and reprinted it (with minor changes) as the lead chapter in his book *The Panda's Thumb* (1980). In it, he explains how biological "imperfections" provide one powerful line of evidence that species arise through an evolutionary process. In the line creationists usually quote from the essay, Gould writes: "Odd arrangements and funny solutions are the proof of evolution—paths that a sensible God would never tread but that a natural process, constrained by history, follows perforce."

The next essay, "The Role of Theology in Current Evolutionary Reasoning," is by Paul Nelson, who has been one of the most important behind-the-scenes leaders of the intelligent design movement from its beginning. Nelson is currently most known for his forthright defense of the "young-earth" creationist view (e.g., in Moreland and Reynolds, eds., *Three Views on Creation and Evolution*, Zondervan, 1999) and will likely soon be more visible after the publication of his forthcoming book that will question the evidence for the evolutionary thesis of common descent. However, here I include his most recent critique of Stephen Jay Gould's essay. Nelson previously argued (under his pseudonym Peter Gordon) that the panda's thumb isn't imperfect after all, but rather efficient. In this article, he argues that Gould is using a theological argument to support evolution, and that this violates the strictures of methodological naturalism that scientists impose against creationists. Nelson calls "foul," arguing that biologists can't have it both ways.

Philosopher of biology Kelly Smith responds to Nelson in "Appealing to Ignorance Behind the Cloak of Ambiguity," written for this volume. He explains that the apparent inconsistency that Nelson points to is "a product of heterogeneity in creationism's arguments, not any difficulty within evolutionary theory itself." Smith identifies a particular ambiguity that he says creationists, including Nelson, regularly hide behind. He argues that Nelson also misstates the nature of methodological naturalism. Smith then analyzes Nelson's arguments about the panda's thumb case both under the assumption that Nelson is speaking of a "Mysterious God" and under the assumption that Nelson is speaking of a "Reasonable God," and shows that neither version helps the creationist case. In the end, Smith concludes that the creationist argument is simply a fallacious appeal to ignorance.

In the final essay in this section, we return to Stephen Jay Gould, writing about what he takes to be the proper relationship between science and religion, particularly in light of the ongoing creationism controversy. In a reflection on the 1995 statement of Pope John Paul II that evolution is not in contradiction to the teachings of the Catholic Church, Gould finds this position to be in line with his own view that science and religion, properly understood, are not in conflict because they occupy different domains. He proposes a principle he calls NOMA, short for "nonoverlaping magisteria," to express their relationship—science deals with matters of fact and theory having to do with the empirical world, while religion covers questions of moral meaning and value. Gould concludes on a personal note: "I believe, with all my heart, in a respectful, even loving concordat between our magisteria.... NOMA represents a principled position on moral and intellectual grounds, not a mere diplomatic stance."

# 29

## Who's Got the Magic?

William A. Dembski

In criticizing Phillip Johnson's "intelligent design creationism," Robert Pennock raises a particularly worrisome legal consequence of Johnson's view. According to Pennock, Johnson insists "that science admit the reality of supernatural influences in the daily workings of the world." But what if the same reasoning that Johnson is trying to import into science were adopted in Johnson's own area of specialization—the law (Johnson is a law professor at UC Berkeley)? Here's the concern as Pennock lays it out in *Tower of Babel* (p. 295):

For the law to take [Johnson's view] seriously as well, it would have to be open to both suits and defenses based on a range of possible divine and occult interventions. Imagine the problems that would result if the courts had to accept legal theories of this sort. How would the court rule on whether to commit a purportedly insane person to a mental hospital for self-mutilation who claims that the Lord told her to pluck out her eye because it offended her? How would a judge deal with a defendant, Abe, accused of attempted murder of his son, Ike, who claims that he was only following God's command that he kill Ike to prove his faith?

Implicit in this passage and throughout Pennock's book is a forced choice between mechanism and magic: Either the world works by mechanisms that obey inviolable natural laws and that admit no break in the chain of natural causation, or all hell breaks loose and the world admits supernatural interventions that make a hash of science and our understanding of the world generally (and legal studies in particular). Pennock is offering his readers mechanism. Johnson is offering them magic. Any reasonable person knows which option to choose.

Originally published in *Metanexus: The Online Forum on Religion and Science* (2000, no. 042) ⟨http://www.metanexus.net⟩.

But as with most forced choices, there's a tertium quid that Pennock has conveniently ignored, and that when properly understood shows that the real magician here is in fact Pennock and not Johnson. The tertium quid here is intelligent design, which is entirely separable from creationism. No doubt, Pennock's constant conjoining of the two serves a useful rhetorical purpose, rallying the troops, giving Darwinists a single common enemy, and keeping biology safe from teleology (indeed, it has become a point of grammatical correctness with Pennock never to use "intelligent design" without "creationism"—"intelligent design" properly being an adjective that only modifies "creationism"). But Pennock, as a trained philosopher, knows that design is an old notion that requires neither miracles nor a creator (F. H. Sandbach's *The Stoics*, for instance, makes this abundantly clear). Intelligent design is detectable; we do in fact detect it; we have reliable methods for detecting it (cf. my *The Design Inference*); and its detection involves no recourse to the supernatural. Design is common, rational, and objectifiable.

How, then, is Pennock a magician? There are at least three forms of magic. One is the art of illusion, where appearance is carefully crafted to distort reality. As entertainment, this form of magic is entirely unobjectionable. Another form of magic is to invoke the supernatural to explain a physical event. To call this magic is certainly a recent invention, since it makes most theists into magicians (Was Thomas Aquinas a magician for accepting as a historical fact the resurrection of Jesus? Was Moses Maimonides a magician for thinking that his namesake had parted the Red Sea?). According to Pennock, intelligent design creationism is guilty of this form of magic. Deep down, though, Pennock must realize that intelligent design (leaving off the creationism) can avoid this charge.

Pennock is guilty of his own form of magic, however. The third form of magic, and the one Pennock and his fellow scientific naturalists are guilty of, is the view that something can be gotten for nothing. This third form of magic can be nuanced. The "nothing" here need not be an absolute nothing. And the transformation of nothing into something may involve minor expenditures of effort. For instance, the magician may need to utter "abracadabra" or "hocus-pocus." The Darwinian just-so stories that attempt to account for complex, information rich biological structures are likewise incantations that give the illusion of solving a problem but in fact merely cloak ignorance.

The great appeal behind this third form of magic is the offer of a bargain—indeed an incredible bargain for which no amount of creative accounting can ever square the books. The idea of getting something for nothing has come to pervade science. In cosmology, Alan Guth, Lee Smolin, and Peter Atkins all claim that this marvelous universe could originate from quite unmarvelous beginnings (a teaspoon of ordinary dust for Guth, black-hole formation for Smolin, and set-theoretic operations on the empty set for Atkins). In biology, Jacques Monod, Richard Dawkins, and Stuart Kauffman claim that the panoply of life can be explained in terms of quite simple mechanisms (chance and necessity for Monod, cumulative selection for Dawkins, and autocatalysis for Kauffman).

We have become so accustomed to this something-for-nothing way of thinking that we no longer appreciate just how deeply magical it is. Consider, for instance, the following evolutionary account of neuroanatomy by Melvin Konner, an anthropologist and neurologist at Emory University: "Neuroanatomy in many species—but especially in a brain-ridden one like ours—is the product of sloppy, opportunistic half-billion year [evolution] that has pasted together, and only partly integrated, disparate organs that evolved in different animals, in different eras, and for very different purposes" (*IEEE Spectrum*, March 2000). And since human consciousness and intelligence are said to derive from human neuroanatomy, it follows that these are themselves the product of a sloppy evolutionary process.

But think what this means. How do we make sense of "sloppy," "pasted together," and "partly integrated," except with reference to "careful," "finely adapted," and "well integrated." To speak of hodge-podge structures presupposes that we have some concept of carefully designed structures. And of course we do. Humans have designed all sorts of engineering marvels, everything from Cray supercomputers to Gothic cathedrals. But that means, if we are to believe Melvin Konner, that a blind evolutionary process (i.e., Richard Dawkins's blind watchmaker) cobbled together human neuroanatomy, which in turn gave rise to human consciousness, which in turn produces artifacts like supercomputers, which in turn are not cobbled together at all but instead carefully designed. Out pop purpose, intelligence, and design from a process that started with no purpose, intelligence, or design. This is magic.

Of course, to say this is magic is not to say it is false. It is after all a logical possibility that purpose, intelligence, and design emerge by purely mechanical means out of a physical universe initially devoid of these. Intelligence, for instance, may just be a survival tool given to us by an evolutionary process that places a premium on survival and that is itself not intelligently guided. The basic creative forces of nature might be devoid of intelligence. But if that is so, how can we know it? And if it is not so, how can we know that? It does no good simply to presuppose that purpose, intelligence, and design are emergent properties of a universe that otherwise is devoid of these.

The debate whether nature has been front-loaded with purpose, intelligence, and design is not new. Certainly the ancient Epicureans and Stoics engaged in this debate. The Stoics argued for a design-first universe: the universe starts with design and any subsequent design results from the outworkings of that initial design (they resisted subsequent novel infusions of design). The Epicureans, on the other hand, argued for a design-last universe: the universe starts with no design and any subsequent design results from the interplay of chance and necessity.

What is new, at least since the Enlightenment, is that it has become intellectually respectable to cast the design-first position as disreputable, superstitious, and irrational; and the design-last position as measured, parsimonious, and alone supremely rational. Indeed, the charge of magic is nowadays typically made against the design-first position, and not against the design-last position, as I have done here.

But why should the design-first position elicit the charge of magic? Historically in the West, design has principally been connected with Judeo-Christian theism. The God of Judaism and Christianity is said to introduce design into the world by intervening in its causal structure. But such interventions cannot be anything but miraculous. And miracles is the stuff of magic. So goes the argument. The argument is flawed because there is no necessary connection between God introducing design into the world and God intervening in the world in the sense of violating its causal structure. Theists like Richard Swinburne, for instance, argue that God front-loads design into the universe by designing the very laws of nature. Paul Davies takes a similar line. Restricting design to structuring the laws of nature precludes design from violating those laws and thus violating nature's causal structure.

Design easily resists the charge of magic. Rather, it's the a priori exclusion of design that has a much tougher time resisting it. Indeed, the design-last position is inherently magical. Consider the following remark by Harvard biologist Richard Lewontin in *The New York Review of Books*:

> We take the side of science *in spite of* the patent absurdity of some of its constructs, *in spite of* its failure to fulfill many of its extravagant promises of health and life, *in spite of* the tolerance of the scientific community for unsubstantiated just-so stories, because we have a prior commitment, a commitment to materialism. It is not that the methods and institutions of science somehow compel us to accept a material explanation of the phenomenal world, but, on the contrary, that we are forced by our *a priori* adherence to material causes to create an apparatus of investigation and a set of concepts that produce material explanations, no matter how counterintuitive, no matter how mystifying to the uninitiated.

If this isn't magic, what is?

Even so, the scientific community continues to be skeptical of design. The worry is that design will give up on science. In place of a magic that derives something from nothing, design substitutes a designer who explains everything. Magic gets you something for nothing and thus offers a bargain. Design gets you something by presupposing something unimaginably bigger and thus asks you to sell your scientific soul. At least so the story goes. But design can be explanatory without giving away the store. Certainly this is the case for human artifacts, which are properly explained by reference to design. Nor does design explain everything: There's no reason to invoke design to explain a random inkblot; but a Dürer woodcut is something else altogether. The point of the intelligent design program is to extend design from the realm of human artifacts to the natural sciences. The program may ultimately fail, but it is only now being tried and it is certainly worth a try.

Just as truth is not decided at the ballot box, so truth is not decided by the price one must pay for it. Bargains are all fine and well, and if you can get something for nothing, go for it. But there is an alternate tendency in science which says that you get what you pay for and that at the end of the day there has to be an accounting of the books. Some areas of science are open to bargain-hunting and some are not. Self-organizing complex systems, for instance, are a great place for scientific bargain-hunters to shop. Bernard cell convection, Belousov-Zhabotinsky reactions, and a host of other self-organizing systems offer complex organized structures apparently for free. But there are other areas of science that frown upon bargain-hunting. The conservation laws of physics, for instance, allow no bargains. The big question confronting design is whether it can be gotten on the cheap or must be paid for in kind. Design theorists argue that design admits no bargains.

Pennock and his fellow scientific naturalists are bargain hunters. They want to explain the appearance of design in nature without admitting actual design. That's why Richard Dawkins begins *The Blind Watchmaker* with "Biology is the study of complicated things that give the appearance of having been designed for a purpose," whereupon he requires an additional three hundred pages to show why it is only an appearance of design. Pennock and his fellow naturalists have my very best wishes for success in their hunt for the ultimate bargain. They may even be right. But they are not guaranteed to be right. And they certainly haven't demonstrated that they are right. They have yet to pull the rabbit out of the hat.

# 30

# The Wizards of ID: Reply to Dembski

Robert T. Pennock

In his article "Who's Got the Magic?" (Dembski, 2000), William Dembski discusses my book *Tower of Babel: The Evidence Against the New Creationism*, and defends Phillip Johnson's intelligent-design creationism (IDC).[1] I had expected that Dembski would also respond to criticisms I had made of his own specific arguments, and was disappointed that he decided not to engage them. For instance, he does not address my criticisms of his "explanatory filter" or of his problematic, idiosyncratic uses of the key terms in his argument—"law," "chance," "information," "specified," and "design." He does not confront examples I gave to show that the Darwinian mechanism can produce functional and novel complexity. He ignores a counter-example that shows how a genetic algorithm can produce complex specified information (which he says is impossible to do), and that avoids what he claims to be a defect in examples Dawkins and Sober have given. He does not engage my criticisms of his SETI analogy. And so on. Indeed, until now, Dembski has had nothing to say in response to my book and has only made a couple of brief ad hominem remarks on his web page (Dembski, 1999b).

I will reserve for a forthcoming article detailed criticisms of what he has published since I finished *Tower of Babel*, and will here confine my reply to issues related to the one substantive issue he does discuss in his current paper, which has to do with whether IDC or evolutionary theory is a form of magic. With the exception of a passing remark in my book about Michael Behe's appeal to supernatural design as a scientific explanation, I have never discussed IDC explicitly in terms of magic, but

Originally published in *Metanexus: The Online Forum on Religion and Science* (2000, no. 089) ⟨http://www.metanexus.net⟩.

Dembski's analogy is apt so I will form my reply along the lines that his analysis suggests.

Dembski distinguishes three kinds of magic. The first is the familiar form of stage magic—"the art of illusion"—an entertainment he says is unobjectionable. It is the second kind of magic—"to invoke the supernatural to explain a physical event"—that he objects to, at least when the label is applied to his and other IDCs' views. He is keen to distance his movement from the term "creationism" and suggests that it is wrong to link their notion of design with the supernatural. There is no magic of the second kind in our design, Dembski appears to say. It is really scientific naturalists who have the magic, he charges: a third type of magic which is the belief "that something can be gotten for nothing."

I shall argue that science is not guilty of this, but that IDC does involve both of these sorts of magic. I'll show that IDCs practice the first form as well, which they use to try to divert attention from the other two. Clever misdirection is what makes the stage show work. Watch closely and see how the trick is done.

## Smoke

Dembski chides me for never using the term "intelligent design" without conjoining it to "creationism." He implies (though never explicitly asserts) that he and others in his movement are not creationists and that it is incorrect to discuss them in such terms, suggesting that doing so is merely a rhetorical ploy to "rally the troops."[2] Am I (and the many others who see Dembski's movement in the same way) misrepresenting their position? The basic notion of creationism is the rejection of biological evolution in favor of special creation, where the latter is understood to be supernatural. Beyond this there is considerable variability. Some creationists think the world is young while a fewer number accept that it is ancient. (Among IDCs, Paul Nelson, Percival Davis, Sigfried Scherer and John Mark Reynolds, to name a few, are among those in the first camp, while Michael Behe is in the second. Most IDCs hide their views on this and other relevant theses.) Creationists typically base their views on the Bible, but there are also, for instance, Native American creationists whose specific alternative creation stories stem from their own religious traditions. So-called scientific creationism or creation-science, which is our main concern here, is a form of apologetics that claims not to base its conclusions on a religious belief or text, but purports that these are supported by scientific research alone. I have previously discussed many varieties of creationism in detail, showing how IDC compares to other forms (Pennock, 1999, ch. 1), and will not repeat that discussion. Here I will focus mostly on Dembski's own views to document how they fit in the overall pattern.

Dembski's other complaint against me arises directly out of the first; he objects to having "design" yoked to the second form of magic that he defined, namely, invoking the supernatural to explain a physical event. He says, "To call this magic is certainly a recent invention, since it makes most theists into magicians" (Dembski, 2000). But Dembski surely knows that etymologists can trace the English word "magic" back through Latin and the Greek (*mágos*) to the Old Persian word *magus*, which dictionaries define as a Zoroastrian priest, or a seer or wizard knowledgeable in sorcery, numerology and astrology. The plural of *magus* is *magi*, and who were the Three Magi, if not theist astrologers who had read the sign of a star? IDCs engage in their own form of numerology, believing that a few probability calculations will reveal signs of the divine. Not all theists are magicians in this sense, but those who are should honestly acknowledge it. Let me be clear that my aim was never to dissuade such theists of their faith in a supernatural designer-creator, but just to show it is not legitimate to accept the creationist contention that this is a scientific conclusion. IDCs, like other creationists, do hold this view. Again, I shall focus here mostly upon establishing that Dembski's position is as I have claimed.

In his response to me, Dembski cites variations of deism to show that God could have created without miraculous violations of natural laws. This response is puzzling, in that I had myself discussed the deist option in my book as one way that a person could accept evolution and scientific methodology while still retaining belief in God as Creator. I gave this as one of several counterexamples to IDCs' *rejection* of such a possibility. Johnson explicitly dismisses deist views throughout his writings. Indeed, to try to set up the (false) dichotomy that he needs to legitimate purely negative argument, he goes much further and dismisses any form of theistic evolution. Dembski's response is all the more puzzling, since he adopts the same position.

IDCs see theistic evolution as, in Dembski's words, "an oxymoron." He writes "*Design theorists are no friends of theistic evolution*" (Dembski, 1995, p. 3; emphasis in original). (I have quoted this line on several occasions, but in one article the last word was printed "evolutionists" instead of "evolution." Let me assure Dembski that the mistake was unintentional and not "by design." I do not know how the error was introduced, but I should have caught it during proofing and I apologize for missing it. Dembski took offense at the error, and was quite correct when he complained that there is a huge difference in meaning between refusing friendship with a group of people rather than with their ideas. More on that shortly.) Theistic evolutionists accept the truth of evolution as science has discovered it, but retain a belief in God, though obviously not in the kind of God that creationists—the large majority of whom are fundamentalist Christians—find worthy of the name. Dembski tells us: "As far as design theorists are concerned, theistic evolution is American evangelicalism's ill-conceived accommodation to Darwinism" (Dembski, 1995, p. 3). Like other IDCs, Dembski thinks that theistic evolution is nonsensical; he says it is equivalent to "purposeful purposelessness" (Dembski, 1995, p. 3).

I recently attended a talk Dembski gave to a group of about fifteen high school and junior high school science teachers in which he tried to explain how they could incorporate intelligent design into their classes.[3] Illustrating the concept of design by reference to violins and other musical instruments, Dembski noted that the early Christian fathers compared the world to a lute, and he advised that this is a better analogy than a watch. Why? A good watch doesn't need winding, but an instrument is meant to be played, he explained.

Anyone familiar with Paley's design argument would have understood exactly what he was implying. If the world is like a watch that God created, then, the deist argues, we should expect that God got it exactly right from the start, and would not thereafter continually intervene to make adjustments. But, as Johnson defines it, the creationist view he advocates holds that "a supernatural Creator not only initiated this process but in some meaningful sense controls it in furtherance of a purpose" (Johnson, 1991, p. 4). Direct control in the form of supernatural intervention is at the very heart of the IDC program, which is aimed not just at overturning evolution, but scientific and metaphysical naturalism as well. In this respect, IDCs are again right in line with the classic creationists, and Dembski is clearly in lockstep with this position. He has written: "I don't believe in fully naturalistic evolution controlled solely by purposeless material processes" (Dembski, 1995, p. 5). Moreover, in his most recent and definitive discussion of ID, he says explicitly that biological forms were created with periodic "discrete insertions" of design and that the complex specified information purportedly required to affect such creation "transcends natural causes" (Dembski, 1999a, p. 171).

Look again now at how Dembski rebutted my challenge regarding IDCs' attempts to bring the supernatural into science. He did so by citing the Stoics and saying that design "requires neither miracles nor a creator." This is not to the point. No one objects to archeologists, anthropologists, or even SETI researchers, who all make use of this completely ordinary and *natural* notion. If that were the only sense of "design" that Dembski and company were talking about, then most of this discussion would be obviated. IDCs often complain that the term "evolution" can be used in many ways and claim that scientists use it to "cover a multitude of sins," but the same can be said, with more justice, about their use of the term "design."

The trick can be seen by examining how Dembski's argument would look against just a slightly different stage setting. For instance, it has become common in this New Age to hear purported explanations of all sorts of physical ailments in terms of blockage of the flow of the body's "energy." Believers may hold that this is a spiritual form of energy that is somehow "infused" into the world, and they may reject the scientific explanation of the same ailments, deriding them as "materialist" and "reductionist". They try to poke holes in medical theory by pointing to ailments that doctors don't know how to explain, and claiming that physicians maintain what is, after all, "just a theory" because of their "naturalistic blinders" and desire to hold on to their cultural prestige and power. We may imagine a New Age Dembski who promotes "energy theory" and defends it by saying that energy need not involve the supernatural but is, as he puts it, "common, rational and objectifiable." Well, perhaps so, but that natural notion is not what we were talking about, was it?

So, let us not be fooled by their stagecraft into thinking that IDCs constitute a real scientific movement.[4] Replace "design theorists" in

Dembski's statement about "ill-conceived accommodation to evolution" quoted above with "evolutionary biologists" or "physicists" or "complexity theorists" and one immediately sees the difference. IDC is a theological movement crafted to win a particular political goal—initially, getting their form of special creation into the public school science classes—in what IDCs take to be the key strategic game in the "culture wars."

Dembski admits as much even in the title he chose for a course he prepared to teach recently at Trinity Graduate School and Trinity Evangelical Divinity School at Trinity International University: "Intelligent Design—The New Player in the Creation-Evolution Controversy." In his syllabus for the course, Dembski describes ID as "an alternative to scientific creationism ... that challenges Darwinian evolution and its naturalistic legacy."[5]

Dembski writes that ID "does not treat the biblical book of Genesis as a scientific text, but instead argues on general scientific and philosophical grounds that Darwinian evolution is a failed scientific research program." This is exactly the strategy that classic creation-science followed in trying to get its views into the public schools—provide only a vague positive thesis and rely upon negative argumentation against evolution in the hope of winning by default. Indeed, as I elsewhere showed in detail (Pennock, 1999, pp. 250–251), all the key elements of Dembski's argument, including the use of Yockey's and Orgel's concept of specified complexity, were previously articulated by Norman Geisler, who was one of the expert witnesses for the creationist side in the famous "balanced treatment of creation-science and evolution-science" case in Arkansas in the early 1980s. Today the ambiguous term they use is "designer" instead of "creator," but otherwise the game plan of the "design theorists" is little different than that of the "creation scientists."

Continuing in his syllabus, Dembski predicts that his movement will have earth-shaking consequences: "Because a vast naturalistic superstructure has been built on Darwinism, its impending collapse at the hands of intelligent design promises to be one of the great upcoming cultural convulsions (witness the recent furor in Kansas)." Here, of course, he is referring to the vote by creationists on the Kansas State Board of Education to expunge evolution, as well as any reference to the ancient age of the earth, from the state's science standards. Creationists on the Board had also tried to insert the following statement into the science standards: "The design and complexity of the design of the cosmos requires an intelligent designer" (NYT 1999). That sentence was excised from the final approved version, but its initial inclusion is telling. In the wake of the Board's decision, IDC leaders—Phillip Johnson, Jonathan Wells, Stephen Meyer, among others—flew to Kansas to take the creationist side in public debates, and to promote it in churches and in talks sponsored by campus Christian groups.

With these quoted passages as background, Dembski concludes by stating that the purpose of his course is to "examine intelligent design as a cultural and intellectual movement and show how Christian theology and apologetics stand to benefit from it."

One might occasionally get the impression that Dembski does not reject evolution, as when he says elsewhere, somewhat vaguely, that he accepts "that organisms have undergone some change in the course of natural history." However, that impression is dispelled when he immediately qualifies this and asserts that "this change has occurred within strict limits." Dembski's position is indistinguishable from that of Henry and John Morris and other creationists, who put the point in exactly the same terms. Also like the Morrises, Dembski specifically insists that "human beings were specially created" (Dembski 1995, p. 5).

We shall see further evidence of Dembski's view in subsequent sections, but this should be sufficient to establish the point. Judge for yourself whether it is unfair to call this position "creationism."

When IDCs use the word "design" as Dembski does, it is the equivalent of the conjurer who assures his audience "I hold in my hand an ordinary deck of cards." Unlike scientists, who lay their cards on the table in peer-reviewed journals, the Wizards of ID do not want anyone to examine their props too closely, because it would then be obvious that their "design theory" is a façade. This is one reason why IDCs are not forthright when asked to state and defend their positive views, and why their standard move is to try to shift the burden of proof. Johnson and Plantinga are masters of this legerdemain and Dembski, their star apprentice, has learned their trick.[6] Rather than confront the challenge that magic of the second kind poses, he first obfuscates the difficulty and then tries to reflect it back, charging that it is actually naturalists who are engaging in a form of magic. In stagecraft this is known as a classic "smoke and mirrors" trick.

## Mirrors

So, let us now turn to Dembski's charge that scientific naturalists are pulling their own magic stunt—the ultimate trick of "getting something from nothing." In his article, he derides the scientific naturalist's method as "bargain hunting," and warns that you get what you pay for. Dembski wishes us luck, but suggests that scientific naturalists will never pull a rabbit out of a hat. (Michael Behe makes a similar complaint, claiming that Darwinians are illegitimately pulling so-called irreducible complexity out of the "black box" of the cell.) Like the creationists on the Board of Education in Kansas, Dembski finds problems not just with biological evolution, but with cosmology and other sciences as well. Here I will retain our focus upon the former. Dembski makes his complaint at a very abstract level, but I will not take up the philosophical chestnut of how, metaphysically, something can come from nothing; that is a fascinating problem, but in this context it is just more smoke to distract from the empirical question of how biological species evolved. Dembski believes his reflected complaint applies specifically to this issue. It really is meant as a challenge to human evolution and to rabbit evolution as well, so let us examine it in that context.

This is obviously not the place to review the vast array of evidential support for evolution: we have abundant evidence that contemporary complex forms evolved from simple ancestral forms, and we know a fair amount about the pathways that that descent with modification took, as well as many of the causal processes that produced those changes.[7] A review of the literature will show that evolutionary biology does not pull rabbits out of a hat. Rabbits (such as the European species *Oryctolagus cuniculus*) and hares (such as *Lepus euopaeus*) belong to the Leporidae family and are grouped with the pikas (of the Ochotonidae family) in the order Lagomorpha. Such nested patterns of features among extant organisms are a potent source of evidence for the evolutionary hypothesis of common descent with modification. Despite some creationists' claims that evolution is just "assumed," evolutionary hypotheses are open to empirical test and to disconfirmation. For instance, J. B. S. Haldane used to point out how significantly we would have to change our understanding of evolutionary history if we were to discover a Precambrian rabbit. But in fact we do not find rabbits where they do not belong; the patterns in the fossil record are another independent source of evidence that further supports the evolutionary picture. Primitive rabbits have been found in Paleocene and Eocene sediments and are abundant by the Oligocene. Such fossil finds have helped sort out the evolutionary relationship of the lagomorphs to the rodents, to which they are closely related (Colbert and Morales 1991, pp. 301–302).

We also have molecular evidence to help refine the evolutionary picture. For instance, scientists have recently investigated mannan-binding protein (MBP) in rabbits and incorporated this new biological and genetic information into what was already known about the genes that code for different types of this protein in humans and other animals. Their analysis showed that one MBP gene was lost not only during the evolution of hominids, but also after the separation of birds and mammals (Kawai et al., 1998). Molecular techniques now allow scientists to precisely compare the sequences of amino acids in proteins across species and to use this to help infer phylogenetic trees (i.e., the pathways of evolutionary divergence). As expected, when such phylogenies are constructed (as they can now be done routinely as an exercise in evolutionary biology lab courses), human beings turn out to be more closely related to rabbits than to frogs, fruit flies or yeast. We retain other clear signs of that evolutionary history. The human appendix, the small and often problematic worm-shaped structure that must often be snipped from the intersection of the large and small intestines, is a vestige of a large organ found in rabbits and other herbivorous mammals that aids their digestion of cellulose that has been imperfectly co-opted to a different function as part of our immune system. One could easily extend this list of evidence indefinitely.

This is not to say that evolutionary science is close to being able to answer all questions. Dembski's criticism of the speculative evolutionary accounts some scientists have given of phenomena like consciousness does contain a kernel of truth; indeed, that criticism and the very term he used—"just-so stories"—was first made by evolutionary biologists against others who drew specific conclusions that went beyond the

Robert T. Pennock

available evidence. However, these are far from representative and, in any case, it is wrong to dismiss such accounts wholesale, since many may be regarded as what philosophers of science call "how-possibly explanations." Even these do not come "from nothing" but are based upon known types of causal processes. As such, they are constrained and perfectly legitimate, serving a useful purpose of illustrating the explanatory range of a theory and suggesting lines for further investigation.

What about design explanations? Here there are fewer constraints, but in certain contexts, if we stick to our ordinary, natural notion of intentional design, we can still make some headway; when archeologists pick out something as an artifact or suggest possible purposes for some unfamiliar object they have excavated they can do so because they already have some knowledge of the causal processes involved and have some sense of the range of purposes that could be relevant. It gets more difficult to work with the concept when speaking of extraterrestrial intelligence, and harder still when considering the possibility of animal or machine intelligence. But once one tries to move from natural to supernatural agents and powers as creationists desire, "design" loses any connection to reality as we know it or can know it scientifically.

In his response, Dembski again declines to confront this problem, and simply claims that design does not explain everything: "There's no reason to invoke design to explain a random inkblot; but a Dürer woodcut is something else altogether." But here, as elsewhere, Dembski's examples are not to the point. His goal is to overturn scientific naturalism, but his examples of inferred design always involve natural objects, natural agents and natural causal processes. Also notice that by specifying that the inkblot is "random" Dembski is simply begging the question about its possible design. The relevant issue is whether and how, given an observation of a particular inkblot, we may infer its design. It is all too easy to *read in* design (purposeful intention) to a random design (pattern), as in Rorschach ink blots. Recently, an oil stain on a glass office building supposedly looked like an image of the Virgin Mary, and was interpreted by hundreds of believers as a divine sign. Moreover, while an inkblot on a sheet of paper may have occurred by chance, it might instead have been drawn by a doodler or an artist. Or perhaps it only appears to be a real inkblot, but is actually a plastic inkblot left by a practical joker. Given a naturalistic method we can often find evidence to distinguish among such possibilities, but how can we tell whether it was designed if we abandon this method and admit supernatural interventions, as IDCs want science to do? Except for a few hints from Johnson about use of "sacred books" and "mystical states of mind," IDCs have proposed no new method to replace the naturalistic one they claim is dogmatically biased (Pennock 1999, p. 197).

Dembski's example of a Dürer print is also problematic. Again, no one disputes the identification of intelligent design in Dürer's renderings of horrifying scenes of the Four Horsemen of the Apocalypse or sublime pictures of nature. Dürer's charming print of the "Young Hare" is a human artifact, and our inference to that conclusion is entirely naturalistic. That inference, however, is not so simple as Dembski would have us believe. Dembski does not explain how his design inference is supposed to work to identify a Dürer print as having been designed, but it must be because in looking at a particular print we supposedly can see that it contains complex specified information (CSI). Presumably, the image of the hare is complex and specified in the relevant technical sense that licenses the design inference. (There are problems with Dembski's technical conditions and how one tells when they apply, but I will reserve criticisms of those issues for another occasion. In this article I am working under the assumption that the specific examples he gives count as CSI as he defines it, and that it is on that basis alone that we are to infer that they were designed.) To make a woodcut print or a painting we know that the artist had to carve or paint every line by hand. Could Dembski inadvertently be smuggling in such background information? He claims that the design inference does not make any reference to such causal knowledge, so to avoid that feature of the example, let us consider a photographic print instead.

A professional photographer may set up in detail all the elements of the photograph and so have "designed" the picture in every particular. An amateur, on the other hand, may simply point and click, and the resulting picture may be *intended*, that is, be "by design" in a weak sense, but not *designed* in the first, strong sense. A novice, unfamiliar with the workings of a camera, may touch the shutter release accidentally and take a picture quite by chance. A robot camera or an automatic "McGregor's garden web-cam," say, may take pictures using a motion detector, or at regular or random intervals. We may suppose for the sake of argument that all of these cases resulted in a photographic image of a young hare; they vary, of course, but are all comparable to Dürer's image in the relevant sense. Each picture "contains CSI" to the appropriate degree. Yet each picture was produced by a different combination of law, chance and design (or lack thereof), and only the first was designed in the strong sense that Dembski claims his inference rule reliably detects. (Dembski actually goes even farther, and says that once his design inference picks out something as designed that designation can never be undermined by further evidence.)

Note that it will not do to dismiss these counterexamples on the grounds that the *camera* was intelligently designed; the same is true of Dürer's tools, of course, and we are not now asking the question of how these acquired their own complex specificity, but that of the images, which is completely different and which Dembski singled out in his example. Furthermore, Dembski's CSI criterion supposedly identifies design without needing to know anything about the causal processes that produced it, so the design of the camera should be irrelevant. Nor may one object that the CSI in the photo was already in the real hare and thus is not a counter-example because it was not new information, but "just moved around"; the same relation holds for Dürer's print, and I am taking Dembski at his word that his criterion does license the inference to design in that case. Moreover, if he were to retract that inference and say that in both cases the CSI came from some prior intelligence (an unspecified "hare designer"), this would beg the question.

Dembski describes his complexity-specification criterion as a net for catching design. He admits that some things that are designed will "occasionally slip past the net," giving false negatives, but claims that it makes no errors in the other direction and catches only things that were indeed designed *in the strong sense*. He writes "if things end up in the net that are not designed, the criterion will be worthless" (Dembski, 1999a, p. 141). As we have seen, his criterion does give such false positives.

Returning now to Dembski's challenge to naturalists, we already have part of the answer. Where did the information that makes up Y come from? It came by being copied with slight "errors" from some ancestor X. Copying ... replication ... reproduction ... and always with new, chance variations: this is the sort of thing that rabbits (and other biological organisms) do very well. This point holds even with inkblots. Additional sheets of paper applied to the blot on the first sheet will pick up a copy of the blot's shape: exact mirror copies or ones with random variations depending upon the conditions (e.g., how wet the ink remains, whether one uses blotting paper or not). Rorschach inkblots have both chance and lawful features. (Note that any of these blots might *also* have been blotted "by design"—the categories of law, chance and design are not mutually exclusive categories, though Dembski defines them as such in his technical treatment.) Suppose now that whether and how often inkblots get to be blotted depended upon their shape. If certain shapes conferred a functional advantage to some inkblots over others in this competition, then we would have the elements of the Darwinian process, and natural selection would kick in to refine the shapes to improve their adaptations. With regard to inkblots this is just a thought-experiment, but for organisms function is indeed dependent on shape. This is so for the organism as a whole, but also at lower levels of organization including organs, cells and even bio-molecules. A protein's function is determined in large measure by its physical shape (the tertiary structure that is caused in part by the primary structure of its sequence of amino acids). We shall come back to this in a moment.

In the question period following Dembski's talk in Lubbock, one teacher asked for an example of CSI she could use to illustrate the design inference to students. Dembski concurred when someone suggested the sentence "John loves Mary" written in the sand—the example used in *Of Pandas and People*, the IDC school textbook. (Such examples are significant in that they show that Dembski is not strict in holding to the arbitrary $10^{-150}$ probability bound that he previously set for the design inference.) Naturally, there are some significant differences we have to keep in mind when using a linguistic case as a model for the biological —the former can occur by "Lamarckian" as well as "Darwinian" processes—but for the present purpose it works very well. Since IDCs do use linguistic strings as an example, I shall first respond to Dembski's challenge with that analogy and then move to a real biological case. I'll use the accidental misquotation I mentioned earlier, since Dembski's comments about it illustrate an important aspect of his misunderstanding. Here, again, are the two sentences, which we are now to consider (like IDCs' *Pandas* example) on analogy with genetic "information" as examples of "specified complexity."

Design theorists are no friends of theistic evolution.

Design theorists are no friends of theistic evolutionists.

Of my error Dembski wrote: "Perhaps as an evolutionist himself, Pennock thinks the evolution of 'evolution' into 'evolutionIST' represents a minor adaptive change. I don't. I think it represents shoddy scholarship" (Dembski, 1999b). It is ironic that Dembski himself here accidentally misquotes my accidental misquotation of him. He leaves out the final "s," which would give a third variation of the sentence:

Design theorists are no friends of theistic evolutionist.

I do not attribute Dembski's error to shoddy scholarship or intentional "mischief making" as he does of mine, but take it as a nice, inadvertent illustration of the kind of small, random error that occurs without intelligent design, as in Darwinian evolution.

All three sentences are cases of CSI, so according to Dembski's inference rule each had to have been designed in the strong sense. (If Dembski were now to insist that his formal probability threshold be met, we could just extend to the adjoining sentence about "ill-conceived accommodation" quoted above, or however much more he might require.) This kind of example also undermines his inference rule. Dembski is correct that the improbability of these two (almost) matching sequences leads us to disconfirm the hypothesis that they both arose independently simply by chance—that is just a standard statistical inference in which we reject the null hypothesis—but he is wrong to leap to design. It was not chance alone, but chance acting in concert with a copying mechanism that produced the result. Are those gaps of but one to four letters really *impossible* to bridge naturally? Why *couldn't* one sequence have arisen from the other or both from a third? Note that we are not assuming that either was specified as a target. There are also a variety of other sequences with fewer differences that might have occurred instead (or served as "intermediates"), including, but not limited to:

Design theorists are no friends of theistic evolutions.

Design theorists are no friends of theistic evolutionism.

To fill out the analogy further, we would also include the force of natural selection, which selects for adaptive functions in specific environments, but I won't take the time to do that again now. When IDCs use examples of a string of letters to illustrate the supposed impossibility of evolution by natural processes they ignore natural selection in an environment. They also invariably focus upon a single specific case in isolation (like "John loves Mary" written in the sand) and fail to make the analogy relevant to the situation in the biological case. Once the existence of multiple copies and variations is included, it is easy to see how a series of small changes can be shaped by natural selection to evolve significant changes in shape and function.

Moreover, we need not assume that every change is a single "point mutation," which seems to be how IDCs conceive that the calculations must be done. A specific genetic string often has utility in multiple settings, and may appear in a new setting by recombination. In our analogy, the suffix "ists" functions like that, since it serves in many settings in a similar fashion; attach it to "material" or "creation" or "clarinet" and so on. Moreover, we need not assume that this specific form is the only possible one that could serve that "purpose." There can be multiple ways to achieve the same effect; the suffixes "ers" and "ians," for instance, can serve as well as "ists" to shift the referent from the "object" to the persons who "use" it. It may simply be a historical accident that English speakers attach "ist" to "clarinet" and "er" to "trumpet" rather than the other way round. Nor need we assume that the sentence must be perfectly formed to be understood—it often can function even if it contains some spelling and grammatical errors. All these processes, and more, operate at the biological level.

So much for the supposed "irreducible complexity" that IDCs claim undermines Darwinian evolution. Over the decades, creationists have gradually been forced to accept microevolution, but they now try to draw a sharp line between it and macroevolution and continue to reject that latter, claiming that small changes within a species has nothing to do with speciation itself. IDCs also take this view, and Dembski was aligning himself with this position in saying that he accepts some natural change but only within "strict limits." I have previously discussed their error in trying to draw a sharp distinction in this way (Pennock 1999, pp. 156–157). Now it should be clear how this error is connected to the so-called information problem.

On the introductory page of his web page article, Dembski wrote: "Small syntactic changes can introduce big semantic changes—Robert

Pennock's misquote of me does just that" (Dembski, 1999b). In grammar, syntax has to do with the specific arrangement of letters and words, whereas semantics has to do with their resulting meaning. Syntax and semantics correspond to two of several different senses of "information" that IDCs regularly conflate or confuse. (Both are different from technical notions of information defined in various ways in information theory.) The meaning of a string of letters depends not just upon their order, but also their context; for instance, the same string can be meaningful in one language, but have a different meaning or be totally meaningless in another, similar to the way that biological function and fitness varies with environment. The salient point for us here, however, is that a small error that leads to a big change in meaning can be introduced not by design but by accident, as was actually what happened in this case. It is ironic that in his complaint, Dembski has inadvertently articulated the very feature of the Darwinian mechanism—that small chance variations can produce novel functional information—that he claims "design theory" rules out in principle.

The confusions in Dembski's information argument deserve more detailed treatment, but this will have to suffice for the moment. Let me conclude this discussion with a brief illustration that shows how the points made in our analogy apply in a real biological case. Rather than strings of letters and words in English that in specific orders produce sentences with different meanings, what we are really after has to do with nucleotides in DNA and sequences of amino acids that form proteins with different functions—molecular differences that make for the differences between species.

Let us look at an important protein—the beta subunit of the hemoglobin molecule—as an example. The specific sequence of amino acids for beta-globin has been determined not only for human beings, but for a variety of other organisms. In figure 30.1, we see an alignment of the human and the rabbit sequences. Exact matches between amino acids in the two sequences are connected by a bar: the human and rabbit sequences are 91% identical. (The sequences are 92% similar, in that the amino acids connected by dots though not identical, share important chemical properties such as charge or hydrophobicity.) By way of comparison, the beta-globin sequence of cows is 85% identical to the human sequence, that of chickens is 69% identical, and that of carp is 53%

```
Human   1 MVHLTPEEKSAVTALWGKVNVDEVGGEALGRLLVVYPWTQRFFESFGDLS 50
            ||||. ||||||||||||||:||||||||||||||||||||||||||||
Rabbit  1 MVHLSSEEKSAVTALWGKVNVEEVGGEALGRLLVVYPWTQRFFESFGDLS 50


       51 TPDAVMGNPKVKAHGKKVLGAFSDGLAHLDNLKGTFATLSELHCDKLHVD 100
          . .||| ||||||||||||| |||:||.||||||||||| ||||||||||
       51 SANAVMNNPKVKAHGKKVLAAFSEGLSHLDNLKGTFAKLSELHCDKLHVD 100


      101 PENFRLLGNVLVCVLAHHFGKEFTPPVQAAYQKVVAGVANALAHKYH 147
          |||||||||||| ||.|||||||| |||||||||||||||||||||
      101 PENFRLLGNVLVIVLSHHFGKEFTPQVQAAYQKVVAGVANALAHKYH 147
```

**Figure 30.1**

identical. On the other hand, that of gorillas is over 99% identical to the human sequence—indeed, it differs by just a single amino acid.[8] The simplest explanation of this pattern of information is that human beings and gorillas descended from a recent common ancestor (compared to less recent common ancestors with the other species) and that the original sequence was modified in one line by a single mutation. Processes of natural law together with chance variation produced the CSI of both. Where did the specific, singularly improbable polymer come from that IDCs cite as the evidence of biological design? It descended with modification from another sequence that perhaps differed by just one or a few elements, and that from another that differed perhaps a little more. Viewed in the opposite temporal direction, we see the mutations, insertions, deletions, recombinations and duplications, shaped by natural selection, adding and molding information, until it is transformed from one species into another. Small "syntactic" changes can indeed produce big "semantic" changes.

IDCs have yet to give any evidence to support their claim that such small microevolution changes cannot produce macroevolutionary differences. Even if we were to agree, for argument's sake, that Dembski's formal design inference was without flaws, for him to show that it applied in a biological case, he would have to provide an example of an actual genetic sequence of the requisite length that not only has no variations in other extant organisms, but also none in recent organic history. Neither Dembski nor any other IDC has proposed such a case and it is hard to imagine how they ever could. And yet Dembski continues to insist that it is impossible to produce CSI except by intelligent design, and that human beings were specially created separately from gorillas and other organisms.

Evolutionary biology does not pull a rabbit out of a hat, nor does it try. Rather it shows how rabbits (and human beings as well) are a twig on the branch of the mammals, which itself is part of the deeply rooted tree of life. It does not say that rabbits came "from nothing," but shows some of the processes by which such complex forms evolve from simpler ones.

Dembski seems to acknowledge this when he briefly qualifies his criticism about naturalists wanting to get something from nothing, noting in passing that "[t]he 'nothing' here need not be an absolute nothing." But then he immediately returns to using the term "nothing" without qualification in the remainder of his complaint. As we have seen, this characterization is far from the truth. This is one more example of how IDCs tend to use ordinary terms in "creative" ways for rhetorical effect, to make a difficult empirical problem appear to be a profound philosophical one. Despite Dembski's attempt to reflect back the criticism, it is not evolutionary biologists, but IDCs who want to get something from nothing. They utter their magic word "design," but offer no account of the processes by which that purported creative activity is supposedly accomplished. No surprise, since the supernatural infusions and insertions of design that creationists want to wedge into science are literally a creation of something from nothing—*Creation ex nihilo*. IDCs hope to pull a rabbit not out of a hat, but out of thin air.

## Apparitions

Dembski quotes with approval Richard Dawkins's comment that "biology is the study of complicated things that give the appearance of having been designed for a purpose." I do not dispute that biologists sometimes speak in design terms. Consider virologist Linda Stannard's comment that "After many, many years of peering at virus particles through the electron microscope, I have still not ceased to be amazed and excited by the precision and intricacy of design in something so very, very small" (Kaiser, 2000, p. 923). However, although the world (and not just the biological world) exhibits all manner of useful arrangements of parts, we cannot simply read off intentional design from such natural objects. On the contrary, the striking thing about genuine artifacts is that the kind of order they exhibit stands out in such stark contrast from the kind of order we observe in the biological world and the rest of nature. There is, of course, a perfectly reasonable sense of evolutionary design and function that biologists legitimately refer to, but the appearance of *agency* that others see in nature is probably the result of a tendency to anthropomorphize the world. As with inkblots, clouds and ceiling cracks, it is all too easy for us to read in a human aspect and to see our own image apparently reflected in natural design.

If the design argument really did allow us to infer agency in the creation of organisms, as IDCs claim, then what else would follow? Taking Stannard's observation together with Dembski's inference rule, it would seem that we must conclude that the intelligent designer lavished far more attention upon the design of the multitude of micro-organisms than upon less fortunate creatures such as ourselves, who suffer as their hosts. Perhaps the theodicy problem is not greater in aggregate for Dembski's view than for other theist views, but it is certainly far more pointed, for we would have to attribute purposeful insertions of design to account for that precise and intricate specified complexity of every different kind of virus and bacterium.

Of course, IDCs will often say that all they are claiming is that biological organisms had to have been created by some "designing intelligence," but that they draw no conclusions about who that agent or agents might be. Perhaps extraterrestrials, they occasionally suggest. In his closing remarks to the science teachers, Dembski called our attention to a significant fact: "Notice that I never used the G-word in this whole talk."

So, what is this name that they dare not speak? Who is this ghostly apparition who supposedly created all the complex specified information of the biological world? In other settings IDCs do not play dumb and are more forthright about the identity of the designer. As one put it with a wink and a grin when introducing another IDC to a campus Christian group, intelligent design is the "politically correct way to talk about God." Well, who ever thought it was otherwise? Even Dembski, when writing for theologians, will sometimes fill in the blank and say that he believes that it is "God [who] created the world with a purpose in mind" (Dembski, 1995, p. 5).

In such discussions, we find further evidence that the term "creation-ist" is fitting to describe his position. When Dembski says that life could not have arisen by purposeless material processes and required inser-tions of design, he is not giving us the conclusions of IDCs' scientific re-search, if such could be found. When he tells us that biological change has occurred only within strict limits and specifies that human beings were specially created, he is doing just what the classic creationists have always done—he is starting with his conclusions already in place. Per-haps as a teleologist himself, Dembski thinks that starting with one's end-point in hand and designing one's theory so it fits one's preferred conclusions is a reasonable method. I don't. I think it represents ideo-logical bias.

Dembski's theistic science begins and ends with a Biblical assumption, read under a particular hermeneutic. He writes: "If we take seriously the word-flesh Christology of Chalcedon (i.e. the doctrine that Christ is fully human and fully divine) and view Christ as the *telos* toward which God is drawing the whole of creation, then any view of the sciences that leaves Christ out of the picture must be seen as fundamentally deficient" (Dembski, 1999a, p. 206). IDCs' theistic science is inherently sectarian, but Dembski tries to assure us that privileging this specific Christian view is epistemically beneficial rather than harmful.

Why won't "privileging Christology," as he puts it, undermine the in-tegrity of science? Because, says Dembski, Christ is not an "addendum" to science, but a "completion" of it (Dembski, 1999a, p. 207). Accord-ing to Dembski, "[T]he validity of the scientist's insights can never be divorced from Christ, who through the incarnation enters, takes on and transforms the world and thus cannot help but pervade the scientist's domain of inquiry" (Dembski 1999a, p. 209). Using a mathematical analogy, he assures us that "Christ, as the completion of our scientific theories, maintains the conceptual soundness of those theories even as the real numbers maintain the conceptual soundness of the applied mathe-matician's calculations. Christ has assumed the fullness of our humanity and entered every aspect of our reality. He thereby renders all our studies the study of himself" (Dembski 1999a, p. 210).

According to Dembski, the key to all this and the reason that the dis-ciplines will be "taken seriously" and supposedly not lose their integrity is that since on the Chalcedonian doctrine Christ is both fully divine and fully human, "Christ can never be less than human" (Dembski 1999a, p. 206). But being not less than human is not the problem. The problem is being more than human: being also superhuman. Dembski still has not solved the problem we began with, but just concealed it in the kind of obscurantist reasoning that gave the Sophists a bad name.

Metaphysically, the doctrine that Christ is at once both fully divine and fully human may be taken as a theological paradox or religious mystery. However, if one were to view it as an epistemic principle it is simply a contradiction (unless, of course, "human" and "divine" are synonymous), and to advocate it as such would be to promote an absurdity that undermines reason itself, for in deductive logic anything whatsoever follows from a contradiction. Similarly, scientific inductive reasoning collapses once one scraps the methodological constraint of lawful causal regularities and tries to introduce supernatural interventions as an epistemic principle, as IDCs do indeed desire in their "theistic science."

This is not meant as an attack upon Dembski's faith or that of others who hold those same specific theological views; rather it is a defense against his attempt to wedge them into science and to his simultaneous denial of the same. Like Johnson, Dembski is attacking metaphysical naturalism and he regularly papers over the distinction between it and the epistemological position that I and others are defending as the basis of scientific reasoning. To be sure, some scientists also fail to recognize the distinction, as Lewontin does in the passage Dembski quoted. But while we may excuse a scientist for not being aware of even such significant distinctions in a subject outside his field, we should expect more from someone like Dembski who is trained as a philosopher and is promoting a philosophical position.

Dembski began his article by quoting from *Tower of Babel* a part of my challenge to Johnson: If you expect science to abandon its naturalistic methods and reintroduce appeals to supernatural interventions, show us first in your own field of the law how to do it. I did not ask Johnson to delete design, in its ordinary natural sense, from the law. No one doubts that criminals may leave a trail of evidence and that careful detective work can sometimes uncover their nefarious designs. But the empirical reasoning involved in such sleuthing relies upon a thoroughly naturalistic method, and without it the notion of empirical evidence itself would collapse. Dembski at first seemed that he was going to address this worry, but with a little sleight-of-hand he avoided confronting the problem entirely.

Despite its denials, IDC does incorporate supernatural intervention and it does so as a central belief, right alongside a belief in creation from nothing. The Wizards of ID practice both of these forms of magic but, as we have also seen, they try to obscure this fact in an elaborate smoke and mirrors show. IDCs are performing a magic act of the first kind, hoping to fool the diverse audiences to which they are playing. Dembski's own definition of this form of magic applies very well; it is "the art of illusion, where appearance is carefully crafted to distort reality" (Dembski, 2000). Thoughtful seekers after truth would do well not to be deceived by their act and, when the unnamed apparition is conjured on stage, to pay close attention to the men behind the curtain.

## Notes

1. For simplicity, I'll also use this acronym as an abbreviation for "intelligent-design creationist." The context will make it clear which is intended.

2. It is interesting to note how often Dembski draws such conclusions about his critics' supposed designs and intentions, though without ever explaining how he uses his "design inference" to infer them.

3. Dembski gave his talk at the annual CAST meetings in Lubbock, Texas, October 29, 1999.

4. In a letter in *Books and Culture* (1999, vol. 5, no. 6, p. 5), IDC John G. West challenged my contention that IDC is top-heavy with philosophers and he listed half a dozen scientists who he said were leading the research. I wrote to each of these asking for reprints or references to their scientific publications on intelligent design. No one will be surprised to hear that not one could provide me with any.

5. All the quotations and information about the course come from the syllabus, which Dembski kindly sent me of his own accord. The syllabus also stated that thirty percent of the student's grade would come from class participation and the remaining seventy percent from a "5,000 to 6,000 word critical review of Rob Pennock's *Tower of Babel*" It continued, "We will not read Pennock's book for class. Pennock's book is supposed to be the 'state-of-the-art' refutation of intelligent design. Your task is to use what you've learned in this course to assess and answer Pennock's critique of intelligent design. If (per impossibile) you happen to agree with Pennock, then you will need to write a joint critical review of my book [i.e. *Intelligent Design*] and Pennock's, showing why intelligent design is a failed intellectual project." I thank Dembski for this friendly warning shot across the bow.

6. On Johnson's attempts to shift the burden of proof see (Pennock 1999, chapter 4); on Plantinga's see Parsons (1990).

7. I have discussed a bit of this evidence in Pennock (1999), chapter 2.

8. Learning to analyze such comparisons to help infer evolutionary relationships is becoming a standard laboratory course exercise now that nucleotide and protein sequence data is easily available on line. The comparisons given come from a typical exercise ⟨www.biochem.mcw.edu/science_ed/Pages/hemoglobin/hem_pages/intro.html⟩. To explore further, one can make any number of sequence similarity comparisons for a wide variety of species using NCBI's BLAST sequence search tool ⟨http://www.ncbi.nlm.nih.gov/BLAST⟩.

# References

Colbert, Edwin H. and Michael Morales (1991). *Evolution of the Vertebrates.* New York: Wiley & Sons.

Dembski, William A. 1995. What Every Theologian Should Know about Creation, Evolution, and Design. *Center for Interdisciplinary Studies Transactions* 3 (2): 1–8.

Dembski, William A. 1999a. *Intelligent Design: The Bridge Between Science and Theology.* Downers Grove, InterVarsity Press.

Dembski, William A. 1999b. Pennock's Convenient Distortion. Published on Dembski's personal home page. ⟨www.baylor.edu/~William_Dembski/critics.htm⟩. Expanded from original letter to the editor, *Books and Culture*, vol. 5, no. 6, Nov/Dec 1999, pp. 4–5.

Dembski, William A. 2000. Who's Got the Magic? *Metanexus: The Online Forum on Religion and Science* (2000, no. 042). Reprinted in this volume, chap. 29.

Johnson, Phillip E. 1991. *Darwin on Trial.* Washington, D.C., Regnery Gateway.

Kaiser, Jocelyn. 2000. Virus Portraits. *Science* 288 (5468): 923.

Kawai, Takao, et al. 1998. Molecular and biological characterization of rabbit mannan-binding protein (MBP). *Glycobiology* 8 (3): 237–244.

New York Times. 1999. Kansas School Officials Monkey With Evolution: Concept deleted from state science curriculum. *New York Times.* August 12.

Parsons, Keith. 1990. *God and the Burden of Proof: Plantinga, Swinburne, and the Analytic Defense of Theism.* Buffalo, New York, Prometheus Books.

Pennock, Robert T. 1999. *Tower of Babel: The Evidence against the New Creationism.* Cambridge, Mass.: MIT Press.

# 31

## The Panda's Thumb

Stephen Jay Gould

Few heroes lower their sights in the prime of their lives; triumph leads inexorably on, often to destruction. Alexander wept because he had no new worlds to conquer; Napoleon, overextended, sealed his doom in the depth of a Russian winter. But Charles Darwin did not follow the *Origin of Species* (1859) with a general defense of natural selection or with its evident extension to human evolution (he waited until 1871 to publish *The Descent of Man*). Instead, he wrote his most obscure work, a book entitled: *On the Various Contrivances by Which British and Foreign Orchids Are Fertilized by Insects* (1862).

Darwin's many excursions into the minutiae of natural history—he wrote a taxonomy of barnacles, a book on climbing plants, and a treatise on the formation of vegetable mold by earthworms—won him an undeserved reputation as an old-fashioned, somewhat doddering describer of curious plants and animals, a man who had one lucky insight at the right time. A rash of Darwinian scholarship has laid this myth firmly to rest during the past twenty years (see essay 2). Before then, one prominent scholar spoke for many ill-informed colleagues when he judged Darwin as a "poor joiner of ideas ... a man who does not belong with the great thinkers."

In fact, each of Darwin's books played its part in the grand and coherent scheme of his life's work—demonstrating the fact of evolution and defending natural selection as its primary mechanism. Darwin did not study orchids solely for their own sake. Michael Ghiselin, a Cali-

Revised (1980) version of the article originally published in *Natural History* (1978, November). Reprinted with permission from the author and *Natural History*. © 1978 the American Museum of Natural History.

fornia biologist who finally took the trouble to read all of Darwin's books (see his *Triumph of the Darwinian Method*), has correctly identified the treatise on orchids as an important episode in Darwin's campaign for evolution.

Darwin begins his orchid book with an important evolutionary premise: continued self-fertilization is a poor strategy for long-term survival, since offspring carry only the genes of their single parent, and populations do not maintain enough variation for evolutionary flexibility in the face of environmental change. Thus, plants bearing flowers with both male and female parts usually evolve mechanisms to ensure cross-pollination. Orchids have formed an alliance with insects. They have evolved an astonishing variety of "contrivances" to attract insects, guarantee that sticky pollen adheres to their visitor, and ensure that the attached pollen comes in contact with female parts of the next orchid visited by the insect.

Darwin's book is a compendium of these contrivances, the botanical equivalent of a bestiary. And, like the medieval bestiaries, it is designed to instruct. The message is paradoxical but profound. Orchids manufacture their intricate devices from the common components of ordinary flowers, parts usually fitted for very different functions. If God had designed a beautiful machine to reflect his wisdom and power, surely he would not have used a collection of parts generally fashioned for other purposes. Orchids were not made by an ideal engineer; they are jury-rigged from a limited set of available components. Thus, they must have evolved from ordinary flowers.

Thus, the paradox, and the common theme of this trilogy of essays: Our textbooks like to illustrate evolution with examples of optimal design—nearly perfect mimicry of a dead leaf by a butterfly or of a poisonous species by a palatable relative. But ideal design is a lousy argument for evolution, for it mimics the postulated action of an omnipotent creator. Odd arrangements and funny solutions are the proof of evolution—paths that a sensible God would never tread but that a natural process, constrained by history, follows perforce. No one understood this better than Darwin. Ernst Mayr has shown how Darwin, in defending evolution, consistently turned to organic parts and geographic distributions that make the least sense. Which brings me to the giant panda and its "thumb."

Giant pandas are peculiar bears, members of the order Carnivora. Conventional bears are the most omnivorous representatives of their order, but pandas have restricted this catholicity of taste in the other direction—they belie the name of their order by subsisting almost entirely on bamboo. They live in dense forests of bamboo at high elevations in the mountains of western China. There they sit, largely unthreatened by predators, munching bamboo ten to twelve hours each day.

As a childhood fan of Andy Panda, and former owner of a stuffed toy won by some fluke when all the milk bottles actually tumbled at the county fair, I was delighted when the first fruits of our thaw with China went beyond ping pong to the shipment of two pandas to the Washington zoo. I went and watched in appropriate awe. They yawned, stretched, and ambled a bit, but they spent nearly all their time feeding on their beloved bamboo. They sat upright and manipulated the stalks with their forepaws, shedding the leaves and consuming only the shoots.

I was amazed by their dexterity and wondered how the scion of a stock adapted for running could use its hands so adroitly. They held the stalks of bamboo in their paws and stripped of the leaves by passing the stalks between an apparently flexible thumb and the remaining fingers. This puzzled me. I had learned that a dexterous, opposable thumb stood among the hallmarks of human success. We had maintained, even exaggerated, this important flexibility of our primate forebears, while most mammals had sacrificed it in specializing their digits. Carnivores run, stab, and scratch. My cat may manipulate me psychologically, but he'll never type or play the piano.

So I counted the panda's other digits and received an even greater surprise: there were five, not four. Was the "thumb" a separately evolved sixth finger? Fortunately, the giant panda has its bible, a monograph by D. Dwight Davis, late curator of vertebrate anatomy at Chicago's Field Museum of Natural History. It is probably the greatest work of modern evolutionary comparative anatomy, and it contains more than anyone would ever want to know about pandas. Davis had the answer, of course.

The panda's "thumb" is not, anatomically, a finger at all. It is constructed from a bone called the radial sesamoid, normally a small component of the wrist. In pandas, the radial sesamoid is greatly enlarged

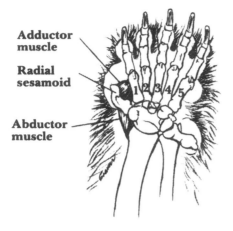

**Adductor muscle**

**Radial sesamoid**

**Abductor muscle**

**Figure 31.1**

and elongated until it almost equals the metapodial bones of the true digits in length. The radial sesamoid underlies a pad on the panda's forepaw; the five digits form the framework of another pad, the palmar. A shallow furrow separates the two pads and serves as a channelway for bamboo stalks.

The panda's thumb comes equipped not only with a bone to give it strength but also with muscles to sustain its agility. These muscles, like the radial sesamoid bone itself, did not arise *de novo*. Like the parts of Darwin's orchids, they are familiar bits of anatomy remodeled for a new function. The abductor of the radial sesamoid (the muscle that pulls it away from the true digits) bears the formidable name *abductor pollicis longus* ("the long abductor of the thumb"—*pollicis* is the genitive of *pollex*, Latin for "thumb"). Its name is a giveaway. In other carnivores, this muscle attaches to the first digit, or true thumb. Two shorter muscles run between the radial sesamoid and the pollex. They pull the sesamoid "thumb" towards the true digits.

Does the anatomy of other carnivores give us any clue to the origin of this odd arrangement in pandas? Davis points out that ordinary bears and raccoons, the closest relatives of giant pandas, far surpass all other carnivores in using their forelegs for manipulating objects in feeding. Pardon the backward metaphor, but pandas, thanks to their ancestry, began with a leg up for evolving greater dexterity in feeding. Moreover, ordinary bears already have a slightly enlarged radial sesamoid.

In most carnivores, the same muscles that move the radial sesamoid in pandas attach exclusively to the base of the pollex, or true thumb. But in ordinary bears, the long abductor muscle ends in two tendons: one inserts into the base of the thumb as in most carnivores, but the other attaches to the radial sesamoid. The two shorter muscles also attach, in part, to the radial sesamoid in bears. "Thus," Davis concludes, "the musculature for operating this remarkable new mechanism—functionally a new digit—required no intrinsic change from conditions already present in the panda's closest relatives, the bears. Furthermore, it appears that the whole sequence of events in the musculature follows automatically from simple hypertrophy of the sesamoid bone."

The sesamoid thumb of pandas is a complex structure formed by marked enlargement of a bone and an extensive rearrangement of musculature. Yet Davis argues that the entire apparatus arose as a mechanical response to growth of the radial sesamoid itself. Muscles shifted because the enlarged bone blocked them short of their original sites. Moreover, Davis postulates that the enlarged radial sesamoid may have been fashioned by a simple genetic change, perhaps a single mutation affecting the timing and rate of growth.

In a panda's foot, the counterpart of the radial sesamoid, called the tibial sesamoid, is also enlarged, although not so much as the radial sesamoid. Yet the tibial sesamoid supports no new digit, and its increased size confers no advantage, so far as we know. Davis argues that the coordinated increase of both bones, in response to natural selection upon one alone, probably reflects a simple kind of genetic change. Repeated parts of the body are not fashioned by the action of individual genes—there is no gene "for" your thumb, another for your big toe, or a third for your pinky. Repeated parts are coordinated in development; selection for a change in one element causes a corresponding modification in others. It may be genetically more complex to enlarge a thumb and *not* to modify a big toe, than to increase both together. (In the first case, a general coordination must be broken, the thumb favored separately, and correlated increase of related structures suppressed. In the second, a single gene may increase the rate of growth in a field regulating the development of corresponding digits.)

The panda's thumb provides an elegant zoological counterpart to Darwin's orchids. An engineer's best solution is debarred by history. The

**Marsh *Epipactis*, lower sepals removed**

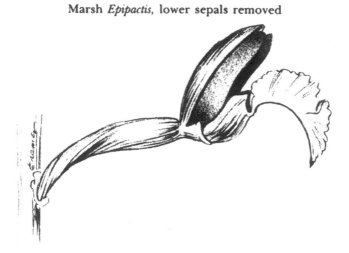

**Figure 31.2**
Runway of labellum depressed after insect lands.

**Figure 31.3**
Runway of labellum raised after insect crawls into cup below.

panda's true thumb is committed to another role, too specialized for a different function to become an opposable, manipulating digit. So the panda must use parts on hand and settle for an enlarged wrist bone and a somewhat clumsy, but quite workable, solution. The sesamoid thumb wins no prize in an engineer's derby. It is, to use Michael Ghiselin's phrase, a contraption, not a lovely contrivance. But it does its job and excites our imagination all the more because it builds on such improbable foundations.

Darwin's orchid book is filled with similar illustrations. The marsh Epipactus, for example, uses its labellum—an enlarged petal—as a trap. The labellum is divided into two parts. One, near the flower's base, forms a large cup filled with nectar—the object of an insect's visit. The other, near the flower's edge, forms a sort of landing stage. An insect alighting on this runway depresses it and thus gains entrance to the nectar cup beyond. It enters the cup, but the runway is so elastic that it instantly springs up, trapping the insect within the nectar cup. The insect must then back out through the only available exit—a path that forces it to brush against the pollen masses. A remarkable machine but all developed from a conventional petal, a part readily available in an orchid's ancestor.

Darwin then shows how the same labellum in other orchids evolves into a series of ingenious devices to ensure cross-fertilization. It may develop a complex fold that forces an insect to detour its proboscis around and past the pollen masses in order to reach nectar. It may contain deep channels or guiding ridges that lead insects both to nectar and pollen. The channels sometimes form a tunnel, producing a tubular flower. All these adaptations have been built from a part that began as a conventional petal in some ancestral form. Yet nature can do so much with so little that it displays, in Darwin's words, "a prodigality of resources for gaining the very same end, namely, the fertilization of one flower by pollen from another plant."

Darwin's metaphor for organic form reflects his sense of wonder that evolution can fashion such a world of diversity and adequate design with such limited raw material:

Although an organ may not have been originally formed for some special purpose, if it now serves for this end we are justified in saying that it is specially contrived for it. On the same principle, if a man were to make a machine for some special purpose, but were to use old wheels, springs, and pulleys, only slightly altered, the whole machine, with all its parts, might be said to be specially contrived for that purpose. Thus throughout nature almost every part of each living being has probably served, in a slightly modified condition, for diverse purposes, and has acted in the living machinery of many ancient and distinct specific forms.

We may not be flattered by the metaphor of refurbished wheels and pulleys, but consider how well we work. Nature is, in biologist François Jacob's words, an excellent tinkerer, not a divine artificer. And who shall sit in judgment between these exemplary skills?

# 32

# The Role of Theology in Current Evolutionary Reasoning

Paul A. Nelson

## Introduction

The theory of evolution was born in a turbulent embrace with theology, and it has yet to relinquish that embrace. By "the theory of evolution," I mean Darwin's theory of the common descent of all organisms via the natural selection of randomly arising variation. By "theology," I mean propositions about what God would (or should) have done in creating the world.

The embrace in question is dialectical. Suppose one wants, as Darwin did, to refute the view that organisms were specially created by God. Or suppose one wants (again, as Darwin did) to reform the practice and content of biology generally, by showing the creationists and natural theologians still in the room politely but firmly to the door. For those ends it is necessary, at least for a time, to take the creationists and natural theologians seriously. That is, it is necessary to assign empirical content to propositions beginning "If God had created organisms ..." Those propositions can then be compared with the evidence; the theological theory can be found wanting; and the creationists and theologians, banished from biology.

But in that dialectic all sorts of things can go awry. The reforming program may find that it is easiest to rid biology of theology by declaring the latter to be a lot of empty nonsense. Meanwhile, the empirical program is busy comparing the predictions of the theory of creation with the evidence, and declaring the predictions falsified. Eventually some-

Originally published in *Biology & Philosophy* (1996, vol. 11, no. 4, pp. 493 517).

one will notice that these programs are markedly incongruent with each other. As Sober (1993, p. 46) observes, many biologists

have taken pains to point out how the hypothesis of evolution by natural selection makes predictions that differ dramatically from those that flow from the design hypothesis.... At the same time and often in the same book, some biologists and philosophers have pursued a quite different line of attack. They have argued that creationism is not a scientific hypothesis because it is untestable.... If creationism cannot be tested, then what was one doing when one emphasized the imperfection of nature? Surely it is not possible to test and find wanting a hypothesis that is, in fact, untestable.

Now evolutionary biologists might acknowledge this incongruity and yet wave it away. Creationism is either false or untestable? *Fascinating*. Let someone else sort it out.

We might ask those same biologists, however, to explain why they think evolution is *true*. This is a task they face regularly, if only for pedagogical reasons. Consider a well-known example. The giant panda (*Ailuropoda melanoleuca*) possesses a pseudothumb built from the radial sesamoid, a wrist bone. The panda uses this structure, somewhat clumsily in the eyes of certain observers, to manipulate its main food, bamboo. Odd structures like the panda's pseudothumb, argues Stephen Jay Gould, "are the primary proofs that evolution has occurred" (1991, p. 61), for

If God had designed a beautiful machine to reflect his wisdom and power, surely he would not have used a collection of parts generally fashioned for other purposes.... Odd arrangements and funny solutions are the proof of evolution— paths that a sensible God would never tread but that a natural process, constrained by history, follows perforce. (1980, pp. 20–21)

This passage, from Gould's essay on the panda's thumb, is an instance of what is frequently called the *imperfection argument* for evolution. God is an optimizing creator. This structure, and hence, organism, is imperfect. Therefore this organism evolved.

Consider another example. It is widely held that all organisms descended from a common ancestor because they share certain biochemical universals, such as the genetic code (Dawkins 1986, p. 270; Ridley 1986, pp. 119–20; Mayr 1991, p. 23). These molecular universals are generally regarded as a strong evolutionary prediction. As Douglas Futuyma puts it,

The only possible reason for these chemical universalities is that living things got stuck with the first system that worked for them. Once the genetic code was established, no species was ever free to try a new one. A mutation that caused the nucleotide sequence UUU to code for glycine instead of phenylalanine would have messed up all the species' proteins. (1983, p. 205)[1]

On the other hand (argues Mark Ridley, developing the same point), if "different species had all been created separately, we should be very surprised if they had all been built with exactly the same genetic code" (1985, p. 10). We should be very surprised—to supply the missing, but implied, premise—because a *freely acting creator* could have constructed many different codes:

Where a Creator would have been free to use different biochemical building blocks, evolution was not free: the history of the earliest organisms determined everything that happened thereafter. (Futuyma 1983, p. 205)

This is a molecular variant of the *homology argument* for evolution. The apparent uniformity of certain biological patterns, such as the genetic code, is inconsistent with the freedom of a creator to act as he wishes. Therefore those patterns evolved.

Arguments of both sorts are common in the recent evolutionary literature. They occur most often where a case for evolution is being made: in the introductory chapters of books on evolution, for instance; in popular or semipopular essays and books; or, in polemical writings, against creationists and other doubters of evolution. The arguments are given as good reasons for thinking that evolution, and not some other theory, best explains the origin and diversity of life.

Now this is a problem of much greater interest. It is widely held that evolutionary theory partakes necessarily of *methodological naturalism*, according to which one cannot in *scientific* reasoning refer to "God," "the Creator," "creation" (understood as the act of a divine intelligence), or other theological concepts (Eldredge and Cracraft 1980, p. 3; Hoffman 1989, pp. 11–12; see also Holton 1993). But the arguments for evolution that we shall consider are formulated in theological terms, usually explicitly so—a practice plainly inconsistent with methodological naturalism. We aren't supposed to be able to *say* anything, scientifically speaking, about God. Whatever we claim to know about God may be true or false, considered theologically or philosophically, but that knowledge isn't the stuff of scientific explanation. How, then, do so many evolutionary biologists speak with confidence about what God would or would not have done?

One can dismiss this problem impatiently. One might argue for example that methodological naturalism is philosophically sound, and necessary for the practice of science. Therefore a *theological* argument for evolution is strictly speaking a *non sequitur*, or an indiscriminate rhetorical lurch into theology.[2] Zoologist Steven Scadding, for instance, after finding the theological premise of the "vestigial organ" argument, concludes

that presented in this way, the vestigial organ argument is essentially a theological rather than a scientific argument, since it is based on the supposed nature of the Creator. (1981, p. 174)

The argument is "based on an assumption about the nature of God," Scadding observes, "and thus should have no place in a scientific presentation of evolution" (Scadding 1982, p. 173). One can be heedlessly scrupulous about method here, however, and force a great deal of evolutionary reasoning out of science. As Mayr (1964, p. xii) points out, Darwin himself

was converted to his new ideas only after he had made numerous observations that were to him quite incompatible with creation. He felt strongly that he must establish this point decisively before his readers would be willing to listen to the evolutionary interpretation. Again and again, he describes phenomena that do not fit the creation theory.

That the phenomena do not fit the creation theory implies of course that they *might* have fit. As it happens, they do not, and thus the theory of creation in question is false. Methodological naturalism however holds that since "God's will" (for instance) is inscrutable to science, the truth or falsity of any theory of creation can never be known. Nonetheless, Darwin piled up phenomena that he thought were plainly inconsistent with "God's will" as usually conceived, and his arguments persuaded most of his peers.

Or one might try to justify theological arguments for evolution pragmatically, as devices for shutting up the creationists. The arguments are indeed theological, this justification holds, but only because of their peculiar context. The arguments take the logical form of *reductio ad absurdum*, where one assumes the truth of an opponent's premises provisionally to derive a contradiction from them. Terms like "God" and "the Creator" appear in the arguments because they were introduced first by creationists, in *their* arguments, into a cultural debate about the truth of evolution. "God" is the principal cause invoked in non-evolutionary theories, and such theories do have genuine observational consequences. "If theology presumes to speak of the natural, material world," argues evolutionary biologist Bruce Naylor (1982, p. 94), "its statements become open to scientific examination and potential falsification." The panda's thumb, in other words, can be stuck in the eye of the creationists. As a polemical tool, therefore, theology is useful. But evolutionary theory *as a natural science* claims nothing for itself theologically. When the debate is over, the theology, borrowed for the evening's *reductio*, goes into the trash bin with the folded programs and coffee cups.

It's a plausible rejoinder. This pragmatic justification collapses completely, however, when one examines the actual context of many of the arguments. An encyclopedia article on the evidence for evolution (see below) might reasonably be expected to be a straightforward summary of the data; likewise, a textbook treatment of the same material. Why use a theological *reductio* in those contexts? The rhetorical setting is that of a lecture, not a debate. The creationists (one might say) have left the building.

But what we see as *the evidence* for evolution exists against an epistemological backdrop where theology of one form or another has always been present. The panda's thumb is a sign of history—i.e., of descent—only when one is certain that "a sensible God" (Gould 1980, p. 20) would not stoop directly to contrive such oddities. Among their possible histories, we can conceive that organisms might have been divinely created at some point in the past (Indeed, this is what creationists maintain). The road to naturalistic common descent passes through the refutation of that possible history. One needs a God with qualities, therefore: a causal entity from which predictions can be derived. Then one can get at the business of refutation.

That borrowed God may remain, however, after the theory in which he served a causal role is gone. Just such a God haunts evolutionary theory today. Biologists have accepted (more or less uncritically, I think) that in justifying evolution, saying what a creator would or would not have done is unproblematical.[5] For this practice, they have the example of the *Origin* itself, and indeed, Darwin's writings generally, where arguments of the sort at issue play an important role in the case for evolution (Gillespie 1979; Kitcher 1985). Yet biologists and philosophers should consider Darwin's theological metaphysics with the same careful gaze they have turned on (for instance) his speculations about heredity. When the case for evolution is made today, the theological and aesthetic criteria at work usually stem directly from Darwin—that is, from his theological metaphysics. In the last section, I speculate briefly about the influence of Darwinian metaphysics on current theory.

In what follows immediately, however, I look critically at a number of the received arguments for evolution, namely, those resting on unjustified theological assumptions. While the arguments are familiar, their fragility is still largely unappreciated.

## The Imperfection Argument

In the *History of Creation*, Haeckel argued that "even if we knew absolutely nothing of the other phenomena of development, we should be obliged to believe in the truth of the Theory of Descent, solely on the ground of the existence of rudimentary organs" (1876, p. 291). Under the heading of "Dysteleology," Haeckel gathered a number of apparently useless or imperfect structures that, he argued, could be reconciled with the theory of creation only by "ludicrous" ad hoc conjectures. In stressing the evidential force of imperfection, Haeckel followed Darwin's lead. Darwin's language is never more bitter than when condemning the failed teleology of theories of creation, which impute imperfect organic design to the direct intent of a rational and benevolent creator (The argument itself is ancient, of course, with roots extending at least to Lucretius).

Imperfection arguments occur widely in the recent literature, in a variety of contexts (e.g., Jacob 1982; Sober 1984, pp. 175–76; Futuyma 1985, p. 6; Dawkins 1986, pp. 91–94; Burian 1986; Williams 1992, pp. 7, 72–76). Doubtless the most influential formulations, however, occur in Stephen Jay Gould's writings. Since many authors draw on Gould's formulations, I consider them here in detail.[4] I have emphasized key words and phrases.

The theory of natural selection would never have replaced the doctrine of divine creation if *evident, admirable design* pervaded all organisms. Charles Darwin understood this, and he focused on features that would be out of place in a world constructed by *perfect wisdom*.... This principle remains true today. The best illustrations of adaptation by evolution are the ones that strike our intuition as *peculiar* or *bizarre*. (1977, p. 91)

*Odd arrangements* and *funny solutions* are the proof of evolution—paths that a sensible God would never tread but that a natural process, constrained by history, follows perforce. (1980, pp. 20–21)

Evolution lies exposed in the *imperfections* that record a history of descent. Why should a rat run, a bat fly, a porpoise swim, and I type this essay with structures built from the same bones unless we all inherited them from a common ancestor? *An engineer, starting from scratch, could design better limbs in each case.* (1983, p. 258)

But how can a scientist infer history from single objects?.... Darwin answers that we must look for *imperfections and oddities*, because any *perfection* in organic design or ecology obliterates the paths of history and *might have been created as we find it*. This principle of imperfection became Darwin's most common guide.... I like to call it the "panda principle".... (1986, p. 63)

It will be useful to formalize Gould's argument:

1.   If *p* is an instance of organic design, then *p* was produced either by a wise creator, or by descent with modification (evolution).

2.   If organic design *p* was produced by a wise creator, then *p* should be perfect (or exhibit no imperfections).

3.   Organic design *p* is not perfect (or exhibits imperfections).

The conclusion follows that

∴   Organic design *p* was not produced by a wise creator, but by descent with modification.

Premises 1 and 2 are theological. Gould's terms for the creator, in the passages cited above and in other instances of this argument, include "a perfect engineer" (1977, p. 91), "a sensible God" (1980, pp. 20–21), "a rational agent" (1983, p. 164) and "a wise creator" (1983, p. 258). Premises 2 and 3 refer also to "perfection," and we may infer that Gould holds that humans can readily discern the presence or absence of perfection when they examine organic designs.

### Some Problems with the Imperfection Argument

The imperfection argument is popular and compelling. Each premise is attended, however, with difficulties.

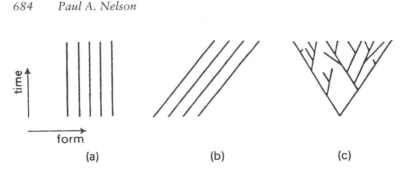

**Figure 32.1**
Three patterns of the history of life: (a) separate creation, (b) transformism, (c) evolution. In (a), species appear at the present very much as they were originally created, and "there were as many origins of species as there have been species" (Ridley 1985, p. 3; figure after Ridley 1985, p. 2).

*If* p *is an instance of organic design, then* p *was produced either by a wise creator, or by descent with modification.*

Assume that the terms "wise creator," "perfection," and "imperfection" are unambiguous, that is, understood in the same way by all observers (assumptions at issue below). Even granting this, the first premise describes a false dichotomy. Consider Figure 1, from Ridley's (1985) *The Problems of Evolution.* Here, (a) depicts what Ridley calls "separate creation," (b) is Lamarckian transformism, and (c) is evolution, or common ancestry. Ridley formulates "separate creation" as stating "that species do not change and that there were as many origins of species as there have been species" (1985, p. 3). Now some creationists may defend this view, although Ridley cites no authority for this interpretation of "separate creation." Pattern (a) represents what I will term a *static* theory of creation, in which designs display (more or less exactly) the form in which they were created. One would be hard pressed to find any expression of that view in the creationist literature, whether recently or within the past several decades. Rather, one will find extended discussions of what I will term *dynamic* theories of creation—as represented, for instance, by figure 32.2. Here, the terminal species are members of basic types, stemming from common ancestors which were themselves created. Considerable—albeit ultimately bounded—change may have occurred between the creation of a design *p* and our observation of *p*. For instance, *p* may have speciated, or undergone genetic changes which gave rise to phenotypic modifications. In short, creationists defend

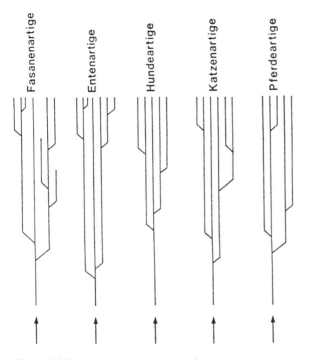

**Figure 32.2**
A "creation model" (after Junker and Scherer 1988, p. 16). The arrows in the figure depict the creation of various "ground types" (*Grundtypen*), which then vary within boundaries by microevolutionary processes. Note that the species observed at the present, within any ground type, may differ considerably from the originally created forms.

the dynamic pattern of figure 32.2, rather than Ridley's static pattern (Murris 1986; Landgren 1993; Scherer 1993).

The imperfection argument however presupposes a static theory, in which organisms appear today largely as they were originally created. Yet, as I have noted, few if any creationists defend that view. In fact, they argue that some designs are biologically "imperfect," but that such imperfection is consistent with their theory. Not all imperfections, therefore, count against creation, or a discontinuous geometry of organic form.

Consider blind cave animals. Futuyma asks, of the functionless lens and retina of the cave salamander, "Do we find evidence here of wise design?" (1983, p. 198). Yet in the same year that Futuyma posed his question, two well-known creationists, independently considering the same phenomenon, saw it as easily understood degenerative change:

Blind cave fish with remnants of eyes ... appear to have true vestigial organs. These and similar degenerations apparently have indeed resulted from typically disadvantageous mutations.... When hereditary changes are small enough to permit survival and reproduction, vestiges may remain. However, these vestigial structures at best are indicative of changes within limits; they are usually degenerative changes within a species. (Frair and Davis 1983, p. 29)

So Futuyma's question has answers other than the one he presupposes. One may be able to explain the apparently poor design of the cave salamander's eye fairly easily within a dynamic theory of creation.

Consider another example, the rudimentary wings of flightless birds, which Naylor (1982, p. 93) regards as true vestigial structures whose existence contradicts the theory of creation. The Dutch creationist Hendrik Murris (1986, pp. 200–201) argues however:

Suppose that (as an oversimplified example) the allele "A" imparts the ability to fly, while "a" signifies flightlessness. If birds with AA and Aa combinations arrive and breed on an island where they have no natural enemies, the flightless aa individuals which will inevitably be hatched will survive. Some generations later, according to our model experiment, the entire population could be flightless!

Murris reasons that known population genetic processes may explain the origin of some, though not all, species of flightless birds. The German creationists Junker and Scherer similarly explain the origin of the rudimentary wings of flightless beetles and insects as cases of degenerative microevolution (1988, p. 126). In these dynamic theories, extant organic designs are the products not just of original creative intent, but also of the perturbing effects of natural causes, e.g., selection or drift (Darnbrough 1986, pp. 252–262). These causes must be separated from original design (if that analysis is possible).

By presupposing a static theory of creation, the first premise of the imperfection argument describes a false dichotomy. Of course, many supposedly imperfect organic designs, such as the human alimentary canal (Williams 1992, p. 7) or retina (Dawkins 1986, p. 93) cannot be explained by a dynamic theory of creation as degenerative changes. A dynamic theory can accommodate only certain limited neutral or degenerative changes without contradicting its tenet that variation is bounded. Most "vestigial" structures, for instance, appear to signify relationships expressly denied by even the most flexibly dynamic theories of creation.

In any event, the imperfection argument need not, indeed should not, assume a static theory of creation. That it does so often presuppose such a theory, however, should alert us that the argument may rest on other doubtful assumptions.

*If* p *was produced by a wise creator, then* p *should be perfect (or exhibit no imperfections).*

Here we come upon the major theological difficulties of the imperfection argument. Terms like "wise creator" must be fixed objectively, so that one knows (1) what a "wise creator" or a "sensible God" *is*, and (2) what a "wise creator" *would do*.

To illustrate the first problem, assume an imperfect organic design *p*. Then suppose (as John Stuart Mill believed) that the creator is benevolent and wise, but not omnipotent. This creator would not be able to avoid occasional design compromises. Some imperfections would necessarily be included in the creation—including, let us say, the imperfect organic design *p*. Here, the conclusion that imperfection of design is evidence of *descent* would not follow in every case. Gould writes that perfection alone cannot demonstrate descent, because "perfection need not have a history" (Gould 1980, p. 28). Given Mill's conception of the creator, however, imperfection need not have a history either. If a stapler that continually jams or a water pitcher with a dribbling spout were designed *de novo*, they have no history in any evolutionary sense—yet both artifacts are manifestly imperfect to anyone knowing their intended functions.

Mill's limited creator is of course heterodox (in the Christian tradition), and some may argue that one either defends the usual omnipotent conception of the creator, or one defends no conception at all. The point however, is that we have no grounds *within evolutionary theory itself* to exclude Mill's creator, or any one of a number of conceivable creators whose natures allow imperfection. The creator's place in the argument cannot be filled by just any conception. The conclusion *imperfection of organic design is evidence of descent* requires *logically* the conventional picture of an omnipotent and beneficent artificer (hereafter, the conventional conception). Far from being theologically neutral, therefore, the imperfection argument has a stake in the truth of a particular theology.

Consider next the problem of what a "wise creator" *would do*. According to the second premise, if a perfect God created the world, we should expect to observe "perfect" organic design—but what do we denote by this adjective? Might biological entities judged imperfect when considered individually combine to form a macrosystem judged perfect? Here, theological difficulties ordinarily ignored in any biological analysis come crowding forward. These difficulties can be avoided only by stipulation.

Take the question of the creator's *proper domain*. Many philosophers and theologians take the creator's proper domain to be the entirety of time and space, and furthermore hold that issues of moral value figure ultimately in any theory of creation. If this is so then the necessary finitude or limits of scientific observation may lead us to infer mistakenly that an organic design (e.g., the panda's thumb) is imperfect, when its imperfection is only *apparent*, that is, *local*. On this view, any judgment of perfection or imperfection must be qualified with a proviso that perfection—defined as divinely created perfection—can be judged only on the scale of the whole creation. And there is no reason for a creator to optimize one part of the universe at the expense of the whole. As one commentator writes:

According to this view, what appears to be evil, when seen in isolation or in a too limited context, is a necessary element in a universe which, viewed as a totality, is wholly good. From the viewpoint of God, who sees timelessly and as a whole the entire moving panorama of created history, the universe is good.... (Hick 1967, p. 137)

Several philosophers (notably Augustine and Leibniz) have articulated just such a global theodicy. In his *Theodicy*, Leibniz argued:

[W]e acknowledge ... that God does all the best possible, in accordance with the infinite wisdom which guides his actions.... But when we see some broken bone, some piece of animal's flesh, some sprig of a plant, there appears to be nothing but confusion, unless an excellent anatomist observe it: and even he would recognize nothing therein if he had not seen like pieces attached to their whole. It is the same with the government of God: that which we have been able to see hitherto is not a large enough piece for recognition of the beauty and order of the whole. ([1710], 1985, pp. 206–207)

Although one may regard such a theodicy with scorn (see *Candide*), the problem remains: how do we judge divinely created perfection? Is it global or local? One can stipulate that only biological optimality matters, but the stipulation is arbitrary.

*Organic design* p *is not perfect (or exhibits imperfections).*

The terms "perfection" and "imperfection" have long been part of the descriptive vocabulary of natural history. Many authors use the terms with little apparent reflection, however, perhaps thinking that, as operational constructs in biology, "perfection" and "imperfection" are perspicuous. They are not. The epistemological difficulties that plague optimality arguments in evolutionary theory (Lewontin 1987) also occur in judgments of perfection (or imperfection). In the latter case, however, the difficulty of determining whether a state of a trait or organism is optimal *is magnified immeasurably by the theological context.*

The second premise says that a "wise creator" will create perfect organic designs. This seems clear enough until we come to cases, such as the panda. Gould argues that we can use optimality theory to designate "ideals for assessing natural departures" (1986, p. 66). It follows that in finding existing pandas to be imperfect, Gould must have some notion of an ideal panda, departure from which evokes a judgment of imperfection. So what is an ideal panda? That's rather hard to say, as Maynard Smith (1978, p. 32) has pointed out generally:

It is clearly impossible to say what is the "best" phenotype unless one knows the range of possibilities. If there were no constraints on what is possible, the best phenotype would live for ever, would be impregnable to predators, would lay eggs at an infinite rate, and so on. It is therefore necessary to specify the set of possible phenotypes, or in some other way describe the limits on what can evolve.

With the imperfection argument, however, the question is not what can possibly evolve, but *what can possibly be created.* Given the conventional conception of the creator, there seem to be *no* limits on what is possible, nor any reason (short perhaps of logical contradiction) why one hypothetically possible panda should be preferred, as a counterfactual ideal, to another. If "perfection" is limited only by one's imagination, then specifying an ideal phenotype, for the panda or any other organism, quickly becomes a fanciful exercise. Why couldn't the creator have given pandas the ability to fly?

We might then turn the problem around, and define a criterion of optimality that a "wise creator" ought to be able to achieve. Real organisms, if they were specially created, should then meet that criterion. The difficulty we hoped to escape, however, now returns in another form. We must now explain why, from all possible criteria, we have chosen one particular criterion (or set of criteria) for placing limits on the perfection expected of the creator. Just as within evolutionary theory, "a proper optimization theory must be capable of explaining why particular constraints on [phenotypic] accessibility are regarded as absolute while others are not" (Lewontin 1987, p. 156), so the imperfection argument must explain why the creator's designs should be constrained in certain instances but not in others. This is exceedingly difficult to do, and may be impossible.

Gould argues that the panda's thumb is "somewhat clumsy" and "wins no prizes in an engineer's derby" (Gould 1980, p. 24). Nevertheless, while watching pandas at the Washington zoo, he was "amazed at their dexterity, and wondered how the scion of a stock adapted for running could use its hands so adroitly" (Gould 1980, p. 21). Indeed, other observers *heap praise* on the panda's use of its forelimbs:

The panda can handle bamboo stems with great precision, by holding them as if with forceps in the hairless groove connecting the pad of the first digit and pseudothumb.... When watching a panda eat leaves ... we were always impressed by its dexterity. Forepaws and mouth work together with great precision, with great economy of motion.... (Schaller, Jinchu, Wenshi, and Zing 1986, pp. 4, 48)

Although the panda's thumb may be suboptimal for many tasks, it does seem suited for what appears to be its usual function, manipulating bamboo. (At any rate the facts of the matter are very much in dispute.)

But even if the pseudothumb were suboptimal for manipulating bamboo, *it might still be the best structure possible*. The creator could have been bound by "compossibility" constraints, which would limit the design possibilities that are *mutually consistent*. One cannot, for instance, expect an electric clock designed to *obtain* its regularity from alternating current to be *more regular* than that current. The thumb may have some primary function for which it was designed, and the panda has co-opted it secondarily to strip bamboo. One may have failed to identify the correct reference situation by which to judge the design, perhaps by observing too little of the panda's life-history. The flippers of marine turtles, for example, strike us as badly designed for digging holes in beach sand to place eggs. The same flippers, however, perform efficiently in the water, where the turtles spend most of their time. Which reference situation do we employ (Lewontin 1984, pp. 234–251)?

If the creator need only "act reasonably," that is, create designs which meet some specific criteria for optimality, then we must say why those (and not some other) criteria obtain if our suboptimality claims are to have any evidential force. This problem is made acute by the bothersome truth that any suboptimal design can be made optimal if we specify the right constraints (Lewontin 1987, pp. 158–159). How, then, do we specify criteria of optimality for an omnipotent creator?

A simple equation illustrates the problem. Suppose we define an optimal organism (design) as scoring 1.0, where the observed and expected design values in the following equation correspond exactly:

$$\frac{\text{observed design}}{\text{expected design}} = \text{optimality (suboptimality) measure}$$

Now suboptimality occurs when the numerator value falls below that of the denominator. If an optimal (created or ideal) panda has an expected design value of, say, 50, but actual pandas score 30, the panda as a species is suboptimal, suffering what we might call a *design shortfall*:

$$\frac{\text{actual (observed) pandas : 30}}{\text{optimal (expected) pandas : 50}} = 0.6 \; (design\; shortfall\; of\; 0.4)$$

We cannot solve this equation, however, without the expected design value. Absent the denominator, the equation has two unknowns and thus is unsolvable.[5] The expected design must be determined by optimality criteria, however, metrics along which design is measured. We have no such metrics for living things *as divinely created*. Thus we have no principled way of assigning the expected design value.

In summary, the premises of the imperfection argument provide a poor grounding at best for any empirical conclusions about the truth of evolution. Gould repeatedly uses the word "proof" for the imperfection argument (1977, p. 91; 1980, pp. 20–21; 1991, p. 61). That, it surely is not.

## The Homology Argument

On opening any moderately advanced biology textbook nowadays one is likely to find, amid the discussion of the evidence for common descent, an illustration showing an array of tetrapod forelimbs (see figure 32.3). The text will state that the pattern of similarity abstracted from the limbs (the *pentadactyl limb*) can be explained only by common descent. Fran

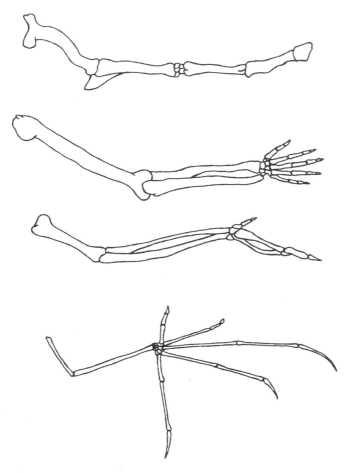

**Figure 32.3**
An array of vertebrate forelimbs. From top: horse, human, bird, and bat. (After Ayla 1988.)

cisco Ayala, for instance, is his *Encyclopedia Britannica* article on evolution (1988, p. 987) writes:

> From a purely practical point of view, it is incomprehensible that a turtle should swim, a horse run, a person write, and a bird or bat fly with structures built of the same bones. An engineer could design better limbs in each case. But if it is accepted that all of these skeletons inherited their structures from a common ancestor and became modified only as they adapted to different ways of life, the similarity of their structures makes sense.

"An engineer could design better limbs in each case" has the ring of an empirical finding. But the story is rather more complicated.

In Chapter XIII of the *Origin*, Darwin argued that it would be "hopeless" to explain the pentadactyl pattern "by utility or by the doctrine of final causes" (1859, p. 435). As Cain (1964, p. 44) observes, Darwin's view of these patterns is now canonical:

Darwin ... originated the evolutionary interpretation which has been followed ever since, that the general plan of the pentadactyl limb is not now adaptive, although it must have been in the common ancestor, but its modifications are adaptive.

But how do we know that the general plan is suboptimal? This claim, after all, drives the inference to descent, or, to put it another way, makes implausible the inference to an optimizing designer. (A designer may have used the same pattern in different organisms precisely because that pattern is optimal for the functions in question.) What, then, grounds this seemingly empirical determination of suboptimality?

Here a brief historical excursus will be helpful. The patterns of homology employed by Darwin were familiar to pre-Darwinian anatomists, having been worked out by them in a non-evolutionary context. "Pre-Darwinian systematics did not profess an evolutionary explanation for homology," writes Ronald Brady (1985, p. 114), "but that privation did not prevent an extensive investigation of comparative anatomy, during which the principles of systematics were developed." Although the "unity of plan" of the tetrapod forelimb was powerfully suggestive of descent, and was so seen by some pre-Darwinians, descent was far from being the only plausible causal account available (Russell [1916] 1982, p. 214). The patterns of similarity evident among major groups of animals suggested similar functional requirements (Cuvier), non-material archetypes (Owen), or the plan of the Creator (Agassiz). Without transitional forms or a mechanism of functional transformation, non-material causes were genuinely competing explanations (Rieppel 1988, pp. 49–51).

Darwin made the patterns themselves the puzzle. Common descent would become the only reasonable explanation if Darwin's readers could be persuaded that, even without other "facts or arguments" (1859, p. 458) for descent, the theory explained patterns before which rival theories stood silent. But the rival theories—in particular, creation— must in fact stand silent, for if they also explained the patterns at hand, descent might remain only a plausible but uncompelling theory, unable to claim broader explanatory promise than its rivals.

In the *Origin* (especially Chapter XIII), therefore, Darwin frames the patterns of comparative natural history in terms favorable to common descent, but uncongenial to any non-material explanation invoking design. In particular, one important avenue of explanation open to the creationist must be cut off, namely, the possibility that homologous patterns, such as the pentadactyl limb, are functionally optimal, and thus, could reasonably have been intended, and realized, by an optimizing creator.

For this task, Darwin finds a ready if unwitting ally in Owen.[6] "What could be more curious," asks Darwin, "than that the hand of a man, formed for grasping, that of a mole for digging, the leg of the horse, the paddle of the porpoise, and the wing of the bat, should all be constructed on the same pattern, and should include the same bones, in the same relative positions?" (1859, p. 434). Four of the five examples given— human hand, mole, horse, and bat—are Owen's, from *On the Nature of Limbs*. (Darwin substitutes a more familiar creature, the porpoise, for Owen's example of an aquatic mammal, the dugong.) It would be "hopeless," Darwin warns, to explain this pattern of similarity by functional utility: "The hopelessness of the attempt has been expressly admitted by Owen in his most interesting work on the 'Nature of Limbs'" (1859, p. 435).

Now it appears from Darwin's phrasing ("expressly admitted") that Owen, having failed to show that the pentadactyl pattern was functionally useful, was conceding as much. But this is "seriously misleading in one respect," Cain notes (1964, p. 44). "The hopelessness of the attempt is not what Owen was driven by the facts to *admit*, but what his whole lecture set out enthusiastically to *proclaim*." Owen was *keen to refute* the notion that the structures of organisms were specifically designed for their functions. He thus makes room for his "legitimate fruit of inductive research," namely, the "higher law of archetypal conformity" (1849, p. 70). In attacking the principle of specific design, and arguing for the constraints of archetypal homology, Owen cannot help supporting Darwin—who understandably then calls on him as an anatomical authority favoring descent, Owen's qualms about that the naturalistic version of that theory notwithstanding.

One will search *On the Nature of Limbs* in vain, however, for anything resembling an *empirical demonstration* that an homologous plan limits functionality, thus rendering an organism suboptimal. Owen's argument rests, rather, on an *a priori principle*:

> The teleologist would rather expect to find the same direct and purposive adaptation of the limb to its office as in the machine [devised by humans]. (1849, p. 10)

Given some functional end, the human engineer "does not fetter himself by the trammels of any common type," says Owen, but uses whatever design is best suited:

> There is no community of plan or structure between the boat and the balloon, between Stephenson's locomotive engine and Brunel's tunneling machinery: a very remote analogy, if any, can be traced between the instruments devised by man to travel in the air and on the sea, through the earth or along its surface. (1849, p. 10)

Yet when we consider organismal structures, Owen argues, a remarkable "unity of plan" is found—"so little to be expected, a priori":

> That every segment and almost every bone which is present in the human hand and arm should exist in the fin of the whale, solely because it is assumed that they were required in such number and collocation for the support and movements of that undivided and inflexible paddle, *squares ... little with our idea of the simplest mode of effecting the purpose* .... (1849, p. 40; emphasis added)

Richard Owen would have designed organisms differently. But of what evidential significance are *Richard Owen's ideas* about "the simplest mode of effecting the purpose"? We want to know if the structures of animals are well-suited to the functions they must perform: a question to be answered—if it can be answered at all—on the grounds, not of any "deep and pregnant" *a priori principle* (1849, p. 10), but by observation and experiment. As Cain argues,

> The fin of the dugong or whale may be of a simple external appearance when compared with the hand of Man, but it is not a simple stiff plate capable only of being waved up and down. The hoof of the horse may merely rest on the ground or beat upon it, and so be simpler than the hand of Man, but each hoof must be picked up and put down without the whole body being raised to correspond, or much energy will be wasted in an intolerably jerky gait. The hoof must adjust itself to some extent to the different angles at which the surface of the uneven ground may meet it, so that it does not slip, and yet it must bear a considerable weight—it cannot be as delicate as the sucker on the tube-foot of a starfish. (1964, p. 43)

Owen never demonstrates that the various mammalian forelimbs he has examined, constructed on a common plan, are functionally less than optimal for being so constructed. Yet this is what Darwin takes away from Owen, and that, in turn, evolutionary biologists have taken away from Darwin. "An engineer could design better limbs in each case" (Ayala 1988, p. 987). *There is no evidence that this is true.*

What *does* ground the perception of suboptimality so widely shared among evolutionary biologists? Here, I would argue, strong theological preconceptions are at work. If the creator is *free to do as he pleases*, the appearance of plan can become the appearance of *limitation* or *constraint*, suggesting an unimaginative or even slavish repetition of structures along some predetermined pattern. "Intelligence and purpose," writes Neal Gillespie (1979, p. 71), interpreting Darwin's arguments against creation, "should be more creative than nature showed itself to be." This theological intuition—that the apparent uniformity of certain biological patterns is inconsistent with the freedom of a creator to act as he wishes—is nowhere better illustrated than in Darwin's book on orchids. After reviewing the homologies of orchids and ordinary flowers, Darwin appeals to our intuitions about *what God would have done in this case*:

> Can we feel satisfied by saying that each Orchid was created, exactly as we now see it, on a certain "ideal type;" that the omnipotent Creator, having fixed on one plan for the whole Order, did not depart from this plan; that he, therefore, made the same organ to perform diverse functions—often of trifling importance compared with their proper function—converted other organs into mere purposeless rudiments, and arranged all as if they had to stand separate, and then made them cohere? Is it not a more simple and intelligible view that all the Orchideae owe what they have in common, to descent from some monocotyledonous plant ... ? ([1877] 1984, pp. 245–246)

Removing the theology from Darwin's argument for the common descent of the Orchideae would eviscerate it. Darwin provides no fossil evidence that orchids evolved from ordinary flowers, nor indeed any experimental evidence that such a transformation is even possible. Rather, Darwin describes patterns of similarity among orchids—which patterns might to a creationist indicate the purposeful workings of a designer. If one accepts, however, the premise that it is *unfitting* to ascribe variations on an "ideal type" to the direct artifice of an omnipotent creator, the same patterns become evidence of common descent. The theology in the passage is thus far more than a rhetorical device. It is the logical pivot of Darwin's entire argument.

In this vein, Futuyma (1985, p. 6) points out that flying vertebrates could have been designed otherwise than they are:

An omnipotent creator could, as we can in imagination, create organisms with wings on their shoulders (e.g., angels, or the mythical Pegasus), but the wings of all flying vertebrates are modifications of the front legs of their ancestors.

*How uncreative to have done things that way* is the principal deliverance of this species of biological imagination. To be told that evidence suggests nevertheless that the structures in question were created (i.e., are transformationally discrete, not derived materially from simpler forms) does not of course answer the question, "but why should they then appear to share the same plan?" One can imagine that organisms could have been constructed in any number of ways.

Here again speculation has free rein. Yet it is precisely here that speculation is likeliest to mislead us. Surely the reason that homology is seen as evidence for descent is *not* because the phenomenon contradicts what one would expect a rational creator to do. An omnipotent creator could have made non-homologous vertebrate limbs, or different genetic codes, or indeed built organisms out of different types of matter entirely. These are not grounds to support an empirical claim about the causal history of homologous patterns. Suppose one argues, *contra* Darwin, that we have every reason for thinking a creator *would* have designed each species of orchid to show homologies with ordinary flowers. How, by everyday scientific methods, would one go about settling this dispute? One may assume or deny the truth of Darwin's particular theological aesthetic, but it is hard to see how that assumption is binding on other observers (or why we should take it as intelligible).

Many formulations of the homology argument, however, rest on similar theological assumptions. It is curious that in glossing the orchid arguments, Gould (1980) and other commentators (e.g., Ghiselin 1984) have not noticed this problem. Perhaps Darwin's theological aesthetic fits so closely with their own intuitions that its role in the argument escapes comment. Nevertheless, these theological assumptions need to be justified, or else it should be admitted that they stand as bare postulates. In analyzing Darwin's argument about homologies—in particular, his claim that a Creator would not use such patterns—Løvtrup (1987, p. 132) observes:

Why not? Even the Creator may use a good device more than once. Yes, why not indeed? Darwin's arguments against this possibility are postulates, unfounded by any evidence.

If homology provides evidence for descent, it must do so *not* because homologies are inconsistent with what a rational creator would have done. A rational creator might have done any number of things. Rather, homologies appear to mark out a pathway of natural transformation characterizing a continuous geometry of organic form, i.e., of descent. Is the appearance of natural transformation more than an appearance? Is the geometry of nature profoundly continuous? These questions want empirical answers. Speculations about the freedom of the creator should be seen for what they are, and abandoned.

### The Influence of Darwinian Metaphysics

Darwin's argument for descent with modification was pressed on many fronts—among them, the theological (Gillespie 1979; Brooke 1985; Cornell 1987; Kohn 1989). Darwin's corpus is permeated by a metaphysical program which was, Cornell (1987, pp. 384–385) argues, "more than useful rhetoric to Darwin, and more than a methodological convention that promoted science." Consider the notion of perfection:

The assumption of perfect adaptation, which Darwin shared with most of the biologists of his generation, was derived from the belief that nature is a created, harmonious, and purposeful whole.... It is a natural, perhaps necessary, corollary of the belief that nature is a harmonious system preplanned in every detail by a wise and benevolent God. (Ospovat 1980, pp. 189–190)

Cornell (1987, p. 396) concurs:

The word "perfect" is an adjective generally reserved for divine action. That is how, for instance, Paley used it, and it was probably what Darwin understood, even when he was criticizing the belief in the perfection of particular forms ... because that belief implied special creation by God.

Now, while Darwin came to reject the idea that organisms were perfectly designed for their environments, he never rejected the theoretical apparatus implied by the very terms "perfection" and "imperfection." Many arguments in the Origin make sense only if one presupposes the creator of early nineteenth century English natural theology—and Darwin does not challenge this conception. Rather, he turns to certain aspects of organic design which appear to fit only awkwardly into the usual schemes of natural theology, and drives these counterexamples back into the machinery of the argument from design. Instead of impiously attacking the nature or existence of the creator (as a skeptic, e.g., Aveling, might do), Darwin offers his theory of descent and secondary causes to explain what would otherwise be intolerable anomalies. *All this incongruity of design could not have been directly created.*

In so doing, of course, Darwin impales his creationist opponents on the horns of a dilemma. Either they deny the benevolence and wisdom of the creator, by making him the authors of "abhorrent" designs, or they retain their conception of the creator, but must greatly circumscribe his actions, for if imperfect designs could be due to secondary causes, then could not many other (in fact, nearly all) organic structures be the products of secondary causes as well?

But note again that little indicates Darwin ever rejected the deep presuppositions which he inherited from English natural theology, namely, perfection as an observable quality of organic design, and the conventional conception of the nature (if not the actions) of the creator. Indeed, a close reading of the Notebooks would suggest that Darwin saw his theory as providing a more sublime conception of the actions of the creator (see, for instance, D 36: "What a magnificent view one can take of the world ..."). *Darwin employed a particular conception of God to judge theories of God's creative activity.* Otherwise, why should the multiple creations scornfully derided in D 37 as a "long succession of vile Molluscous animals" be beneath the "dignity" of God? Cornell (1987, p. 397) argues, of this and other passages from later notebooks:

As always, Darwin's idea of "perfection" refers to the nice relationship of organisms to their physical surroundings. But it also refers to the overall design of the world, from a divine viewpoint.... Darwin's sense of a comprehensive system, the invocation of divine perfection, and his new theory are thus all closely related.

And, as Brooke (1985, p. 46) argues:

The fact is that there are several entries in the transmutation notebooks which indicate that Darwin was discovering a philosophy of nature which he genuinely believed conferred a new grandeur on the deity, despite—or rather because of—the fact that it superseded Paley.

While current evolutionists may be indifferent or opposed to Darwin's theology, their use of the imperfection and homology arguments for evolution presupposes the intelligibility of notions rooted in Darwin's theological metaphysics: perfection as an observable quality of organic design, and the intuition at the heart of Darwin's metaphysics—that a rational and benevolent God would have created an organic world different from the one we observe. Both continue to inform evolutionary theory.[7]

Yet many hold that the Darwinian revolution entailed the surrender of theological speculation in biology (Mayr 1983, p. 25). Indeed, many scientists and philosophers would argue that natural science and theology view each other across a largely (if not completely) impassable epistemological gulf (Kolakowski 1982; Gilkey 1985). Science, on this view, is by its very nature committed to a thoroughgoing methodological naturalism. Hence, the problem which opened this essay: the persistence of Darwinian theological *themata* in evolutionary theory is inconsistent with the doctrine of methodological naturalism.

But should natural science necessarily be committed to methodological naturalism? The shortcomings of theological arguments for evolution may be evidence enough that science has no business meddling in theology (or vice versa). I draw a different moral, however. Science will have to deal with theological problems if science is a truth-seeking enterprise; theology must confront the patterns of scientific experience if it hopes to speak to all of reality. What this essay helps to show, I think, is how very easy it will be to do both theology and science badly. That is not a brief for methodological naturalism, however. It is a tale of caution about how we should go about explaining the origin of the world's creatures.

### Acknowledgements

Many persons have helped me in thinking about theology and evolution; too many, in fact, to acknowledge individually. I mention as instrumental, however, Bill Wimsatt, Bob Richards, and Leigh Van Valen. Each disagrees with some of the preceding paper. This work was supported in part by grants from the Pascal Centre for Advanced Studies in Faith and Science, Ancaster, Ontario, Canada.

# Notes

1. Since 1985 several variant nuclear codes have been discovered, leading many workers to doubt the theory that once established the code must be invariant (Jukes and Osawa 1991).

2. "Pure wind," a biologist called the arguments. "I never took Gould's suggestions seriously," a philosopher wrote, "that is, I never thought he meant me to take it seriously. If you find that philosophers or biologists do take the thing seriously I am profoundly disappointed in the intelligence of my colleagues." Most did take the arguments seriously. And, as pure wind, the arguments blow pretty steadily in many evolutionary biology textbooks.

3. In the spring of 1979, I attended a series of lectures on evolution at the Carnegie Museum of Natural History, in Pittsburgh. The first lecture in the series was given by Leonard Krishtalka, a vertebrate paleontologist on the museum staff. For his opening illustration, Krishtalka had borrowed a peccary (a pig-like mammal) from the museum's collection, which he placed on the dais at the front of the auditorium. Pointing to the peccary's "dew claws" (so-called because these toes, on the rear of the limb just above the hoof, appear to touch only the "dewy" surface of the ground), Krishtalka asked, "Now why would God have created this animal with non-functional structures like the dew claw?" But of course God didn't create the peccary, he continued, natural selection did. What strikes me now about this illustration was how utterly clear its theological content seemed to Krishtalka, that is, as evidence supporting the causal story he was about to tell.

4. Some readers argued that I ought not to cite Gould's general and semi-popular essays, on the grounds that the essays *were* general and semi-popular. The more I thought about this argument, however, the odder it seemed. The argument means either (a) Gould has knowingly misrepresented evolutionary theory to his lay readers, or (b) in presenting evolutionary theory to his lay readers, Gould has used language so "analogous, rhetorical, sometimes imprecise" (to quote one of my correspondents) that he has—despite his best intentions—misrepresented evolutionary theory. Those who make this argument can hardly have meant (a), but is (b) any more credible? On the topic of biological imperfection, Gould's popular (and technical) writings are strikingly consistent. It is hard to believe that a scientist renowned for his prose abilities would explain his own theory so poorly that he could, for more than a decade, continually mislead his readers. And Gould takes his popular writing far more seriously than many of his colleagues. "The concepts of science, in all their richness and ambiguity," he argues, "can be presented without any compromise, without any simplification counting as distortion, in language accessible to all intelligent people" (1989, p. 16). The reading list for Gould's introductory science course at Harvard (Science B–16) includes, out of a total of 62 readings, 33 of his essays from *Natural History* magazine.

5. Gabriel Nelson pointed this out to me.

6. In the section of Chapter XIII headed "Morphology," only two authors are cited by name: Geoffroy St. Hilaire (once) and Owen (four times).

7. Some have urged me to note that Gould, like many (most?) evolutionary theorists, is himself an agnostic, or, as Gould has put it, a "nontheist." Jeffrey Levinton observed that Gould "is about as 'theological' as anyone calling himself an atheist." True, and immaterial. The theological convictions of a person have no bearing on the truth or falsity of the propositions, theological or otherwise, that he puts forward publicly as worthy of assent.

## References

Ayala, F.: 1988, "Evolution, The Theory of," *Encyclopedia Britannica*, 15th ed., Encyclopedia Britannica, Chicago.

Brady, R. H.: 1985, "On the Independence of Systematics," *Cladistics* 1, 113–126.

Brooke, J. H.: 1985, "The Relations Between Darwin's Science and his Religion," in J. Durant (ed.), *Darwinism and Divinity*, Basil Blackwell, London, pp. 40–75.

Burian, R.: 1986, "Why the Panda Provides no Comfort to the Creationist," *Philosophica* 37, 11–26.

Cain, A. J.: 1964, "The Perfection of Animals," in J. D. Carthy and C. L. Duddington (eds.), *Viewpoints in Biology*, Vol. 3, Butterworths, London, pp. 36–63.

Cornell, J.: 1987, "God's Magnificent Law: The Bad Influence of Theistic Metaphysics on Darwin's Estimation of Natural Selection," *Journal of the History of Biology* 20, 381–412.

Darnbrough, C.: 1986, "Genes—Created but Evolving," in E. H. Andrews, W. Gitt, and W. J. Ouweneel (eds.), *Concepts in Creationism*, Evangelical Press, Herts, England, pp. 241–266.

Darwin, C.: 1859 [1964], *On the Origin of Species*, Harvard University Press, Cambridge, Mass.

Darwin, C.: 1877 [1984], *The Various Contrivance by Which Orchids Are Fertilized by Insects*, University of Chicago Press, Chicago.

Dawkins, R.: 1986, *The Blind Watchmaker*, W. W. Norton, New York.

Eldredge, N. and J. Cracraft.: 1980, *Phylogenetic Patterns and the Evolutionary Process*, Columbia University Press, New York.

Frair, W. and Davis, P.: 1983, *A Case for Creation*, Moody Press, Chicago.

Futuyma, D.: 1983, *Science on Trial*, Pantheon, New York.

Futuyma, D.: 1985, "Evolution as Fact and Theory," *Bios* 56, 3–13.

Ghiselin, M.: 1984, *The Triumph of the Darwinian Method*, 2nd ed., University of Chicago Press, Chicago.

Gilkey, L.: 1985, *Creationism on Trial*, Winston Press, Minneapolis.

Gillespie, N.: 1979, *Charles Darwin and the Problem of Creation*, University of Chicago Press, Chicago.

Gould, S. J.: 1977, *Ever Since Darwin*, W. W. Norton, New York.

Gould, S. J.: 1980, *The Panda's Thumb*, W. W. Norton, New York.

Gould, S. J.: 1983, *Hen's Teeth and Horse's Toes*, W. W. Norton, New York.

Gould, S. J.: 1986, "Evolution and the Triumph of Homology, or Why History Matters," *American Scientist* 74, 60–69.

Gould, S. J.: 1989, *Wonderful Life*, W. W. Norton, New York.

Gould, S. J.: 1991, *Bully for Brontosaurus*, W. W. Norton, New York.

Haeckel, E.: 1876, *The History of Creation*, D. Appleton, New York.

Hick, J.: 1967, "Evil, The Problem of," in P. Edwards (ed.), *The Encyclopedia of Philosophy*.

Hoffman, A.: 1989, *Arguments on Evolution*, Oxford University Press, Oxford.

Holton, G.: 1993, *Science and Anti-Science*, Harvard University Press, Cambridge, Mass.

Jacob, F.: 1982, *The Possible and the Actual*, Pantheon, New York.

Jukes, T. and Osawa, S.: 1991, "Recent Evidence for Evolution of the Genetic Code," in S. Osawa and T. Honjo (eds.), *Evolution of Life*, Springer-Verlag, Berlin, pp. 79–95.

Junker, R. and Scherer, S.: 1988, *Entstehung und Geschichte der Lebewesen*, Weyel Lehrmittelverlag, Giessen.

Kohn, D.: 1989, "Darwin's Ambiguity: The Secularization of Biological Meaning," *British Journal for the History of Science* 22, 215–239.

Kolakowski, L.: 1982, *Religion*, Oxford University Press, Oxford.

Kitcher, P.: 1985, "Darwin's Achievement," in N. Rescher (eds.), *Reason and Rationality in Natural Science*, University Press of America, Lanham, Maryland, pp. 127–189.

Landgren, P.: 1993, "On the Origin of "Species": Ideological Roots of the Species Concept," in S. Scherer (ed.), *Typen des Lebens*, Pascal Verlag, Berlin, pp. 47–64.

Leibniz, G. W.: [1710] 1985, *Theodicy*, Open Court, La Salle, Illinois.

Lewontin, R. C.: 1984, "Adaptation," in E. Sober (ed.), *Conceptual Issues in Evolutionary Biology*, MIT Press, Cambridge, Mass., pp. 234–251.

Lewontin, R. C.: 1987, "The Shape of Optimality," in J. Dupre (ed.), *The Latest on the Best*, MIT Press, Cambridge, Mass., pp. 151–159.

Løvtrup, S.: 1987, *Darwinism: The Refutation of a Myth*, Croom Helm, London.

Maynard Smith, J.: 1978, "Optimization Theory in Evolution," *Annual Review of Ecology and Systematics* 9, 31–56.

Mayr, E.: 1964, Introduction to the facsimile reprint of *On the Origin of Species* by Charles Darwin, Harvard University Press, Cambridge, Mass.

Mayr, E.: 1983, "Darwin, Intellectual Revolutionary," in D. S. Bendall (ed.), *Evolution from Molecules to Men*, Cambridge University Press, Cambridge, pp. 23–41.

Mayr, E.: 1991, *One Long Argument*, Harvard University Press, Cambridge, Mass.

Murris, H.: 1986, "The Concept of the Species and its Formation," in E. H. Andrews, W. Gitt and W. J. Ouweneel (eds.), *Concepts in Creationism*, Evangelical Press, Herts, England, pp. 175–207.

Naylor, B. G.: 1982, "Vestigial Organs are Evidence of Evolution," *Evolutionary Theory* 6, 91–96.

Ospovat, D.: 1980, "God and Natural Selection: The Darwinian Idea of Design," *Journal of the History of Biology* 13, 169–194.

Owen, R.: 1849, *On the Nature of Limbs*, John Van Voorst, London.

Ridley, M.: 1985, *The Problems of Evolution*, Oxford University Press, Oxford.

Ridley, M.: 1986, *Evolution and Classification*, Longman, London.

Rieppel, O.: 1988, *Fundamentals of Comparative Biology*, Birkhauser Verlag, Basel.

Russell, E. S.: [1916] 1982, *Form and Function*, University of Chicago Press, Chicago.

Scadding, S.: 1981, "Do 'Vestigial Organs' Provide Evidence for Evolution?" *Evolutionary Theory* 5, 173–176.

Scadding, S.: 1982, "Vestigial Organs do not Provide Scientific Evidence for Evolution," *Evolutionary Theory* 6, 171–173.

Schaller, G., Jinchu, H., Wenshi, P., and Jing, Z.: 1986, *The Giant Pandas of Wolong*, University of Chicago Press, Chicago.

Scherer, S.: 1993, "Basic Types of Life," in S. Scherer (ed.), *Typen des Lebens*, Pascal Verlag, Berlin, pp. 11–30.

Sober, E.: 1984, *The Nature of Selection*, MIT Press, Cambridge, Mass.

Sober, E.: 1993, *Philosophy of Biology*, Westview Press, Boulder, Colorado.

Williams, G. C.: 1992, *Natural Selection: Domains, Levels, and Challenges*, Oxford University Press, Oxford.

# 33

# Appealing to Ignorance Behind the Cloak of Ambiguity

Kelly C. Smith

One should always keep an open mind, but not so open that one's brains fall out.
—Bertrand Russell

## I. Introductory Comments

In "The Role of Theology in Current Evolutionary Reasoning," Paul Nelson (1996) fires a fresh salvo in the renewed skirmishing between evolutionary theory and creationism. His arguments are, in some ways, quite different from those we are all accustomed to seeing from traditional creationism. In fact, some of issues raised in this "new" creationism are quite thoughtful and deserve to be taken very seriously. On the other hand, Nelson does not shy away from some of the same sorts of rhetorical "rude tricks" creationism has always resorted to. Thus there's something here to please everyone.

A charitable reconstruction of the *logic* of Nelson's basic approach involves these points:

(1) Evolutionists are often careless when making claims about testability and imperfection.

(2) There are (creationist) theological positions that are consistent with some of the evidence amassed by evolutionary theory.

(3) Evolutionary theory must make certain sorts of theological commitments.

(4) Evolution's basic *methodological naturalism* is inconsistent with its theological commitments.

These points run the gamut from true to misleading to false. The first point is perfectly true. The second point is the most interesting—it too is true in a sense, though not in any way that represents a problem for evolution exclusive of any other rational pursuit of knowledge. The third point is also true, but to employ it as Nelson intends is to endorse a hideous concoction of creationist ambiguity and appeal to ignorance. Finally, the fourth point is simply false. The difficulty is that formulating a response must necessarily take us rather deep into philosophy of science and issues theological—ground many evolutionary biologists fear to tread. Fortunately, this is precisely the kind of thing at which philosophers of science excel.

Nelson's *rhetorical* strategy can not be viewed in so favorable a light. In fact, it can be summarized in four basic dicta:

(1) Revealeth not a specific claim about God, for to do so will invite the wrath of your colleagues and the tribulations of stale theological disputes!

(2) Relyeth always on the bare possibility of a God "behind" the data, for verily shall ye be burned should ye make specific predictions!

(3) Maketh your own problems seem as problems of your foes, for then they will be sore afflicted indeed!

(4) Obscureth your position whenever possible, in order to frustrate thine enemies' response!

The new creationism deserves a fair hearing, hence this paper. In fact, it's something of a professional duty for junior philosophers to provide such critiques as each new variation on the old creationist tune comes along. It's doubtful, of course, whether any critique, no matter how careful, will fundamentally alter creationist tactics. Darwin's observation is no less true today than in 1859: "Great is the power of steady misinterpretation."

## II.   On Testability and the Cloak of Ambiguity

Nelson's first barrage centers on the observation that evolutionists sometimes argue that creationism is untestable, and sometimes that it's testable, but false. It seems it can't be both, so one is left to infer that evolution needs to get the very foundations of its logical house in order before it can engage in any serious critique of the alternatives. However, the apparent inconsistency here is a product of heterogeneity in *creationism's* arguments, not any difficulty within evolutionary theory itself. When one's opponent hides behind a *cloak of ambiguity*, one can be forgiven the occasional imprecise thrust.

It should be granted that many of evolution's defenders *have* been a bit sloppy on this point—for example, being overly quick to label creationism as untestable. Nelson makes a perfectly legitimate point here that we should take seriously, though he is not the first to do so (see, e.g., Burian 1986; Sober 1993). He also provides the basic ammunition for an appropriate response by pointing out some of the great variety of positions the creationist platform encompasses. Like any major intellectual movement, creationism is full of disagreements over even the most basic elements of belief; some variants of creationism are untestable, while others are testable but falsified. Thus there is really no inconsistency in saying that creationism is either untestable or false, providing one is careful to specify which variant of creationism fits which description.

So, what precisely is the nature of the ambiguity behind which Nelson hides? It has to do with what *kind* of God we are positing in creationism. In particular, it matters a great deal whether we believe that God's motives and procedures, and thus the patterns we finds in His creations, are in any meaningful way transparent to human reason. It would be impossible, of course, to do justice to every position that has actually been defended concerning God's nature, much less all positions that might conceivably be defended.[1] However, for the present purposes it will suffice to set up and examine the two endpoints of the continuum of possible positions:

(1) The Mysterious God (MG): God and His ways are utterly mysterious. In particular, humans are incapable of ever comprehending in a meaningful way the purpose(s) and mechanism(s) underlying the universe— indeed, there may not even *be* such things as we understand them.

(2) The "Reasonable" God (RG): God reasons in a manner at least analogous to human reason. Thus, although finite human intellect may never *fully* comprehend His ways, it is perfectly legitimate for us to *attempt* to explain them. We can, at the very least, understand the *essence* of God and know something about what that essence probably entails.

We could, of course, simply believe in an MG and not make any claims about the operation of the natural universe. However, suppose we believe in an MG yet still want to be able to make claims about the divine purpose(s) and mechanism(s) used to create the universe (call this

MG-creationism). We will not get very far, because we do not have, nor could we ever have, any grounds for preferring one story about God's creation over another. If God need not even conform to our mortal concepts of right and wrong, logical and illogical, what could we possibly learn about the universe He created?[2] *Any* explanation of His actions is arbitrary and wholly unsupported. Thus we should simply avoid speculating about such matters—all we can say is that He did what He did for reasons and in ways He alone can comprehend. Nelson certainly seems to endorse exactly this view in places, as when he argues (p. 512 [698]) that "Speculations about the freedom of our creator should be seen for what they are [bare postulates], and abandoned."

At a minimum, it seems patently disingenuous for a creationist to make such an argument. In order to find creationism attractive, one must believe that some sort of creation account of the natural universe is correct. But to do this requires, in turn, a very specific view of God's nature and actions. Granted, Nelson does not discuss God's creation of the universe, but he must believe in something like this, else it's not clear how his position really differs from evolution in the first place.[3] How can a creationist possibly argue that we should not speculate about God's nature?

Nelson himself speculates with abandon, unconstrained even by basic principles of consistency. He endorses an MG, but also implies that rational analysis of the universe is perfectly appropriate on various interpretations of an RG. Hypocrisy aside, endorsing the possibility of an MG is a very dangerous strategy for a creationist to pursue, since it threatens to undermine the basis of his *own* position. Of course, if one takes the possibility of an MG as something that must be refuted, then the speculations of evolutionists about such things as "optimal design" seem entirely unsupported. After all, who are we mere mortals to try to decipher God's mysteries? But how could Nelson fail to realize that such a position makes discussion of *creationism's* explanation of the universe equally unsupported? An MG is just as likely to have created the universe in a manner consistent with *evolution* as with creationism—for that matter, He could have created the universe in a way completely different from *either* sort of theory and totally incomprehensible to human reason.

Whatever we may think of an MG as a strictly theological position,[4] when it makes claims about the natural universe, such a theory is clearly untestable in principle and should never be mistaken for science. Why is this? It's untestable because there is absolutely no evidence we could ever adduce, under any conditions, that would show an MG either did or did not exist. We can never even assess the *likelihood* of such a God in any sort of straightforward fashion, since any rational tools we might employ to do so would be highly suspect if such a God did in fact exist. One can point to the explanatory power of evolutionary biology or the biblical account of Genesis, but in vain. Nothing whatsoever is *inconsistent* with an MG, nor is anything to be *expected* of an MG because we can not, in principle, speculate in even the most minimal way about His motives, powers, etc. On such a view, the answer to any question like, "Why would God have made the human appendix?" or "Is this the best of all possible worlds?" is always, "Who could ever know?"

So, if one endorses both an MG and creationism (MG-creationism), then:

(1) One undermines one's own position in a deep and fundamental way. (2) Evolutionists are perfectly justified in saying that this position is *untestable* and therefore not scientific. It might even be true that God *is* an MG, but MG-creationism certainly isn't scientific (and thus has no place in a science classroom).

On the other hand, if we make the methodological assumption[5] that God is an RG, this empowers us to reason about matters where theology intersects the natural universe. Thus we can argue that we are justified in making some general claims about God (e.g., that He is omnipotent, omniscient, all good, etc.) and thus are also justified in making predictions about His actions based on those claims. Since God's reason is held to be at least analogous to human reason, this sort of analysis is not a category error. To the extent that a particular creationist is an RG-creationist, then, we can hold his claims to be *testable* (at least in principle).

For example, a young-earth creationist might argue that the entire fossil record can be explained as the product of a single worldwide flood some 5,000 years ago. This hypothesis encounters so much contrary evidence, however, that it seems to have been decisively falsified.[6] One could also argue that creationism in general, taken as a research program encompassing several related points of view, has been so consistent in producing these sorts of bad hypotheses that any creationist hypothesis is therefore suspect to some extent. This does not mean, as many have been careful to allow (see Burian 1986), that creationism is *necessarily* untestable. It does not even mean that the scientific falsification of past RG-creationist hypotheses is infallible, since science does not claim infallibility.[7]

By accepting an RG, the RG-creationist agrees to play by an established set of rules governing human reasoning. In particular, he indicates his willingness to take up a certain *burden of proof* whenever one of his claims seems inconsistent with these rules. It is perfectly legitimate to ask an RG-creationist to provide *specific* hypotheses concerning why and how God created the universe. We can subject those hypotheses to empirical analysis and, if they come up lacking, we can argue that this presents a prima facie *problem* for creationism. It is not sufficient for such a person to simply point out that there *might* be an *unspecified* way an RG creation is *consistent* with the empirical data. To do this would be to refuse to provide a rationale that one allows both exists and is accessible, in spite of the fact that the burden of proof now rests squarely on the creationist's shoulders (see section V).

Nelson seems perfectly content to waffle between an MG and an RG as it suits his purposes. Not only that, he alludes to many different possible positions that rely on an RG without committing himself to any one. This allows him to cite every strength of every possible alternative position when *critiquing* evolution, while not having to *defend* the particular weaknesses each position entails. Hiding behind a cloak of ambiguity like this may sometimes be an effective rhetorical technique—making it seem to the uncritical that evolution dies the death of a thousand cuts. But, however effective it may be as a courtroom technique, it's hardly the approach endorsed by anyone who genuinely wishes to *further* human knowledge. Consider the consequences if we were to endorse this technique as a general means of reasoning:

*Laura:*   What was that thump under the bed?

*Jessie:*   Oh, that's the gremlin who lives under there.

*Laura:*   Really, how do you know there's a gremlin living under the bed?

*Jessie:*   Well, I see him from time to time moving around under there.

*Laura:*   So he's not invisible or magic or anything?

*Jessie:*   I don't think so.

*Laura:*   Do you mind if I look under the bed? (looking) I don't see any gremlin, but I do see our baby brother Sam playing with his toys. Don't you think it's more likely to be Sam than the gremlin who caused the bump?

*Jessie (puzzled):*   No, I'm quite sure there's a Gremlin there.

*Laura:*   Why?

*Jessie (brightening):*   Oh! Maybe the gremlin is magic after all and altered himself to look just like Sam.

*Laura:*   But you said the gremlin wasn't magic.

*Jessie:*   No, I said I didn't *think* he was magic. He *could* be magic—one can never tell about gremlins …

*Laura:*   Ok, let's assess that. If he's magic and turned into a semblance of Sam, then I should be able to find the real Sam somewhere in the house (goes off to look). Nope. No Sam, so I think there's evidence against your position.

*Jessie:*   Well, the gremlin could have made the real Sam disappear … or maybe he just turned invisible when Sam came along.

*Laura (sighing):*   Look, can't you just commit to some position so I can see if it's correct?

*Jessie:*   Well, I have no good idea what gremlins are really like, only what they *might* be like. He could have any sort of power you'd care to imagine—maybe even some we *can't* imagine …

As you can see, this discussion could go on forever, given a sufficiently stubborn defender of the gremlin hypothesis. And, of course, it's certainly logically *possible* that Jessie is right about the gremlin. At some point, however, Laura is justified in deciding that Jessie is not being scientific about the whole matter, since (1) there doesn't seem to be any possibility of ever proving such an amorphous claim wrong and (2) Jessie refuses to provide any positive evidence of her contention, settling instead for establishing the mere *possibility* that the evidence agrees with her views. The reason this seems such a silly example is because we reject this type of *appeal to ignorance* intuitively as completely fruitless—but this is precisely the style of reasoning Nelson is using in his paper (though the issues involved are, admittedly, much more complex).

As a good scientist, I certainly do not want to claim infallibility here. Perhaps I have even been unfair to Nelson and his allies—perhaps they really do have a definite position between the two extremes I describe. Of course, I see no evidence whatsoever for such a position in the paper, but it's certainly *possible*. At the very least, however, my confusion should serve as an urgent call for Nelson to clarify his position. Creationists cannot hide behind the cloak of ambiguity forever and, once their views are exposed to the light of day, they will be subject to critical analysis.

## III.   Methodological Naturalism

Now we are in a position to address Nelson's concerns about science's *methodological naturalism* (MN) and related muddles. There is a great deal of confusion about the meaning of this critical phrase. When used by philosophers of science to describe scientific practice, methodological naturalism simply means that scientists are committed to a set of empirical methods for investigation of the world, especially the methods of the natural sciences. In the hands of creationists like Phillip Johnson, however, methodological naturalism is often taken to be a much stronger claim:

Naturalism assumes the entire realm of nature to be a closed system of material causes and effects, which cannot be influenced by anything "outside." Naturalism does not explicitly deny the mere existence of God, but it does deny that a supernatural being could in any way influence natural events, such as evolution, or communicate with natural creatures like ourselves. (Johnson 1993, pp. 114–15)

Nelson seems to go even further when he writes (p. 496): "Methodological naturalism however holds that since 'God's will' (for instance) is inscrutable to science, the truth or falsity of creation can never be known." This is a much more sophisticated argument than one is used to seeing from the creationists, and it seems to be gaining favor within the creationist community.[8] It certainly has rhetorical appeal, because it forces scientists to go rather far out of their accustomed realm—into philosophy of science and even theology—to craft a reply. There is a crucial falsehood concealed in this claim that needs to be examined more closely, but there is also an interesting and important element of truth here—that to the extent that God really is inscrutable, science cannot *justify* (in some sense of the term) any claims about the universe. However, this point is not as telling as Nelson thinks, though the reason for this is complex.

## Methodological Naturalism

The first question that arises is, "Does MN really prohibit any discussion of God's nature?" Not really. The falsehood in Nelson's claims is the implication that science is *necessarily* opposed to theological speculations. MN is, after all, *methodological* (see Pennock 1999). It is part of the very nature of science to be open to new possibilities, and it is not in the business of ruling things impossible. Science *is* in the business of trying to figure out which explanations—out of all those (including theological ones, at least in principle) that *might* be true—are *more likely* to be true. To make this even more interesting, science must always do this based on limited and imperfect evidence. Strictly speaking, science is neutral on the question of God's existence: He might exist or He might not.[9] Science does tend to shy away from theological explanations, but on purely methodological grounds. That is, theological explanations of the natural universe (and, in particular, creationist explanations) have a long and sordid history. In general, they tend to be very poor explanations that don't survive critical scrutiny for long. Moreover, the research traditions they form are unusually barren of novel hypotheses. The rule "Don't invoke divine mechanisms in a scientific explanation" is simply a rule of thumb (though a good one)—it does *not* say that such explanations are unacceptable *in principle*, much less that it's *impossible* they are correct. Thus Nelson is misstating the basic nature of the methodological naturalist's position.

## Inscrutability, Science, and Mysticism

The second question Nelson's discussion of MN raises is this: "Is it really the case that the truth or falsity of creation could never be known *if* God's plan *were* inscrutable?" We will examine this issue more deeply in subsequent sections, but the short answer is, "Yes, if we have very strong rules of evidence." However, this critique cuts just as deeply into the arguments of friend as of foe—if we take it to its logical conclusion, we could argue that evolution is untenable, but the same rough treatment will spill over to all other areas of science and much of nonscience as well. If this is a victory for creationism, it is the very paradigm of a Pyrrhic victory.

In fact, it may even turn out that theology itself is not possible if God's ways are entirely inscrutable. It is probably no accident that Nelson claims God's ways are not open to *scientific* investigation, thus implying that other methods may fare better. Perhaps religious mysticism or some brand of biblical interpretation holds the key to deciphering God's ways. If we attribute this sort of view to Nelson, it would certainly explain how he could criticize science's theological speculation while still maintaining that his own theology is not similarly problematic. The difficulty is that, once rid of science's restraint, it is entirely unclear how we are to *evaluate* competing claims concerning God's universe. Pennock (1999) paints a vivid picture of the consequences of adopting just such a position, which includes such tantalizing "new" approaches as teaching the theory of demonic possession (and associated techniques for exorcism) in medical school. I assume the difficulties with such a "theistic science" are sufficiently clear and sufficiently unappealing that I need not dwell on them. If Nelson wishes to maintain he has insight into God's ways that science does not, however, he will have to take up this problem.

## IV.   On Evolutionary Theory as Theology

Nelson contends that evolutionary theory is inherently theological in character. Since MN supposedly prohibits this, evolutionary explanations are held to be deeply inconsistent. As with so many of the things Nelson says, this makes sense superficially, but it's very unclear precisely what it *means*.

The issues arise primarily in the context of Nelson's discussion of imperfection arguments for evolution. He provides (p. 499 [683]) a schematization of Stephen J. Gould's classic imperfection argument for evolution (the Panda's thumb):[10]

1. If $p$ is an instance of organic design, then $p$ was introduced either by a wise creator, or by descent with modification.
2. If organic design $p$ was introduced by a wise creator, then $p$ should be perfect (or exhibit no imperfections).
3. Organic design $p$ is imperfect (or exhibits imperfections).

The conclusion follows that
∴  Organic design $p$ was not produced by a wise creator, but by descent with modification.

The difficulty Nelson raises is that, in order for this argument to work, it must assume that God would produce *perfect* traits. But this is necessarily to speculate concerning what *kind* of trait God *would* produce if He existed—and that is a patently theological claim for which evolutionary theory provides no framework. The problem here, again, is one of ambiguity. Why *exactly* does Nelson think this situation is problematic? There seem to be three basic claims he might be making: one is simply confused, one is false, and the other is a radical claim with extremely unpalatable consequences.

### Internal Consistency of Evolution

Let's deal with the confused possibility first. Nelson could be arguing that evolutionary explanations are *internally inconsistent*. As a science, evolution can not make theological claims, yet it must imply such claims in order to make sense of its arguments; therefore evolution is deeply illogical. There are two responses to offer here.

First, evolution is not making a theological hypothesis at all. It is making a perfectly naturalistic hypothesis, which happens to be *opposed* to some sorts of theological hypotheses. If this is all that Nelson means when he says that evolutionary theory is "theological," then evolution is guilty as charged. It is certainly true that endorsing evolution as an explanation of the natural universe is necessarily to "reject" many alternative, logically possible, explanations.[11] Whether scientists take the time to be explicit on this point or not, if one argues for some particular explanation in science, one is arguing that it is *superior* to any other available. Scientific endorsement of an hypothesis is an inherently *comparative* enterprise.

Second, even if evolution were making an explicitly theological hypothesis, there is nothing inconsistent about this. MN provides methodological *advice*, not prohibitions. If the evidence really seemed to favor the creationist hypothesis over natural alternatives (which it clearly does not), we might be justified in making a theological hypothesis.[12] So there is nothing logically inconsistent within the structure of evolutionary theory.

### Scientific Theology

The next claim Nelson might be making is that science is inappropriately arguing in favor of one theological hypothesis over another. Maybe science is trying to argue that God must be conceived of in the traditional way, as opposed to an alternative account like John Stuart Mill's hypothesis that God is not omnipotent. But this is exactly what science is *not* doing. It is therefore ironic that Nelson (p. 503 [687]) uses science's circumspection on religious matters against them when he says of alternative accounts of God: "The point however, is that we have no grounds *within evolutionary theory itself* to exclude Mill's creator, or any one of many conceivable creators whose natures allow imperfection."

If biologists routinely engaged in theological disputes about which specific concepts of God are preferable to others, they could justly be accused of straying from their area of expertise. However, they don't do this.[13] Therefore, the worst someone like Gould can be accused of is accepting (not defending) an incomplete hypothesis of God's nature as the one against which to pit evolutionary explanations. Yet the choice of something like Paley's design hypothesis as a pedagogical foil seems perfectly straightforward—since it's perfectly natural to pick the most common alternative view among the audience to which one is making a presentation.

Strangely enough, Nelson (p. 497 [681]) considers this possibility—stranger still is his reason for rejecting it: "This pragmatic justification collapses completely however, when one examines the actual context of many of the arguments. An encyclopedia article on the evidence for evolution might reasonably be expected to be a straightforward summary of the data; likewise, a textbook treatment of the same material. Why use a theological reductio in these contexts?"

Now, let's examine the argument Nelson is making here:

Premise 1: If the mention of a creationist hypothesis in evolutionary contexts were simply for pedagogical reasons, then one would expect the occurrence of the hypothesis to vary with the intended audience.
Premise 2: But the creationist hypothesis is constant across different audiences.

Therefore: It is not true that the mention of the creationist hypothesis in evolutionary theory is simply for pedagogical reasons. Something more fundamental is involved.

Students in introductory logic would (one hopes) recognize this as a valid *modus tollens* form of argument. Fine, but is it *sound*—are the premises actually *true*? Premise 1 seems fine, but premise 2 is false. In fact, the theological content of evolutionary presentations *does* vary with the makeup of the audience, precisely as expected. For example, it would be exceptional indeed to mention God's plans in a presentation on evolution to an audience of biologists at a professional conference. This is because the audience can be assumed to be biologically sophisticated and not very inclined toward creationist accounts. Theological comparisons are also rare, even for general presentations, in countries like Russia where few in the audience will ever even have heard of creationism.

So why is it so common for evolutionary presentations in *this* country to discuss creationist hypotheses explicitly, even in encyclopedias? The answer is incredibly obvious: because in this country, a general audience is certain to contain a large percentage of people (perhaps even a majority) who believe in creationism. Any American professor can tell you that it is a simple matter to elicit the most unsophisticated creationist accounts of the natural world from college students—indeed, even from biology majors in a senior-level biology course! Since it seems inconceivable that Nelson would be genuinely unaware of this basic fact about American intellectual life, he is guilty here of the basest sort of sophism.

## Let's Get Radical

Now let's consider the radical interpretation of Nelson's complaint. Perhaps he is saying that one can only have confidence in the truth of a hypothesis to the extent that one has *conclusively* refuted *every* alternative hypothesis. This odd sort of "Cartesian" claim would certainly cause great trouble for science if it were adopted. Science simply can't prove that God does not exist; in fact such a proof is impossible in principle under most conceptions of empirical induction. Since God can't be ruled out and could be any sort of mysterious God we choose to think up, no scientific hypothesis that relies on refuting theological claims could be established. What's more, since theological concepts are so very general, to endorse almost *any* natural explanation of the material universe is necessarily to reject some kind of theological belief. This is entirely unavoidable and is part of the reason science and theology have not always been easy bedfellows.

Wow! So much for evolution, since it is clearly in opposition to some theological claims that can't be conclusively refuted. So much for biology in general, for that matter—it could always be, for example, that the circulation of blood is caused not by the hydrodynamics and cardiovascular action we describe in our favorite theories but rather by the continual exertion of force from a divine being. "But that's so unlikely!" one might be tempted to retort. However, a determined defender of this position could argue, this judgement of likelihood is based on some very poorly defined sense of what the universe is expected to be. The bottom line is that it's certainly *possible*. So much for chemistry as well, since we *could* explain chemical bonding in terms of God's matchmaking. In fact, so much for science in general, as there will *always* be alternate theological hypotheses for any scientific one. Oh well, it was fun while it lasted . . .

In short, this kind of genie can't be let out of the bottle without a great deal of damage to other disciplines, perhaps every discipline. In particular, theology may fare no better than science. After all, if we are required to establish *theological* claims by rejecting all possible *scientific* ones, progress will be, to put it mildly, slowed considerably. Perhaps theology can substitute its own rules of reasoning, since it isn't a science (though as we discussed earlier, it's entirely mysterious and more than a little worrisome exactly what those rules might be). However, at the very least, all progress in any discipline that fancies itself a science will halt, never to resume until all theological disputes are settled. That's a rather stiff price to pay, I should think.

It's difficult to avoid the idea that Nelson has something like this radical critique in mind, though he doesn't seem aware of all the implications. It certainly might play well to an uncritical audience: on the one hand, if science *fails* to wade into the theological quagmire, it can be accused of "covering up" its theological assumptions; on the other hand, if science *does* enter the theological arena, it can be accused of violating its own principle of MN. Science seems boxed in. Of course, the way science *actually* proceeds is to compare each new hypothesis, not to every single possible alternative, but to the "leading" alternatives. What counts as a "leading hypothesis" may be determined in a number of ways (historical popularity, strength of empirical support, theoretical considerations, etc.), but science wisely never took on the task of decisively defeating every single challenger.

In light of this, Nelson's incredulity at the choice of theological hypothesis to which evolution compares itself is misplaced. Mill might be right when he speculates that God is not omnipotent.[14] However, since the heterodox view of God in many different traditions for thousands of years has been something like the omnipotent, omniscient, omnibenevolent God referred to in Paley, and since this is *still* the view espoused by *most* of those today who consider themselves religious, is it really surprising that Gould and others choose to discuss this conception rather than the "new" ideas put forward by the intellectual creationists?

## V.  The Appeal to Ignorance

Reading Nelson's paper, one gets the impression that a great many difficulties might be avoided if evolution simply admitted that a creationist God is possible. Very well, on behalf of evolution, I give him that admission: *It is logically possible that God (however we wish to interpret that term) exists and created the universe according to one's favorite creation account.* This of course implies that it's possible for everything evolutionists have ever said to be false. I admit that as well. How can I say this without giving up on evolution and opening the door to creationism?

Think of it this way: if the only criterion we employ to screen hypotheses is their bare possibility, we will cast a wide net indeed. For example, it's logically *possible* that the universe was designed by white lab rats in order to conduct excruciatingly subtle experiments on human behavior (see Adams 1980). It's *possible* that chemical bonding is caused by sexual attraction on the part of the constituent atoms rather than anything having to do with electrical charge, etc., etc. If we accept the mere possibility of an alternative explanation as sufficient grounds to abandon an hypothesis, we will never commit to any hypothesis whatsoever, because the alternatives to be ruled out are limited only by our imaginations. Surely, the proper approach is to view the mere logical possibility of an alternative hypothesis as a *necessary* but not *sufficient* condition for its acceptance. Put another way, logical possibility is the *minimum* standard any hypothesis must meet, but hypotheses could certainly meet this minimum standard without being very *good*. To confuse the *mere possibility* of a proposition with *evidence that it is true* is such a bad style of reasoning, in fact, that logic has a special term for it: it's the fallacy of *appeal to ignorance*.

A careful scientist should admit, therefore, that God may exist (in whatever form). He is under no pressure to prove this hypothesis false, which is fortunate because such a task would be impossible. Rather, all the scientist needs to do is to establish that the positive evidence for evolution seems to exceed the positive evidence for other possible explanations like creationism—that, based on the evidence we have at the moment, evolution seems more likely to be true. The question, therefore, is not, "Is the creationist hypothesis *possible*?" but rather, "Is there good *reason* to believe the creationist hypothesis is more likely to be true than the evolution hypothesis?"

If creationism is to ever accumulate positive evidence, then it must abandon the cloak of ambiguity and take a particular position with particular hypotheses (which implies an RG). When it does this, it can be held accountable for failure to provide an explanation that should be available. Every time creationism has done this in the past, evolutionary hypotheses have won by the overwhelming weight of the evidence. As Darwin puts it: "We can plainly see that nature is prodigal in variety, though niggard in innovation. But why this should be a law of nature if each species has been independently created, no man can explain" (1964, p. 471).

It's not a question of whether such a pattern is possible, for it clearly is. Rather, it's a question of which hypothesis explains the pattern better. The answer almost every scientist who has studied the question since Darwin has given is that the evidence seems very strongly (but, of course, not conclusively) in favor of evolution by natural selection. Evolution is a bloody good theory, but it's always *possible* (however low the estimate of the associated probability) that it might be wrong. We might wish that things were more certain, but they aren't. Fortunately, in spite of the uncertainty, science works very well.

## VI.   On Imperfection and Optimality

### Imperfection Arguments in General

There are two remaining points I'd like to make concerning Nelson's treatment of the imperfection argument before discussing the related concept of optimality. First is a very obvious point, but one often overlooked in all the logical maneuvering. Even if we were to admit that any particular instance of imperfection put forward in the past (e.g., the panda's thumb) is poorly supported, this does little to undermine evolution (Eldredge and Cracraft 1980). The reason is that, as Nelson freely admits, there are any number of examples of imperfection that are *clearly inconsistent* with the standard creationism accounts.[15] Therefore, Nelson's argument against the panda's thumb, even if it went through without a hitch, would be telling only to the extent that it represents a *general* problem in evolution. He implies that it does, yet he never addresses the relative frequency of the two kinds of cases. This is a curious omission, since the mere fact of a mismatch between any one example, even one as popular as the Panda's thumb, and evolutionary theory is largely irrelevant unless the generality of the problem is considered.

Second, Nelson accuses Gould of offering a false alternative in the first premise of the imperfection argument:

If *p* is an instance of organic design, then *p* was introduced either by a wise creator, or by descent with modification.

One fork of this dilemma concerns the various theories of God that might be employed in cashing out Gould's "wise God" phrase,[16] but these have already been addressed. The other fork concerns the fact that Gould overlooked alternative biological accounts such as Lamarkianism and "dynamic creationism." Leaving the discussion of dynamic creationism for the next section, it is puzzling that Nelson accuses Gould of being ignorant of Lamarck. In fact, it seems simply laughable, since Gould is one of the relatively few biologists to take the history of biology seriously.

So, if Gould knew Lamarckianism was a possibility, why did he not cite it explicitly? One obvious response is that we must be able to take some things for granted in order to discuss anything at all without resorting to a Ph.D. dissertation. A popular reading of Lamarck says that traits can be altered through use and disuse, then passed on to progeny. Since this has been proven false, it's no more surprising that Gould fails to discuss this possibility than that he fails to discuss the possibility that heredity is accounted for by pangenetic gemmules (as Darwin famously, and falsely, believed). On a more charitable reading of Lamarck, his theory is simply one variant of the descent with modification theme. Nelson's attack here illustrates a consistent failure of the creationists: the inability to distinguish debate over the *mechanisms* of evolution from debate over the *occurrence* of evolution.[17] The question creationists want to take up largely concerns whether or not evolution occurred, not what mechanism was employed (provided, of course, we are talking about a natural mechanism). What consolation would it be to the creationist if evolution turned out to be the result of a Lamarkian mechanism? There would still be heritable variation, competition, and natural selection, with the same pattern of common ancestry. Indeed, as Sober argues (1993), Darwin proposes two entirely separate hypotheses: *Darwinian natural selection* and *the tree of life*, the former being only one possible mechanism to explain the latter. Darwin himself got the mechanism of inheritance wrong, but it doesn't make his general theory one whit less convincing.

## Optimality and the RG Assumption

Nelson's strategy for defusing optimality arguments is the same as the one employed with imperfection arguments—namely, to show that they are inextricably tangled in theological difficulties. Thus, "the difficulty of determining whether a state of a trait or organism is optimal is *magnified immeasurably by the theological context*" (p. 504 [689]).[18] In order to label a given trait as "best," it is first necessary to establish some range of possible options over which comparison will be made.[19] However, Nelson argues, ultimately we cannot know the range of *possible* traits unless we know a great deal about God's nature, and scientists are barred from speculation on this point by their own methodology. Therefore, we simply cannot reason about matters of optimality without making questionable theological assumptions.

Strictly speaking, Nelson is right about this. The fact of the matter is that science *does* tend to *assume* that the natural universe is explicable via human reason (i.e., that God, if He exists, is an RG). But just because this assumption cannot be conclusively established does not mean there aren't good reasons for it. In fact, there are at least two very good reasons. First, without such an assumption, *any* investigation (not just in biology) into the natural universe will be, at best, extremely difficult and, at worst, utterly impossible. Since scientists want to explain the natural universe, it is entirely understandable they make assumptions that, if true, explain how this is possible.[20] Of course, this is a pragmatic justification and might be mistaken—an MG could still have made the universe along creationist lines.

It might be profitable to reexamine the difficulties facing the alternative assumption of an MG, since scientific evaluation is essentially comparative. First, an MG-creationism will have to justify itself in the face of the extreme difficulties it poses for scientific explanation. In light of the overwhelming success of science, this will be an *extremely* heavy burden to carry. Second, an MG is every bit as much of an assumption as an RG.[21] To label the scientific position that the universe is intelligible an "assumption" or a "bare postulate" is to imply something that is simply not the case—namely, that the alternative one is proposing does *not* suffer from this same difficulty.

## RG and Optimality

As long as we are forced to seriously consider the possibility of appeal to an MG as an explanation of natural events, we will not make any progress in science whatsoever (or in creationism, for that matter). Therefore, science must assume that God, if He exists, is an RG. *Given that assumption*, we can ask: "Do we have good grounds for our evaluations of the optimality of a trait?"

Nelson is quite right that we often make such claims without sufficient evidence.[22] Aesthetic judgments are not the kinds of things about which there's agreement among *humans*, for example, so extrapolating from a human judgment about a matter of taste to divine sensibilities is clearly inappropriate. It is equally true that many designs that appear poor on first inspection actually have advantages revealed only after prolonged study.

So, should we even discuss biological optimality? Yes. For one thing, there are models of optimality that are much more highly developed and precise than the sort of morphological example (Panda's thumb) Nelson discusses. For example, we can very precisely calculate the optimal sex ratio for a particular organism, given relatively accessible facts about the population in which it lives. Even in situations where our knowledge of the relevant variables is not as precise, the use of optimality models can still provide very useful *estimates* of optimality. Moreover, such estimates will suffice perfectly well for deciding between creationist and evolutionary explanations. Consider the following rule of thumb:

If God is an RG, then He could reasonably be expected to produce designs *at least as good* as those a human engineer could produce.

This does not necessarily mean that a design that appears to be less than humanly possible is inconsistent with an RG, of course. However, the burden of proof would at least shift to the RG-creationist, who must *explain why* his expectations were not met. Now, Nelson makes much of the fact that even the most obvious numerical analysis of optimality in terms of the ratio of actual to potential optimality is impossible in principle. This is because one cannot solve an equation with two variables (one for actual utility, one for highest possible utility), and one cannot make numerical predictions about the range of potential utility without venturing into theology (e.g., speculating on the limits of God's power, etc.). In light of the RG assumption, is this really a problem?

It's perfectly feasible (even common) in such situations to determine, not the precise value for the two variable expression, but rather a maximum or minimum value. This at least allows us to compare data to the limiting case and, sometimes, draw conclusions in spite of our uncertainty. On the very reasonable assumption that an RG would be at least as competent as a human designer, we can plug in the optimality figure for the best design a *human* could achieve as the *minimum* optimality we can reasonably expect of God's designs. The possible optimality may well be higher than this minimum expectation—indeed, we would expect this to be the case quite often. However, if we find an optimality *lower* than this minimum, it constitutes a clear prima facie problem for the RG-creationist. This difficulty may not, of course, be inescapable, but the creationist is now obliged to explain himself on this point.

Put this way, it is not as difficult as Nelson thinks to get some handle on expected optimality. We look at the natural universe and simply ask, "Is this design better or worse than what a human engineer could achieve?" If a natural design is no worse than a human design, the evidence doesn't cut either way. However, if a design in nature is clearly *inferior* to what a human engineer could produce, then we are entitled to request an explanation of this deviation from the RG-creationist prediction. Refusal to provide such an explanation takes us right back to the appeal to ignorance.

There are many examples that do pose such problems for RG-creationism. Consider the bizarre fact that, in mammals, the recurrent laryngeal nerve does not travel straight from the cranium to the larynx. Rather, it travels down the neck to the chest, where it loops around a pulmonary ligament and then travels back up the neck to the larynx. In animals like the giraffe, this can mean a twenty-foot length of nerve where twelve inches or so would suffice. Is this a good design?

A human engineer would be hard pressed to come up with *any* explanation for such a structure in terms of defensible design principles. At a minimum, the nerve performs its duties in a demonstrably inefficient manner when an obvious (and cheaper) alternative design, using the materials at hand, could do much better. This case is therefore problematic for the RG-creationist. Of course, as always, we *could* have missed something. It might be that there is some basic aspect of neurophysiology that we understand imperfectly, or that this is an instance of local imperfection that is somehow required for God's realization of global perfection. As we have established, however, the fact that these are *possible* should not be confused with evidence that they are *true*.

So it's *possible* there's an adequate creationist explanation for this example, but we might be forgiven if we don't hold our breath waiting for one. RG-creationism doesn't seem able to explain this sort of example. On the other hand, this weird assembly makes perfect sense if we explain it from the evolutionary perspective. The reason the recurrent laryngeal nerve takes this odd path, says evolution, is because it is descended from a vagus nerve branch that passed through an arterial arch to enervate the gills in the giraffe's ancestor, the fish.[23] The sixth arterial arch around which the vagus nerve branch traced its original development in the gills became modified into the pulmonary ligament, while the nerve itself was diverted to the larynx. In other words, the nerve was part of a developmental system that, for whatever reason, was not changed in the course of evolution—even as the relative position and function of the players in the system changed radically. Thus the nerve must still develop in the proximity of the pulmonary ligament (the remnant of the sixth arterial arch) because that's what it was "programmed" to do so many millions of years ago when this was still adaptive (Strickberger 1996, p. 47).

So, given these two facts—(1) that the structure seems, on non-arbitrary grounds, a very poor design for its intended purpose—far below the minimum expectation for an RG, and (2) evolution provides a perfectly sensible explanation in terms of historical constraints—we are perfectly justified in asserting that this example is highly problematic for the RG-creationist account. It is now up to the RG-creationist to attempt an explanation of this pattern in some other fashion.

## VII.   On Dynamic Creationism

Assuming one can get the creationist to endorse an RG and abandon the cloak of ambiguity, it might seem a simple matter to test the creationist hypotheses. When scientists do this, however, Nelson still has one final defense. He makes the surprising claim that evolutionists are arguing with a historically outmoded straw man when they debate, as they usually do, the theory of *static* creationism (fixity of species). He asserts that "few if any creationists defend that view" (p. 500 [685]).

As with his protestations concerning the "surprising" theological content of evolutionary presentations, on this point as well it is very difficult to be charitable to Nelson. Even if we grant that dynamic creationism is the current darling of those who publish on creationism (which I suspect is not the whole story), the position espoused by the vast majority of those who believe in creationism is closer to the "old-fashioned" static view than the dynamic theories Nelson discusses. Moreover, it is the height of irony that creationism, having adopted the basic mechanisms of evolutionary theory only after 150 years of vigorous protest, then claims to have no association with its past worthy of critique! The spectacular success of evolutionary theory finally forced a change in creationism's most basic tenets, but this is not to be used as evidence of the two theories' relative fruitfulness!

However, on the principle of charity, I will put history aside and give the position the benefit of the doubt, treating dynamic creationism to its own separate critique. Nelson cites two examples of dynamic creation theory with evident approval. In one, Frair and Davis (1983) take up Futuyma's (1983) question as to whether blind cave animals show evidence of wise design. They explain that cave animals with vestigial eyes are the result of mutations that, because they "are small enough to permit survival and reproduction," manage to accumulate over time. However, such structures are "at best indicative of change within limits" (p. 500 [685]). The second example involves the rudimentary wings of some flightless birds. Murris (1986) explains that a recessive "flightless gene" could propagate through a population of birds in circumstances, as on a deserted island, where "they have no natural enemies."

First, note that neither example addresses the question posed by Futuyma. At best, they explain *how God could*, not *why God would* design such structures. In other words, the creationist is simply pointing out that such structures are consistent with natural mechanisms. What a surprise! This is a claim any good evolutionist would endorse as a matter of course, particularly since the mechanisms in question seem to be lifted straight out of evolutionary theory. It certainly is *not* evidence that creationist explanations are in fact *true*.

Second, as was pointed out earlier, these sorts of cases need not establish a *general* problem, as indeed they do not. Even if we were to allow that the creationist account of vestigial traits is just as good as the evolutionary account (which seems extremely dubious), this is certainly *not* the case for imperfect traits in general, which is a much more inclusive set of data. Evolution can explain all sorts of examples here that the creationist cannot—like the giraffe's laryngeal nerve. So the inference Nelson intends—that creationism and evolution stand on the same evidentiary basis with respect to imperfection arguments—is false.[24]

Third, to the extent the vestigial traits in question are being explained at all, they are being explained by standard *evolutionary* mechanisms like drift and selection. These are clearly not concepts developed independently by creationists to bolster their case. Of course, as long as we restrict discussion to the points where creationism has borrowed from evolution, we will find little but agreement. Even here, however, there is one point on which evolution and creationism might disagree. The dynamic creationism theory shows that vestigial traits are *consistent* with natural processes, but it provides no reason to *expect* them. Evolution, on the other hand, can argue that selection may actively *favor* these traits under the right conditions, and thus that they are to be *expected*. All else being equal, one should prefer a theory that renders the pattern *likely* to one that merely establishes its *possibility*.

Perhaps dynamic creationists mean to admit natural selection as well.[25] I was actually quite surprised to see a creationist admitting the truth of mutation and drift, so why not natural selection as well? Fine. At this point, however, we need to highlight the differences in the predictions made by the two theories. If dynamic theory encompasses fitness differentials, mutation, population genetics, and drift, then what exactly *distinguishes* it from evolution in terms of its specific predictions? The answer, of course, is the concept of *bounded change*—evolution argues that there is no inherent limit to the changes wrought by these mechanisms, while creationists argue that change is always "bounded".[26] The problem is that creationism, again, does not offer any *evidence* for such bounds, nor does Nelson even discuss them as salient in this way.

Consider, however, the implications of adopting (as Murris clearly does) population genetic methods. Population genetics predicts that a given type of gene (an allele) can take over a population entirely (go to fixation) under the right circumstances—the situation Murris describes is just such a case. Further, we know that a great many traits, perhaps even most, are influenced by a large number of genes simultaneously (polygeny). It seems a clear implication, therefore, that any number of genes could be fixated in succession over a long period of time, producing as much change as could possibly be desired.[27]

If dynamic creationists are to adopt the basic concepts of population genetics while rejecting its conclusions, they need to offer a good reason for this. Such a move seems justified only if we have evidence for one of two basic possibilities:

(1) Some essential ingredient in the population genetic formulae, which is present for initial changes, is no longer present for changes beyond some limit. Population genetics still applies, but no longer predicts change after some critical threshold has been passed.

(2) The components of evolution by natural selection remain in place, yet change is prevented by some outside force (e.g., God's will) at some threshold. Population genetics has been rendered ineffective by an outside force.

What's interesting to note here is that both of these possibilities are perfectly *testable*. To test the first, we could simply assay island populations of birds for features such as genetic variance, fitness differentials, and the like. Not only are there well-worked-out methods for doing this, but an excellent and comprehensive study is currently underway for populations of finches in the Galapagos (Grant and Grant 1995, 1997). A dynamic creationist might predict that some essential variable becomes "subcritical" as selection produces more and more morphological change. At present, however, there doesn't seem to be any such pattern, which is prima facie evidence against the dynamic creationist hypothesis. The alternative creationist claim, that population genetics is thwarted by an outside force, is subject to at least an indirect test. If we were to routinely encounter populations that did meet the population genetic criteria for evolution by natural selection, yet where there was no evidence of such change for long periods of time,[28] then that would be prima facie evidence against the evolutionary story (and thus, indirectly, for the dynamical creationist theory). To the extent that we have made a general survey of such situations and failed to encounter this pattern, this seems unlikely to be the case.

Despite painting a gloomy picture of their prospects, I welcome careful scientific investigations of these sorts of situations by any creationist with an empirical bent. If the creationists who back dynamic theory really want to make the point that creationism is a testable science, then they should fund the sorts of investigations I outline above. Evolution has embarked on many investigations of its own and found a wealth of data consistent with evolutionary theory, yet inconsistent with dynamic creationism. Thus it seems perfectly sensible for biology to withhold credence from dynamic creationism until such time as evidence in its favor can be produced and vetted. Of course, as I have consistently been at pains to emphasize, the lack of creationist evidence need not imply that creationism is wrong, just that the evidence is currently (very much) in evolution's favor. It is simply not true, despite Nelson's arguments, that the arguments for evolution work only when applied to static creation.

## VIII.   On the Homology Argument

### Pedagogical Reflections
The basic logic of the homology argument as Nelson presents it is not really so different from the imperfection argument. We can interpret the widespread occurrence of common design elements across taxa (e.g., the pentadactyl limb structures) as evidence either of descent with modification or of some divine design conservatism. To some extent, which we prefer hinges on how strong the evidence is for *imperfect* common

designs, since these would tend to cast doubt on the divine design hypothesis. However, asks Nelson, who are we to analyze the optimality of God's plan, having such limited information?

First, however, I must say a word about the pedagogical style in this section, which seems a particularly egregious example of creationist hypocrisy. Nelson begins by telling us he will engage in "a brief historical excursus" to illuminate the homology argument. Then, he details the arguments employed by Darwin and Owen *for four pages*. This emphasis leaves the clear implication that nothing new has been done in the intervening century and a half to show that traits may be designed imperfectly. Thus there is nothing new to support the argument that homology is best explained via common descent rather than creation. Even if we allowed without argument that he's right about Darwin and Owen (which is hardly the case), this is nothing more than an attack on an "outmoded" position—exactly the sort of complaint *he himself* lodges against evolution's analysis of the static view of creation. In fact, Nelson has far less justification for his own historical indulgences, since virtually everyone, even in the general population, has *long* since abandoned concepts like Owen's archetypes.

Yet never does he consider any evidence from the field of biomechanics, which is dedicated to just the kind of enterprise he says is lacking in Darwin and Owen—scientific analysis of trait function and design (see, e.g., Vogel 1988 and Wainwright 1988). Nelson responds to Ayala's (1988) claim that an engineer could have improved on the design of the pentadactyl limb with the flat statement that "*there is no evidence this is true*" (p. 510 [696]). Given his temporal lapse, how can he possibly justify such a claim? At the very least, cases like the giraffe's laryngeal nerve seem to throw the ball squarely back into the creationists' court.

### Statistical Parsimony

There is also a statistical sort of argument that could be made as to why evolutionary explanation of common "design" is superior to the possibility that God favors common elements of design. Suppose we accept Nelson's claim that we should make no stipulations whatsoever about God's powers or motives. However, we think it desirable to proceed with some attempt to formulate scientific explanations (on pain of an end to science). We must ask the question, "What exactly is the null hypothesis concerning the expected distribution of 'design elements' across the taxa?"

Well, if we make no assumptions about God's motives or powers, then the obvious null hypothesis is that the distribution of design elements is expected to be *random*. This is, of course, perfectly testable. Even the most superficial analysis of the data will quickly reveal that morphologies are distributed in anything but a random array. It is still possible, of course, that even such a clear pattern results from a random process. However, since the probability of such an occurrence in this case is vanishingly low, we have excellent grounds to seek an explanation that makes the observed pattern an *expected* outcome—evolution's theory of common descent does precisely this. Moreover, the evolutionary explanation is consistent with a wide array of other theories inside and outside of biology and explains a wide variety of data, not just this instance of pattern.

Of course, creationism is welcome to provide an alternative explanation. However, in order to explain the pattern, creationists will have to offer some *specific* hypothesis—perhaps that God is not omnipotent (à la Mill), or perhaps that this "microimperfection" is a piece of a larger system that is "macroperfect." Both of these are possibilities Nelson mentions, but he fails to commit to either in any clear way. If Nelson ever were to commit to any specific theological position, then he would immediately open a theological can of worms with which he would have to deal. And, even if Nelson were to commit to a particular theological position and could show that it is in fact *consistent* with the data, such consistency is merely a *minimum* requirement for a scientific hypothesis. A good hypothesis is one that is not only consistent with the available evidence, but that offers novel testable hypotheses. Evolutionary theory clearly does this, but creationism either says nothing specific (à la Nelson) or makes specific claims that are falsified.

Of course, we always have the option of declaring the search for any explanation fruitless, as we might if we have a prior belief in God's mysterious ways, for example. However, since this approach will mean an end to scientific inquiry altogether, the cure seems much worse than the disease.

## IX.  Concluding Remarks

It seems that the situation is something like the following:

(1) An MG provides no true comfort to the creationist, since He would either undermine the creationists' case by robbing them of the justification for their own creationist account, or make scientific reasoning of any kind impossible.

(2) There is some (albeit imperfect and fallible) evidence that seems inconsistent with *any* version of RG-creationism (e.g., giraffe nerves, panda's thumb). Moreover, *specific* hypotheses concerning RG-creationism will have to deal with all sorts of major difficulties, both theological and biological. At best, these render creationism deeply problematic and, at worst, they falsify creationist theory altogether.

(3) There is no evidence *independently* favoring RG-creationism. While *some* of the evidence supporting evolution is consistent with *some* versions of RG-creationism, this is simply not a reason to believe that creationism is *true*.

(4) Evolutionary explanation is supported in ways RG-creationism is not (more general explanations, expectations of observed patterns, etc.).

The bottom line is this: the mere fact of consistency with data is clearly not adequate grounds for belief—the appeal to ignorance is considered fallacious for good reason. This approach is particularly pernicious when combined with a willingness to hide behind a cloak of ambiguity. The combination makes the creationist critique of evolution seem far stronger than it really is, and it exposes its proponent as someone unconcerned with furthering human knowledge.

## Acknowledgments

This paper has benefited from the insightful comments of many people, but most especially Robert Pennock, Elliot Sober and Eugenie Scott. Any "imperfections" that remain are due entirely to my own efforts.

## Notes

1. Hence the seeming strength of the cloak of ambiguity.

2. Indeed, many theologians have rejected the MG on the grounds that it threatens to make aspects of heterodox Judeo-Christian-Islamic thinking nothing more than elaborate fiction.

3. This is particularly true given his endorsement of the evolution-like "dynamic creationism" (see section VII).

4. I certainly do not mean to take a position on this issue. Unless MG theorists attempt to offer hypotheses concerning the natural universe, their view is perfectly compatible with science in general and evolution in particular.

5. And this is clearly an assumption of sorts—at least in the sense that a proponent of an RG could never definitively refute the alternative possibility of an MG (see section V).

6. For example, a flood of sufficient magnitude to have deposited all the fossil strata known to exist would require a volume of water far greater than has ever existed on Earth (Natl. Acad. of Sciences 1984). We could always be mistaken, but given what we know, we're justified in holding that the claim is falsified.

7. Being human, individual *scientists* may occasionally make this mistake, but this is generally recognized as inappropriate.

8. See Pennock 1999, chapter 4 for a detailed discussion of these and similar points.

9. Individual scientists, of course, have their own personal views. In fact, polls tend to show that a majority of scientists do believe in God.

10. A similar schema is also used by Shanahan (1997).

11. This is "rejection" only in the sense that one considers the alternatives *less likely* to be true, not logically *impossible*.

12. Of course, given their methodological commitments, scientists would accept such a hypothesis only if it were suitably *empirical*. For the sake of argument, suppose that (1) there were a detailed theological theory that made a number of highly specific predictions about the future, (2) these predictions all came true, and (3) there seemed no possibility of a more natural explanation. Then scientists would be justified in further investigation of the theory. We could debate the *likelihood* of such an event, of course, but there seems no reason to reject this possibility in principle.

13. Except in the sense that they assume an RG (see sections III and VI).

14. Of course, there's no reason to stop here—God might be an inanimate atom that just so happens to have some peculiar properties like necessary existence.

15. Nelson admits it, but it's all too common to see denials of this in the literature. For example, Frair and Davis (1983) claim that *most* vestigial traits are degenerative changes within species, which is clearly false.

16. Actually, it seems fairly clear what Gould has in mind—that this is not the *only* interpretation (Nelson's argument) is another point entirely.

17. Interestingly, the tendency to conflate debate over details with debate over a general conclusion occurs in other contexts as well. Holocaust revisionists, for example, sometimes try to show that there is less consensus in the historical community concerning the *occurrence* of the holocaust by pointing out that historians disagree over the exact *number* of jews killed, etc. (see Tolson 2000).

18. As discussed in section IV, it's precisely the *immeasurable* scope of this complication that makes it unacceptable, not just for evolution, but for science in general.

19. Note that it is also necessary to specify precisely what a trait is best *for* (gene, organism, group, etc.)—a problem into which creationism offers no insights whatsoever.

20. This is another issue that is far too complex for this paper to explore in detail. Suffice it to say that there has been a great deal of thought concerning issues like the principle of sufficient reason (e.g., Spinoza, Leibniz) and transcendental assumptions (e.g., Kant). Scientists too have endorsed principles like this, as with Einstein's famous rejection of the Copenhagen interpretation of quantum mechanics.

21. In fact, depending on how one views the process of induction and confirmation, it could be argued that there is at least some evidence for the scientific assumption, but could never be (by definition) for the alternative MG assumption.

22. Ironically, Gould and Lewontin (1979) make this exact point forcefully in their famous critique of the adaptationist program.

23. Note that Nelson cannot claim here that the nerve served some adaptive function in a *recent* ancestor. I doubt even the most liberal of dynamic creation theories would be willing to accept the transformation of fish to giraffe.

24. And, in any event, imperfection arguments are not the only evidence biology can adduce for evolution (despite Nelson's intimation to that effect). Gould may have been right that evolution would never have been accepted (when it was) without imperfect traits, but science has managed to accumulate a great deal of additional evidence in the 150 years since Darwin's original work. Indeed, Darwin himself goes to great lengths to point out other evidence in 1859.

25. Whether they actually use the "N word" (natural selection) or not, Frair and Davis's discussion of morphological changes "small enough to permit survival and reproduction" and Murris's reference to the lack of natural predators is clearly stipulating something about the selective environment; thus they must have some concept of differential fitness.

26. Exactly *how* bounded change must be is, of course, extremely difficult to pin down. It is usually expressed by saying that variants within a species are allowed, but transspecific change is not.

27. There are often genes with very large effects that might produce such change more rapidly. Since the rate of change is not critical here, I will not deal with this complication.

28. There are of course populations that remain relatively unchanged for long periods of time, but there have always been plausible evolutionary explanations for why this would occur.

## References

Adams, D. 1980. *The Hitchhiker's Guide to the Galaxy*. Harmony Books, New York.

Ayala, F. 1988. The Theory of Evolution. In *Encyclopedia Britannica*, 15th ed., Britannica, Chicago.

Burian, R. M. 1986. Why the Panda's Thumb Provides no Comfort to the Creationist. *Philosophica* 37 (1): 11–26.

Darwin, C. 1964. *On the Origin of Species* (facsimile of the first edition), Harvard University Press, Cambridge.

Eldredge, N. and J. Cracraft. 1980. *Phylogenetic Patterns and the Evolutionary Process: Method and Theory in Comparative Biology.* Columbia University Press, New York.

Frair, W. and P. Davis. 1983. *A Case for Creation.* Moody Press, Chicago.

Futuyma, D. 1983. *Science on Trial.* Pantheon, New York.

Gordon, P. 1984. The Panda's Thumb Revisited. *Origins Research* 7 (1): 12–14.

Gould, S. J. and R. C. Lewontin. 1979. The Spandrels of San Marco and the Panglossian Paradigm: a critique of the adaptationist program. *Proceedings of the Royal Society of London* B205: 581–598.

Grant, P. R. and R. B. Grant. 1995. The Founding of a New Population of Darwin's Finches. *Evolution* 49 (2): 229ff.

———. 1997. Genetics and the Origin of New Bird Species. *Proceedings of the National Academy of Sciences* 94 (15): 7768ff.

Johnson, P. E. 1993. *Darwin on Trial*, second ed. Regnery Gateway, Washington, D.C.

National Academy of Sciences, Committee on Science and Creationism. 1984. *Science and Creationism: A view from the national Academy of Sciences.* National Academy Press, Washington, D.C.

Nelson, P. A. 1996. The Role of Theological Reasoning in Current Evolutionary Theory. *Biology & Philosophy* 11 (4): 493–517. Reprinted in this volume, chap. 32.

Murris, H. 1986. The Concept of the Species and its Formation. In *Concepts in Creationism*, E. H. Andrews, W. Gitt, and W. J. Ouweneel (eds.). Evangelical Press, Herts, England, pp. 175–207.

Pennock, R. T. 1999. *Tower of Babel: The Evidence against the New Creationism.* The MIT Press, Cambridge.

Shanahan, T. 1997. Darwinian Naturalism, Theism, and Biological Design. *Perspectives on Science and Christian Faith* 49 (3): 170–178.

Sober, E. 1993. *Philosophy of Biology.* Westview, Boulder.

Strickberger, M. W. 1996. *Evolution.* Jones & Bartlett, Boston.

Tolson, Jay 2000. New Debates over Old Horrors: The Holocaust and the Rewriting of History. *U.S. News and World Report*, February 14, 2000, 44–45.

Vogel, S. 1988. *Life's Devices: The Physical World of Animals and Plants.* Princeton University Press, Princeton.

Wainwright, S. A. 1988. *Axis and Circumference.* Harvard University Press, Cambridge.

# 34

## Nonoverlapping Magisteria

Stephen Jay Gould

Incongruous places often inspire anomalous stories. In early 1984, I spent several nights at the Vatican housed in a hotel built for itinerant priests. While pondering over such puzzling issues as the intended function of the bidets in each bathroom, and hungering for something other than plum jam on my breakfast rolls (why did the basket only contain hundreds of identical plum packets and not a one of, say, strawberry?), I encountered yet another among the innumerable issues of contrasting cultures that can make life so interesting. Our crowd (present in Rome for a meeting on nuclear winter sponsored by the Pontifical Academy of Sciences) shared the hotel with a group of French and Italian Jesuit priests who were also professional scientists.

At lunch, the priests called me over to their table to pose a problem that had been troubling them. What, they wanted to know, was going on in America with all this talk about "scientific creationism"? One asked me: "Is evolution really in some kind of trouble; and if so, what could such trouble be? I have always been taught that no doctrinal conflict exists between evolution and Catholic faith, and the evidence for evolution seems both entirely satisfactory and utterly overwhelming. Have I missed something?"

A lively pastiche of French, Italian, and English conversation then ensued for half an hour or so, but the priests all seemed reassured by my general answer: Evolution has encountered no intellectual trouble; no new arguments have been offered. Creationism is a homegrown

Originally published in *Natural History* (1997, March). Reprinted with permission from the author and *Natural History*. © 1997 The American Museum of Natural History.

phenomenon of American sociocultural history—a splinter movement (unfortunately rather more of a beam these days) of Protestant fundamentalists who believe that every word of the Bible must be literally true, whatever such a claim might mean. We all left satisfied, but I certainly felt bemused by the anomaly of my role as a Jewish agnostic, trying to reassure a group of Catholic priests that evolution remained both true and entirely consistent with religious belief.

Another story in the same mold: I am often asked whether I ever encounter creationism as a live issue among my Harvard undergraduate students. I reply that only once, in nearly thirty years of teaching, did I experience such an incident. A very sincere and serious freshman student came to my office hours with the following question that had clearly been troubling him deeply: "I am a devout Christian and have never had any reason to doubt evolution, an idea that seems both exciting and particularly well documented. But my roommate, a proselytizing Evangelical, has been insisting with enormous vigor that I cannot be both a real Christian and an evolutionist. So tell me, can a person believe both in God and evolution?" Again, I gulped hard, did my intellectual duty, and reassured him that evolution was both true and entirely compatible with Christian belief—a position I hold sincerely, but still an odd situation for a Jewish agnostic.

These two stories illustrate a cardinal point, frequently unrecognized but absolutely central to any understanding of the status and impact of the politically potent, fundamentalist doctrine known by its self-proclaimed oxymoron as "scientific creationism"—the claim that the Bible is literally true, that all organisms were created during six days of twenty-four hours, that the earth is only a few thousand years old, and that evolution must therefore be false. Creationism does not pit science against religion (as my opening stories indicate), for no such conflict exists. Creationism does not raise any unsettled intellectual issues about the nature of biology or the history of life. Creationism is a local and parochial movement, powerful only in the United States among Western nations, and prevalent only among the few sectors of American Protestantism that choose to read the Bible as an inerrant document, literally true in every jot and tittle.

I do not doubt that one could find an occasional nun who would prefer to teach creationism in her parochial school biology class, or an occasional orthodox rabbi who does the same in his yeshiva, but creationism based on biblical literalism makes little sense in either Catholicism or Judaism, for neither religion maintains any extensive tradition for reading the Bible as literal truth rather than illuminating literature, based partly on metaphor and allegory (essential components of all good writing) and demanding interpretation for proper understanding. Most Protestant groups, of course, take the same position—the fundamentalist fringe notwithstanding.

The position that I have just outlined by personal stories and general statements represents the standard attitude of all major Western religions (and of Western science) today. (I cannot, through ignorance, speak of Eastern religions, although I suspect that the same position would prevail in most cases.) The lack of conflict between science and religion arises from a lack of overlap between their respective domains of professional expertise—science in the empirical constitution of the universe, and religion in the search for proper ethical values and the spiritual meaning of our lives. The attainment of wisdom in a full life requires extensive attention to both domains—for a great book tells us that the truth can make us free and that we will live in optimal harmony with our fellows when we learn to do justly, love mercy, and walk humbly.

In the context of this standard position, I was enormously puzzled by a statement issued by Pope John Paul II on October 22, 1996, to the Pontifical Academy of Sciences, the same body that had sponsored my earlier trip to the Vatican. In this document, entitled "Truth Cannot Contradict Truth," the pope defended both the evidence for evolution and the consistency of the theory with Catholic religious doctrine. Newspapers throughout the world responded with front-page headlines, as in the *New York Times* for October 25: "Pope Bolsters Church's Support for Scientific View of Evolution."

Now I know about "slow news days," and I do admit that nothing else was strongly competing for headlines at that particular moment. (The *Times* could muster nothing more exciting for a lead story than Ross Perot's refusal to take Bob Dole's advice and quit the presidential race.) Still, I couldn't help feeling immensely puzzled by all the attention paid to the pope's statement (while being wryly pleased, of course, for we need all the good press we can get, especially from respected outside sources). The Catholic Church had never opposed evolution and had no reason to do so. Why had the pope issued such a statement at all? And why had the press responded with an orgy of worldwide, front-page coverage?

I could only conclude at first, and wrongly as I soon learned, that journalists throughout the world must deeply misunderstand the relationship between science and religion, and must therefore be elevating a minor papal comment to unwarranted notice. Perhaps most people really do think that a war exists between science and religion, and that (to cite a particularly newsworthy case) evolution must be intrinsically opposed to Christianity. In such a context, a papal admission of evolution's legitimate status might be regarded as major news indeed—a sort of modern equivalent for a story that never happened, but would have made the biggest journalistic splash of 1640: Pope Urban VIII releases his most famous prisoner from house arrest and humbly apologizes, "Sorry, Signor Galileo ... the sun, er, is central."

But I then discovered that the prominent coverage of papal satisfaction with evolution had not been an error of non-Catholic Anglophone journalists. The Vatican itself had issued the statement as a major news release. And Italian newspapers had featured, if anything, even bigger headlines and longer stories. The conservative *Il Giornale*, for example, shouted from its masthead: "Pope Says We May Descend from Monkeys."

Clearly, I was out to lunch. Something novel or surprising must lurk within the papal statement, but what could it be?—especially given the accuracy of my primary impression (as I later verified) that the Catholic Church values scientific study, views science as no threat to religion in general or Catholic doctrine in particular, and has long accepted both the legitimacy of evolution as a field of study and the potential harmony of evolutionary conclusions with Catholic faith.

As a former constituent of Tip O'Neill's, I certainly know that "all politics is local"—and that the Vatican undoubtedly has its own internal reasons, quite opaque to me, for announcing papal support of evolution in a major statement. Still, I knew that I was missing some important key, and I felt frustrated. I then remembered the primary rule of intellectual life: when puzzled, it never hurts to read the primary documents—a rather simple and self-evident principle that has, nonetheless, completely disappeared from large sectors of the American experience.

I knew that Pope Pius XII (not one of my favorite figures in twentieth-century history, to say the least) had made the primary statement in a 1950 encyclical entitled *Humani Generis*. I knew the main thrust of his message: Catholics could believe whatever science determined about the evolution of the human body, so long as they accepted that, at some time of his choosing, God had infused the soul into such a creature. I also knew that I had no problem with this statement, for whatever my private beliefs about souls, science cannot touch such a subject and therefore cannot be threatened by any theological position on such a legitimately and intrinsically religious issue. Pope Pius XII, in other words, had properly acknowledged and respected the separate domains of science and theology. Thus, I found myself in total agreement with *Humani Generis*—but I had never read the document in full (not much of an impediment to stating an opinion these days).

I quickly got the relevant writings from, of all places, the Internet. (The pope is prominently on-line, but a Luddite like me is not. So I got a computer-literate associate to dredge up the documents. I do love the fracture of stereotypes implied by finding religion so hep and a scientist so square.) Having now read in full both Pope Pius's *Humani Generis* of 1950 and Pope John Paul's proclamation of October 1996, I finally understand why the recent statement seems so new, revealing, and worthy of all those headlines. And the message could not be more welcome for evolutionists and friends of both science and religion.

The text of *Humani Generis* focuses on the magisterium (or teaching authority) of the Church—a word derived not from any concept of majesty or awe but from the different notion of teaching, for *magister* is Latin for "teacher." We may, I think, adopt this word and concept to express the central point of this essay and the principled resolution of supposed "conflict" or "warfare" between science and religion. No such conflict should exist because each subject has a legitimate magisterium, or domain of teaching authority—and these magisteria do not overlap (the principle that I would like to designate as NOMA, or "nonoverlapping magisteria"). The net of science covers the empirical universe: what is it made of (fact) and why does it work this way (theory). The net of religion extends over questions of moral meaning and value. These two magisteria do not overlap, nor do they encompass all inquiry (consider, for starters, the magisterium of art and the meaning of beauty). To cite the arch clichés, we get the age of rocks, and religion retains the rock of ages; we study how the heavens go, and they determine how to go to heaven.

This resolution might remain all neat and clean if the nonoverlapping magisteria (NOMA) of science and religion were separated by an extensive no man's land. But, in fact, the two magisteria bump right up against each other, interdigitating in wondrously complex ways along their joint border. Many of our deepest questions call upon aspects of both for different parts of a full answer—and the sorting of legitimate domains can become quite complex and difficult. To cite just two broad questions involving both evolutionary facts and moral arguments: Since evolution made us the only earthly creatures with advanced consciousness, what responsibilities are so entailed for our relations with other species? What do our genealogical ties with other organisms imply about the meaning of human life?

Pius XII's *Humani Generis* is a highly traditionalist document by a deeply conservative man forced to face all the "isms" and cynicisms that rode the wake of World War II and informed the struggle to rebuild human decency from the ashes of the Holocaust. The encyclical, subtitled "Concerning some false opinions which threaten to undermine the foundations of Catholic doctrine," begins with a statement of embattlement:

*Disagreement and error among men on moral and religious matters have always been a cause of profound sorrow to all good men, but above all to the true and loyal sons of the Church, especially today, when we see the principles of Christian culture being attacked on all sides.*

Pius lashes out, in turn, at various external enemies of the Church: pantheism, existentialism, dialectical materialism, historicism, and of course and preeminently, communism. He then notes with sadness that some well-meaning folks within the Church have fallen into a dangerous relativism—"a theological pacifism and egalitarianism, in which all points of view become equally valid"—in order to include people of wavering faith who yearn for the embrace of Christian religion but do not wish to accept the particularly Catholic magisterium.

What is this world coming to when these noxious novelties can so discombobulate a revealed and established order? Speaking as a conservative's conservative, Pius laments:

*Novelties of this kind have already borne their deadly fruit in almost all branches of theology.... Some question whether angels are personal beings, and whether matter and spirit differ essentially.... Some even say that the doctrine of Transubstantiation, based on an antiquated philosophic notion of substance, should be so modified that the Real Presence of Christ in the Holy Eucharist be reduced to a kind of symbolism.*

Pius first mentions evolution to decry a misuse by overextension often promulgated by zealous supporters of the anathematized "isms":

*Some imprudently and indiscreetly hold that evolution ... explains the origin of all things.... Communists gladly subscribe to this opinion so that, when the souls of men have been deprived of every idea of a personal God, they may the more efficaciously defend and propagate their dialectical materialism.*

Pius's major statement on evolution occurs near the end of the encyclical in paragraphs 35 through 37. He accepts the standard model of NOMA and begins by acknowledging that evolution lies in a difficult area where the domains press hard against each other. "It remains for US now to speak about those questions which, although they pertain to the positive sciences, are nevertheless more or less connected with the truths of the Christian faith."[1]

Pius then writes the well-known words that permit Catholics to entertain the evolution of the human body (a factual issue under the magisterium of science), so long as they accept the divine Creation and infusion of the soul (a theological notion under the magisterium of religion).

*The Teaching Authority of the Church does not forbid that, in conformity with the present state of human sciences and sacred theology, research and discussions, on the part of men experienced in both fields, take place with regard to the doctrine of evolution, in as far as it inquires into the origin of the human body as coming from pre-existent and living matter—for the Catholic faith obliges us to hold that souls are immediately created by God.*

I had, up to here, found nothing surprising in *Humani Generis*, and nothing to relieve my puzzlement about the novelty of Pope John Paul's recent statement. But I read further and realized that Pope Pius had said more about evolution, something I had never seen quoted, and that made John Paul's statement most interesting indeed. In short, Pius forcefully proclaimed that while evolution may be legitimate in principle, the theory, in fact, had not been proven and might well be entirely wrong.

One gets the strong impression, moreover, that Pius was rooting pretty hard for a verdict of falsity.

Continuing directly from the last quotation, Pius advises us about the proper study of evolution:

*However, this must be done in such a way that the reasons for both opinions, that is, those favorable and those unfavorable to evolution, be weighed and judged with the necessary seriousness, moderation and measure.... Some, however, rashly transgress this liberty of discussion, when they act as if the origin of the human body from pre-existing and living matter were already completely certain and proved by the facts which have been discovered up to now and by reasoning on those facts, and as if there were nothing in the sources of divine revelation which demands the greatest moderation and caution in this question.*

To summarize, Pius generally accepts the NOMA principle of non-overlapping magisteria in permitting Catholics to entertain the hypothesis of evolution for the human body so long as they accept the divine infusion of the soul. But he then offers some (holy) fatherly advice to scientists about the status of evolution as a scientific concept: the idea is not yet proven, and you all need to be especially cautious because evolution raises many troubling issues right on the border of my magisterium. One may read this second theme in two different ways: either as a gratuitous incursion into a different magisterium or as a helpful perspective from an intelligent and concerned outsider. As a man of good will, and in the interest of conciliation, I am happy to embrace the latter reading.

In any case, this rarely quoted second claim (that evolution remains both unproven and a bit dangerous)—and not the familiar first argument for the NOMA principle (that Catholics may accept the evolution of the body so long as they embrace the creation of the soul)—defines the novelty and the interest of John Paul's recent statement.

John Paul begins by summarizing Pius's older encyclical of 1950, and particularly by reaffirming the NOMA principle—nothing new here, and no cause for extended publicity:

*In his encyclical "Humani Generis" (1950), my predecessor Pius XII had already stated that there was no opposition between evolution and the doctrine of the faith about man and his vocation.*

To emphasize the power of NOMA, John Paul poses a potential problem and a sound resolution: How can we reconcile science's claim for physical continuity in human evolution with Catholicism's insistence that the soul must enter at a moment of divine infusion:

*With man, then, we find ourselves in the presence of an ontological difference, an ontological leap, one could say. However, does not the posing of such ontological discontinuity run counter to that physical continuity which seems to be the main thread of research into evolution in the field of physics and chemistry? Consideration of the method used in the various branches of knowledge makes it possible to reconcile two points of view which would seem irreconcilable. The sciences of observation describe and measure the multiple manifestations of life with increasing precision and correlate them with the time line. The moment of transition to the spiritual cannot be the object of this kind of observation.*

The novelty and news value of John Paul's statement lies, rather, in his profound revision of Pius's second and rarely quoted claim that evolution, while conceivable in principle and reconcilable with religion, can cite little persuasive evidence, and may well be false. John Paul states—and I can only say amen, and thanks for noticing—that the half century between Pius's surveying the ruins of World War II and his own pontificate heralding the dawn of a new millennium has witnessed such a growth of data, and such a refinement of theory, that evolution can no longer be doubted by people of good will:

*Pius XII added ... that this opinion [evolution] should not be adopted as though it were a certain, proven doctrine.... Today, almost half a century after the publication of the encyclical, new knowledge has led to the recognition of more than one hypothesis in the theory of evolution. It is indeed remarkable that this theory has been progressively accepted by researchers, following a series of discoveries in various fields of knowledge. The convergence, neither sought nor fabricated, of the results of work that was conducted independently is in itself a significant argument in favor of the theory.*

In conclusion, Pius had grudgingly admitted evolution as a legitimate hypothesis that he regarded as only tentatively supported and potentially (as I suspect he hoped) untrue. John Paul, nearly fifty years later, reaffirms the legitimacy of evolution under the NOMA principle—no news here—but then adds that additional data and theory have placed the factuality of evolution beyond reasonable doubt. Sincere Christians must now accept evolution not merely as a plausible possibility but also as an effectively proven fact. In other words, official Catholic opinion on evolution has moved from "say it ain't so, but we can deal with it if we have to" (Pius's grudging view of 1950) to John Paul's entirely welcoming "it has been proven true; we always celebrate nature's factuality, and we look forward to interesting discussions of theological implications." I happily endorse this turn of events as gospel—literally *good news*. I may represent the magisterium of science, but I welcome the support of a primary leader from the other major magisterium of our complex lives. And I recall the wisdom of King Solomon: "As cold waters to a thirsty soul, so is good news from a far country" (Prov. 25:25).

Just as religion must bear the cross of its hard-liners, I have some scientific colleagues, including a few prominent enough to wield influence by their writings, who view this rapprochement of the separate magisteria with dismay. To colleagues like me—agnostic scientists who welcome and celebrate the rapprochement, especially the pope's latest statement—they say: "C'mon, be honest; you know that religion is addlepated, superstitious, old-fashioned b.s.; you're only making those welcoming noises because religion is so powerful, and we need to be diplomatic in order to assure public support and funding for science." I do not think that this attitude is common among scientists, but such a position fills me with dismay—and I therefore end this essay with a personal statement about religion, as a testimony to what I regard as a virtual consensus among thoughtful scientists (who support the NOMA principle as firmly as the pope does).

I am not, personally, a believer or a religious man in any sense of institutional commitment or practice. But I have enormous respect for religion, and the subject has always fascinated me, beyond almost all others (with a few exceptions, like evolution, paleontology, and baseball). Much of this fascination lies in the historical paradox that throughout Western history organized religion has fostered both the most unspeakable horrors and the most heart-rending examples of human goodness in the face of personal danger. (The evil, I believe, lies in the occasional confluence of religion with secular power. The Catholic Church has sponsored its share of horrors, from Inquisitions to liquidations—but only because this institution held such secular power during so much of Western history. When my folks held similar power more briefly in Old Testament times, they committed just as many atrocities with many of the same rationales.)

I believe, with all my heart, in a respectful, even loving concordat between our magisteria—the NOMA solution. NOMA represents a principled position on moral and intellectual grounds, not a mere diplomatic stance. NOMA also cuts both ways. If religion can no longer dictate the nature of factual conclusions properly under the magisterium of science, then scientists cannot claim higher insight into moral truth from any superior knowledge of the world's empirical constitution. This mutual humility has important practical consequences in a world of such diverse passions.

Religion is too important to too many people for any dismissal or denigration of the comfort still sought by many folks from theology. I may, for example, privately suspect that papal insistence on divine infusion of the soul represents a sop to our fears, a device for maintaining a belief in human superiority within an evolutionary world offering no privileged position to any creature. But I also know that souls represent a subject outside the magisterium of science. My world cannot prove or disprove such a notion, and the concept of souls cannot threaten or impact my domain. Moreover, while I cannot personally accept the Catholic view of souls, I surely honor the metaphorical value of such a concept both for grounding moral discussion and for expressing what we most value about human potentiality: our decency, care, and all the ethical and intellectual struggles that the evolution of consciousness imposed upon us.

As a moral position (and therefore not as a deduction from my knowledge of nature's factuality), I prefer the "cold bath" theory that nature can be truly "cruel" and "indifferent"—in the utterly inappropriate terms of our ethical discourse—because nature was not constructed as our eventual abode, didn't know we were coming (we are, after all, interlopers of the latest geological microsecond), and doesn't give a damn about us (speaking metaphorically). I regard such a position as liberating, not depressing, because we then become free to conduct moral discourse—and nothing could be more important—in our own terms, spared from the delusion that we might read moral truth passively from nature's factuality.

But I recognize that such a position frightens many people, and that a more spiritual view of nature retains broad appeal (acknowledging the factuality of evolution and other phenomena, but still seeking some intrinsic meaning in human terms, and from the magisterium of religion). I do appreciate, for example, the struggles of a man who wrote to the *New York Times* on November 3, 1996, to state both his pain and his endorsement of John Paul's statement:

*Pope John Paul II's acceptance of evolution touches the doubt in my heart. The problem of pain and suffering in a world created by a God who is all love and light is hard enough to bear, even if one is a creationist. But at least a creationist can say that the original creation, coming from the hand of God was good, harmonious, innocent and gentle. What can one say about evolution, even a spiritual theory of evolution? Pain and suffering, mindless cruelty and terror are its means of creation. Evolution's engine is the grinding of predatory teeth upon the screaming, living flesh and bones of prey. . . . If evolution be true, my faith has rougher seas to sail.*

I don't agree with this man, but we could have a wonderful argument. I would push the "cold bath" theory; he would (presumably) advocate the theme of inherent spiritual meaning in nature, however opaque the signal. But we would both be enlightened and filled with better understanding of these deep and ultimately unanswerable issues. Here, I believe, lies the greatest strength and necessity of NOMA, the nonoverlapping magisteria of science and religion. NOMA permits—indeed enjoins—the prospect of respectful discourse, of constant input from both magisteria toward the common goal of wisdom. If human beings are anything special, we are the creatures that must ponder and talk. Pope John Paul II would surely point out to me that his magisterium has always recognized this distinction, for *in principio erat verbum*—"In the beginning was the Word."

## Postscript

*Carl Sagan organized and attended the Vatican meeting that introduces this essay; he also shared my concern for fruitful cooperation between the different but vital realms of science and religion. Carl was also one of my dearest friends. I learned of his untimely death on the same day that I read the proofs for this essay. I could only recall Nehru's observations on Gandhi's death—that the light had gone out, and darkness reigned everywhere. But I then contemplated what Carl had done in his short sixty-two years and remembered John Dryden's ode for Henry Purcell, a great musician who died even younger: "He long ere this had tuned the jarring spheres, and left no hell below."*

*The days I spent with Carl in Rome were the best of our friendship. We delighted in walking around the Eternal City, feasting on its history*

*and architecture—and its food! Carl took special delight in the ano-*
*nymity that he still enjoyed in a nation that had not yet aired* Cosmos,
*the greatest media work in popular science of all time.*

*I dedicate this essay to his memory. Carl also shared my personal sus-*
*picion about the nonexistence of souls—but I cannot think of a better*
*reason for hoping we are wrong than the prospect of spending eternity*
*roaming the cosmos in friendship and conversation with this wonderful*
*soul.*

## Note

1. Interestingly, the main thrust of these paragraphs does not address evolution in general but lies in refuting a doctrine that Pius calls "polygenism," or the notion of human ancestry from multiple parents—for he regards such an idea as incompatible with the doctrine of original sin, "which proceeds from a sin actually committed by an individual Adam and which, through generation, is passed on to all and is in everyone as his own." In this one instance, Pius may be transgressing the NOMA principle—but I cannot judge, for I do not understand the details of Catholic theology and therefore do not know how symbolically such a statement may be read. If Pius is arguing that we cannot entertain a theory about derivation of all modern humans from an ancestral population rather than through an ancestral individual (a potential fact) because such an idea would question the doctrine or original sin (a theological construct), then I would declare him out of line for letting the magisterium of religion dictate a conclusion within the magisterium of science.

# IX

## Creationism and Education

This final set of essays deals with the thorny issue of whether creationism belongs in science classes. The Association for Philosophy of Education organized a special symposium at the American Philosophical Association annual Eastern Division meetings in December 1998 to deal with the question. They asked Alvin Plantinga and me to address the question: should creationism be taught in the public schools?

I took the negative side. I had previously written on this topic, and in my earlier article I based my argument on the distinction between private faith and public knowledge, arguing that to teach creationism in the public schools would violate a variety of important epistemic, political, and religious values. I took the opportunity in this new paper to cast a wider net. I discuss the main legal arguments that have ruled in the public school case, as well as creationists' arguments from academic freedom, fairness, censorship, parental rights, and majority rule. I also look at epistemological issues regarding competing claims of truth, and the contention that excluding "what Christians know" (Plantinga) amounts to "viewpoint discrimination" (Johnson), and argue that religious protection arguments actually favor excluding creationism more than including it.

Alvin Plantinga takes the affirmative side in his critical commentary on my essay. He first turns the tables and asks whether *evolution* should be taught, given that it is anathema to many people who hold a Christian "epistemic base." Appealing to a Rawlsian notion of justice, he argues for what he calls a basic right, namely, that "Each of the citizens party to the contract has the right not to have comprehensive beliefs taught to her children that contradict her own comprehensive beliefs." He concludes that we cannot in fairness teach either evolution or creationism as the settled truth in the public schools, and modestly proposes that the just solution is to teach them both *conditionally*, that is, relative to different epistemic bases.

In my reply, I focus on Plantinga's Rawlsian appeal to justice. I argue that Plantinga has missed a crucial element of the Rawlsian procedure by considering the question only from the point of view of parents, and that when one also takes into account other points of view (including that of children), it becomes clear that Plantinga's "modest proposal" is not the just solution. I conclude by trying to find some common ground— while evolution should be taught, it should not be taught dogmatically.

# 35

# Why Creationism Should Not Be Taught in the Public Schools

Robert T. Pennock

## I.  The Question

I would like to thank the Association for Philosophy of Education for inviting me to address the question: should creationism be taught in the public schools? The full range of issues that are the subject of debate in the creation/evolution controversy are too numerous to be covered here, so in this first section I begin by analyzing the question so that we might better focus our discussion. I'll take this opportunity to point out related questions that have so far not been considered in the debate, but that deserve the attention of philosophers of education. In subsequent sections I will consider a variety of legal arguments, creationists' extralegal arguments, epistemological arguments, religious protection arguments, and arguments from educational philosophy that are relevant to our specific question.

## (a)  The Public Schools

The controversies about educational policies having to do with creationism have almost exclusively involved the teaching of evolution in the public schools, so this is a natural locus for us to consider the problem. However, several key elements of the controversy would be quite different if we looked at the issue in other educational settings, such as private and parochial schools, home schooling, and in higher education.

There has been very little consideration of the issues for private rather than public schools, no doubt because the governance of the former is not subject to public control and review in the same way. However, there is still a measure of external oversight in that private schools must meet certain standards in order to get and maintain accreditation. It is a reasonable question to ask whether a school deserves to be accredited if it teaches creationism rather than evolution in its science classes. In general, however, we tend to take for granted that private schools, if they do not receive public funds or certification, are not subject to public standards and may teach whatever private, esoteric doctrines they choose.

For parochial schools we fully expect that religious views will be taught. Indeed, this is the most natural setting for creationist views, and it is fair to say that it is primarily in parochial schools that we find creationism taught in science classes. This is not to say, of course, that all or even most parochial schools teach creationism. Based on informal assessment of my undergraduate students, those who studied at Catholic high schools typically have had the best education on evolution, often better than their public school counterparts. Fundamentalist and evangelical "Bible schools," on the other hand, often cite the creationist orientation of their science curriculum as a major selling point. As they see it, all true knowledge has a biblical basis. Gil Hansen, of the Fairfax Baptist Temple school, explains his school's educational philosophy, which seems to be representative on this point: "What we do here is base everything on the Bible. This becomes really the foundation, the word of God is the foundation from which all academics really spring" (Duvall 1995). How does the school teach evolutionary theory? Hansen is clear about the school's position on this as well: "We expose it as a false model."

Unless the government begins to significantly fund parochial schools with tax dollars such as through a voucher program, parochial schools can probably expect to remain free to teach creationism or whatever religious doctrines they choose. Moreover, fundamentalist and evangelical schools often choose to forego secular educational accreditation, and may be accredited only within their own independent system. I would argue that there are serious issues of educational philosophy to consider even in this setting. Is it right to teach children that something that is known to be true is false? Is it not bad faith to misrepresent the findings of science in what is purported to be a science class? If the basis for knowledge is taken to be biblical revelation, isn't it intellectually dishonest to put such revelations forward as science?

Most of these same considerations apply if we move to consideration of home schooling. There have always been parents who, for one reason or other, chose to teach their own children at home rather than send them to school, but currently the vast majority of home schoolers consists of religious conservatives who do not want their children to be exposed to what they take to be the evils of the public schools, be it sex education or evolution. Home schooling raises some unique issues, since basic education is compulsory and parents must demonstrate that they are providing their children with an education that meets state standards. Oversight of parents who teach their own children is inconsistent, and it seems to be fairly common that fundamentalist home schoolers teach the bare minimum of what they have to of subjects they object to, and then regularly supplement the required curriculum with the religious education—in Bible study, creationism, and so on—that they desire. Are parents doing an educational disservice to their children in teaching them creationism on the sly? One might ask whether stricter oversight is necessary in such cases.

Considering the issue in the setting of higher education overlaps some of the previous considerations, but with some relevant differences. One of the most important is the age and maturity of the students. Most undergraduates are at a more advanced developmental stage then they were in high school, so some new educational goals begin to apply. Certainly one of the most significant is that we expect undergraduates to begin to hone their critical and evaluative thinking skills, and to develop (disciplined) independence of mind. At this stage, it can be quite appropriate and instructive to discuss creationist views, so that students can come on their own to see what is wrong with them. Of course, there are any number of other topics that also could serve the same end, but a professor might legitimately choose to dissect creationism in the same way that one might choose to have students dissect a snake rather than a frog in anatomy class. One question we will have to address is whether this might not be a reasonable educational goal in secondary school public education as well.

For the most part, we must leave consideration of these other venues aside, and focus on the public schools, which historically have been (and for the most part remain) the central locus of the controversy. The general causes of the creationism controversy—perceived conflicts between evolution and some Christian views about Creation—have remained fairly constant over the decades, but these have manifested themselves differently in different periods. In the early decades of the public school system in the United States, few textbooks incorporated evolution, and once they did begin to do so many states responded by passing legislation that banned the teaching of evolution altogether. The 1925, antievolutionary Butler Act in Tennessee led to the first legal battle over creation and evolution in the schools—the famous *Scopes* trial. Such antievolutionary laws remained in effect until they were finally overturned by the U.S. Supreme Court in 1968. Creationists countered at first by passing laws in the early 1970s to give "equal emphasis" to the Biblical account. Since these and similar state laws were struck down in the 1980s, creationist activists have turned to other tactics and other venues, getting laws passed that require, for example, "disclaimers" to be read before biology classes in which evolution would be covered. Alabama public school students found a disclaimer pasted in their biology textbooks that began:

This textbook discusses evolution, a controversial theory some scientists present as a scientific explanation for the origin of living things, such as plants, animals and humans. No human was present when life first appeared on earth. Therefore, any statement about life's origins should be considered as theory, not fact.

In other states, creationist "stealth candidates" got themselves elected to local School Boards and to State Boards of Education and then worked to change science curriculum standards to include creationism or to gut any evolution component. A few go further and require that "evidence against evolution" be presented. Some creationist teachers sometimes simply ignore the law and go ahead and teach their views in their individual classrooms. These and other examples of creationist activism in the public schools keeps this venue at the center of the controversy.

### (b)   Kinds of Creationism

The next element of the question before us that we need to examine is the notion of creationism itself.

The most common form of "creation-science" is what is known as "young-earth creationism" (YEC). In the law they got passed in 1982 in Arkansas, creationists proposed the following as an outline of what they wanted to have taught:

(1) Sudden creation of the universe, energy, and life from nothing; (2) The insufficiency of mutation and natural selection in bringing about the development of

all living kinds from a single organism; (3) Changes only within fixed limits of originally created kinds of plants and animals; (4) Separate ancestry of humans and apes; (5) Explanation of the earth's geology by catastrophism, including the occurrence of a worldwide flood; and (6) A relatively recent inception of the earth and living kinds. (La Follette 1983, p. 16)

A more complete outline of the "Creation model" may be found in Aubrey (1998 [1980]). It is also important to understand that creationism does not end with its rejection of biological evolution, though this is the main thesis that so far has been at issue in the public controversy. As we see in the list above, and as I have shown in more detail elsewhere (Pennock 1999, ch. 1), creationism also rejects scientific conclusions of anthropology, archeology, astronomy, chemistry, geology, linguistics, physics, psychology, optics, and so on.

We must also be aware that there is now considerable factionalism among creationists. Disagreements about the details of Christian theology, partial acceptance of scientific views, and different political strategies have given rise to splinter groups that question one or another of the standard views. Old-earth creationists, for example, do not insist that the world is only six to ten thousand years old and accept something closer to the scientific chronology. A few creationists doubt that there was a single, catastrophic worldwide flood, and hold that the Noachian Deluge may have been local to the Mediterranean, or, if global, then "tranquil" rather than catastrophic. Adherents of the YEC view far outnumber members of other factions, but it is important that we recognize that creationism is not a monolithic view and is split by deep divisions.

Though the traditional creationists remain the most active in their political and educational work, there was a significant evolution of creationism in the 1990s, beginning with the publication of *Darwin on Trial*, by Berkeley law professor Phillip Johnson. Johnson neither endorses nor denies the young-earth view, and he argues the we should understand creationism as belief in the process of creation in a more general sense. People are creationists, according to Johnson's definition, "if they believe that a supernatural Creator not only initiated this process but in some meaningful sense *controls* it in furtherance of a purpose" (Johnson 1991, p. 4). Rather than speaking of "creation-science," Johnson and others among these new creationists call their view "intelligent design theory" and advocate a "theistic science." Intelligent design creationists include both young-earth and old-earth creationists, but for the most part they keep their specific commitments hidden and speak only of the generic thesis of "mere creation." As do other creationists, they oppose accommodation to evolution and take it to be fundamentally incompatible with Christian theism. In another way, however, they go further than creation-science does; they reject scientific methodology itself, arguing that scientific naturalism itself must be tossed out and replaced by their theistic science (though they are never clear about what its distinctive methods might be).

Although it has been fundamentalist and evangelical Christian creationists, especially the YECs, who have been the most active in opposing the teaching of evolution in the schools and pressing for the inclusion of their view of Creation, we cannot fairly evaluate the question without also taking into account non-Christian creationist views. For the most part, adherents of these views have not been as politically active in the United States so these have not reached the public attention to the same degree. It is impossible to even begin to canvas these numerous views, but I will mention by way of example two recent cases that have made the news.

In Kennewick, Washington the 1996 discovery of a fossil human skeleton led to a very public legal battle between science and religion. The 9,000-plus-year-old bones were claimed by a coalition of Northwest Indians who wanted to immediately bury them. However, features of the skull seemed to be more Caucasian than Indian, and scientists questioned whether it was really a tribal ancestor and suggested that further analysis could help reveal something of early human history in the area. Armand Minthorn, of Oregon's Umatilla Tribe, said his people were not interested in the scientist's views: "We already know our history. It is passed on to us through our elders and through our religious practices." Their history says that their God created them first in that place. Many Amerindian tribes have origin stories that, on their face, are antithetical to evolutionary theory and other scientific findings. (In parts of northern Canada, an alliance of Native Indians and Christian creationists has formed to oppose teaching evolution in the schools. It is an uneasy union, of course, because the groups differ sharply in what story of Creation they would put in its stead.) The controversy over the Kennewick skeleton also involves another religious group, the Asatru Folk Assembly, an Old Norse pagan group. The members of this pre-Christian faith revere

Viking-era Scandinavian gods and goddesses, and believe that their ancestors were the first inhabitants of the region. They expect that scientific study of the skull would support their claim of priority, though their other religious beliefs would certainly put them at odds with other aspects of the scientific picture.

Another religious anti-evolutionary view became newsworthy in 1995, when NBC broadcast a program entitled *Mysterious Origins of Man* that purported to reveal scientific evidence that human beings had lived tens of millions of years ago. Creationists were at first elated by this prime-time repudiation of evolution, but quickly withdrew their endorsement when they learned that the program was based on the 1993 book *Forbidden Archeology* (Cremo and Thompson 1993), which advances purportedly scientific evidence for a position that mirrors a Hindu view of creation and reincarnation.

Finally, let me mention one more type of view that is relevant to the controversy. The Raëlian Movement, which had its start in France in the 1970s, calls itself a "scientific religion." Raëlians reject evolution and believe that life on earth is the result of purposeful, intelligent design, but they also reject creationism, in the sense that they believe that the creator was not supernatural. Instead, they believe that life on earth was genetically engineered from scratch by extraterrestrials. The founder of the Raëlian Movement, Claude Vorilhon, claims that he knows this because the truth was revealed to him by an extraterrestrial, who anointed him as the Guide of Guides for our age. Some 70,000 adherents worldwide, many in the U.S. and Canada, share this faith.

In considering whether creationism should be taught in the public schools, we must always keep in mind that all these (and many more) anti-evolutionary views would have to be included as "alternative theories" to the scientific conclusions.

## (c)  Taught How?

The third element of the question that requires preliminary discussion involves the *kind* of academic course in which, and the *way* in which creationism might be taught.

The issues change dramatically, for instance, if the different forms of creationism were to be taught in a comparative religion class. The Constitution is not taken to bar discussion of religion in a course of this sort.

One could imagine a course that surveyed the splendid variety of views of creation of different religions, and could make a good case that such a course might serve a useful educational purpose in fostering an appreciation of American and global cultural diversity. Many opponents of teaching creationism in the schools would be willing to compromise if the topic were to be introduced in such a course. The controversy arises mostly because creationists insist that their view be included as part of the science curriculum, and that it replace or be given equal weight as evolution.

Moreover, creationists want evolution to be revealed as a false model. Creationist textbooks that are used in fundamentalist schools often go further and teach that it is as an evil view as well, promoted by atheist scientists who want to lure people away from God. The textbooks they have lobbied for use in public schools, however, keep the sermonizing about evolutionary evils to a minimum. The most common creationist proposals have followed what is known as the "dual model" approach, whereby "the two theories" are presented and contrasted. The Arkansas "Balanced Treatment" Act specified that the schools should give equal consideration to "creation-science" and "evolution-science." Since all such legislation has been found unconstitutional, creationists now try to argue that science classes should simply present "alternative theories" besides the scientific view. In practice, however, the proposals are essentially unchanged. For example, the textbook *Of Pandas and People* (Davis and Kenyon 1993), which presents intelligent design creationism and is by far the most carefully crafted creationist offering to date, follows the same misleading framework of presenting the views of those who hold "the two" theories of biological origins, natural evolution and intelligent design, neglecting the variety of other views. The terms may have changed, but the dubious strategy it adopts is the same: evolutionary theory is claimed to be riddled with holes, with creationism left as the only alternative. Students are told that the textbook will allow them to do what no other does, namely, let them decide for themselves which theory is true.

One cannot, however, judge simply from such creationist textbooks how creationism would likely be taught were it allowed in the classroom, since these are currently written more for political than for pedagogical purposes, to present an innocent face and get a foot in the door. (*Pandas*, for example, which is offered as a biology text, contains a long section that gives philosophical arguments for why intelligent design should not be disqualified as a scientific hypothesis, and why it purportedly does not violate the court rulings that have found teaching creationism in the public schools to be unconstitutional. This is hardly standard fare for secondary school textbooks.) One can get a more realistic sense of what might happen in classrooms by looking at cases in which teachers have gone ahead and taught creationism despite the laws prohibiting it. In a middle school in Harrah, Oklahoma, a suburb of Oklahoma City, a teacher took away students' textbooks and distributed creationist material, teaching the students that a person who believes in evolution cannot believe in God (RNCSE 1998, vol. 18, no. 2, p. 5).

Finally, let me suggest one further way that creationism might be taught that has not been considered in the literature: creationist views could be used as illustrations of how *not* to do science. Specific creationist tenets could be presented and then the evidence reviewed, showing how scientists came to see that they were false. This might turn out to be a useful educational exercise, in that examination of some of the many errors of so-called creation-science or intelligent design could help teach students how real science is done. As John Dewey pointed out, science education is a failure if it consists of nothing more than the recitation and memorization scientific facts (Dewey 1964 [1910]; 1964 [1938], p. 19). To teach science well is to teach the methods of scientific reasoning, and a critical examination of creationism could serve very well for this purpose. It is because of the possible pedagogic utility of this approach that I actually find myself of two minds about whether teachers should introduce creationism into their science lesson plans. If the courts were to permit the teaching of creation in the public school classrooms I expect that this would soon become a common educational exercise in many science classrooms. Indeed, this is the only intellectually responsible way that it could be taught in a science class.

I'll return to some of these issues shortly, but having delimited the focus of our discussion, let me now turn to the relevant arguments, beginning with a brief review of the legal reasoning that has excluded creationism from the public schools.

## II. Legal Arguments

Because our focus is on the public schools, the legal arguments involving the teaching of creationism have been and continue to be the most significant. Both sides have argued that this is primarily a Constitutional question, with some proponents of teaching creationism claiming that it should actually be protected by the First Amendment, under the free exercise of religion clause. Opponents respond that creationists are not being prevented from exercising their beliefs in their churches, homes, or private schools, but that teaching Creation in the public schools would violate the First Amendment's establishment clause. In a long series of cases, the courts have consistently ruled against the antievolutionists' arguments.

Antievolution laws were struck down by the U.S. Supreme Court in the 1968 *Epperson v. Arkansas* case, on the grounds that the Constitution does not permit a state to tailor its requirements for teaching and learning to the principles or prohibitions of any particular religious sect or doctrine. Subsequent rulings have helped define the boundaries of this ruling. In a 1981 case, a parent sued California, claiming that classes in which evolution was taught prohibited his and his children's free exercise of religion. The Sacramento Superior Court ruled (*Segraves v. California*) that teaching evolution did not infringe on religious freedom, and that a 1972 antidogmatism policy of the School Board—which said that statements about origins should be presented conditionally, not dogmatically, and that class discussions on the topic should emphasize that scientific explanations focus on how things occur, not on ultimate causes—was an appropriate compromise between state science teaching and individual religious beliefs.

Creationists next tried to argue that their view should *not* be excluded on grounds of separation of Church and State, because it was not religion but science. The Arkansas legislature passed a bill requiring "balanced treatment" of what they called "creation-science" and "evolution-science." The court struck down the law in the 1982 *McLean v. Arkansas* case, finding that "creation-science" was not a science. The U.S. Supreme Court came to the same conclusion in the 1987 *Edwards v. Aguillard* case, striking down Louisiana's "Creationism Act," which required the teaching of creationism whenever evolution was taught. The court found that, by advancing the religious belief that a supernatural being created humankind, the act impermissibly legislated the teaching of religion, and that a comprehensive science education is undermined when it is forbidden to teach evolution except when creation-science is also taught.

The court has also ruled, as in the 1994 *Peloza v. Capistrano School District* case, that a teacher's First Amendment right to free exercise of religion is not violated by a school district's requirement that evolution be taught in biology classes, rejecting creationists' contention that "evolutionism" is a religion. In another recent case, in Louisiana, the 1997 *Freiler v. Tangipahoa Parish Board of Education* case, the court overturned a policy that would require teachers to read aloud a disclaimer whenever they taught about evolution, and also found that making curriculum proposals in terms of "intelligent design" is no different from the legal standpoint than earlier proposals for teaching creation-science.

When considering questions about what we ought or ought not do, however, philosophers are never content with answers that stop with the law, if only because we may have ethical duties that require more of us than the law does. Moreover, we must always consider the possibility that current law is itself unjust or unwise. In the sections that follow, I'll examine several other sorts of arguments that address these points.

## III. Creationist Extralegal Arguments

In pressing for the reintroduction of their views into the schools, creationists most often argue that the legal rulings are themselves unjust in one or another way, appealing to a handful of arguments from fairness, majority rule, parental rights, academic freedom, and censorship. Here I'll briefly review and then respond to these as they have been put forward by creationist lawyer Wendell Bird in a video that creationists air on public access television stations. It was Bird who laid out the legal strategy in the early 1980s of promoting creationism as though it were a science. Bird argued the creationist side in the *Edwards v. Aguillard* case in which the Supreme Court overturned laws that had been based on that strategy. Since that defeat, Bird rarely speaks of "creation-science" and instead uses the term "abrupt appearance theory." Bird's arguments here are representative of the main creationist arguments, but they are not unique to him. Indeed, most of these arguments were originally made by the Great Commoner, William Jennings Bryan, himself in his anti-evolution crusade that led up to the *Scopes* trial.

By far the most common argument creationists make is to say that it is *unfair* for the law to exclude their view from the public school science classroom. Isn't it biased and one-sided, they challenge, to teach evolution to the exclusion of creationism? Bird argues that "the only fair approach is to let the children hear all the scientific information and make up their own minds." Phillip Johnson makes the argument in stronger terms, claiming that excluding creationism amounts to "viewpoint discrimination." Bird tries to bolster the argument by appealing to *majority rule*, saying "The fact is that, contrary to all of the smoke, the great majority of the American public feels that it is unfair to teach just the theory of evolution." He cites polls indicating that a large percentage of Americans believe that "the scientific theory of Creation" should be taught alongside evolution in the schools.[1] This majoritarian argument was the main plank of Bryan's position, and he too cited figures about Americans' beliefs. As Bryan saw it, "The hand that writes the pay check rules the schools" (Larson 1997, p. 44). A related argument involves claimed *parental rights* to determine what one's children will study in the public schools. If it is granted that parents have such a right, then a creationist parent should be able to insist that creationism be taught or that evolution excluded.

Creationists also appeal to what is properly taken to be a prime educational value, *academic freedom*. Bird says: "To me the basic issue is academic freedom, because no one is trying to exclude evolution from public schools while teaching a theory of creation. Instead, the evolutionists are trying to exclude alternatives, while, in general, defending the exclusive teaching of evolutionism." This way of putting the argument is somewhat disingenuous, since creationists have indeed tried to exclude evolution from the public schools, and were very successful in keeping it out until just the past few decades. Moreover, they have subsequently tried to exclude it unless it is taught in conjunction with their view, and they continue to work to diminish or undermine its place in the science curriculum in whatever way they can. Bird is correct, however, that science educators do now usually defend the exclusive teaching of the scientific view, and this leads to his final objection. Mentioning organizations that oppose teaching creationism, he concludes that what they are doing amounts to *censorship*: "They have a very specific desire to preserve the exclusive teaching of evolution and to exclude any teaching of a scientific theory of creation or a scientific theory of abrupt appearance. That's censorship in my view."

We should agree that, at first glance at least, some of these charges exert a powerful pull upon us. No one wants to be seen as engaging in censorship or in unfair, discriminatory exclusion of a popular viewpoint. It is certainly incumbent on professionals to take such charges seriously, and to examine them carefully to see whether they have merit. When we do this, however, we find that the charges do not apply or are irrelevant to the issue before us.

The notion of parental rights to determine what one's children are to be taught may sound attractive at first, but parents typically have no special expertise about specific subject matter, and they certainly do not have a right to demand that teachers teach what is demonstrably false. A recent poll showed that 44 percent believe the creationist view that "humans were created pretty much in their present form about 10,000 years ago," but it is not relevant that a large number of Americans reject the scientific findings and don't believe that evolution occurred.[2] It does not matter what the figures are, because matters of empirical fact are not appropriately decided by majority rule. Nor is it "unfair" to teach what is true even though many people don't want to hear it. Neither are the schools "censoring" creationism; they are simply and properly leaving out what does not belong in the curriculum.

The charge that such a policy violates academic freedom is not so easily dismissed. One might reasonably dispute whether academic freedom applies in the public elementary and secondary schools in the same *way* that it does in higher education, but prima facie there seems to be no good reason to think that this important protection should be afforded to university professors and not to others of the teaching profession who serve in other educational settings. Isn't this good enough reason to allow teachers to exercise their professional judgment about whether to include creationism or not? However, academic freedom is not a license to teach whatever one wants. Along with that professional freedom come special professional responsibilities, especially of objectivity and intellectual honesty. Neither "creation-science" nor "intelligent design" (nor any of the latest euphemisms) is an actual or viable competitor in the scientific field, and it would be irresponsible and intellectually dishonest to teach them as though they were.[3]

In the previous century, the situation in science with regard to the question of the origin of species was quite different, but it cannot now fairly be said that the basic theses of evolution are scientifically controversial. There currently are no "alternative theories" to evolution that scientists take seriously, since the evidence has gone against previous contenders (including the forms of creationism held by nineteenth-century scientists) and continues to accrue in favor of evolutionary theory. Evolution is in no sense "a theory in crisis," as creationists purport. This is not to say that there are not problems that remain to be solved, but that is true of every science, and such issues are of sufficient complexity that they are properly reserved for consideration at professional meetings, in the primary literature, and in graduate programs. Unresolved issues at the cutting-edge of science are well above the level that would be likely to be included in secondary school textbooks. What is included at these lower levels that concern us here is well confirmed and scientifically uncontroversial.

## IV. Epistemological Arguments

Creationists of course refuse to accept the evidence that supports the various hypotheses of evolutionary theory. As they see it, one must look to revelation to determine what is absolutely true, rather than believing the "mere theories" of science. The basic issue, as they see it, is whose truths are to be taught. God help us, they say, if we fail to teach our children God's Truth. Phillip Johnson has outlined a new legal strategy for reintroducing creationism into the public schools, arguing that excluding the religious perspective amounts to "viewpoint discrimination." Citing the 1993 *Lamb's Chapel* case, in which the court found that a school could not bar an evangelical Christian perspective in a class that discussed the subject of family relationships, Johnson claims that it is similarly improper to bar consideration of intelligent design when the topic of biological origins is discussed in science class.

One major problematic assumption behind this kind of argument is thinking that questions about empirical fact are simply a matter of one's peculiar point of view, so that excluding one or another is "discrimination," in the sense of subjective prejudice rather than the sense of objective assessment of differences. But there is a real difference between what is true and what is false, what is well confirmed and what is disconfirmed, and surely it is a good thing for science to discriminate the true empirical hypotheses from the false by empirical tests that can tell which is which. Creationists also are interested in truth, but they believe that they already know what the truths are, indeed, as Johnson puts it, with what the truths "with a capital 'T' " are. However, one cannot ascertain truths except by appropriate methods, and creationists are typically unwilling to even say what their special methods are, let alone show that they are reliable.

A second major questionable assumption is that it makes sense to talk about "the religious" or even "the creationist" viewpoint. Professor Plantinga recognizes that truths must be ascertained by a justifiable method, but he argues that different epistemological assumptions may be taken to be properly basic. On that ground, he argues for what he calls an "Augustinian science," in which scientists pursue their research along parallel epistemological tracks. Christians, he tells us, should do their science starting with "what Christians know" (Plantinga 1997). The problem with this is that there is little that Christians can say univocally that they know.

Christians disagree, sometimes violently, about what it is they supposedly know. With the exception perhaps of Roman Catholicism, Fundamentalist Christianity actually provides perhaps the broadest general consensus to be found, since it traces its roots to a series of publications at the beginning of the twentieth century the explicit purpose of which was to try to distill just "the fundamentals" of the faith. But even our brief view of the factions among creationists reveals the splinters that nevertheless form among even fundamentalists over disagreements about the smallest points of theology. (Intelligent design creationists try to paper over this problem by remaining silent about the details and promoting that minimal positive thesis that God creates for a purpose. Though even this vaguely stated view may resonate with religious meaning, it is devoid of empirical content—it certainly neither opposes nor supports any particular view about the truth of evolutionary theory—and provides no method of investigation.) The problems increase exponentially once we look beyond Christianity, and bring Hindu, Pagan, Amerindian, and other creationist viewpoints into the classroom, taking into account what these other religions take to be properly basic beliefs. If these theologically perspectival epistemologies are taken to stand on a par with ordinary natural science, then what we will be left with is a Balkanized science where specific private revelations that one or another group professes to "know" and take as given will vie with one another with no hope of public resolution.

The knowledge that we should impart in public schools is not this private esoteric "knowledge," but rather public knowledge—knowledge that we acquire by ordinary, natural means. The methodological constraints that science puts on itself serve to provide just this sort of knowledge, and thus it is scientific knowledge that is appropriate to teach in the public schools.[4]

## V.   Religious Protection Arguments

It is in part a recognition that the esoteric knowledge claims of religions are of a different sort than the conclusions of scientific investigation that the special Constitutional protection of freedom of religion is needed. This leads to several additional arguments for the exclusion of creationism from the public schools.

Probably the main reason typically offered against the teaching of creationism is that it improperly promotes one religious view over others. We need not dig into the theological soils within which creationism is rooted to see that this is so. In their literature, creationists write as though they are defending the Christian faith and that the enemy consists solely of godless evolutionists, but in reality it is the religious who are more often in the forefront of the opposition to seeing creationism taught in the schools. The plaintiffs opposing the "Balanced Treatment" Act in the Arkansas case included Episcopal, Methodist, A.M.E., Presbyterian, Roman Catholic, and Southern Baptist officials, and national Jewish organizations. Though creationists attempt to portray their views as purely scientific and nonsectarian, other religious groups are not taken in by the disguise, and quite understandably argue that to sanction the teaching of creationism would indeed be to privilege one religious viewpoint over others.

One might argue that this unfair singling out of one view could be avoided by allowing all religious views into the science class. So should we then, following the creationists' pedagogic philosophy, teach them all and let the children decide which is the true one? This is hardly a wise course of action, in that it would make the classroom a place where different religious were inevitably pitted against one another.

Creationists and other conservative Christians often take issue with the Court's interpretation of the Constitution that set up the "wall of separation" between Church and State in the 1947 *Everson v. Board of Education* case. However, the idea of tossing all religions into the science classroom to see which wins would actually violate what some took to be the original intent of the establishment clause of the First Amendment. The influential nineteenth-century Supreme Court Justice, Joseph Story, wrote that the amendment's main object was not just to prevent the exclusive patronage of some particular religion that would result from any national ecclesiastical establishment, but also "to exclude all rivalry among Christian sects ..." (Larson 1989, p. 93). Religious rivalry and outright persecution was all too common in the American colonies when theocracy was the norm, and it seems reasonable to thank the secularization of the government and the Constitutional policy of religious neutrality in large measure for the fact that the United States has been relatively free of the sectarian violence that continues in other parts of the world. In a comparative religion class, religious differences could be respectfully described and studied, but in the setting of a science class, where the point is to seek the truth by submitting differences to the rigors of crucial tests, it is hard to see how conflict could be avoided.

Creationist law professor Phillip Johnson argues that this rationale of neutrality is "wearing thin" (1995, p. 28) in that teaching evolution is tantamount to governmental endorsement of naturalism, which he says is the "established religion" (ibid. p. 35) of the West. Here he is giving a variation of a complaint that creationists have made over the decades, that "evolutionism" is a religion, but this argument has already been tested in court, and evolution has been properly found not to be a religion. Neither is scientific naturalism religious.[5] Scientific organizations might be tempted to get the religious tax exemptions if the court were to rule otherwise, but they could not in good conscience accept them.

Let me mention one further reason to oppose the introduction of creationism into the schools under this heading that I have not seen discussed before, and that some religious people might find to be compelling, namely, that introducing creationism in the science classroom would necessarily place their religious beliefs under critical scrutiny. Creationists typically teach that Christianity stands or falls with the truth or falsity of each and every specific claim in the Bible interpreted literally, or at least "robustly." Fundamentalist and evangelical Christian parents who are familiar only with creationist literature (which invariably describes evolution as "a theory in crisis," and obviously seen to be false to anyone not "blinded" by "naturalistic biases") have no idea how vast is the amount of evidence that supports evolutionary theory, and how weak are the specific claims of creationists. They also do not recognize that a few of what one might call "missionary atheists" are as eager as creationists to have the "Creation hypothesis" included in the public school curriculum, being confident that a side-by-side examination of the claims and evidence would destroy any student's naïve beliefs in the religious view. In my experience, science teachers who teach evolution currently go out of their way to be respectful of their students' religious views. However, should the curriculum change so that they had to discuss the various "Creation" or "design" hypotheses—the Hindu, Amerindian, Pagan, and Raëlian versions, as well as the multiple Christian ones[6]—as though these were simply "alternative theories," they could not avoid a direct confrontation. Given that we expect the government to neither help nor hinder religion, it does not seem wise policy to open the door to having children's religious beliefs explicitly analyzed and rebutted in the public schools in this way.

## VI.   Educational Arguments

In this final section, let us set aside the above considerations and simply ask whether it would be a good educational policy to teach creationism if there were no other factors to consider. To answer the question put this way we must turn our attention to philosophy of education more generally.

The choice of what to teach in the public schools must be made in light of the goals of public education. I take it for granted that one of the basic goals of education is to provide students a true picture of the natural world we share. Another is to develop the skills and instill the civic virtues that they will require to function in harmony in society. While there are several common purposes of this kind toward which all public education aims, the more specific goals, of course, will vary depending on, among other things, the age of the student. It makes little sense, for example, to confront students with material that is beyond their developmental level. We also need to ask what is to be included when one teaches a discipline. What should be taught under the particular subject heading of *science*? In particular, does creationism belong within that subject area?

If we think of science in terms of its set of conclusions, then it is clear that creationism does not belong with it. That creationism and science both have things to say about "the subject of origins" is not sufficient to say that the views of former are a part of the subject matter we ought to teach. The specific hypotheses of creation-science have been rejected by science as the evidence accumulated against them, and the general thesis that "God creates" is not a hypothesis that science considers or can treat of at all. Some scientists do discuss their theological musings—some theistic, some atheistic—in their popular writings, but research on the questions of the existence, or possible activities and purposes of a Creator simply is not to be found in the primary scientific literature. The only proper way to treat the specific empirical claims of creation-science in a science textbook is, thus, as an interesting historical footnote about hypotheses that have been long overturned.

But now let me return to Dewey's important contention that to teach science properly is to teach not a collection of facts but a way of thought. Science education that focuses only on scientific conclusions, and omits teaching scientific methods, misrepresents the nature of scientific inquiry and fails in its basic mission of preparing students with the best skills to function in the natural world. "Theistic science," despite its name, rejects science's methodology and therefore does not belong within the subject. "Creation-science," "abrupt appearance theory," "intelligent design theory," and so on are the creationists' cuckoo eggs that they hope will pass unnoticed, enabling them to garner the resources (and cultural prestige) of science and the forum of the science classroom for their own religious ends.

We should not haggle over mere terminology, but it remains the case that neither the conclusions nor the methods of creationism are properly described as "science." Disciplinary boundaries may not be sharply defined, nor should we expect them to be, but they are generally distinguished by a characteristic order. It is because practitioners must adhere to constraints—be they the precedents of (tentatively) accepted conclusions or the procedures of inquiry themselves—that the notion of a "discipline" makes sense at all.

Could creationism could be taught under a different heading, rather than as a science? If "theistic science" were to prove its value as an independent discipline, then educators might have to consider whether it would be worthwhile in relation to the educational goals of the public schools to include in the curriculum. Historically, theistic science had many centuries to prove itself, but in the end scientists concluded that they had no need of that hypothesis, and contemporary creationists have nothing to show for their attempts to revive the view that theology is the queen of the sciences. Intelligent design creationists plead that they are only beginning their researches and ask for patience when asked for concrete results of their approach, and at present there is no sign that they will succeed in developing a fruitful discipline. It does not make sense to create a separate class to teach a discipline that does not exist.

The fact that intelligent design and other versions of creationism have nothing positive to offer accounts for the pattern that we find in all creationist literature and in proposed texts such as *Of Pandas and People*, namely, that they consist almost exclusively of pointing out purported explanatory gaps in evolutionary theory. The "Creation theory" or the "design hypothesis" is supposed to win by default. As we have seen, this dual model strategy has appeared in various forms, but the current favorite is in the creationist proposals to teach their view under the heading of "critical thinking."

As mentioned above, I find some merit in the idea of considering the creation/evolution controversy as a case study to develop critical thinking. At the university level this can work, but there are several practical obstacles to implementing it at lower levels. The main problem is simply that quantity of material that would have to be covered. It takes a semester-long college course just to give undergraduates an introduction to evolutionary theory, and one needs at least that much background to be able to begin to judge the evidence for oneself. It is also questionable whether high school students have developed cognitively to the degree that would make such an exercise worthwhile. Even some honors-level college freshmen are not sufficiently mature intellectually to begin that sort of evaluative project.

But there is a more important reason for not following the creationists' proposal to teach critical thinking by criticizing evolutionary theory in the way they desire. Creationists are ideologues who "know" in advance what is the "absolutely true" answer to the question of origins, and they want the critical tools to be used against evolutionary theory rather than turned on their own views. But it is simply not intellectually honest or professionally responsible to teach as though scientific conclusions were simply a matter of opinion, or that the creationist views of the origin of species are on a par with the findings of evolutionary theory.

Consider what the effect would be if we were to buy into the curricular framework that creationists propose; it would not be only evolutionary biology that would have to be put under critical scrutiny. Take the subject of world history. There are any number of advocacy groups that, for religious, political, or other ideological reasons, advance some idiosyncratic version of history that is at odds with the findings of historians. If we accept the creationists' proposals regarding evolution, we should also be sure to present alternative theories, attach disclaimers to the standard accounts, and give equal time to the "evidence against" the conclusions of historical research so that the students can judge for themselves. In studying the assassination of President Kennedy, for example, we might begin by screening Oliver Stone's movie *JFK*, and then go on to consider the other theories, such as that Vice President Johnson masterminded his death, or that the CIA was behind it. When studying the landing of American astronauts on the moon, we should probably issue a disclaimer and respectfully consider the views of those who believe that the whole event was filmed by the government in a secret Hollywood studio as part of an elaborate charade. When teaching about World War II we would need to give balanced treatment to those who hold that the Holocaust never happened and was just a Zionist propaganda ploy to gain sympathy for Jews.[7] But such a notion of "fairness" and "balance" is absurd. It is certainly not sound educational philosophy.

We should be no less diligent in teaching the results of careful investigation of the history of life on earth as we are in teaching the history of our nation and the other nations of the world. Critical thinking does not mean indiscriminate thinking, but thinking governed by the rules of reason and evidence.

## VII. The Answer

In reviewing the arguments, we find many good reasons for excluding creationism from the schools, and few good reasons for not doing so. On balance, it seems the wiser course not to allow the conflict into the classroom. Should creationism be taught in the public schools? The answer is that it should not.

## Notes

An article that combines this paper and my reply to Plantinga is to be published in *Science & Education* (in press).

1. Such a poll question is misleading and biased, however, in that it assumes what is false: that there is just *one* theory of Creation and that it is a science.

2. The 44 percent figure comes from a November 1997 poll. There has been no statistically significant change in the percentage of Americans who accept this creationist view since 1982 (45%), when the Gallup poll began to track beliefs about human origins. The social, political, and demographic breakdown of the poll figures showed that the "those most likely to believe in the *creationist version* were older Americans, the less well-educated, southerners, political conservatives (the New Religious Right?), biblical literalists, and Protestants, particularly in fundamentalist denominations such as Baptists" (Bishop 1998).

3. It should be obvious, but let me nevertheless state explicitly that in making these arguments I am taking it for granted that evolutionary theory is true. I mean this, of course, in the standard scientific sense of approximate, revisable truth; no one thinks that evolutionary theory is complete or that one or another of its specific elements might not have to be modified should new, countervailing evidence be found. Creationism is false in the basic sense that, whatever its specific positive commitments, it by definition rejects evolution. Most of creationists' specific claims about the processes of the origins of cosmological, geological, and biological phenomena (among others) have been shown to be false as well, provided that we are able to judge these by ordinary scientific means and standards. Note, however, that I do not assume that this means that God does not exist; the larger question of whether a supernatural designer created the world is not answerable simply by appeal to scientific methods. I have discussed some of the evidence for these conclusions elsewhere (Pennock 1999, chs. 2–3) and will not review them here.

4. I previously developed this position in a paper entitled "Creationism in the Science Classroom: Private Faith vs. Public Knowledge," delivered at the 1995 Conference on Value Inquiry, the proceedings of which have yet to be published. I expand the argument in Pennock (1999, ch. 8) and so will not rehearse it here.

5. I have previously rebutted Johnson's claims about scientific naturalism and shown how it is a methodological and not a dogmatic view in Pennock (1996).

6. Creationists want to include only their own view, of course. In 1973, they got the Tennessee legislature to pass a law that would require public school textbooks to give equal emphasis to the Genesis view and to explicitly identify the evolutionary view of human origins as "merely a theory" and not a scientific fact, and attached an amendment so that the Bible did not also have to carry that disclaimer, and another amendment to expressly exclude the "teaching of all occult or satanic beliefs of human origins." In the 1975 *Daniel v. Waters* case Federal Court of Appeals immediately struck down the law as "patently unconstitutional," holding that claims to the contrary would be "obviously frivolous" and did not merit review, citing the amendments in particular and noting that no law could give such preferential treatment to the Biblical view of Creation over "occult" ones (Larson 1989, p. 136).

7. I mention such examples of conspiracy theories intentionally. As one reads creationist literature, from Morris to Johnson and beyond, one is struck by the regularity with which creationists describe scientists as being engaged in a deliberate conspiracy to deceive everyone into accepting evolution so that they might maintain their cultural authority, promote atheism, and spread immorality. Such implicit and explicit accusations are more than irresponsible.

## References

Aubrey, Frank. 1998 (1980). Yes, Virginia, There Is a Creation Model. *Reports of the National Center for Science Education* 18 (1): 6.

Bishop, George. 1998. What Americans Believe about Evolution and Religion: A Cross-National Perspective. Paper read at 53rd Annual conference of the *American Association for Public Opinion Research*, at St. Louis, Missouri.

Cremo, Michael A. and Richard L. Thompson. 1993. *Forbidden Archeology: The Hidden History of the Human Race*. Alachua, Fla.: Govardhan Hill Publishing.

Davis, Percival and Dean H. Kenyon. 1993. *Of Pandas and People*. Dallas, Texas: Haughton Publishing Co.

Dewey, John. 1964 (1910). Science as Subject Matter and as Method. In *John Dewey on Education: Selected Writings*, edited by R. D. Archambault. Chicago: University of Chicago Press.

Dewey, John. 1964 (1938). The Relation of Science and Philosophy as a Basis of Education. In *John Dewey on Education: Selected Writings*, edited by R. D. Archambault. Chicago: University of Chicago Press.

Duvall, Jed. 1995. School Board Tackles Creationism Debate. *CNN Interactive (WWW)*, November 5.

Johnson, Phillip E. 1991. *Darwin on Trial*. Washington, D.C.: Regnery Gateway.

Johnson, Phillip E. 1995. *Reason in the Balance: The Case against Naturalism in Science, Law, and Education*. Downers Grove, Ill.: Inter Varsity Press.

La Follette, Marcel Chotkowski, ed. 1983. *Creationism, Science, and the Law: The Arkansas Case*. Cambridge, Mass.: The MIT Press.

Larson, Edward J. 1989. *Trial And Error: The American Controversy over Creation and Evolution*. Updated ed. New York and Oxford: Oxford University Press.

Larson, Edward J. 1997. *Summer for the Gods: The Scopes Trial annd America's Continuing Debate over Science and Religion*. New York, NY: Basic Books.

Pennock, Robert T. 1996. Naturalism, Evidence, and Creationism: The Case of Phillip Johnson. *Biology and Philosophy* 11 (4): 543–559. Reprinted in this volume, chap. 3.

Pennock, Robert T. 1999. *Tower of Babel: The Evidence against the New Creationism*. Cambridge, Mass.: MIT Press.

Plantinga, Alvin. 1997. Methodological Naturalism? *Perspectives on Science and Christian Faith* 49 (3): 143–154. Reprinted in this volume, chap. 13.

# 36

## Creation and Evolution: A Modest Proposal

Alvin Plantinga

The topic of our meeting is the question, should Creationism be taught in the (public) schools? That is an excellent question, and Professor Pennock has interesting things to say about it. I want to begin, however, by asking a complementary question, after which I shall return to this one: should evolution be taught in the public schools? I'm not asking whether it is legally permissible to teach evolution in the public schools; that matter has been long settled. I'm asking instead whether it should be taught. Given that it is permissible, is it also the right thing to do? But why should that even be a question? Daniel Dennett thinks it is a foolish question: "Should evolution be taught in the schools?" he asks? Well, "Should arithmetic? Should history?" Isn't it utterly obvious that evolution should be taught in public schools? I don't think so; the answer isn't nearly so simple. But we must initially specify the question a bit more closely. First, I am asking whether evolution should be taught in the public schools of a country like the United States, one that displays the pluralism and diversity of opinion our country presently displays. And second, I am asking whether evolution should be taught as the sober truth of the matter, rather than as, for example, the best current scientific hypothesis, or what accords best or is most probable (epistemically probable) with respect to the appropriate scientific evidence base. The question is whether evolution should be taught in the way arithmetic and chemistry and geography are taught: as the settled truth.

Still another need for specification: the term "evolution" can expand and contract upon demand: it covers a multitude of sins, as some might put it. First, there is the idea that at least some evolution has occurred, that there have been changes in gene frequencies in populations. I suppose everyone accepts this, so we can put it to one side. Second, there is the claim that the earth is very old—billions of years old, and that life has been present on earth for billions of years. Third, there is the progress thesis, as we humans like to think of it: first there were prokaryotes, then single-celled eukaryotes, then increasingly more complex forms of life of great diversity, achieving a contemporary maximum in us. Fourth, there is the claim of universal common ancestry: the claim that any two living things you pick, you and the poison ivy in your backyard, for example, share a common ancestor. Fifth, there is what I will call "Darwinism," the thesis that the cause of the diversity of forms of life is natural selection working on a source of genetic variation like random genetic mutation. Sixth and finally, there is the idea that life itself arose by way of purely natural means, just by way of the workings of the laws of physics and chemistry on some set of initial conditions, or just by way of the workings of those laws together with what supervenes on their workings; this thesis is part of the contemporary scientific picture of the origin of life, although at present all such accounts of the origin of life are at best enormously problematic.[1] Since the first thesis is accepted by everyone, we can set it aside, and use the term "evolution" or "the theory of evolution" to refer to the conjunction of the remaining five theses, or occasionally to the conjunction of the first four.

So why is there a question as to whether evolution should be taught in the public schools? And if there is such a question, what sort of question is it? I believe there *is* a question here, and it is a question of justice or fairness. First, our society is radically pluralistic; and here I am thinking in particular of the plurality of religious and quasi-religious views our citizenry displays. I say "quasi-religious": that is because I mean the term to cover, not only religious belief, as in Christianity, Islam, Judaism, Hinduism, Buddhism, and the like, but also other deep ways of understanding ourselves and our world, other deep ways of interpreting ourselves and our world to ourselves. Thus consider philosophical naturalism, the idea that there is no such person as God or anything or anyone at all like him: on this use, naturalism is or can be a quasi-religious view. Following John Rawls, let's call beliefs of this sort "comprehensive" beliefs. Now for many, perhaps most citizens, these comprehensive beliefs are of enormous importance; for some they are the most important beliefs of all. And it is natural for these citizens to want their children to be educated into what they take to be the true and correct comprehensive beliefs; they think it is a matter of great importance which comprehensive beliefs their children adopt, some even thinking that one's eternal welfare depends on their accepting the true comprehensive beliefs.

Next, we must think for a moment about the purpose of public schools. This purpose is somehow determined by or supervenes on the purposes of the citizens who support and employ these schools. It is as if we are all party to a sort of implicit contract: we recognize the need to train and educate our children, but don't have the time or competence to do it individually. We therefore get together to hire teachers to help instruct and educate our children, and together we pay for this service by way of tax money. But what should we tell these teachers to teach? Of course all the citizens party to the contract would prefer that their children be educated into their own comprehensive beliefs—be taught that those comprehensive beliefs are the sober truth. But that isn't feasible, because of the plurality of comprehensive beliefs. It would clearly be unfair, unjust, for the school, which we all support, to teach one set of religious beliefs as opposed to another—to teach that evangelical Christianity, for example, is the truth. This would be unfair to those citizens who are party to the contract and whose comprehensive beliefs —Judaism, naturalism, Islam, whatever—are incompatible with evangelical Christianity. The teacher can't teach all or even more than one of these conflicting sets of beliefs as the truth; therefore it would be unfair to select any particular one and teach *that* one as the truth. More generally, fairness dictates that no belief be taught as the settled truth that conflicts with the comprehensive beliefs of some group of citizens party to the contract. We can put this in terms of what I'll call "the basic right" (BR):

(BR)   Each of the citizens party to the contract has the right not to have comprehensive beliefs taught to her children that contradict her own comprehensive beliefs.

Our society is a pluralistic society; there are many mutually inconsistent sets of comprehensive beliefs. But then no particular set of comprehensive beliefs can be taught without infringing on that basic right. It is therefore unfair and unjust to teach one religious belief as opposed to others in the schools; it is improper and unjust to teach, for example, Protestant beliefs, as opposed to Catholic, or Christian as opposed to

Jewish or Hindu, or religious beliefs as opposed to naturalism and atheism. More generally, take any group of citizens who are party to the contract: it would be unfair for the public schools to teach beliefs inconsistent with their religious or comprehensive beliefs: unfair, because it would go against (BR). Of course (BR) is a prima facie right. It is at least possible that special circumstances should arise, perhaps as in wartime, in which this right would be overridden by other desiderata, for example national security. The majority might also insist on teaching the denial of certain comprehensive views, Naziism, for example, in which case the fair thing to do would be to exclude the Nazis from the contract (and also exclude them from the tax liability).

But then it is also easy to see how an issue of justice or fairness can arise with respect to the teaching of evolution. As Professor Pennock points out, many American Indian tribes, for example, "have origin stories that, on their face, are antithetical to evolutionary theory and other scientific findings" (this volume, p. 760). Now consider a public school in an Indian village of this sort, one where many or most of the citizens hold and are deeply committed to comprehensive beliefs that are contradicted by contemporary evolutionary theory. Perhaps they believe that the first human beings were specially created by God a hundred miles or so from their village, some thousands of years ago. Would it be fair or just to teach their children, in this public school, that these religious accounts of human origins are false? Would it be right to teach their children that their ancestors emerged on the plains of Serengetti more than a million years ago, and that they were not specially created at all, but descended from earlier, nonhuman forms of life? Would it be just to teach their children accounts of human origins that contradict their religious accounts? I think we can see that this would be unfair and unjust. These citizens are party to the implicit contract by which public education is founded; they support and help finance these schools. By virtue of (BR), then, they have a right not to have their children taught, in public schools, the denials of their cherished religious beliefs. If their children are taught the denials of these beliefs, these citizens' rights are being violated. They are being violated just as surely as if their children were taught, for example, that their religion is merely superstition and evangelical Christianity is the truth of the matter.

Now the fact is there is a substantial segment of the population, at least in certain states and certain parts of our country, whose comprehensive beliefs are indeed contradicted by the theory of evolution. There are fundamentalist Christian, Jewish, and Muslim parents, and quite a few of them, who think the earth is very young, perhaps only 10,000 years old. This is not a casual opinion with them, as might be their opinion that there are mountains on the far side of the moon. It is a part of their comprehensive belief: it is one of their religious beliefs that the Bible (or the Koran) contains the truth on all the matters on which it speaks, and on this matter what it says is that the earth is young. There are others who believe that the first human beings were created specially by God, so that the theory of universal common ancestry is false. We may disagree with their beliefs here, or even think them irrational; but that doesn't change matters. Even if their beliefs are irrational from our point of view, (BR) still applies: they have the right to require that public schools not teach as the settled truths beliefs that are incompatible with their comprehensive beliefs.[2]

So there is therefore a clear prima facie question of justice here: these citizens are party to the implicit contract; they pay their taxes; they support these schools, and send their children to them. But then they have a prima facie right to have their children taught, as settled fact, only what is consistent with their comprehensive beliefs. And this means that it is unfair or unjust to teach evolution—universal common ancestry, for example—in the public schools, at any rate where there is a substantial segment of the population whose comprehensive beliefs are incompatible with evolution. In the very same way, of course, it would be unjust to teach creationism as the settled truth. Both doctrines conflict with the comprehensive beliefs of some of the parties to the contract.

But now for a reply, a reply suggested by some of Professor Pennock's comments. Doesn't *truth* have any rights here? Perhaps (BR) is a prima facie right, so runs the reply, but this right is overridden by the demands of truth. Pennock "takes it for granted that one of the goals of education is to provide students a true picture of the natural world we share" (this vol., p. 772), which seems fair enough; he also takes it for granted that evolutionary theory is true (p. 776); but then he concludes that evolutionary theory ought to be taught in the public schools. "Matters of empirical fact," he says, "are not appropriately decided by majority rule, nor is it unfair to teach what is true, even though many people don't want to hear it" (p. 767). That seems to suggest that if a proposition is true, then it is fair to teach it in public schools, even if it goes contrary to the comprehensive beliefs of the citizens who are party to the contract and support the public schools. But the reasoning seems deeply flawed. Suppose Christianity is in fact true, as indeed I believe it is: would that mean that it is fair to teach it in public schools where most of the citizens, citizens who support those schools, are not Christians and reject Christian comprehensive beliefs? I should think not; that would clearly be unfair, and the fact that the system of beliefs in question is true would not override the unfairness. We can't sensibly just insist that what is true can properly be taught, even if it contradicts the comprehensive beliefs of others party to the contract. After all, *they* also believe that *their* comprehensive beliefs are true: that is why they hold them.[3]

But other things he says suggest a different objection. According to Stephen J. Gould, there is the realm of values, and there is the realm of fact; religion and comprehensive beliefs occupy the realm of values (hence the expression "religious values"); science occupies the realm of fact. Hence when things are done properly there can be no conflict. There are no properly religious beliefs on matters of fact. Of course this is much too strong: clearly *most* religions make factual claims: that there is such a person as God, that the world was created, that Mohammed was God's prophet and spokesman. A slightly (but only slightly) more nuanced view, one that seems to me to be suggested by some of the things Pennock says can be put as follows: when it comes to matters of empirical fact (however that phrase is to be understood) scientific consensus trumps comprehensive belief. These questions of the origin of human beings and of life are factual questions, questions of empirical fact. The proper way to deal with them, then, is by way of science; it is simply a matter of trespass for someone in the name of religion to propose an answer to these factual questions. And this fact of trespass means that (BR) is overridden in some cases. If it's a factual question that's at issue, then the way to deal with it is by way of science. If you happen to have mistaken opinions about algebra or prime numbers (perhaps it is part of your comprehensive belief, somehow, that there is a greatest prime) that is your problem; you can't require the public schools to respect your comprehensive opinion here, and refrain from demonstrating to your children that in fact there is no greatest prime. Citizens do not have the right to object, on the ground of religious or comprehensive beliefs, to any scientific teaching. When it comes to issues that are dealt with by science, the prima facie claims of (BR) are overridden, and it is entirely right to teach the denials of comprehensive beliefs in the public schools, if those comprehensive beliefs are in fact contrary to contemporary scientific consensus.

But again, this seems entirely mistaken. First, why should we think scientific consensus overrides (BR)? Perhaps because we think science is our best bet with respect to the discovery of the truth or the approximate truth on the subjects on which it speaks. But if it is the truth we want taught to our children, then it's far from clear that current science should be treated with this much deference. We all know how often scientific opinion has changed over the years; there is little reason to think that now it has finally arrived at the unrevisable truth, so that its current proposals are like the claim that there is no greatest prime. According to Bryan Appleyard, "At Harvard University in the 1880's John Trowbridge, head of the physics department, was telling his students that it was not worthwhile to major in physics, since all the very important discoveries in the subject had now been made. All that remained was a routine tidying up of loose ends, hardly a heroic task worthy of a Harvard graduate."[4] Twenty years later the same opinion seemed dominant: for example, in 1902 Albert Michelson, of Michelson-Morley fame, declared that "the most important fundamental laws and facts of physical science have all been discovered and these are now so firmly established that the possibility of their ever being supplanted on consequences of new discoveries is remote."[5] And of course we all know of the scientific theories that once enjoyed consensus but are now discarded: caloric theories of heat, effluvial theories of electricity and magnetism, theories involving the existence of phlogiston, vital forces in physiology, theories of spontaneous generation of life, the luminiferous ether, and so on.

But there is another and even more important consideration. Pennock, we are supposing, thinks the way to approach questions of empirical fact is by way of science, not by way of religion; thus scientific consensus trumps religious or comprehensive belief in such a way that the prima facie requirements of (BR) are overridden; and hence it is fair to teach evolution as settled fact, even if it does conflict with the religious beliefs of some of the citizens party to that implicit contract. But now consider this claim, that is:

(PC)    The right way to answer questions of empirical fact—for example, questions about the origin of life, the age of the earth, whether human beings have evolved from earlier forms of life—is by way of science, or scientific method.

Note first that (PC) is not, of course, itself a question of empirical fact. Science itself does not decide between (PC) and other possibilities—for example, the claim that the right way to approach certain empirical questions is not by way of scientific inquiry but by way of consulting the Bible, or the elders of the tribe. The question whether the scientific epistemic or evidential base is the right way to settle these issues is not itself to be settled with respect to the scientific epistemic base; this dispute is philosophical or religious rather than scientific. Note second that there are many others, of course, who do not share Pennock's opinion: they do not accept (PC). Indeed, there are many others such that a proposition incompatible with this opinion is part of their religious or comprehensive beliefs. Perhaps (PC) is part of Pennock's comprehensive beliefs; but its denial is part of the comprehensive beliefs of others who are party to the contract. But then clearly it would be unfair to act on (PC), as opposed to these other comprehensive beliefs that are incompatible with it. Suppose in fact fundamentalists are right: the truth is the correct way to determine the age of the earth is by way of consulting Scripture under a certain literal construal of early Genesis: would it follow that it was right in public schools to teach as the settled truth that the age of the earth is some 10,000 years or so? I should think not; and the same goes with respect to (PC). (PC) may be true and (more likely, in my opinion) it may be false; either way it is just one comprehensive belief among others. It would be unfair to teach comprehensive beliefs that entailed the denial of (PC); but by the same token, it would also be unfair to teach (PC).

What we have seen so far, therefore, is that it is improper, unfair, to teach either creationism or evolution in the schools—that is so, at any rate, for areas where a substantial proportion of the parents hold religious or comprehensive beliefs incompatible with either. But then what *can* be taught, in public schools, about this crucial topic of origins, a topic deeply connected with our sense of ourselves, our sense of where we come from, what our prospects are, what is good for us, and the like? If we can't teach either Creationism or evolutionism, what can we teach in the public schools?

Well, possibly nothing. One answer is to say: in a pluralistic society like ours, there is no fair way to teach anything about origins; hence public schools ought not to teach anything on that subject. They should instead stick to subjects where there isn't disagreement at the level of religious or comprehensive beliefs. This would be just a reflection of a more general difficulty in having public schools of our sort in a pluralistic society. Perhaps, when the citizens get together to found a system of education, what they discover is that there is too much diversity of opinion to make it feasible. But this is a counsel of dispair; I think perhaps we can do better.

We can see a bit more deeply into this question by turning to a bit of epistemology. We have already noted that different people accept different religious or comprehensive beliefs. More generally, for each person P there is an epistemic base, EB$_P$, with respect to which the probability or acceptability of proposed beliefs is to be evaluated. This epistemic base includes, first, P's current beliefs. Since some beliefs are held more strongly than others, it includes, second, an index of degree of belief. Some beliefs, furthermore, are of the form *probably S*. An epistemic base also includes, third, prescriptions as to how to conduct inquiry, how to learn more about the world, under what conditions to change belief, and the like. And finally, an epistemic base includes comprehensive beliefs. These comprehensive beliefs are not, of course, frozen in stone; nor are they impervious to argument and reasoning; nor are they irrational just as such, or held in an irrational way. A person's epistemic base is not static, of course; it constantly changes under the pressure of experience, what we are told by others, and the like. An epistemic base can also undergo sudden and drastic revision, as in a religious conversion, for example. A proper characterization of the notion of an epistemic base would take us far afield, and would certainly require an entire paper on its own. But I think the basic idea is fairly clear.

Now what parents want, presumably, is that their children be taught the truth—which, of course, they take to be what is in accord with their own epistemic bases. What is in accord with their own epistemic bases, of course, is not just the propositions they themselves happen to accept. I may know or believe that there are people who hold lots of beliefs I don't hold—about, for example, mathematics; I may also believe that their beliefs, whatever they are, are true, or likely to be true, or more likely to be true than the beliefs I actually hold; and I may therefore want my children taught *those* beliefs, even though they are not parts of my own epistemic base, and even though some may conflict with beliefs in my own epistemic base. These beliefs, we may say, are in accord with my epistemic base, although not contained in it. This can happen with respect to religious or comprehensive beliefs too: I may be an American Indian who holds that the tribal elders know the truth about important matters of origin, or whatever; then I may want my children taught what these elders believe, even if I don't myself know precisely what it is that they do believe.

We must note next that science has its own epistemic base. This base is presumably not identical with that of any of the citizens, although it overlaps in complex ways with those of some of the citizens. It is not important, here, to say precisely what goes into the scientific epistemic base (or how it is related to those of the citizens); presumably logic goes into it, together with prescriptions as to how to conduct various kinds of inquiry, together with a host of common sense beliefs, together with a good bit of firmly established current science. But it is important to note certain beliefs that do *not* go into EB$_P$, at least with respect to science as currently practiced. Among these would be the belief that there is such a person as God, that God has created the world, and that God has created certain forms of life specially—human beings, perhaps, or the original forms of life, or for that matter sparrows and horses. That is because science commonly respects what is often called "methodological naturalism," the policy of avoiding hypotheses that mention or refer to God or special acts on the part of God, or other supernatural phenomena, or hypotheses whose only support is the Bible, or some other alleged divine revelation. There is dispute as to whether science by its very nature involves methodological naturalism, and there is also dispute as to whether science *has* a nature. But as commonly practiced, science does seem to involve methodological naturalism. This means that EB$_P$ does not include any propositions of the above sort. It is not entirely clear whether EB$_P$ includes the *denials* of some propositions about God, or rather just fails to include those propositions. It is also worth noting that a person could think that EB$_P$ is the proper epistemic base from which to conduct scientific inquiry, even if her own epistemic base contains some of those propositions excluded from EB$_P$ by methodological naturalism. Indeed she might hold that a given proposition is a good scientific hypothesis even if it conflicts with one of her comprehensive beliefs and is therefore, as she sees it, false. Thus someone might think that a given scientific hypothesis—Darwinism, for example—is in fact false, but is nevertheless a source of fertile and useful hypotheses.

To return to our subject, then: we can't in fairness teach evolution as the settled truth in public schools in a pluralistic society like ours, and of course we can't in fairness teach creationism either. But there is something else we can do: we can teach evolution *conditionally*. That is, we can teach, as the sober truth, that from the vantage point of EB$_P$ the most satisfactory hypothesis is the ancient earth thesis, or universal common ancestry, or Darwinism, or even some hypothesis entailing naturalistic origins. We can also distinguish between the likelihoods of these hypotheses, on EB$_P$; the ancient earth thesis is very nearly certain on this basis, universal common ancestry much less certain, but still a very good bet, Darwinism still less certain, and naturalistic origins, or rather any particular current theory of naturalistic origins, unlikely, at least with respect to the current EB$_P$. There is one further complication we must note: given plausible views about EB$_P$, it might be that a hypothesis is the best scientific hypothesis from the point of view of EB$_P$, even though it is not, from that point of view, more probable than not. This might be first just because there are several conflicting hypotheses in the field, all of which enjoy substantial probability with respect to EB$_P$, but none of which enjoys a probability as great as $1/2$. But second, it might be that a conjecture is a fine, fertile hypothesis such that inquiry pursued under its aegis is fruitful and successful, even though the hypothesis in question is unlikely with respect to EB$_P$. Many more questions arise about EB$_P$; there is no time to explore them now.

Now consider the claim that evolution is the best hypothesis (the one most likely to be true), or even that it is much more likely than not with respect to EB$_P$: that claim, I take it, will be compatible with everyone's religious and comprehensive beliefs. There would then be no objection from the point of view of fairness to teaching this claim as the settled truth—while refraining, of course, from teaching evolution itself as the settled truth. Perhaps this is something like what the court had in mind in *Segraves v. California* when it declared that "any speculative statements concerning origins, both in texts and in classes, should be presented conditionally, not dogmatically" (p. 8). And the same would go for creationism: with respect to certain widely shared epistemic bases, the most likely or satisfactory hypothesis will be the claim that God created human beings specially, or even the claim that the earth is only 10,000 years old. Of course the public schools will not, under this proposal, teach that one epistemic base—either that of evangelical Christianity, for example, or scientific naturalism—is in fact the correct or right or true epistemic base. The question of which epistemic base is the correct one is not a question on which public schools should pronounce, at least in areas where there is relevant religious disagreement. What the public schools should teach as the sober truth is what is in accord with all the relevant epistemic bases; this is what should be taught unconditionally.

To return to our original question then: should creationism be taught in the public schools? Should evolution? The answer is in each case the same: no, neither should be taught unconditionally; but yes, each should be taught conditionally.[6]

### Notes

1. See, e.g., *Origins* by Robert Shapiro (New York: Summit Books, 1986).

2. A slightly different issue: there are still others who believe, as part of their comprehensive belief, that God created the world and humankind one way or another, where one possibility is that he did it by way of an evolutionary process. These parents may very well believe that it is possible (epistemically possible) that human beings are geneologically related to earlier forms of life, but that this suggestion is far from certain. They may therefore quite properly resist having it taught as settled truth, on a par with arithmetic and the proposition that there has been an American Civil War.

3. In the same vein, Pennock also holds that "parents certainly don't have a right to demand that teachers teach what is false" (this vol., p. 767). But this too seems to me mistaken: didn't parents at the end of the last century have the right to demand that science teachers teach, e.g., Newtonian mechanics, even though as a matter of fact it is false? They also had a right to demand that science teachers teach that there is such a thing as the luminiferous ether, although that too is false, at least by our current lights. Earlier parents had a similar right to demand

that science teachers teach the caloric theory of heat, that there is such a thing as phlogiston, that electricity is a kind of fluid, that the sun goes around the earth, and so on.

4. *Understanding the Present* (New York: Doubleday, 1992), p. 110.

5. Quoted in Hanbury Brown, *The Wisdom of Science: Its Relevance to Culture and Religion* (Cambridge: Cambridge University Press, 1986), p. 66.

6. I'm grateful to Tom Crisp, Marie Pannier, and David VanderLaan for comments and criticism.

## References

Pennock, Robert T. (2000). Why Creationism Should Not Be Taught in the Public Schools. This volume, chap. 35.

# 37

# Reply to Plantinga's "Modest Proposal"

Robert T. Pennock

I would like to thank Professor Plantinga for his comments, and for providing us with much food for thought. I must confess, however, that after chewing over his offering, I still find some of his claims and conclusions hard to swallow. He uses idiosyncratic and shifting definitions of evolution, seriously underestimates the evidential support for the various evolutionary hypotheses, and makes some misleading statements about these.[1] But rather than pick out these and other pits from the salad, I will turn directly to problems with the main course. Professor Plantinga turns the table and asks whether *evolution* should be taught in the public schools, and his central argument overlaps the earlier creationists' arguments claiming "unfairness" and "parental rights" that I discussed above. He links these by appealing to a Rawlsian notion of justice, claiming that every party to the social contract has a basic right (BR), "not to have comprehensive beliefs taught to her children that contradict her own comprehensive beliefs" (this vol., p. 781). When Jonathan Swift made his own "modest proposal" that the Irish might solve the problem of starvation during the potato famine by devouring children, it was clear that he was being satirical. It seems, however, that Professor Plantinga means for us to take (BR) seriously, even though its effects on children's education would be much the same as Swift's. My argument is not, as Professor Plantinga supposes, that science overrides (BR). I would not accept (BR) in the first place, nor should anyone who is concerned with justice.

Professor Plantinga argues for (BR) as follows: "The teacher can't teach all or even more than one of ... conflicting sets of beliefs as the truth; therefore it would be unfair to select any particular one and teach *that* one as the truth" (this vol., p. 781). But this argument is seriously flawed; even if there is no way for everyone to eat the whole pie, it does not follow that there is no fair way to divide it or even to pick just one person who will get it. Plantinga appeals to the Rawlsian analysis of justice, but does not describe the reasons that he thinks would lead free and equal people in Rawls's Original Position to agree to such a basic right. We need to see this reasoning before we can evaluate the justice of the proposal from a Rawlsian perspective.[2] Perhaps he thinks that it would go something like this: One knows that people are likely to disagree about "comprehensive beliefs" such as those profound religious beliefs about God and God's methods of Creation, but under the veil of ignorance one does not know what one's own view will be on such matters or whether, say, one would be in a position see to it that only those beliefs were taught in the schools. Plantinga must think that under such conditions rational persons would agree to institute (BR) as a defensive safeguard—since no one could be sure that their own preferred view would be taught, they all would want at least the right to see to it that another view that opposes it not be taught either. In other words, I supposedly would reason as follows: since I might be a creationist whose epistemology tells me that true knowledge comes only from the Bible and that evolution is contrary to God's Truth, I would want the right to exclude evolution from the schools to protect my children from what I would take to be its evil influence. But this reasoning is surely mistaken. Rational agents would never agree to such a gag rule as a basic right for a variety of reasons.

First, to do so would be to gut the curriculum, for even the most well-established facts may threaten some element of some person's comprehensive beliefs. As we saw, creationists' comprehensive beliefs put them at odds with not just the facts of evolutionary biology, but also with the findings of most other sciences. Furthermore, because of the interconnections among scientific theories, opposition to what might appear to be just a single finding necessarily involves opposition to many other ones as well. Nor are creationists the only ones that would make use of (BR); there are thousands of special interests groups that would use such a right to prohibit the teaching of specific facts or even whole subjects they objected to. One does not have to look far to find parents who would object to teaching about racial equality, the facts of reproductive health, or that even that the earth is round.[3] Only the utterly trivial could have a chance of escaping the gag of (BR). No rational person would agree to such a situation.

Second, under the veil of ignorance one should not limit one's deliberations to the scenario in which one was a parent with a comprehensive belief that he or she did not want challenged. The rational person would also have to think about being a child growing up in the household of such a parent. Sadly, we all know parents who are bigots or ideologues and others who are simply narrow-minded or ignorant. A good education may be a child's only window to a clear picture of the world and to an open future. To agree to (BR) would be to close that window. This would be a serious harm for the children of such parents. It would harm other children as well, all of whom could be deprived of a decent education, because the public schools could be forced to omit even basic scientific facts, simply because *some* parent found them offensive to his or her private religious beliefs. Indeed, not one of the facts that Professor Plantinga agrees we may teach unconditionally as settled truths would pass muster if we grant with him that each person may appeal to (BR) from the warrant of her personal "epistemic base" (EB$_P$). Judging the question after considering the points of view of both parents and children, rational persons in Rawls's Original Position would reject (BR). Thus a school policy that ignores (BR) is not unjust, for (BR) itself is unjust.

So, what *would* rational agents agree to? They would assent, I believe, to something very similar to our current system, along the lines I mentioned above—a separation of the public and the private. They would require that public institutions, like the public schools, not teach views based on "private epistemologies" such as special revelation, because one cannot rationally adjudicate among beliefs that different persons purport "to know" simply "as a Christian" or a Hindu, a Raëlian, a Pagan, or whatever. Rational agents would agree to a basic right to hold such beliefs privately, but as a defensive measure, would not want the government to support beliefs of this sort even "conditionally," as in Professor Plantinga's proposal, since it would be impossible for schools to teach the myriad of private views without favoring some over others. Instead they would require that public institutions be constrained to public knowledge. In teaching about the empirical world, this means that the schools should limit their science curriculum to scientific findings—testable conclusions that we can rationally draw on the basis of observational evidence and the methodological assumption of natural law.

Although for such reasons I am afraid I cannot stomach the main entrée that Professor Plantinga offers, I hope we may still at least break bread together in another way. While I cannot agree that we should teach creationism as true even conditionally, I can agree that we should not teach science (including evolutionary theory) dogmatically. Scientific truths are never more than approximate, and always come attached with some greater or lesser degree of likelihood that depends on their evidential support. While it may be reasonable to teach well-confirmed findings as "sober" truths, scientists must always remain cognizant that even these might have to be revised should new evidence appear that undermines their support; even "settled" truths may not legitimately be put forward as being absolute. The good science teacher will point this out. This is not the same as teaching science "conditionally," as though its epistemic base were no better or worse than any private epistemology; rather it is a straightforward acknowledgment of features of its methodology. Knowing the rational limits of science, the rational person would agree not to have it portrayed as something it is not. It is not a religion and should not be taught as one. Its findings are trustworthy, but they are not dogma. Teaching science simply as a set of facts to be memorized can make it seem like a catechism, but if we teach science properly, by also teaching its methods of investigation, dogmatism will not be an issue. Creationists and others who hold to some private epistemology may still teach in their homes and churches that man does not live by bread alone, but surely the reasonable and just compromise is to agree that in the public schools we should content ourselves with sharing science's simple, basic fare.

## Notes

1. For instance, while it is true that scientists still know rather little about *how* life originated from nonlife, *that* organic life arose from the inorganic is well confirmed, and progress continues to be made in understanding the chemical processes that were involved, such as the discovery that RNA can act as its own catalyst.

2. According to John Rawls's influential view, justice in a pluralistic democracy is a function of fairness, with fairness understood in terms of fair procedures (Rawls 1971). The "Original Position" is an abstraction of such a procedure in which we are to imagine ourselves as rational persons who are contracting to form what will be the basic social institutions, while under a "veil of ignorance" that prevents our knowing in advance what position we will hold in that society once it is constituted.

3. Around the time of the *Scopes* trial, schools in Zion, Illinois rejected on biblical authority not only evolution, but also "modern astronomy" and its "infidel theories" of a moving, round earth. See Schadewald (1989) for a description of how the flat-earth view persisted in Zion schools until the mid-1930s and continues to have a few adherents even today.

## References

Plantinga, Alvin. 1998. Creation and Evolution: A Modest Proposal. (This volume.)

Rawls, John. 1971. *A Theory of Justice*. Cambridge, Mass.: The Belknap Press of Harvard University Press.

Schadewald, Robert. 1989. The Earth Was Flat in Zion. *Fate* (May): 70–79.

# Index